GRAPHS, DIOIDS AND SEMIRINGS
New Models and Algorithms

OPERATIONS RESEARCH/COMPUTER SCIENCE INTERFACES

Professor Ramesh Sharda
Oklahoma State University

Prof. Dr. Stefan Voß
Universität Hamburg

Greenberg /*A Computer-Assisted Analysis System for Mathematical Programming Models and Solutions: A User's Guide for ANALYZE*
Greenberg / *Modeling by Object-Driven Linear Elemental Relations: A Users Guide for MODLER*
Brown & Scherer / *Intelligent Scheduling Systems*
Nash & Sofer / *The Impact of Emerging Technologies on Computer Science & Operations Research*
Barth / *Logic-Based 0-1 Constraint Programming*
Jones / *Visualization and Optimization*
Barr, Helgason & Kennington / *Interfaces in Computer Science & Operations Research: Advances in Metaheuristics, Optimization, & Stochastic Modeling Technologies*
Ellacott, Mason & Anderson / *Mathematics of Neural Networks: Models, Algorithms & Applications*
Woodruff / *Advances in Computational & Stochastic Optimization, Logic Programming, and Heuristic Search*
Klein / *Scheduling of Resource-Constrained Projects*
Bierwirth / *Adaptive Search and the Management of Logistics Systems*
Laguna & González-Velarde / *Computing Tools for Modeling, Optimization and Simulation*
Stilman / *Linguistic Geometry: From Search to Construction*
Sakawa / *Genetic Algorithms and Fuzzy Multiobjective Optimization*
Ribeiro & Hansen / *Essays and Surveys in Metaheuristics*
Holsapple, Jacob & Rao / *Business Modelling: Multidisciplinary Approaches — Economics, Operational and Information Systems Perspectives*
Sleezer, Wentling & Cude/*Human Resource Development And Information Technology: Making Global Connections*
Voß & Woodruff / *Optimization Software Class Libraries*
Upadhyaya et al / *Mobile Computing: Implementing Pervasive Information and Communications Technologies*
Reeves & Rowe / *Genetic Algorithms—Principles and Perspectives: A Guide to GA Theory*
Bhargava & Ye / *Computational Modeling And Problem Solving In The Networked World: Interfaces in Computer Science & Operations Research*
Woodruff / *Network Interdiction And Stochastic Integer Programming*
Anandalingam & Raghavan / *Telecommunications Network Design And Management*
Laguna & Martí / *Scatter Search: Methodology And Implementations In C*
Gosavi/ *Simulation-Based Optimization: Parametric Optimization Techniques and Reinforcement Learning*
Koutsoukis & Mitra / *Decision Modelling And Information Systems: The Information Value Chain*
Milano / *Constraint And Integer Programming: Toward a Unified Methodology*
Wilson & Nuzzolo / *Schedule-Based Dynamic Transit Modeling: Theory and Applications*
Golden, Raghavan & Wasil / *The Next Wave in Computing, Optimization, And Decision Technologies*
Rego & Alidaee/ *Metaheuristics Optimization via Memory and Evolution: Tabu Search and Scatter Search*
Kitamura & Kuwahara / *Simulation Approaches in Transportation Analysis: Recent Advances and Challenges*
Ibaraki, Nonobe & Yagiura / *Metaheuristics: Progress as Real Problem Solvers*
Golumbic & Hartman / *Graph Theory, Combinatorics, and Algorithms: Interdisciplinary Applications*
Raghavan & Anandalingam / *Telecommunications Planning: Innovations in Pricing, Network Design and Management*
Mattfeld / *The Management of Transshipment Terminals: Decision Support for Terminal Operations in Finished Vehicle Supply Chains*
Alba & Martí/ *Metaheuristic Procedures for Training Neural Networks*
Alt, Fu & Golden/ *Perspectives in Operations Research: Papers in honor of Saul Gass' 80th Birthday*
Baker et al/ *Extending the Horizons: Adv. In Computing, Optimization, and Dec. Technologies*
Zeimpekis et al/ *Dynamic Fleet Management: Concepts, Systems, Algorithms & Case Studies*
Doerner et al/ *Metaheuristics: Progress in Complex Systems Optimization*
Goel/ *Fleet Telematics: Real-time management and planning of commercial vehicle operations*

GRAPHS, DIOIDS AND SEMIRINGS

New Models and Algorithms

Michel Gondran
University Paris-Dauphine

and

Michel Minoux
University Paris VI

Springer

Michel Gondran
University Paris-Dauphine
France

Michel Minoux
University Paris VI
France

Series Editors
Ramesh Sharda
Oklahoma State University
Stillwater, Oklahoma, USA

Stefan Voß
Universität Hamburg
Germany

Library of Congress Control Number: 2007936570

ISBN 978-0-387-75449-9 e-ISBN 978-0-387-75450-5
DOI: 10.1007/978-0-387-75450-5

Printed on acid-free paper.

9 8 7 6 5 4 3 2 1

springer.com

Preface

During the last two or three centuries, most of the developments in science (in particular in Physics and Applied Mathematics) have been founded on the use of classical algebraic structures, namely groups, rings and fields. However many situations can be found for which those usual algebraic structures do not necessarily provide the most appropriate tools for modeling and problem solving. The case of arithmetic provides a typical example: the set of nonnegative integers endowed with ordinary addition and multiplication does not enjoy the properties of a field, nor even those of a ring.

A more involved example concerns *Hamilton–Jacobi* equations in Physics, which may be interpreted as optimality conditions associated with a *variational principle* (for instance, the Fermat principle in Optics, the 'Minimum Action' principle of Maupertuis, etc.). The discretized version of this type of variational problems corresponds to the well-known *shortest path problem in a graph.* By using Bellmann's optimality principle, the equations which define a solution to the shortest path problem, which are *nonlinear* in usual algebra, may be written as a *linear system* in the algebraic structure ($\mathbb{R} \cup \{+\infty\}$, Min, +), i.e. the set of reals endowed with the operation Min (minimum of two numbers) in place of addition, and the operation + (sum of two numbers) in place of multiplication.

Such an algebraic structure has properties quite different from those of the field of real numbers. Indeed, since the elements of $E = \mathbb{R} \cup \{+\infty\}$ do not have inverses for $\oplus$ = Min, this internal operation does not induce the structure of a group on E. In that respect (E, $\oplus$, $\otimes$) will have to be considered as an example of a more primitive algebraic structure as compared with fields, or even rings, and will be referred to as a *semiring*.

But this example is also representative of a particular class of semirings, for which the monoid (E, $\oplus$) is *ordered* by the order relation $\propto$ (referred to as 'canonical') defined as:

$$a \propto b \Leftrightarrow \exists c \in E \quad \text{such that} \quad b = a \oplus c.$$

In view of this, (E, $\oplus$, $\otimes$) has the structure of a *canonically ordered semiring* which will be called, throughout this book, a *dioid*.

More generally, it is to be observed here that the operations Max and Min, which give the set of the reals a structure of canonically ordered monoid, come rather naturally into play in connection with algebraic models for many problems, thus leading to as many applications of dioid structures. Among some of the most characteristic examples, we mention:

- The dioids (ℝ, Min, +) and (ℝ, Max, Min) which provide natural models for the *shortest path problem* and for the *maximum capacity path problem* respectively (the latter being closely related to *the maximum weight spanning tree* problem). Many other path-finding problems in graphs, corresponding to other types of dioids, will be studied throughout the book;
- The dioid ({0,1}, Max, Min) or *Boolean Algebra*, which is the algebraic structure underlying *logic*, and which, among other things, is the basis for modeling and solving *connectivity problems* in graphs;
- The dioid (P(A*), ∪, o), where P(A*) is the set of all languages on the alphabet A, endowed with the operations of union ∪ and concatenation o, which is at the basis of the theory of languages and automata.

One of the primary objectives of this volume is precisely, on the one hand, to emphasize the deep relations existing between the semiring and dioid structures with graphs and their combinatorial properties; and, on the other hand, to show the capability and flexibility of these structures from the point of view of *modeling and solving problems* in extremely diverse situations. If one considers the many possibilities of constructing new dioids starting from a few reference dioids (vectors, matrices, polynomials, formal series, etc.), it is true to say that the reader will find here an almost unlimited source of examples, many of which being related to applications of major importance:

- Solution of a wide variety of optimal path problems in graphs (Chap. 4, Sect. 6);
- Extensions of classical algorithms for shortest path problems to a whole class of nonclassical path-finding problems (such as: shortest paths with time constraints, shortest paths with time-dependent lengths on the arcs, etc.), cf. Chap. 4, Sect. 4.4;
- Data Analysis techniques, hierarchical clustering and preference analysis (cf. Chap. 6, Sect. 6);
- Algebraic modeling of fuzziness and uncertainty (Chap. 1, Sect. 3.2 and Exercise 2);
- Discrete event systems in automation (Chap. 6, Sect. 7);
- Solution of various nonlinear partial differential equations, such as: Hamilton–Jacobi, and Bürgers equations, the importance of which is well-known in Physics (Chap. 7).

And, among all these examples, the alert reader will recognize the most widely known, and the most elementary mathematical object, the dioid of natural numbers: ***At the start, was the dioid N!***

Besides its emphasis on models and illustration by examples, the present book is also intended as an extensive overview of the mathematical properties enjoyed by these "nonclassical" algebraic structures, which either extend usual algebra (as for the case of pre-semirings or semirings), or (as for the case of dioids) correspond to a new branch of algebra, clearly distinct from the one concerned with the classical structures of groups, rings and fields.

Indeed, a simple, though essential, result (which will be discussed in the first chapter) states that a monoid cannot simultaneously enjoy the properties of being a group and of being canonically ordered. Hence the algebra for sets endowed with two internal operations turns out to split into two disjoint branches, according to which of the following two (incompatible) assumptions holds:

- The "additive group" property, which leads to the structures of ring and of field;
- The "canonical order" property, which leads to the structures of dioid and of lattice.

For dioids, one of the immediate consequences of dropping the property of invertibility of addition to replace it by the canonical order property, is the need of considering *pairs of elements* instead of individual elements, to avoid the use of "negative" elements. Modulo this change in perspective, it will be seen how many basic results of usual algebra can be transposed. Consider, for instance, the properties involving the determinant of a square $n \times n$ matrix. In dioids (as well as in general semirings), the standard definition of the determinant cannot be used anymore, but we can define the *bideterminant* of $A = (a_{i,j})$ as the pair $(\det^+(A), \det^-(A))$, where $\det^+(A)$ denotes the sum of the weights of even permutations, and $\det^-(A)$ the sum of the weights of odd permutations of the elements of the matrix. For a matrix with a set of linearly dependent columns, the condition of zero determinant is then replaced by equality of the two terms of the bideterminant:

$$\det^+(A) = \det^-(A).$$

In a similar way, the concept of characteristic polynomial $P_A(\lambda)$ of a given matrix A, has to be replaced by the *characteristic bipolynomial*, in other words, by a pair of polynomials $(P_A{}^+(\lambda), P_A{}^-(\lambda))$. Among other remarkable properties, it is then possible to transpose and generalize in dioids and in semirings, the famous Cayley–Hamilton theorem, $P_A(A) = 0$, by the matrix identity:

$$P_A{}^+(A) = P_A{}^-(A).$$

Another interesting example concerns the classical Perron–Frobenius theorem. This result, which states the existence on $\mathbb{R}_+$ of an eigenvalue and an eigenvector for a nonnegative square matrix, may be viewed as a property of the dioid $(\mathbb{R}_+, +, \times)$, thus opening the way to extensions to many other dioids. Incidentally we observe that it is precisely this dioid $(\mathbb{R}_+, +, \times)$ which forms the truly appropriate underlying structure for measure theory and probability theory, rather than the field of real numbers $(\mathbb{R}, +, \times)$.

One of the ambitions of this book is thus to show that, as complements to usual algebra, based on the construct "Group-Ring-Field", other algebraic structures based on alternative constructs, such as "Canonically ordered monoid- dioid- distributive lattice" are equally interesting and rich, both in terms of mathematical properties and of applications.

Acknowledgements

Many people have contributed in some way to the "making of" the present volume. We gratefully acknowledge the kind and professional help received from Springer's staff, and in particular, from Gary Folven and Concetta Seminara-Kennedy.

Many thanks are due to Patricia and the company Katex for the huge text-processing work in setting-up the initial manuscript. We also appreciate the contribution of Sundardevadoss Dharmendra and his team in India in working out the final manuscript.

We are indebted to Garry White for his most careful and professional help in the huge translation work from French to English language.

We want to express our gratitude to Prof. Didier Dubois and to Prof. Michel Gabisch for their comments and encouragements, and also for pointing out fruitful links with fuzzy set theory and decision making under uncertainty.

The "MAX-PLUS" group in INRIA, France, has provided us along the years with a stimulating research environment, and we thank J.P. Quadrat, M. Akian, S. Gaubert and G. Cohen for their participation in many fruitful exchanges. Also we acknowledge the work of P.L. Lions on the viscosity solutions to Hamilton–Jacobi equations as a major source of inspiration for our research on MINPLUS and MINMAX analyses. And, last but not least, special thanks are due to Professor Stefan Voß from Hamburg University, Germany, for his friendly encouragements and help in publishing our work in this, by now famous, series dedicated to Operations Research/Computer Science Interfaces.

Paris,
January 15th 2008

M. Gondran
M. Minoux

Contents

List of Notation

Sets

$\mathbb{R}$	Set of reals.
$\mathbb{N}$	Set of natural numbers.
$\mathbb{Z}$	Set of integers.
$\mathbb{R}_+$	Set of nonnegative reals.
$\mathbb{C}$	Set of complex numbers.
$\hat{\mathbb{R}}$	The set $\mathbb{R} \cup \{+\infty\}$.
$\check{\mathbb{R}}$	The set $\mathbb{R} \cup \{-\infty\}$.
$\bar{\mathbb{R}}$	The set $\mathbb{R} \cup \{+\infty\} \cup \{-\infty\}$.
$\mathbb{R}_*$	The set of nonzero real numbers: $\mathbb{R}\backslash\{0\}$.
$(E, \oplus)$	Set E endowed with the internal operation $\oplus$.
$(E, \oplus, \otimes)$	Set E endowed with two internal operations $\oplus$ and $\otimes$.
$\overset{\oplus}{\leq}$	The canonical order relation of a canonically ordered monoid $(E, \oplus)$ (often simply denoted $\leq$ when there is no ambiguity).
a^k	For $k \in \mathbb{N}$, $a \in (E, \oplus, \otimes)$, $a^k = a \otimes a \otimes \cdots \otimes a$ (k times).
$a^{(k)}$	$a^{(k)} = e \oplus a \oplus a^2 \oplus \cdots \oplus a^k$, where e is the neutral element for $\otimes$.
a^*	Quasi-inverse of $a \in (E, \oplus, \otimes)$ (when it exists): limit of $a^{(k)}$ for $k \to +\infty$.
$\emptyset$	The empty set.
$\{a, b, c\}$	The set formed by the three elements, a, b, c.
$x \in X$	x belongs to the set X.
$x \notin X$	x does not belong to the set X.
$A \subset X$	A is strictly included in X.
$A \subseteq X$	A is included in X, and possibly $A = X$.
$A \not\subset X$	A is not included in X.
$\lvert A \rvert$	Cardinal of A, number of elements of A.
$X\backslash A$	The set of elements in X which do not belong to A.
$A_1 \cup A_2$	Union of the two subsets A_1 and A_2.
$A \cup \{e\}$	The set obtained by adding element e to A.

$\bigcup_{i \in I} A_i$	Union of the family of subsets A_i for all i in the subset of indices I.
$A_1 \cap A_2$	Intersection of the two subsets A_1 and A_2.
$\bigcap_{i \in I} A_i$	Intersection of the family of subsets A_i, for all i in I.
$A_1 \Delta A_2$	Symmetric difference of A_1 and A_2 ($= (A_1 \cup A_2) \backslash (A_1 \cap A_2)$).
$\{x : x \text{ such that} \ldots\}$	The set of elements x such that ...
$\exists x$:	There exists x such that ...
$\forall x \in X$	For all x in X.
$A \times B$	The Cartesian product of A and B (i.e. the set of pairs (a, b) with $a \in A$ and $b \in B$).
$\mathcal{P}(A)$	The power set of A (the set of all subsets of A).
$\binom{p}{q} = \frac{p!}{q!(p-q)!}$	The binomial coefficient p choose q.
$\lfloor x \rfloor$	The greatest integer less than or equal to $x \in \mathbb{R}$.
$\lceil x \rceil$	The smallest integer greater than or equal to $x \in \mathbb{R}$.
$(P) \Rightarrow (Q)$	Property (P) implies property (Q).
$\downarrow x$	In an ordered set E: the set of $y \in E$ such that $y \leq x$ (ideal).
$\downarrow P$	In an ordered set E, and for $P \subseteq E$: $\bigcup_{x \in P} (\downarrow x)$.
$\uparrow x$	In an ordered set E, the set of $y \in E$ such that $x \leq y$ (filter).
$\uparrow P$	In an ordered set E, and for $P \subseteq E$: $\bigcup_{x \in P} (\uparrow x)$.
$x \vee y$	In an ordered set E, least upper bound of x and y.
$x \wedge y$	In an ordered set E, greatest lower bound of x and y.
Per(n)	The set of permutations of the set $\{1, 2, \ldots, n\}$, $n \in \mathbb{N}$.
$\text{Per}^+ n$, $\text{Per}^-(n)$	The set of even, the set of odd permutations of $\{1, 2, \ldots, n\}$.
$\text{sign}(\sigma)$	Signature of the permutation $\sigma \in \text{Per}(n)$.
$\text{char}(\sigma)$	Characteristic of the permutation $\sigma \in \text{Per}(n)$ (see definition in Chap. 2, Sect. 4.1).
Part(n)	Set of partial permutations of the set $\{1, 2, \ldots, n\}$.
$\text{Dom}(\sigma)$	Domain of the partial permutation $\sigma \in \text{Part}(n)$.
$\text{Part}^+(n)$, $\text{Part}^-(n)$	Set of partial permutations of $\{1, \ldots, n\}$ with characteristic $+1, -1$.

Matrices and vectors

E^n	The Cartesian product of the set E with itself (n times): the set of vectors with n components in E.
$x = \begin{pmatrix} x_1 \\ x_2 \\ \vdots \\ x_n \end{pmatrix}$	A vector $x \in E^n$.
x^T	The transpose of $x \in E^n$.
$x \leq y$	For $x \in E^n$, $y \in E^n$, each component of x is less than or equal to the corresponding component of y.
$A = (a_{ij})$	Matrix with entries a_{ij}.
A^k	The k-th power of matrix A.

$[A^k]_{ij}$	The entry of matrix A^k corresponding to row i and column j.
A_I^J	The submatrix of A, the rows (resp. columns) of which correspond to the subset of rows (resp. columns) I (resp. J).
$A \otimes B$	Product of matrices A and B.
A^T	Transpose of matrix A.
$M_n(E)$	Set of square $n \times n$ matrices with entries in E.
$\det(A)$	Determinant of a real matrix $A \in M_n(\mathbb{R})$.
$(\det^+(A), \det^-(A))$	Bideterminant of a matrix $A \in M_n(E)$.
$\mathrm{Perm}(A)$	Permanent of $A \in M_n(E)$.
$A^{(k)}$	For $k \in \mathbb{N}$, $A \in M_n(E)$, $A^{(k)} = I \oplus A \oplus A^2 \oplus \cdots \oplus A^k$, where I denotes the identity matrix of $M_n(E)$.
A^*	Quasi-inverse of matrix $A \in M_n(E)$ (when it exists): limit of $A^{(k)}$ for $k \to +\infty$.
$A^{[k]}$	For $k \in \mathbb{N}$, $A \in M_n(E)$: $A^{[k]} = A \oplus A^2 \oplus \cdots \oplus A^k$.
A^+	Limit (when it exists) of $A^{[k]}$ when $k \to +\infty$.
$\rho(A)$	Spectral radius of a matrix A.
$\mathcal{V}(\lambda)$	For a given matrix, the set of eigenvectors associated with the eigenvalue λ.
$\mathrm{Sp}(x_1, \ldots, x_p)$	Subspace or semi-module generated by the family of vectors $X = \{x_1, x_2, \ldots, x_p\}$.

Graphs

$G = [X, U]$	Graph with vertex set X and arc (or edge) set U.
Γ_i	Set of immediate successors of vertex i in a given graph.
Γ_i^{-1}	Set of immediate predecessors of vertex i in a given graph.
$G = [X, \Gamma]$	Graph represented by its associated point-to-set map Γ.
$\hat{\Gamma}$	Transitive closure of the mapping Γ.
$\hat{\Gamma}_i$	Set of successors of vertex i: the set of vertices j such that there is at least one path from i to j.
$d_G(i)$	Degree of vertex i in graph G.
$d_G^+(i), d_G^-(i)$	Out-degree, in-degree of vertex i in graph G.
$\omega(A)$	The set of arcs or edges having one endpoint in A and the other in $X \backslash A$.
$\omega^+(A), \omega^-(A)$	The set of arcs in $\omega(A)$ having initial endpoint, terminal endpoint in A.
U_A	The set of arcs or edges having both endpoints in A.
$G_A = [A, U_A]$	Subgraph of $G = [X, U]$ induced by the subset of vertices $A \subseteq X$.
G/Y	Shrunken graph deduced from $G = [X, U]$ by shrinking the subset of vertices Y.
P_{ij}^k	In a directed graph, the set of paths from i to j composed of exactly k arcs.
$P_{ij}^{(k)}$	The set of i–j paths composed of at most k arcs.
$P_{ij}^k(p)$	The subset of paths in P_{ij}^k taking no more than p times each elementary circuit in the graph.

$P_{ij}^{k}(0)$	The subset of elementary paths in P_{ij}^{k}.
$P_{ij}^{(k)}(p)$	The subset of paths in $P_{ij}^{(k)}$ taking no more than p times each elementary circuit.
$P_{ij}^{(k)}(0)$	The subset of elementary paths in $P_{ij}^{(k)}$.

Functions

$f : E \to F$	Function which, to each $x \in E$, lets correspond $y = f(x) \in F$.
$\dfrac{\partial f}{\partial x_i}(\bar{x})$	Partial derivative of f in $\bar{x}$ with respect to variable x_i (assuming f differentiable).
$\dfrac{\partial^2 f}{\partial x_i \partial x_j}(\bar{x})$	Second partial derivative of f in $\bar{x}$ w.r.t. x_i and x_j (assuming f twice differentiable).
$\nabla f(\bar{x})$	Gradient of f in $\bar{x}$, i.e. the vector in $\mathbb{R}^n$ with components $\dfrac{\partial f}{\partial x_i}(\bar{x})$ $(i = 1, \ldots, n)$.
$\partial^- f(\bar{x})$	Subdifferential of f in $\bar{x}$, for f: $\mathbb{R}^n \to \mathbb{R}$ not everywhere differentiable.
$\partial^+ f(\bar{x})$	Super-differential of f in $\bar{x}$, for f: $\mathbb{R}^n \to \mathbb{R}$ not everywhere differentiable.
f_*	l.s.c. closure of f: $\mathbb{R}^n \to \mathbb{R}$.
f^*	u.s.c. closure of f: $\mathbb{R}^n \to \mathbb{R}$.
F(f)	Legendre–Fenchel transform of f: $X \to \mathbb{R}$ (often denoted $\hat{f}$).
$\partial_\theta^{1,-} f(\bar{x})$	Order-1 subdifferential of f: $\theta \to \overline{\mathbb{R}}$ in $\bar{x} \in \theta$. Denoted $\partial^- f(\bar{x})$ when θ is an open set.
$\partial_\theta^{1,+} f(\bar{x})$	Order-1 super-differential of f: $\theta \to \overline{\mathbb{R}}$ in $\bar{x} \in \theta$. Denoted $\partial^+ f(\bar{x})$ when θ is an open set.
$\partial_\theta^{2,-} f(\bar{x})$	Order-2 subdifferential of f: $\theta \to \overline{\mathbb{R}}$ in $\bar{x} \in \theta$.
$\partial_\theta^{2,+} f(\bar{x})$	Order-2 super-differential of f: $\theta \to \overline{\mathbb{R}}$ in $\bar{x} \in \theta$.
(f, g)	MINPLUS (or MINMAX) scalar product of two functions f and g.
f_λ	Inf-Moreau–Yosida transform of f with parameter λ.
f^λ	Sup-Moreau–Yosida transform of f with parameter λ.
f_{**}	Convex l.s.c. closure of a function f.
$f_{\circledast}$	Quasi-convex l.s.c. closure of a function f.
$LSC(\theta)$	The set of lower semi-continuous functions: $\theta \to \mathbb{R}$.
$USC(\theta)$	The set of upper semi-continuous functions: $\theta \to \mathbb{R}$.
$LSD^1(\theta)$	The set of order-1 lower semi-differentiable functions: $\theta \to \mathbb{R}$.
$USD^1(\theta)$	The set of order-1 upper semi-differentiable functions: $\theta \to \mathbb{R}$.
$SD^1(\theta)$	The set of order-1 semi-differentiable functions: $\theta \to \mathbb{R}$.
$LSD^2(\theta)$	The set of order-2 lower semi-differentiable functions: $\theta \to \mathbb{R}$.
$USD^2(\theta)$	The set of order-2 upper semi-differentiable functions: $\theta \to \mathbb{R}$.
$SD^2(\theta)$	The set of order-2 semi-differentiable functions: $\theta \to \mathbb{R}$.

Algorithms and pseudocode

Conditional instructions:

```
If (logical condition) then
      {block of instructions B}
Endif

If (logical condition) then
      {block of instructions B1}
else
      {block of instructions B2}
Endif
```

loops:

$$\text{For} \begin{pmatrix} \text{index i} = \text{val 1,} \\ \text{val 2,} \ldots \text{val k} \end{pmatrix} \text{do}$$

```
      {block of instruction B}
Endfor

Repeat
      {block of instructions B}
Until (logical condition)

While (logical condition) do
      {block of instructions B}
Endwhile
```

$f(n) \in o(g(n)) : \dfrac{f(n)}{g(n)} \to 0$ when $n \to \infty$.

$f(n) \in \mathcal{O}(g(n)) : \exists k > 0$ and $n_0 \geq 0$ such that: $\forall n \geq n_0, f(n) \leq k\, g(n)$.

Chapter 1
Pre-Semirings, Semirings and Dioids

As an introduction to this first chapter, we show, by discussing four characteristic examples, that even with internal operations with limited properties – in particular those are not invertible – there exist nonetheless algebraic structures in which it is possible to solve fixed-point type equations and obtain eigenvalues and eigenvectors of matrices. It will be seen throughout this book that it is possible to reconstruct, in such structures, a major part of classical algebra.

This first chapter is composed of two parts. The first is devoted to some basic properties and to a typology of algebraic structures formed by a set endowed with a single internal operation: semigroups and monoids in Sect. 2, ordered monoids in Sect. 3.

The second part is devoted to the basic properties and typology of algebraic structures formed by a set endowed with two internal operations: pre-semirings in Sect. 4, semirings in Sect. 5 and dioids in Sect. 6.

For each of these structures, the most important subclasses are pointed out and the basic terminology to be used in the following chapters is introduced.

1. Founding Examples

Example 1.1. Let us denote by $\overline{\mathbb{R}}$ the set of reals to which we have added the elements $-\infty$ and $+\infty$. In the algebraic structure ($\overline{\mathbb{R}}$, Max, Min), composed of the set $\overline{\mathbb{R}}$ endowed with operations Max (denoted $\oplus$) and Min (denoted $\otimes$), the equations:

$$\mathrm{a} \oplus \mathrm{x} = \mathrm{b}$$

$$\mathrm{a} \otimes \mathrm{x} = \mathrm{b}$$

do not have solutions if a > b (resp. b > a).

On the other hand, the equation:

$$\mathrm{x} = (\mathrm{a} \otimes \mathrm{x}) \oplus \mathrm{b}$$

has solutions for all the values of a and b: infinitely many solutions, including a *minimal solution* b (minimality being understood in the sense of the usual order relation on $\overline{\mathbb{R}}$) if $b < a$. A unique solution $x = b$ if $a \leq b$. Thus, even if the operations $\oplus$ and $\otimes$ are not invertible (nor symmetrizable), it is possible to solve equations of the fixed-point type as above.

The algebraic structure ($\overline{\mathbb{R}}_+$, Max, Min) is a *distributive lattice* which appears as a special case of the more general *dioid* structure studied in the present work. ||

Example 1.2. In the algebraic structure ($\overline{\mathbb{R}}_+$, Min, +), that is to say in the set of positive real numbers to which we have added $+\infty$, endowed with operations Min (denoted $\oplus$) and + (denoted $\otimes$), the equations:

$$a \oplus x = b$$
$$a \otimes x = b$$

do not have solutions if $a < b$ (resp. $a > b$).

On the other hand, the equation

$$x = (a \otimes x) \oplus b$$

has, here again, solutions for any a and b: infinitely many solutions (the whole segment [0, b]), including a maximum solution $x = b$ if $a = 0$; a unique solution $x = b$ if $a > 0$. The structure ($\overline{\mathbb{R}}_+$, Min, +) is a dioid. ||

Example 1.3. In the algebraic structure ($\mathbb{R}_+$, +, $\times$), that is to say the set of positive real numbers endowed with ordinary addition and multiplication, the equation $a + x = b$ only has a solution in $\mathbb{R}_+$ if $a \leq b$.

On the other hand, for any value of b, the equation $x = ax + b$, has a solution in $\mathbb{R}_+$, $x = \dfrac{1}{1-a} b = (1 + a + a^2 + \cdots)\, b$ as soon as $a < 1$.

The Perron–Frobenius theorem (see Chap. 6, Sect. 4 and Exercise 1) also ensures that a square matrix A with elements in $\mathbb{R}_+$ has an eigenvalue in $\mathbb{R}_+\backslash\{0\}$ and an eigenvector with coordinates in $\mathbb{R}_+\backslash\{0\}$.

We will see in Chap. 6 that this theorem extends to a great number of dioids, and in particular to the above two dioids ($\overline{\mathbb{R}}$, Max, Min) and ($\overline{\mathbb{R}}_+$, Min, +).

The algebraic structure ($\mathbb{R}_+$, +, $\times$) also appears fundamental, because it is subjacent to measure theory and probability theory. On the one hand, measures, or densities of probability, are functions (or distributions) with values in $\mathbb{R}_+$. On the other hand, to define the measurability of a function f with values in $\mathbb{R}$, one must decompose f in the form $f = f^+ - f^-$, where f^+ and f^- are positive and measurable functions. The basic mathematical object corresponding to the concept of measurable function is therefore the pair (f^+, f^-), made up of two functions with values in the dioid ($\mathbb{R}_+$, +, $\times$).

The integral of a function (of a distribution) may be viewed as a *linear form* on the dioid ($\mathbb{R}_+$, +, $\times$). We will see in Chap. 7 that by substituting the dioid ($\mathbb{R}_+$, +, $\times$) with other dioids (such as ($\overline{\mathbb{R}}_+$, Max, Min) or ($\overline{\mathbb{R}}_+$, Min, +) we can define new linear forms on these dioids. These forms are *nonlinear* with respect to ordinary addition and multiplication and thus lead to *nonlinear analyses* which therefore can be studied with tools of "linear analysis". ||

Example 1.4. Let A be a finite set, referred to as an *alphabet*, whose elements are referred to as *letters*. Any finite sequence of letters is called a *word*. The set of words, denoted A^*, is called the *free monoid* (see Sect. 2, Example 2.1.13). We call *language* on A any subset of A^*. We can endow the set $E = \mathcal{P}(A^*)$ of languages on A with two operations: union, denoted by the sign +, and the Cauchy product, denoted $\cdot$:

$$L_1 \cdot L_2 = \{m_1\, m_2 / m_1 \in L_1, m_2 \in L_2\}$$

In this algebraic structure $(\mathcal{P}(A^*), +, \cdot)$, the equations

$$L_1 + X = L_2$$
$$L_1 \cdot X = L_2$$

generally do not have a solution.

On the other hand the system of equations:

$$X = L_1 \cdot X + L_2$$

has, for any L_1 and L_2, an infinity of solutions including a minimal solution:

$X = L_1^* \cdot L_2 = L_2 + L_1 \cdot L_2 + L_1^2 \cdot L_2 + \cdots$. The algebraic structure $(\mathcal{P}(A^*), +, \cdot)$ is the basis of Kleene's theory of *regular languages*. (see e.g. Eilenberg, 1974) ||

2. Semigroups and Monoids

After having presented semigroups and monoids through a number of examples, we recall some combinatorial properties of finite semigroups before introducing regular monoids and groups.

2.1. Definitions and Examples

Definition 2.1.1. *We call* semigroup *a set* E *endowed with an internal* associative *(binary) law denoted* $\oplus$:

$$a \oplus b \in E \quad \forall a, b \in E$$
$$(a \oplus b) \oplus c = a \oplus (b \oplus c) \quad \forall a, b, c \in E.$$

Example 2.1.2. $(\mathbb{R}, \text{Min})$, the set of reals endowed with the operation Min is a semigroup. The same applies to $(\mathbb{Z}, \text{Min})$, $(\mathbb{N}, \text{Min})$, $(\mathbb{R}, \text{Max})$, $(\mathbb{Z}, \text{Max})$, $(\mathbb{N}, \text{Max})$. ||

Example 2.1.3. $\mathbb{R}_+\backslash\{0\}$ the set of strictly positive reals, endowed with addition or multiplication is a semigroup. The same applies to $\mathbb{N}_*$., the set of strictly positive integers. ||

Example 2.1.4. $\mathbb{R}_+\backslash\{0\}$ endowed with the law $\oplus$ defined as:

$$a \oplus b = \frac{a+b}{1+ab}$$

is a semigroup (for the associativity, note the $a \oplus b \oplus c = \dfrac{a+b+c+abc}{1+ab+ac+bc}$).

The same applies to $\mathbb{R}_+$ endowed with the law

$$a \oplus b = a(1+b^2)^{1/2} + b(1+a^2)^{1/2}.$$

(for the associativity, note that $(a \oplus b) \oplus c = a(1+b^2)^{1/2}(1+c^2)^{1/2} + b(1+a^2)^{1/2}$ $(1+c^2)^{1/2} + c(1+a^2)^{1/2}(1+b^2)^{1/2} + abc$). (see Exercise 1 at the end of the chapter.) ||

Example 2.1.5. $\mathbb{C}_+$, the set of complex numbers z with strictly positive real component $\{z \in \mathbb{C}, \operatorname{Re}(z) > 0\}$, endowed with addition is a semigroup. ||

Example 2.1.6. $\mathbb{C}_+$, endowed with the law $\oplus$ defined as

$$a \oplus b = \frac{a+b}{1+ab}$$

is a semigroup. ||

Example 2.1.7. The complex numbers of the form $x + iy$ with $x > |y|^\rho$, $0 < \rho \le 1$, endowed with addition, form a semigroup (indeed a sub-semigroup of $(\mathbb{C}_+, +)$). ||

Example 2.1.8. If we consider for a set E, the set of mappings $\mathcal{F}$ of E onto itself, the set $\mathcal{F}$ endowed with the law of composition of mappings is a semigroup. ||

Example 2.1.9. Let $E = C[0, \infty]$, the Banach space of continuous functions defined on the closed interval $[0, +\infty]$ with the norm $||f|| = \sup_t |f(t)|$. We define for every $\alpha > 0$:

$$T(\alpha)[f] = f(t+\alpha)$$

The family $T(\alpha)$ is a semigroup with one parameter of linear transformations in $C[0, \infty]$ with $||T(\alpha)|| = 1$.

It is the prototype of semigroups with one parameter upon which a great part of functional analysis is based, see, e.g. Hille and Phillips (1957). For examples, see Exercise 1. ||

The *neutral element* of a semigroup $(E, \oplus)$, denoted ε, is defined by the property:

$$\varepsilon \oplus x = x \oplus \varepsilon = x \quad \forall x \in E.$$

If a semigroup has a neutral element ε, then this neutral element is *unique*. Indeed, if ε' was another neutral element, we would have $\varepsilon \oplus \varepsilon' = \varepsilon = \varepsilon'$. As a result, if the neutral element does not exist, we can add one to the set E. Thus, in the case of the semigroup $(\mathbb{R}, \text{Min})$ (Example 2.1.2), a neutral element in $\mathbb{R}$ does not exist. However, we can add one, denoted $+\infty$, which augments $\mathbb{R}$ to $\hat{\mathbb{R}} = \mathbb{R} \cup \{+\infty\}$.

Definition 2.1.10. *We call* monoid *a set* E *endowed with an associative internal law and a neutral element.*

Examples 2.1.3, 2.1.4, 2.1.5 and 2.1.6 become monoids by adding the neutral element 0.

Remark 2.1.11. The terms *semigroup* and *monoid* seem more or less stabilized today. Bourbaki applied the term *magma* to what we have referred to as a semigroup and restricted the term semigroup (which suggest that the set is "almost a group"), to a monoid for which $\oplus$ is simplifiable and which, consequently, is extendable to a group via symmetrization. This is what we will refer to henceforth as a *cancellative monoid* (see Sect. 2–3). ||

If the operation $\oplus$ is commutative, then the monoid $(E, \oplus)$ is said to be *commutative*.

An element a is *idempotent* if $a \oplus a = a$.

If all the elements of E are *idempotent*, the monoid $(E, \oplus)$ is said to be *idempotent*.

An even more special case is when the operation $\oplus$ satisfies the property of *selectivity*:

$$a \oplus b = a \textit{ or } b \quad \forall a, b \in E.$$

In this case, the monoid $(E, \oplus)$ is said to be *selective*.

Selectivity obviously implies idempotency, but the converse is not true, as shown by the operation below (mean-sum) defined on the set of real numbers

$$a \oplus b = \frac{a + b}{2} \quad \forall a, b \in R.$$

This law is commutative and idempotent, but not selective (and not associative either).

An element $w \in E$ is said to be *absorbing* if and only if

$$x \oplus w = w \oplus x = w \quad \forall x \in E.$$

Example 2.1.12. $(E, \oplus)$, where $E = \mathcal{P}(X)$ is the power set of a set X, endowed with $\oplus$ = union of sets, is a *commutative monoid.* It has a neutral element $\varepsilon = \emptyset$ (the empty set) and an absorbing element $w = X$. ||

Example 2.1.13. (the free monoid)

Let A be a set (called "alphabet") whose elements are referred to as *letters*.

We take for E the set of finite sequences of elements of A which we call *words*, and we define the operation $\oplus$ as *concatenation*, that is to say:

$$\begin{aligned}
\text{If} \quad m_1 \in E: &\quad m_1 = s_1\, s_2 \ldots s_p \\
m_2 \in E: &\quad m_2 = t_1\, t_2 \ldots t_q \\
m_1 \oplus m_2 &= s_1\, s_2 \ldots s_p\, t_1\, t_2 \ldots t_q \\
m_2 \oplus m_1 &= t_1\, t_2 \ldots t_q\, s_1\, s_2 \ldots s_p
\end{aligned}$$

We see that, in general, $m_1 \oplus m_2 \neq m_2 \oplus m_1$, the operation $\oplus$ is therefore not commutative. The set denoted A^* of finite words on A endowed with the operation of concatenation is called *the free monoid on the alphabet* A. ||

Example 2.1.14. $\overline{\mathbb{R}} = \mathbb{R} \cup \{-\infty\} \cup \{+\infty\}$ endowed with the operation $\oplus = \text{Min}$ ($a \oplus b = \text{Min}\ \{a, b\}$) is a commutative monoid. It has a neutral element $\varepsilon = +\infty$ and an absorbing element $w = -\infty$. ||

Example 2.1.15. R and [0, 1] endowed with the operation $\oplus$ defined as $a \oplus b = a + b - ab$ are commutative monoids. They have a neutral element $\varepsilon = 0$ and an absorbing element $w = 1$. ||

Example 2.1.16. [0, 1] endowed with the operation $\oplus$ defined as $a \oplus b = \text{Min}(a + b, 1)$ is a commutative monoid with neutral element $\varepsilon = 0$ and absorbing element $w = 1$. Similarly, [0, 1] endowed with the operation $\oplus$ defined as $a \oplus b = \text{Max}(0, a + b - 1)$ is a commutative monoid with neutral element $\varepsilon = 1$ and absorbing element $w = 0$. ||

2.2. Combinatorial Properties of Finite Semigroups

We recall some classical properties of finite semigroups, see for example Lallement (1979) and Perrin and Pin (1997). We present them by noting the associative internal operation in a multiplicative form. x^k denotes the kth power of x, that is to say the product $x \cdot x \cdot \ldots$ (k times).

Proposition 2.2.1. *Any element of a finite semigroup* (E, ·) *has an idempotent power.*

Proof. Let S_x be the sub-semigroup generated by an element x. Since S_x is finite there exist integers $i, p > 0$ such that:

$$x^i = x^{i+p}$$

If i and p are chosen to be minimal, we say that i is the *index* of x and p its *period.* The semigroup S_x then has $i + p - 1$ elements and its multiplicative structure is represented on the figure below:

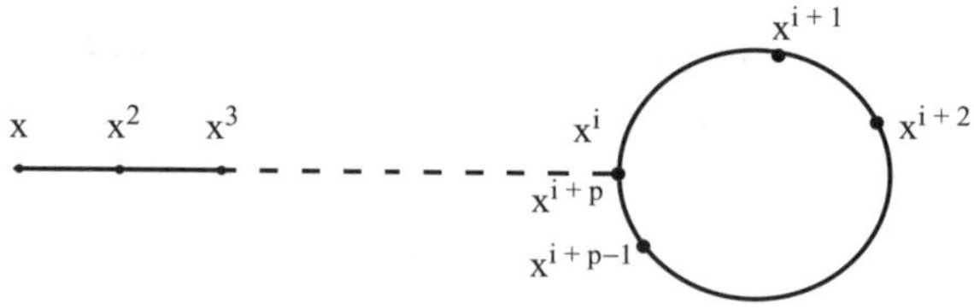

The sub-semigroup $\{x^i, x^{i+1}, \ldots, x^{i+p-1}\}$ of S_x then has an idempotent x^{i+r} with $r \geq 0$ and $r \equiv -i \pmod p$. □

Corollary 2.2.2. *Every non-empty finite semigroup contains at least one idempotent element.*

Proposition 2.2.3. *For every finite semigroup* E*, there exists an integer* q *such that, for every* $x \in E$*,* x^q *is idempotent.*

Proof. Following from Proposition 2.2.1, any element x of E has an idempotent power x^{n_x}. Let s be the least common multiple of n_x, for $x \in E$. Then x^s is idempotent for every $x \in E$. The smallest integer q satisfying this property is called the *exponent* of E. (s not being necessarily the smallest integer k such that x^k is idempotent $\forall x$). □

Proposition 2.2.4. *Let* E *be a finite semigroup and* $n = |E|$*. For every finite sequence* $x_1, x_2, \ldots, x_n$ *of elements of* E*, there exists an index* $i \in \{1, \ldots, n\}$ *and an idempotent* $e \in E$ *such that* $x_1 x_2 \cdots x_i\, e = x_1 x_2 \cdots x_i$.

Proof. Let us consider the sequence $\{x_1\}, \{x_1 \cdot x_2\}, \ldots, \{x_1 \cdot x_2 \cdots x_n\}$. If all the elements of this sequence are distinct, all the elements of E show up in it and one of them, let us say $x_1 x_2 \cdots x_i$, is idempotent (Corollary 2.2.2). The result in this case is immediate. Otherwise, two elements of the sequence are equal, let us say $x_1 \cdot x_2 \cdots x_i$ and $x_1 \cdot x_2 \cdots x_j$ with $i < j$. We then have $x_1 \cdots x_i = x_1 \cdots x_i\, (x_{i+1} \cdots x_j) = x_1 \cdots x_i\, (x_{i+1} \cdots x_j)^q$ where q is the exponent of E. The proposition follows from this, since $(x_{i+1} \cdots x_j)^q$ is idempotent (Proposition 2.2.3). □

With every idempotent e of a semigroup E, we associate the set

$$e\,E\,e = \{e\,x\,e / x \in E\}.$$

This is a sub-semigroup of E, referred to as the *local semigroup* associated with e, and which has e as neutral element. It is therefore a monoid and we easily verify that e E e is the set of elements x of E which have e as a neutral element, that is to say such that $e\,x = x\,e = e$.

2.3. Cancellative Monoids and Groups

Let us complete this section with the definition of cancellative elements, cancellative monoids and groups.

Let $(E, \oplus)$ be a monoid. We say that $a \in E$ is a cancellative element if and only if:

$$\forall x, y \in E: \quad x \oplus a = y \oplus a \Rightarrow x = y$$
$$\text{and} \quad a \oplus x = a \oplus y \Rightarrow x = y$$

When a only satisfies the first (resp. the second) of these conditions, we say that it is only right-cancellative (resp. left-cancellative). When the monoid is commutative, the two concepts coincide.

Definition 2.3.1. *(cancellative monoid)*

We call cancellative monoid a monoid $(E, \oplus)$ *endowed with a neutral element* ε *and in which all elements are cancellative.*

In a cancellative monoid, the internal law $\oplus$ is said to be *cancellative.*

Example 2.3.2. (free monoid for concatenation)

Let us return to the example of the free monoid A^* on an alphabet A (see Example 2.1.13).

It is easy to verify that every word $m \in A^*$ is right-cancellative and left-cancellative for the operation of concatenation. It is therefore a cancellative monoid. ||

Example 2.3.3. On $\mathbb{R}_+$, we consider the law $\oplus$ defined as $a \oplus b = \dfrac{a+b}{1+ab}$.

We verify that $(\mathbb{R}_+, \oplus)$ is a commutative monoid (see Example 2.1.4) with a neutral element 0, that 1 is an absorbing element and that every element different from 1 is cancellative. It then follows that $(\mathbb{R}_+\backslash\{1\}, \oplus\}$ is a cancellative monoid. ||

We say that an element a of a monoid E with neutral element ε has a left inverse (resp. right inverse) if there exists an element a′ (resp. a″) such that

$$a' \oplus a = \varepsilon$$
$$(\text{resp. } a \oplus a'' = \varepsilon)$$

An element a has an inverse if there exists an element a′ such that

$$a \oplus a' = a' \oplus a = \varepsilon$$

Definition 2.3.4. *(group)*

A monoid (E, ⊕) *in which every element* x *has an inverse is called a group.*

Proposition 2.3.5. *Every cancellative commutative monoid is isomorphic to the "nonnegative" elements of a commutative group.*

Proof. From the cancellative commutative monoid (E, ⊕), endowed with the neutral element ε, construct the set S whose elements are ordered pairs of elements of E:

$$S = \{(a, b)/a \in E, b \in E\}$$

By definition, the elements of S of the form (a, ε) are referred to as the nonnegative elements of S, and the elements of the form (ε, b) the nonpositive elements. We observe that there is a one-to-one correspondence between E and S^+, the set of nonnegative elements of S.

By defining on S the following *equivalence relation* $\mathcal{R}$:

$$(a_1, a_2)\, \mathcal{R}(b_1, b_2) \Leftrightarrow a_1 \oplus b_2 = a_2 \oplus b_1$$

and by endowing $G = S/\mathcal{R}$ with the law ⊕ defined as:

$$(a_1, a_2) \oplus (b_1, b_2) = (c_1, c_2)$$

where $(c_1, c_2)\, \mathcal{R}(a_1 \oplus b_1, a_2 \oplus b_2)$.

We easily verify that $G = S/\mathcal{R}$ is a commutative group and that E is isomorphic to the nonnegative elements of G. □

We observe that the concept of *nonnegative element* used in the above proof did not require the existence of an order relation on E (for the study of ordered monoids, see Sect. 3.2).

Remark 2.3.6. Even in the case where the commutative monoid (E, ⊕) is not cancellative, we can construct the set S whose elements are ordered pairs of elements of E: $S = \{(a, b)/a \in E, b \in E\}$, and define the following *equivalence relation* $\mathcal{R}$:

$$(a_1, a_2)\ \mathcal{R}(b_1, b_2) \Leftrightarrow \begin{cases} a_1 \neq b_1, a_2 \neq b_2 & \text{and} \quad a_1 \oplus b_2 = b_1 \oplus a_2 \\ (a_1, a_2) = (b_1, b_2) & \text{otherwise} \end{cases}$$

We then distinguish between three types of equivalence classes: the "nonnegative" elements corresponding to the classes $\overline{(a, \varepsilon)}$, the "nonpositive" elements corresponding to the classes $\overline{(\varepsilon, a)}$ and the "balanced" elements corresponding to the classes $\overline{(a, a)}$. ||

3. Ordered Monoids

The aim of this section is to study the monoids endowed with an order relation compatible with the monoid's internal law.

In Sect. 3.1, we recall some basic definitions concerning ordered sets. Then, in Sect. 3.2, we introduce the concept of ordered monoid, illustrating it through some examples. We next introduce, in Sect. 3.3, the *canonical preorder* relation in a monoid, followed by canonically ordered monoids in Sect. 3.4. Theorem 1 (stating that a monoid cannot both be a group and be canonically ordered) introduces an initial typology of monoids. The subsequent sections further expand the typology of canonically ordered monoids which may be divided into *semi-groups* (Sect. 3.5) *idempotent monoids* and *semi-lattices* (Sect. 3.6).

3.1. Ordered Sets

We recall that an order relation on E, denoted ≤, is a binary relation featuring:

$$\begin{aligned} &\text{reflexivity} \quad (\forall a \in E\colon a \le a), \\ &\text{transitivity} \quad (a \le b \quad \text{and} \quad b \le c \Rightarrow a \le c), \\ &\text{antisymmetry} \quad (a \le b \quad \text{and} \quad b \le a \Rightarrow a = b). \end{aligned}$$

Let E be an ordered set, that is to say a set endowed with an order relation ≤.

Two elements $a \in E, b \in E$ are said to be *non-comparable* if neither of the two relations $a \le b$ and $b \le a$ are satisfied.

If there exist non-comparable elements, we say that E is a *partially ordered set* or a *poset*.

On the other hand, if for any pair $a, b \in E$, either $a \le b$ or $b \le a$ holds, we say that we have a *total order* and E is called a *totally ordered set*.

Remark. Since the relation $\leq$ is reflexive, the set of the elements $x \in E$ satisfying $x \leq a$ contains the element a itself, we say that it is an order relation in the wide sense.

With every order relation $\leq$ in the wide sense it is possible to associate a strict order relation $<$ defined as:

$$a < b \Leftrightarrow a \leq b \quad \text{and} \quad a \neq b.$$

Observe that this relation is *irreflexive* ($a < a$ is not satisfied), *asymmetric* and *transitive*.

Conversely, with every strict order relation $<$ that is irreflexive, asymmetric and transitive, we can associate a symmetric, transitive and antisymmetric order relation $\leq$ defined as:

$$a \leq b \Leftrightarrow a < b \quad \text{or} \quad a = b. \quad ||$$

For a subset $A \subseteq E$, an element $a \in E$ satisfying

$$\forall x \in A\colon x \leq a$$

is called an *upper bound* of A.

An upper bound of A which belongs to A is called the *largest element of* A.

When $A \subseteq E$ has a largest element a, it is necessarily unique. Let us in fact assume that a and a' are two upper bounds of A belonging to A.

We have: $\forall x \in A, \quad x \leq a,$ and, in particular: $a' \leq a$.
Similarly: $\forall x \in A, \quad x \leq a',$ and, in particular: $a \leq a'$.

Through the antisymmetry of $\leq$ we then deduce $a = a'$.

Similarly $b \in E$ is a *lower bound* of A if and only if:

$$\forall x \in A \quad b \leq x.$$

A lower bound of A which belongs to A is called *the smallest element of A*. If A has a smallest element, it is unique.

A subset $A \subseteq E$ is said to be *bounded* if it has an upper bound and a lower bound.

When the set of the upper bounds of $A \subseteq E$ has a smallest element, this smallest element is called the *supremum* of A. It is denoted sup (A). Similarly, when the set of the lower bounds of A has a largest element, we call it the *infimum* of A (denoted inf (A)).

We say that the ordered set $(E, \leq)$ is *complete* if every subset A of E has a supremum, which can be denoted $\vee A$ or $\underset{a \in A}{\vee} a$.

It is said to be *complete for the dual order* if every subset A of E has an infimum, which can be denoted $\wedge A$ or $\underset{a \in A}{\wedge} a$.

We say that a set $S \subset E$ is a *lower set* if $x \in S$ and $y \leq x$ implies $y \in S$. Given a subset P, we denote:

$$\downarrow(P) = \{x \in E \mid \exists\, p \in P, x \leq p\}.$$

This is the smallest lower set containing P.

Furthermore, if a lower set S satisfies, for all a, b $\in$ S, a $\vee$ b $\in$ S, then S is called an *ideal*. If it satisfies, for every a, b $\in$ S, a $\wedge$ b $\in$ S, then S is called a *filter*.

We observe that for x $\in$ E, $\downarrow(\{x\})$ is a ideal. The ideals of this form are referred to as *principal ideals* and denoted $\downarrow(x)$. Exercise 7 at the end of the chapter is concerned with the properties of ideals and filters.

We call *maximal element* of A $\subseteq$ E every a $\in$ A satisfying:

$$\nexists\, x \in A,\ x \neq a, \quad \text{such that:} \quad a \leq x.$$

Similarly, we call *minimal element* of A $\subseteq$ E every a $\in$ A satisfying:

$$\nexists\, x \in A,\ x \neq a, \quad \text{such that:} \quad x \leq a.$$

When a subset A $\subseteq$ E has a maximal (resp. minimal) element, the latter is not necessarily unique.

It is easy to show that every finite subset of a (partially) ordered set E has at least one maximal element (resp. one minimal element).

3.2. *Ordered Monoids: Examples*

Definition 3.2.1. *(ordered monoid)*

We say that a monoid (E, $\oplus$) *is ordered when we can define on* E *an order relation* $\leq$ *compatible with the internal law* $\oplus$*, that is to say such that:*

$$\forall a, b, c \in E \quad a \leq b \Rightarrow (a \oplus c) \leq (b \oplus c).$$

Example 3.2.2. The monoid ($\mathbb{R}_+$, +) is ordered for the order relation "less than or equal to" ($\leq$) on $\mathbb{R}_+$. ||

Example 3.2.3. The monoid ($\hat{\mathbb{R}}$, Min) is ordered for the order relation "less than or equal to" ($\leq$) on $\hat{\mathbb{R}} = \mathbb{R} \cup \{+\infty\}$. ||

Example 3.2.4. The monoid ($\mathbb{R}$, +) is ordered for the order relation "less than or equal to" ($\leq$) on $\mathbb{R}$. ||

Example 3.2.5. (a few algebraic models useful in fuzzy set theory)

An infinite class of ordered monoids can be deduced through isomorphisms from ($\mathbb{R}_+$, +). More precisely, for every one-to-one correspondence φ between M $\subset \mathbb{R}$ and $\mathbb{R}_+$, we can associate with every a $\in$ M, b $\in$ M, the real value:

$$a \oplus b = \varphi^{-1}\, [\varphi(a) + \varphi(b)].$$

This class of ordered monoids arises in connection with many algebraic models in fuzzy set theory (see e.g. Dubois and Prade 1980, 1987).

For instance, considering the parameter $h \in \mathbb{R}_+$, we obtain a family of ordered monoids associated with the following functions:

$$\begin{aligned}
\varphi_h(x) &= x^h && (x \in \mathbb{R}_+)\\
\varphi_h(x) &= x^{-h} && (x \in \mathbb{R}_+)\\
\varphi_h(x) &= e^{-\frac{x}{h}} && (x \in \mathbb{R})\\
\varphi_h(x) &= e^{\frac{x}{h}} && (x \in \mathbb{R})
\end{aligned}$$

Observe that the operation $\oplus^h$ defined as

$$a \oplus^h b = h \ln\left(e^{\frac{a}{h}} + e^{\frac{b}{h}}\right)$$

"tends" towards Max{a, b} when h "tends" towards 0^+, and that the operation $\oplus_h$ defined as $a \oplus_h b = -h \ln\left(e^{-\frac{a}{h}} + e^{-\frac{b}{h}}\right)$ "tends" towards Min{a, b} when h "tends" towards 0^+.

In the same way, the operation $\oplus^h$ defined as $a \oplus^h b = \left(a^h + b^h\right)^{1/h}$ "tends" towards Max{a, b} when h "tends" towards $+\infty$ and the operation $\oplus_h$ defined as $a \oplus_h b = \left(a^{-h} + b^{-h}\right)^{-1/h}$ "tends" towards Min(a, b) when h "tends" towards $+\infty$.

Similarly, we can consider a one-to-one correspondence φ relative to the multiplication (on $\mathbb{R}$) by setting: $a \oplus b = \varphi^{-1}\left[\varphi(a) \cdot \varphi(b)\right]$.

For a detailed study of some of these ordered monoids, refer to Exercise 2 at the end of the chapter. For the study of the asymptotic behavior of the operations $\oplus_h$ and $\oplus^h$, refer to Exercise 3. ||

3.3. *Canonical Preorder in a Commutative Monoid*

Given a commutative monoid (E, ⊕) with neutral element ε, it is always possible, thanks to the internal law ⊕, to define a *reflexive* and *transitive* binary relation, denoted ≤, as:

$$a \le b \Leftrightarrow \exists\, c \in E \quad \text{such that} \quad b = a \oplus c.$$

The reflexivity ($\forall a \in E$: $a \le a$) follows from the existence of a neutral element ε ($a = a \oplus \varepsilon$) and the transitivity is immediate because:

$$\begin{aligned}
a \le b &\Leftrightarrow \exists\, c: b = a \oplus c\\
b \le d &\Leftrightarrow \exists\, c': d = b \oplus c'
\end{aligned}$$

hence: $d = a \oplus c \oplus c'$, which implies $a \le d$.

Since the antisymmetry of ≤ is not automatically satisfied, we can see that ≤ is only a *preorder* relation. We call it the *canonical preorder relation* of (E, ⊕).

We observe that $\oplus$ being assumed to be commutative, the canonical preorder relation of (E, $\oplus$) is *compatible with the law* $\oplus$ because:

$$a \le b \Rightarrow \exists c\colon b = a \oplus c$$

therefore, $\forall d \in E$:

$$\begin{aligned} & b \oplus d = a \oplus c \oplus d = a \oplus d \oplus c \\ & \Rightarrow \ a \oplus d \le b \oplus d. \end{aligned}$$

When (E, $\oplus$) is a *noncommutative monoid* having a neutral element ε, we can define two canonical preorder relations, denoted $\underset{R}{\le}$ (right canonical preorder relation) and $\underset{L}{\le}$ (left canonical preorder relation) as follows:

$$a \underset{R}{\le} b \Leftrightarrow \exists c \in E \quad \text{such that:} \quad b = a \oplus c$$

$$a \underset{L}{\le} b \Leftrightarrow \exists c' \in E \quad \text{such that:} \quad b = c' \oplus a.$$

Here again, the properties of reflexivity (ε being a neutral element on the right and on the left) and transitivity are easily checked.

Example 3.3.1. The free monoid A^* on an alphabet A is not a commutative monoid (see Example 2.1.13). Two words $m_1 \in A^*$, $m_2 \in A^*$ satisfy $m_1 \underset{R}{\le} m_2$ if and only if there exists a word m_3 such that: $m_2 = m_1 \cdot m_3$, in other words if and only if m_1 is a *prefix* of m_2. Similarly: $m_1 \underset{L}{\le} m_2$ if and only if there exists a word m_3 such that: $m_2 = m_3 \cdot m_1$, in other words if and only if m_1 is a *suffix* of m_2. ||

3.4. Canonically Ordered Monoids

Definition 3.4.1. *A commutative monoid* (E, $\oplus$) *is said to be* canonically ordered *when the canonical preorder relation* $\le$ *of* (E, $\oplus$) *is an* order relation, *that is to say also satisfies the property of antisymmetry:* $a \le b$ *and* $b \le a \Rightarrow a = b$.

The Examples 3.2.2 ($\mathbb{R}_+$, +), 3.2.3 ($\hat{\mathbb{R}}$, Min) and 3.2.5 correspond to canonically ordered monoids. The monoid ($\mathbb{R}$, +) in Example 3.2.4 is not canonically ordered.

This property of canonical order with respect to the internal law $\oplus$ ***is precisely the one which will be involved in the basic definition of dioids in Sect. 6.***

The following is an important property on which the typology of monoids (see Sect. 3.9) and the distinction between *dioids* and *rings* (see Sect. 6) will be based:

Theorem 1. *A monoid cannot both be a group and canonically ordered.*

Proof. Let us assume that (E, $\oplus$) is a group (we denote a^{-1} the inverse of $a \in E$) and is canonically ordered. Let a and b be two arbitrary elements $a \neq b$.

Since (E, ⊕) is a group:

there exists c such that $a = b \oplus c \Rightarrow a \geq b$ (take $c = b^{-1} \oplus a$)

there exists d such that $b = a \oplus d \Rightarrow b \geq a$ (take $d = a^{-1} \oplus b$)

If (E, ⊕) is canonically ordered, we deduce $a = b$, which gives rise to a contradiction. □

Thus the group $(\mathbb{R}, +)$ is *not canonically ordered*, and the canonically ordered monoid $(\mathbb{R}_+, +)$ *is not a group*. Let us give some further examples of canonically ordered monoids.

Example 3.4.2. (qualitative addition)

On the set of the signs, together with the indeterminate ?, E = {+, −, 0, ?}, we consider the operation denoted ⊕ defined by the table:

⊕	+	−	0	?
+	+	?	+	?
−	?	−	−	?
0	+	−	0	?
?	?	?	?	?

(E, ⊕) is a canonically ordered idempotent monoid with 0 as neutral element. We have: ? ≥ + ≥ 0 and ? ≥ − ≥ 0, which may be represented by the following diagram:

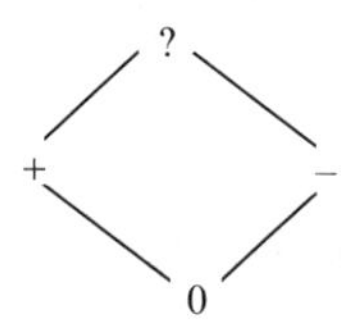

||

Example 3.4.3. (qualitative multiplication)

On the set of signs E = {+, −, 0}, we consider the product of signs ⊗(+⊗− = −, + ⊗ + = +, − ⊗ − = +, 0 ⊗ a = 0 ∀a ∈ E). (E, ⊗) is not a canonically ordered monoid.

We can add to the set E the indeterminate sign (denoted: ?) satisfying: ? ⊗ + =?, ? ⊗ − =?, ? ⊗ 0 = 0, ? ⊗ ? =?. Still, the resulting monoid is not canonically ordered. ||

Examples 3.4.2 and 3.4.3 define a *qualitative physics* where the various signs of E can have the following interpretation: + corresponds to the set]0, +∞[, − to the set]−∞, 0[, ? to the set]−∞, +∞[and 0 to the set {0}.

Example 3.4.4. (order of magnitude monoid)

We consider the set E formed of the pairs (a, α) with $a \in \mathbb{R}_+ \backslash \{0\}$ and $\alpha \in \mathbb{R}$, to which we add the pair $(0, +\infty)$.

We then define the $\oplus$ operation as:

$$(a, \alpha) \oplus (b, \beta) = (c, \min(\alpha, \beta)) \quad \text{with} \quad c = a \text{ if } \alpha < \beta, \quad c = b \text{ if } \alpha > \beta,$$
$$c = a + b \text{ if } \alpha = \beta.$$

We verify that $(E, \oplus)$ is a canonically ordered monoid with neutral element $(0, +\infty)$.

The elements (a, α) of this monoid correspond to the numbers of the form $a\,\varepsilon^\alpha$ when $\varepsilon > 0$ tends to 0^+.

By setting $p = -\ln(\varepsilon)$ and $A = e^{-\alpha}$, we have $\varepsilon^\alpha = A^p$. We can therefore define a new set F formed by the pairs $(a, A) \in (\mathbb{R}_+ \backslash \{0\})^2$ to which we add the pair $(0, 0)$.

F is endowed with the law $\oplus$ defined as $(a, A) \oplus (b, B) = (c, \max(A, B))$ with $c = a$ if $A > B$, $c = b$ if $A < B$, $c = a + b$ if $A = B$.

The elements (a, A) of this monoid correspond to the numbers of the form $a\,A^p$ when p tends to $+\infty$. ||

An important special case arises when the $\oplus$ law is commutative and *idempotent* (i.e. $\forall a \in E, a \oplus a = a$); the antisymmetry of $\le$ can then be directly deduced, without further assumption.

Proposition 3.4.5. *If* $\oplus$ *is commutative and idempotent, then the canonical preorder relation* $\le$ *is an order relation.*

Proof.

$$a \le b \Rightarrow \exists\, c\colon b = a \oplus c$$
$$b \le a \Rightarrow \exists\, c'\colon a = b \oplus c'$$

hence we deduce: $a = a \oplus c \oplus c'$

and

$$b = a \oplus c = a \oplus c \oplus c' \oplus c = a \oplus c \oplus c' = a$$

which proves antisymmetry. □

A slightly more general case of Proposition 3.4.5 arises when the $\oplus$ law is commutative and *m-idempotent*,

$$\text{i.e.} \quad \underbrace{a \oplus a \oplus \cdots \oplus a}_{m+1 \text{ times}} = \underbrace{a \oplus a \oplus \cdots \oplus a}_{m \text{ times}};$$

here again, the anti-symmetry of $\le$ is directly deduced.

We denote $m \times a$ the sum $\underbrace{a \oplus a \oplus \cdots \oplus a}_{m \text{ times}}$.

An example of 2-idempotency corresponds to the following law $\oplus$, defined for elements $a = \begin{pmatrix} a_1 \\ a_2 \end{pmatrix} \in R^2$ with $a_1 \leq a_2$ as:

$$\begin{pmatrix} a_1 \\ a_2 \end{pmatrix} \oplus \begin{pmatrix} b_1 \\ b_2 \end{pmatrix} = \begin{pmatrix} \min(a_1, b_1) \\ \min_2(a_1, a_2, b_1, b_2) \end{pmatrix}$$

where $\min_2(A)$ corresponds to the second smallest element of the set A.

We verify that $\oplus$ is 2-idempotent; indeed

$$2 \times \begin{pmatrix} a_1 \\ a_2 \end{pmatrix} \oplus \begin{pmatrix} a_1 \\ a_2 \end{pmatrix} = \begin{pmatrix} a_1 \\ a_1 \end{pmatrix} \quad \text{and} \quad 3 \times \begin{pmatrix} a_1 \\ a_2 \end{pmatrix} = 2 \times \begin{pmatrix} a_1 \\ a_1 \end{pmatrix}.$$

Proposition 3.4.6. *If $\oplus$ is commutative and m-idempotent, then the canonical preorder relation $\leq$ is an order relation.*

Proof.

$$a \leq b \Rightarrow \exists\, c: b = a \oplus c$$
$$b \leq a \Rightarrow \exists\, c': a = b \oplus c'$$

hence we deduce:

$$a = a \oplus c \oplus c' = b \oplus 2\,c' \oplus c = a \oplus 2\,c \oplus 2\,c' = \cdots = a \oplus m\,c \oplus m\,c',$$
$$b = b \oplus c \oplus c' = b \oplus m\,c \oplus m\,c' = a \oplus (m+1)\,c \oplus m\,c' = a \oplus m\,c \oplus m\,c'. \quad \square$$

Observe that, in a canonically ordered monoid, the relation:

$$\forall a \in E: a \oplus \varepsilon = a \quad \text{implies: } \varepsilon \leq a$$

which shows that ε is (the unique) smallest element of E.

Proposition 3.4.7. *If $\oplus$ is* selective *and* commutative ($a \oplus b = a$ *or* b) *then $\leq$ is a* total *order relation.*

Proof. Selectivity implies idempotency, therefore $\leq$ is an order relation.

Furthermore, $a \oplus b = a$ *or* b implies for every $a, b \in E$:

$$\begin{aligned} &\text{either} \quad a \leq b \\ &\text{or} \qquad b \leq a \end{aligned}$$

which proves that $\leq$ is a total order. $\square$

A selective operation is not necessarily commutative. As an example, the $\oplus$ operation defined as:

$\forall a, b \in E: a \oplus b = a$ (the result is the first of the two elements added) is clearly selective but not commutative (because $a \oplus b = a$ and $b \oplus a = b$).

Proposition 3.4.8. *In a canonically ordered monoid, the following so-called* positivity *condition is satisfied:*

$$a \in E, b \in E \quad \text{and} \quad a \oplus b = \varepsilon \Rightarrow a = \varepsilon \quad \text{and} \quad b = \varepsilon.$$

Proof. $a \oplus b = \varepsilon$ implies $a \leq \varepsilon$ and $b \leq \varepsilon$ but since: $\varepsilon \oplus a = a$ and $\varepsilon \oplus b = b$ we also have: $\varepsilon \leq a$ and $\varepsilon \leq b$.

From the antisymmetry of $\leq$ we then deduce $a = \varepsilon$ and $b = \varepsilon$. $\square$

3.5. Hemi-Groups

Definition 3.5.1. *(hemi-group)*

We call hemi-group *a monoid which is both canonically ordered and cancellative.*

The set $(\mathbb{N}, +)$ is a canonically ordered monoid in which every element is cancellative. It is therefore a hemi-group. The same applies to $(\mathbb{R}_+, +)$, see Example 3.2.2. On the other hand, the set of reals $\mathbb{R}$ endowed with addition and the usual (total) order relation (see Example 3.2.4) is a cancellative ordered monoid but not a canonically ordered one. It is therefore not a hemi-group.

Property 3.5.2. A cancellative commutative monoid $(E, \oplus)$ is a hemi-group if it satisfies the so-called *zero-sum-free* condition: $a \oplus b = \varepsilon \Rightarrow a = \varepsilon$ and $b = \varepsilon$.

Proof. It suffices to show that $(E, \oplus)$ is canonically ordered.

Let us then assume that: $a \le b$ and $b \le a$, So:

$$\exists\, c \quad \text{such that:} \quad b = a \oplus c$$
$$\exists\, d \quad \text{such that:} \quad a = b \oplus d$$

hence:

$$a \oplus b = a \oplus b \oplus \varepsilon = a \oplus b \oplus c \oplus d.$$

Since $a \oplus b$ is a cancellative element, we deduce $c \oplus d = \varepsilon$

The condition of positivity then implies

$$c = d = \varepsilon \quad \text{hence} \quad a = b.$$

The canonical preorder relation is therefore clearly an order relation. □

The *zero-sum-free* condition involved in the previous result is satisfied by many algebraic structures investigated in the present work, e.g. the boolean algebra, distributive lattices, and inclines (see Cao, Kim & Roush, 1984).

3.6. Idempotent Monoids and Semi-Lattices

The concepts of semi-lattice (sup-semi-lattice, inf-semi-lattice) may be defined, either in terms of sets endowed with (partial) order relations, or in algebraic terms. We recall the set-based definitions below, then we show that algebraically, semi-lattices are in fact idempotent monoids.

Definition 3.6.1. *(idempotent monoid)*

A monoid $(E, \oplus)$ *is said to be* idempotent *if the law* $\oplus$ *is commutative, associative and idempotent, that is to say satisfies*:

$$\forall a \in E, \quad a \oplus a = a.$$

Observe here that a cancellative monoid not reduced to its neutral element ε cannot be idempotent. Indeed, for every $a \neq \varepsilon$, $a \oplus a = a = a \oplus \varepsilon$ implies, since a is regular, $a = \varepsilon$, which gives rise to a contradiction. Hemi-groups and idempotent monoids therefore correspond to two disjoint sub-classes of canonically ordered monoids (see Fig. 1 Sect. 3.7).

Proposition 3.6.2. *If* $(E, \oplus)$ *is an idempotent monoid, then the canonical order relation* $\leq$ *can be characterized as*:

$$a \leq b \Leftrightarrow a \oplus b = b.$$

Proof. $a \leq b$ is by definition equivalent to:

$$\exists c \in E \quad \text{such that} \quad a \oplus c = b.$$

We can then write:

$$a \oplus b = a \oplus a \oplus c = a \oplus c = b.$$

We therefore clearly have $a \leq b \Leftrightarrow a \oplus b = b$. □

Definition 3.6.3. *(sup- and inf-semi-lattices)*

We call sup-semi-lattice *a set* E, *endowed with an order relation* $\leq$, *in which every pair of elements* (x, y) *has a least upper bound denoted* $x \vee y$.

Similarly, we call inf-semi-lattice *a set* E, *endowed with an order relation, in which every pair of elements* (x, y) *has a greatest lower bound denoted* $x \wedge y$.

A sup-semi-lattice (resp. inf-semi-lattice) is said to be complete *if every finite or infinite set of elements has a least upper bound (resp. a greatest lower bound).*

Theorem 2. *Every sup-semi-lattice (resp. inf-semi-lattice)* E *is an idempotent monoid for the internal law* $\oplus$ *defined as:*

$$\forall x, y \in E: x \oplus y = x \vee y \quad (\text{resp. } x \oplus y = x \wedge y).$$

Conversely if $(E, \oplus)$ *is an idempotent monoid, then* E *is a sup-semi-lattice for the canonical order relation* $\leq$.

Proof. Let E be a sup-semi-lattice, where $\forall x, y \in E$, $x \vee y$ denotes the least upper bound of x and y; then $(E, \vee)$ is an idempotent monoid.

Conversely, let $(E, \oplus)$ be an idempotent monoid, and let $\leq$ be the canonical order relation. We have $a \oplus b \geq a$ and $a \oplus b \geq b$. This is also the least upper bound of the set $\{a, b\}$ because for every other upper bound x of a and of b, $a \leq x$ and $b \leq x$, we have:

$$a \oplus b \leq x \oplus x = x. \quad \square$$

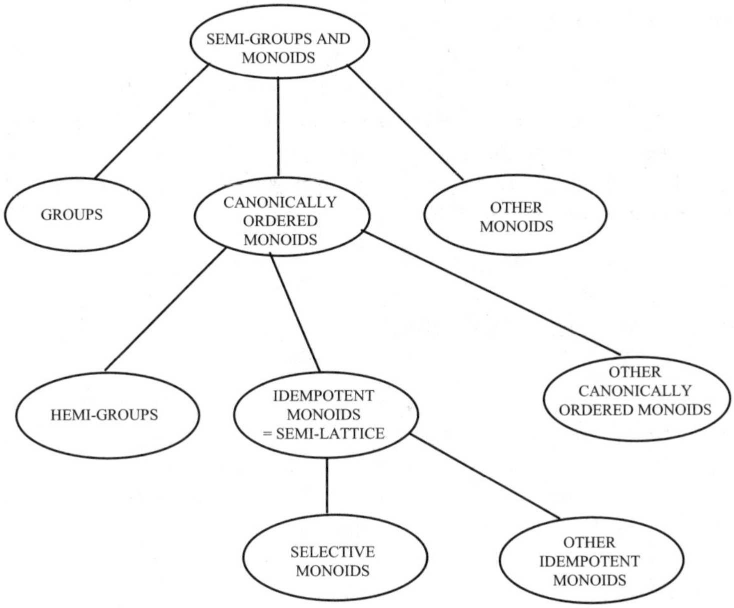

Fig. 1 Classification of monoids

Table 1 The various types of monoids and their basic properties

	Properties of $\oplus$	Canonical preorder relation $\leq$	Additional properties and comments
Monoid	Associative	Preorder	
Commutative monoid	Commutative	Preorder	
Cancellative monoid	Commutative, neutral element, every element is cancellative	Preorder	
Ordered monoid		Preorder	Monoid endowed with an order relation different from the canonical preorder relation
Group	Neutral element ε, every element has an inverse		
Commutative group	Invertible commutative		
Canonically ordered monoid		Order	Monoid in which the canonical preorder relation is an is an order
Idempotent monoid (semi-lattice)	Idempotent	Order	
Selective monoid	Selective	Total order	
Hemi-Group	Cancellative monoid (every element is cancellative)	Order	The zero-sum-free condition is satisfied (see Sect. 3.5)

3.7. Classification of Monoids

Table 1 sums up the main properties of the various types of monoids.

Figure 1 provides a graphic representation of the classification of monoids. Observe on the first level the disjunction between the class of groups and that of canonically ordered monoids and, on the second level, the disjunction between idempotent monoids and hemi-groups.

4. Pre-Semirings and Pre-Dioids

The term of *dioid* was initially suggested by Kuntzmann (1972) to denote the algebraic structure composed of a set E endowed with two internal laws $\oplus$ and $\otimes$ such that (E, $\oplus$) is a commutative monoid, (E, $\otimes$) is a monoid (which is not necessarily commutative) with a property of right and left distributivity of $\otimes$ with respect to $\oplus$. In the absence of additional properties for the laws $\oplus$ and $\otimes$, such a structure is quite limited and here we refer to it as a *pre-semiring*, thus keeping the name of *semi-ring* and of *dioid* for structures with two laws endowed with a few additional properties as explained in Sects. 5 and 6.

4.1. Right, Left Pre-Semirings

Definition 4.1.1. *We call* left pre-semiring *an algebraic structure* (E, $\oplus$, $\otimes$) *formed of a ground set* E *and two internal laws* $\oplus$ *and* $\otimes$ *with the following properties:*

(i) $a \oplus b = b \oplus a \quad \forall a, b \in E$ *(commutativity of* $\oplus$*)*
(ii) $(a \oplus b) \oplus c = a \oplus (b \oplus c) \quad \forall a, b, c \in E$ *(associativity of* $\oplus$*)*
(iii) $(a \otimes b) \otimes c = a \otimes (b \otimes c) \quad \forall a, b, c \in E$ *(associativity of* $\otimes$*)*
(iv) $a \otimes (b \oplus c) = (a \otimes b) \oplus (a \otimes c) \quad \forall a, b, c \in E$
(left distributivity of $\otimes$ *relative to* $\oplus$*)*

The concept of right pre-semiring *is defined similarly, by replacing left distributivity with right distributivity:*

(iv)′ $(a \oplus b) \otimes c = (a \otimes c) \oplus (b \otimes c) \quad \forall a, b, c \in E.$

We observe that in the above definitions, we do not assume the existence of neutral elements. If they do not exist (neither on the right nor on the left), we can easily add them. In the case where ε, the neutral element added for $\oplus$, is absorbing for $\otimes$, we have a semiring structure, see Sect. 5.

Example 4.1.2. There exist many cases where there is neither right distributivity nor left distributivity. As an example, the structure (E, $\oplus$, $\otimes$) with $E = [0, 1]$, $a \oplus b = a + b - ab$ and $a \otimes b = ab$ does not enjoy distributivity and is therefore not a pre-semiring. The same applies to the structure (E, $\oplus$, $\otimes$) with $E = [0, 1]$, $a \oplus b = \mathrm{Min}(1, a + b)$, $a \otimes b = \mathrm{Max}(0, a + b - 1)$. ||

The reason why it is of interest not to assume both right and left distributivity in the most basic structure (the pre-semiring structure) is that there exist interesting applications which do not enjoy both properties. This is the case, in particular, of Example 4.1.3 below.

Example 4.1.3. Left pre-semiring of the set of mappings of a monoid onto itself.

Let $(E, \overset{\circ}{+})$ be a commutative monoid, and H the set of mappings: $E \to E$.

We define on H the following operations $\oplus$ and $\otimes$:

For every $f \in H$, $g \in H$ we denote $f \oplus g$ the mapping which associates with every $a \in E$ the value $f(a) \overset{\circ}{+} g(a)$.

The properties of the $\overset{\circ}{+}$ operation on E induce similar properties for $\oplus$ on H.

If $\overset{\circ}{+}$ has a neutral element ε in E, then we can define the neutral element of $\oplus$ on H as the mapping h^ε $(E \to E)$ given by:

$$h^\varepsilon(a) = \varepsilon, \forall a \in E.$$

$(H, \oplus)$ is therefore a commutative monoid with neutral element h^ε.

For every $f, g \in H$ we denote $f \otimes g$ the mapping which, with every $a \in E$ associates $g \circ f(a) = g(f(a))$ ($\otimes$ is therefore directly deduced from the law of composition for mappings).

We observe that $\otimes$ is associative and has a neutral element which is the identity mapping h^e defined as:

$$h^e(a) = a, \ \forall a \in E.$$

We check the property of left distributivity because $\forall f, g, h \in H$, and $\forall a \in E$:

$$\begin{aligned} f \otimes [g \oplus h](a) &= ([g \oplus h] \circ f)(a) \\ &= [g \oplus h]\ (f(a)) \\ &= g(f(a)) \oplus h\ (f(a)) \\ &= [g \circ f \oplus h \circ f](a) \\ &= [(f \otimes g) \oplus (f \otimes h)](a). \end{aligned}$$

On the other hand, the property of right distributivity is not satisfied without additional assumptions (see Sect. 4.2, Example 4.2.2). The structure $(H, \oplus, \otimes)$ above is therefore a *left pre-semiring*. ||

Particular instances of the above structure have been proposed and studied by many authors in the area of computer program analysis, specifically through data flow analysis. For example, in the case of the data flow analysis models referred to as *monotone* (see e.g. Kam and Ullman 1977), E is taken as an idempotent monoid (= sup semi-lattice) and H is the set of monotone functions: $E \to E$. For further detail, refer to Chap. 8, Sects. 2.1 and 2.2.

4.2. Pre-Semirings

Definition 4.2.1. *We call* pre-semiring *an algebraic structure* (E, ⊕, ⊗) *which is both a right pre-semiring and a left pre-semiring.*

Example 4.2.2. Pre-semiring of the endomorphisms of a commutative monoid.

Let us return to Example 4.1.3 above, but now we assume that we are studying a particular subset of mappings $H' \subset H$ satisfying:

$$h(a \overset{\circ}{+} b) = h(a) \overset{\circ}{+} h(b) \quad \forall h \in H', \forall a, b \in E$$

in other words, we are dealing with *endomorphisms* of E (it is easily checked that H′ is closed for ⊕ and for ⊗).

In addition to left distributivity, which is always present (see Example 4.1.3), right distributivity is now satisfied because:

$$\begin{aligned}\forall a \in E\colon (g \oplus g') \otimes f(a) &= f(g \oplus g'(a)) \\ &= f(g(a) \overset{\circ}{+} g'(a)) \\ &= f(g(a)) \overset{\circ}{+} f(g'(a)) \\ &= g \otimes f(a) \overset{\circ}{+} g' \otimes f(a)\end{aligned}$$

From this we deduce that (H′, ⊕, ⊗) defined above is a *pre-semiring.* ||

A structure of this kind has interesting applications in the area of program analysis (problems of *continuous data flow*, see Kildall 1973) and the study of non-classical path-finding problems in graphs (see Minoux 1976). Examples of such pre-semirings are described in Chap. 8, Sect. 2.2.

4.3. Pre-Dioids

Definition 4.3.1. *We call right pre-dioid (resp. left pre-dioid) a right pre-semiring (resp. left pre-semiring) canonically ordered with respect to ⊕.*

We call pre-dioid a canonically ordered pre-semiring.

Example 4.3.2. The set $\mathbb{R}_+$ endowed with the internal laws Max and + is a pre-dioid (see Chap. 8 Sect. 2.4). ||

In a pre-dioid, the neutral element ε is not necessarily absorbing (see Example 5.3.1). On the other hand, the following result shows that ε is always *nilpotent.*

Proposition 4.3.3. *Let* (E, ⊕, ⊗) *be a pre-dioid in which* ε *and e are the neutral elements of* ⊕ *and* ⊗ *respectively. Then* ε *is* nilpotent ($\varepsilon^k = \varepsilon, \forall k \in \mathbb{N}$).

Proof. It suffices to show that $\varepsilon^2 = \varepsilon$.

Let ≤ be the canonical order relation of (E, ⊕). We have, $\forall a \in E\colon \varepsilon \oplus a = a$ hence: $\varepsilon \le a$.

We deduce $\varepsilon \le \varepsilon^2 = \varepsilon \otimes \varepsilon$, then $\varepsilon \le \varepsilon^2 \le \varepsilon^3$.

By using right and left distributivity, we can also write:

$$
\begin{aligned}
e = e \otimes e = (e \oplus \varepsilon) \otimes (e \oplus \varepsilon) &= (e \oplus \varepsilon) \otimes e \oplus (e \oplus \varepsilon) \otimes \varepsilon \\
&= e \oplus \varepsilon \oplus \varepsilon \oplus \varepsilon^2 \\
&= e \oplus \varepsilon^2
\end{aligned}
$$

We deduce: $\varepsilon^2 \le e$, then $\varepsilon^3 \le \varepsilon$.

From the above it follows:

$$\varepsilon = \varepsilon^2 = \varepsilon^3. \quad \square$$

Proposition 4.3.4. *Let* $(E, \oplus, \otimes)$ *be a pre-dioid where* ε*, the neutral element for* $\oplus$*, is non-absorbing. Then* ε E *is an idempotent pre-dioid.*

Proof. For every $a \in E$ we have:

$$a = (e \oplus \varepsilon) \otimes (a \oplus \varepsilon) = a \oplus \varepsilon \oplus (\varepsilon \otimes a) \oplus \varepsilon^2 = a \oplus (\varepsilon \otimes a)$$

($\varepsilon^2 = \varepsilon$ from Proposition 4.3.3).

By multiplying on the left by ε, we obtain: $\varepsilon \otimes a = (\varepsilon \otimes a) \oplus (\varepsilon^2 \otimes a) = (\varepsilon \otimes a) \oplus (\varepsilon \otimes a)$, which shows that $\oplus$ is idempotent for all elements of the form $\varepsilon \otimes a$ and thus ε E is an idempotent pre-dioid. $\square$

5. Semirings

5.1. Definition and Examples

Definition 5.1.1. *(semiring, right semiring, left semiring)*

A semiring is a pre-semiring $(E, \oplus, \otimes)$ *which satisfies the following additional properties:*

(i) $\oplus$ *has a neutral element* ε
(ii) $\otimes$ *has a neutral element* e
(iii) ε *is absorbing for* $\otimes$*, that is to say:*

$$\forall a \in E: a \otimes \varepsilon = \varepsilon \otimes a = \varepsilon.$$

A right semiring (resp. left) is a right pre-semiring (resp. left) satisfying property (i) *and properties* (ii)′ *and* (iii)′ *below.*

(ii)′ $\otimes$ *has e as a right neutral element* ($a \otimes e = a, \forall a$) (*resp. left:* $e \otimes a = a, \ \forall a$)
(iii)′ ε *is a right absorbing element* ($a \otimes \varepsilon = \varepsilon, \ \forall a$) (*resp. left:* $\varepsilon \otimes a = \varepsilon, \ \forall a$)

A semiring in which the operation $\otimes$ *is commutative is said to be* commutative.

Example 5.1.2. Let us return to Example 4.2.2 where $(E, \oplus)$ is a commutative monoid and H the set of endomorphisms of E. We have seen that $(H, \oplus, \otimes)$ is a pre-semiring. The neutral element h^ε of H for $\oplus$ does not satisfy the absorption property in general.

On the other hand, if we consider the subset $H' \subseteq H$ of endomorphisms having the additional property:

$$h \in H' \Leftrightarrow h \in H \quad \text{and} \quad h(\varepsilon) = \varepsilon$$

then the absorption property:

$\forall h \in H'$: $h^{\varepsilon} \otimes h = h \otimes h^{\varepsilon} = h^{\varepsilon}$ is satisfied and the structure $(H', \oplus, \otimes)$ is a *semiring*. ||

Example 5.1.3. Let us return to Example 4.1.3 of Sect. 4.1 relative to the set H of the mappings of a commutative monoid E onto itself. The neutral element h^{ε} of H for $\oplus$ does not satisfy the absorption property in general. On the other hand, if we limit ourselves to the subset $H' \subseteq H$ of mappings: $E \to E$ having the additional property:

$$h \in H' \Leftrightarrow h \in H \quad \text{and} \quad h(\varepsilon) = \varepsilon$$

then the absorption property is satisfied and the structure $(H', \oplus, \otimes)$ is a *left semiring*. ||

The class of ***semirings*** can be naturally subdivided into ***two disjoint sub-classes*** depending on whether the law $\oplus$ satisfies one of the following two properties:

(1) The law $\oplus$ endows the set E with a group structure;
(2) The law $\oplus$ endows the set E with a canonically ordered monoid structure.

In view of Theorem 1 of Sect. 3.4, (1) and (2) cannot be satisfied simultaneously. In case (1), we are led to the well-known ***Ring*** structure, whose definition is recalled in Sect. 5.2; in case (2) we are lead to the ***Dioid*** structure, (see Sect. 6 below) the in-depth study of which is one of the main objectives of the present volume.

Apart from Dioids and Rings, the other classes of semirings appear to have less potential interest. For example, we can mention the semirings obtained as products of a ring and a dioid. Some other examples are given in Chap. 8, Sect. 2.

5.2. Rings and Fields

Definition 5.2.1. *(ring)*

We call ring *a semiring in which the basic set* E *has a* commutative group *structure for the addition* $\oplus$. *A ring* $(E, \oplus, \otimes)$ *is said to be* commutative *if the operation* $\otimes$ *is commutative.*

Example 5.2.2. The set $(\mathbb{Z}, +, \times)$ of signed integers endowed with the standard operations $+$ and $\times$ is a commutative ring. Similarly, the set of square $n \times n$ matrices with real entries is a (non commutative) ring. ||

An important special case of the ring structure is obviously the ***field*** structure in which the basic set E has a group structure (not necessarily a commutative one) with respect to the law $\otimes$. When $\otimes$ is commutative, we refer to a commutative field.

Definition 5.2.3. *(semi-field)*

We call semi-field *a semiring in which every element other than* ε *has an inverse for the multiplication* $\otimes$.

We will see that many dioids, in particular idempotent-invertible dioids such as $(\mathbb{R}, \text{Min}, +)$ and $(\mathbb{R}, \text{Max}, +)$ are semi-fields (see Sect. 6.7 and Exercise 8).

Hereafter, we will use rarely the term of semi-field, because the resulting classification (based on the properties of "multiplication") would not be directly comparable to that of semirings (based on the properties of "addition"). Classifying with respect to the properties of the first law appears to be more fundamental insofar, in that it is with respect to the first law that the distributivity of the second law is defined.

5.3. The Absorption Property in Pre-Semi-Rings

In order for a pre-semiring to be a semiring, ε (the neutral element for $\oplus$) must be absorbing for $\otimes$. This is not always the case as seen in the following example.

Example 5.3.1. Let us take for E the set of intervals of the real line $\mathbb{R}$ of the form $[\underline{a}, \overline{a}]$ with $\underline{a} \leq 0$ and $\overline{a} \geq 0$.

Let us define the $\oplus$ law as:

$$[\underline{a}, \overline{a}] \oplus [\underline{b}, \overline{b}] = [\text{Min}\{\underline{a}, \underline{b}\}, \text{Max}\{\overline{a}, \overline{b}\}]$$

$\oplus$ is commutative and idempotent with the interval [0, 0] as a neutral element Furthermore, let us define the $\otimes$ law as:

$$[\underline{a}, \overline{a}] \otimes [\underline{b}, \overline{b}] = [\underline{a} + \underline{b}, \overline{a} + \overline{b}].$$

The $\otimes$ law has as neutral element [0, 0].

The distributivity of $\otimes$ with respect to $\oplus$ follows from the immediate properties:

$$\text{Min}\{\underline{a}, \underline{b}\} + \underline{c} = \text{Min}\{\underline{a} + \underline{c}, \underline{b} + \underline{c}\}$$
$$\text{Max}\{\overline{a}, \overline{b}\} + \overline{c} = \text{Max}\{\overline{a} + \overline{c}, \overline{b} + \overline{c}\}.$$

Finally, the canonical preorder relation is an order relation because of the idempotency of $\oplus$.

On the other hand, $(E, \oplus, \otimes)$ *is not a semiring* because ε is not absorbing for $\otimes$. Indeed, for an arbitrary element $[\underline{a}, \overline{a}] \neq \varepsilon$ we have $[\underline{a}, \overline{a}] \otimes [0, 0] = [\underline{a}, \overline{a}] \neq \varepsilon$. The structure $(E, \oplus, \otimes)$ defined above is therefore a pre-semiring (in fact a pre-dioid, due to the canonical order relation, see. Sect. 4.3) but it is not a semiring. ||

The above example shows, moreover, that assuming that $\leq$ is an order relation is not sufficient to guarantee the absorption property.

The following result provides a sufficient condition to guarantee this property in a canonically ordered pre-semiring, that is to say, in a pre-dioid.

Proposition 5.3.2. *If* $\leq$ *(the canonical preorder) is an order relation and if,* $\forall a \in E$: $a \otimes \varepsilon \leq \varepsilon$, *then we have the absorption property:*

$$\forall a \in E\colon a \otimes \varepsilon = \varepsilon.$$

Proof. For every $b \in E$ we have: $b = \varepsilon \oplus b$, therefore $\varepsilon \leq b$.

In particular, if we consider an arbitrary element $a \in E$ and we apply the above property to $b = a \otimes \varepsilon$, we obtain:

$$\forall a \in E\colon \varepsilon \leq a \otimes \varepsilon.$$

With the assumption of the proposition, we therefore have $\forall a \in E$: $a \otimes \varepsilon = \varepsilon$. □

Observe that the above proposition applies in particular when e, the neutral element for $\otimes$, is the largest element of E, that is to say when, for every a, $a \leq e$.

5.4. Product of Semirings

Given p semirings $(E_i, \oplus_i, \otimes_i)$ the *product* semiring is defined as the set $E = E_1 \times E_2 \times \cdots \times E_p$ endowed with the "product" laws $\oplus$ and $\otimes$ defined as:

$$\forall x = \begin{pmatrix} x_1 \\ \vdots \\ x_p \end{pmatrix} \in E, \quad \forall y = \begin{pmatrix} y_1 \\ \vdots \\ y_p \end{pmatrix} \in E\colon \quad x \oplus y = \begin{pmatrix} x_1 \oplus_1 y_1 \\ x_2 \oplus_2 y_2 \\ \vdots \\ x_p \oplus_p y_p \end{pmatrix}$$

and:

$$x \otimes y = \begin{pmatrix} x_1 \otimes_1 y_1 \\ x_2 \otimes_2 y_2 \\ \vdots \\ x_p \otimes_p y_p \end{pmatrix}$$

We easily verify that the laws $\oplus$ and $\otimes$ enjoy the same basic properties as the laws $\oplus_i$ and $\otimes_i$, and that, consequently, $(E, \oplus, \otimes)$ clearly has a semiring structure.

In particular, we verify that the product of a ring and a dioid is a semiring (see also Chap. 8, Sect. 3.1).

5.5. Classification of Pre-Semirings and Semirings

Table 2 below sums up the main properties of the various types of pre-semirings and semirings.

Figure 2 provides a graphic representation of the classification. In the semiring class, it shows two main disjoint sub-classes, *rings* (see Sect. 5.2) and *dioids* which are studied in Sect. 6. In the first case, $(E, \oplus)$ is a *group*, in the second case $(E, \oplus)$ is *canonically ordered*, these two properties being incompatible in view of Theorem 1 (Sect. 3.4).

Table 2 Pre-semirings, semirings and dioids and their basic properties

	Properties of (E, ⊕)	Properties of (E, ⊗)	Relation ≤	Additional properties and comments
Right (resp. left) pre-semiring	Commutative monoid	Monoid	Preorder	Right (resp. left) distributivity of ⊗ with respect to ⊕
Pre-semiring	Commutative monoid	Monoid		Right and left distributivity of ⊗ with respect to ⊕
Semiring	Commutative monoid, neutral element ε	Monoid, neutral element e		Right and left distributivity of ⊗ with respect to ⊕, ε absorbing for ⊗
Ring	Commutative group	Monoid, neutral element e		
Dioid	Canonically ordered monoid	Monoid, neutral element e	Order	

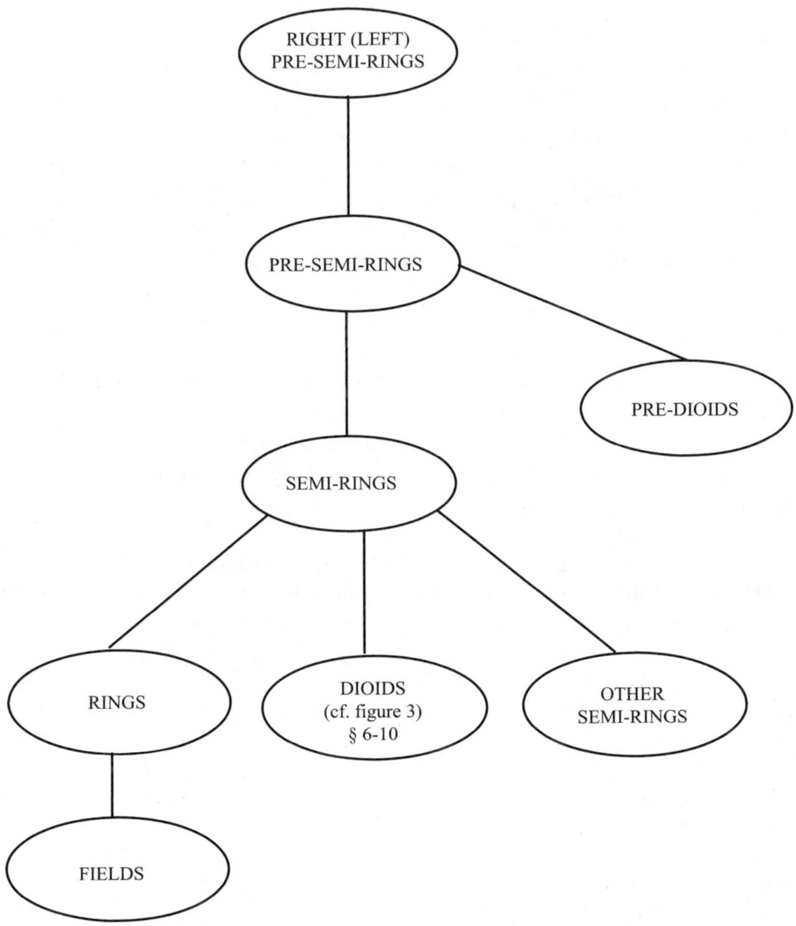

Fig. 2 Classification of pre-semirings, semirings and dioids

6. Dioids

6.1. Definition and Examples

Definition 6.1.1. *(dioid, right dioid, left dioid)*

We call dioid *a set* (E, ⊕, ⊗) *endowed with two internal laws* ⊕ *and* ⊗ *satisfying the following properties:*

(i) (E, ⊕) *is a commutative monoid with neutral element* ε*;*
(ii) (E, ⊗) *is a monoid with neutral element e;*
(iii) The canonical preorder relation relative to ⊕ *(defined as:* $a \le b \Leftrightarrow \exists$ c: $b = a \oplus c$*) is an order* relation, *i.e. satisfies:* $a \le b$ *and* $b \le a \Rightarrow a = b$*;*
(iv) ε *is absorbing for* ⊗*, i.e.:* $\forall a \in E$: $a \otimes \varepsilon = \varepsilon \otimes a = \varepsilon$*;*
(v) ⊗ *is right and left distributive with respect to* ⊕.

We call right dioid (resp. left dioid) a set (E, ⊕, ⊗) satisfying the properties (i) to (iv) above and where ⊗ is only right distributive (resp. only left distributive) with respect to ⊕. (We observe that for a right dioid, it in fact suffices for e to be a right neutral element ($a \otimes e = a, \forall a$) and for ε to be right-absorbing only ($a \otimes \varepsilon = \varepsilon, \forall a$)).

The fundamental difference between a ring and a dioid lies in property (iii). ***In a ring, addition induces a group structure, whereas in a dioid, it induces a canonically ordered monoid structure. From Theorem* 1 *(Sect.* 3.4*) this implies a disjunction between the class of rings and the class of dioids.***

Example 6.1.2. $\mathbb{Z}$ endowed with the standard operations + and ×, is a *ring* but it is not a *dioid*.

Indeed, in this structure, we always have, for every pair of signed integers a, b: $a \le b$ and $b \le a$ for the canonical preorder relation ($a \le b \Leftrightarrow \exists$ c: $b = a + c$).

It is therefore not an order relation. On the other hand, the semiring $\mathbb{N}$ (the set of natural integers) is a dioid because the canonical preorder relation coincides with the standard (total) order relation. ||

It is therefore the presence of an order relation intrinsically linked to the addition ⊕ which constitutes the main distinction between rings and dioids. This order relation will naturally lead to define topological properties. These will be studied in Chap. 3.

Remark. In the Definition 6.1.1, we can replace (iv) by the weaker assumption: $\varepsilon \ge a \otimes \varepsilon$ which suffices to guarantee $a \otimes \varepsilon = \varepsilon$, according to Proposition 5.3.2. ||

Apart from the well-known dioids ($\mathbb{N}$, +, ×) and ($\mathbb{R}$, Min, +) let us give a few other examples of interesting dioids.

Example 6.1.3. Qualitative algebra

On the set of signs E = {+, −, 0, ?}, we consider the law ⊕ (qualitative addition, see Example 3.4.2) and the law ⊗ (qualitative multiplication, see Example 3.4.3). We verify that (E, ⊕, ⊗) is a dioid. (see Chap. 8, Sect. 4.5.3) ||

Example 6.1.4. Right dioid and shortest path with gains or losses

On the set $E = \mathbb{R} \times (\mathbb{R}_+ \backslash \{0\})$ we define the following operations $\oplus$ and $\otimes$:

$$\begin{pmatrix} a \\ k \end{pmatrix} \oplus \begin{pmatrix} a' \\ k' \end{pmatrix} = \begin{cases} \begin{pmatrix} a \\ k \end{pmatrix} & \text{if} \quad \dfrac{a}{k} < \dfrac{a'}{k'} \quad \text{or if} \\ \dfrac{a}{k} = \dfrac{a'}{k'} & \text{and} \quad k = \max\{k, k'\} \\ \left(\dfrac{a'}{k'}\right) & \text{if} \quad \dfrac{a}{k} > \dfrac{a'}{k'} \quad \text{or if} \\ \dfrac{a}{k} = \dfrac{a'}{k'} & \text{and} \quad k' = \max\{k, k'\} \end{cases}$$

$$\begin{pmatrix} a \\ k \end{pmatrix} \otimes \begin{pmatrix} a' \\ k' \end{pmatrix} = \begin{pmatrix} a + ka' \\ kk' \end{pmatrix}$$

$\oplus$ has as neutral element ε any element of the form $\binom{+\infty}{k}$ with $k \in \mathbb{R}_+ \backslash \{0\}$ and $\otimes$ has as neutral element $e = \binom{0}{1}$.

We easily verify all of the properties (i)–(iv) as well as the right distributivity of $\otimes$ with respect to $\oplus$. On the other hand, $\otimes$ is not left distributive as shown in the following example:

$$\binom{4}{1} \otimes \left[\binom{1}{1} \oplus \binom{4}{2}\right] = \binom{4}{1} \otimes \binom{1}{1} = \binom{5}{1}$$

$$\left[\binom{4}{1} \otimes \binom{1}{1}\right] \oplus \left[\binom{4}{1} \otimes \binom{4}{2}\right] = \binom{5}{1} \oplus \binom{8}{2} = \binom{8}{2}$$

The structure $(E, \oplus, \otimes)$ is therefore a *right dioid*. This is the algebraic structure underlying the resolution of the shortest path problem with gains or losses (see Chap. 4, Exercise 4). ||

Example 6.1.5. Order of magnitude dioid

On the set E of pairs (a, α) with $a \in \mathbb{R}_+ \backslash \{0\}$ and $\alpha \in \mathbb{R}$, to which we add the pair $(0, +\infty)$, we define the two laws $\oplus$ and $\otimes$ as:

$$(a, \alpha) \oplus (b, \beta) = (c, \min(\alpha, \beta))$$

with $c = a$ if $\alpha < \beta$, $c = b$ if $\alpha > \beta$, $c = a + b$ if $\alpha = \beta$, (see Example 3.4.4)

$$(a, \alpha) \otimes (b, \beta) = (a\,b, \alpha + \beta).$$

We check that $(E, \oplus, \otimes)$ is a non idempotent dioid.

This dioid is isomorphic to the set of elements of the form $a\,\varepsilon^{\alpha}$ endowed with ordinary addition and multiplication when $\varepsilon > 0$ tends towards 0^+.

We obtain a dioid that is isomorphic to the above by setting $A = e^{-\alpha}$, and by taking for E the set of pairs $(a, A) \in (\mathbb{R}_+ \backslash \{0\})^2$ to which we add the pair (0, 0), endowed with the operations $\oplus$ and $\otimes$ defined as:

$$(a, A) \oplus (b, B) = (c, \operatorname{Max}(A, B))$$

with $c = a$ if $A > B$, $c = b$ if $A < B$, $c = a + b$ if $A = B$,

$$(a, A) \otimes (b, B) = (ab, AB).$$

Moreover, the elements (a, A) of this dioid can be interpreted as the set of elements of the form $a\,A^p$ endowed with ordinary addition and multiplication when p tends towards $+\infty$.

We can interpret (a, A) as the coding of an asymptotic expansion of the form $a\,A^p + \sigma(A^p)$ when $p \to +\infty$.

The latter dioid was introduced by Finkelstein and Roytberg (1993) to calculate the asymptotic expansion of distribution functions in the study of biopolymers. It was also used by Akian et al. (1998) to calculate the eigenvalues of a matrix with coefficients of the form $\exp(-a_{ij}/\varepsilon)$ where ε is a small positive parameter. ||

Example 6.1.6. Non standard number dioid

On the set E of ordered triples $(a, b, \alpha) \in (\mathbb{R}_+ \backslash \{0\})^3$ to which we add the ordered triples $(0, 0, +\infty)$ and $(1, 0, +\infty)$, we define the two laws $\oplus$ and $\otimes$ by:

$$(a_1, b_1, \alpha_1) \oplus (a_2, b_2, \alpha_2) = (a_1 + a_2, b, \min(\alpha_1, \alpha_2))$$

with $b = b_1$ if $\alpha_1 < \alpha_2$, $b = b_2$ if $\alpha_1 > \alpha_2$, $b = b_1 + b_2$ if $\alpha_1 = \alpha_2$,

$$(a_1, b_1, \alpha_1) \otimes (a_2, b_2, \alpha_2) = (a_1\,a_2, b, \min(\alpha_1, \alpha_2))$$

with $b = a_2\,b_1$ if $\alpha_1 < \alpha_2$, $b = a_1\,b_2$ if $\alpha_1 > \alpha_2$,

$$b = a_1\,b_2 + a_2\,b_1 \text{ if } \alpha_1 = \alpha_2.$$

We verify that $(E, \oplus, \otimes)$ is a dioid. This dioid is isomorphic to the set of non standard numbers of the form $a + b\,\varepsilon^\alpha$, $(a > 0, b > 0)$, endowed with ordinary addition and multiplication, when $\varepsilon > 0$ tends towards 0^+. ||

The concept of *positive semiring* and *positive dioid* is studied in Exercise 5.

Proposition 6.1.7. *In a dioid* $(E, \oplus, \otimes)$, *the canonical order relation* $\le$ *is compatible with the laws* $\oplus$ *and* $\otimes$.

Proof. The fact that $\le$ is compatible with $\oplus$ was already proved in Sect. 3.3. Let us show that $\le$ is compatible with $\otimes$. We have: $a \le b \Leftrightarrow \exists\, c \in E$: $a \oplus c = b$, hence:

$$(a \oplus c) \otimes x = b \otimes x$$

thus, using distributivity:

$$a \otimes x \oplus c \otimes x = b \otimes x$$

hence we deduce $a \otimes x \le b \otimes x$.

We would similarly prove that $x \otimes a \le x \otimes b$. □

Definition 6.1.8. *(complete dioid)*

A dioid (E, ⊕, ⊗) *is said to be* complete *if it is complete as an ordered set for the canonical order relation, and if, moreover, it satisfies the two properties of "infinite distributivity":*

$$\forall A \subset E, \forall b \in E \quad \left(\bigoplus_{a\in A} a\right) \otimes b = \bigoplus_{a\in A} (a \otimes b)$$

$$b \otimes \left(\bigoplus_{a\in A} a\right) = \bigoplus_{a\in A} (b \otimes a)$$

From this definition it follows that, for every $A \subset E$ and $B \subset E$:

$$\left(\bigoplus_{a\in A} a\right) \otimes \left(\bigoplus_{b\in B} b\right) = \bigoplus_{(a,b)\in A\times B} (a \otimes b)$$

In a complete dioid, we define the *top-element* T as the sum of all the elements of the dioid

$$T = \bigoplus_{a\in E} a$$

We observe that T satisfies, $\forall x \in E$:

$$T \oplus x = T \quad \text{and} \quad T \otimes \varepsilon = \varepsilon.$$

As an illustration the dioids ($\mathbb{R}$, Max, +) and ($\mathbb{R}$, Min, +) are not complete. To make them complete, a *top-element* must be added:

$$T = +\infty \quad \text{for} \quad (\mathbb{R}, \text{Max}, +)$$
$$T = -\infty \quad \text{for} \quad (\mathbb{R}, \text{Min}, +).$$

In the same way as for other algebraic structures, we say that a subset $F \subset E$ is a *sub-dioid* of (E, ⊕, ⊗) if and only if: $\varepsilon \in F$, $e \in F$, and F is *closed* with respect to the laws ⊕ and ⊗. Thus for instance the dioid ($\mathbb{N}$, Max, +) is a sub-dioid of the dioid ($\mathbb{R}_+$, Max, +).

In the following sections, we discuss some particularly important sub-classes of dioids.

6.2. Dioid of Endomorphisms of a Canonically Ordered Commutative Monoid

Let (E, ⊕) be a canonically ordered commutative monoid with neutral element ε. As in the Examples 4.2.2. and 5.1.2., we then consider the set H of endomorphisms on E satisfying, $\forall h \in H$:

$$h(a \oplus b) = h(a) \oplus h(b) \quad \forall a, b \in E$$
$$h(\varepsilon) = \varepsilon$$

endowed with the laws $\oplus$ and $\otimes$ defined as: $\forall h, g \in H$:

$$(h \oplus g)(a) = h(a) \oplus g(a) \quad \forall a \in E$$
$$(h \otimes g)(a) = g \circ h(a) \quad \forall a \in E$$

where $\circ$ is the law of composition of mappings.

We verify that $(H, \oplus, \otimes)$ is a dioid.

This is a very important class of dioids underlying a wide variety of problems, in particular many non classical path-finding problems in graphs (see Minoux 1976, the two following examples and Example 4.2.3 in Chap. 8). Solution algorithms will be discussed in Chap. 4, Sect. 4.4.

Example 6.2.1. Shortest path with time-dependent lengths on the arcs

Let us consider the following problem. With each arc (i, j) of a graph G we associate a function h_{ij} giving the time t_j of arrival in j when we leave i at the instant t_i: $t_j = h_{ij}(t_i)$.

Leaving vertex 1 at the instant t_1, we seek the earliest time to reach vertex i.

For this problem, we take

$$E = \hat{\mathbb{R}} = \mathbb{R} \cup \{+\infty\}, \oplus = \min, \varepsilon = +\infty.$$

The set H is taken to be the set of *nondecreasing* functions h: $E \to E$ such that $h(t) \to +\infty$ when t tends towards $+\infty$.

These functions are indeed endomorphisms because we have:

$$h(\min(t, t')) = \min(h(t), h(t'))$$
$$h(+\infty) = +\infty$$

$(H, \oplus, \otimes)$ is therefore a dioid.

For a detailed study of this problem and solution algorithms, see Cooke and Halsey (1966) and Minoux (1976). ||

Example 6.2.2. Shortest path with discounting (Minoux 1976)

With each arc (i, j) of a graph G, we associate a length which depends, in a path, on the number of arcs taken previously. If we interpret, for example, the path along the arc (i, j) as the realization of an annual investment program, the cost of the arc (i, j) is $C_{ij}/(1+\tau)^t$ if t is the number of arcs previously taken by the path, that is to say the year of the expenditure c_{ij} (τ being the discounting rate).

We seek the shortest path in terms of discounted value from vertex 1 to the other vertices.

If T is the final time period, we take for S the set of (T + 1) - vectors with components in $\mathbb{R}_+ \cup \{+\infty\}$. If $a = (a_0, a_1, \dots, a_T)$ and $b = (b_0, b_1, \dots, b_T)$, we define $d = a \oplus b = (d_0, d_1, \dots, d_T)$ by setting $d_t = \min(a_t, b_t)$, t from 0 to T. $\varepsilon = (+\infty, \dots, +\infty)$. Then we define the endomorphism h_{ij} as:

$$h_{ij}(a) = b$$

$$\text{with:} \quad b_0 = +\infty \; b_t = a_{t-1} + \frac{c_{ij}}{(1+\tau)^{t-1}} \text{ for } t = 1, \dots, T.$$

We observe that such endomorphisms are T-nilpotent (see. Chap. 4, Sect. 3.3).

After obtaining the optimal label $a \in S$ of a vertex, it is possible to deduce the shortest path with discounted value from vertex 1 to this vertex, the value of which is equal to $\min_{0 \le t \le T} (a_t)$. ||

Many other examples can be constructed on this model (see Chap. 4 Sect. 4.4 and Chap. 8, Sect. 4.2).

6.3. Symmetrizable Dioids

Definition 6.3.1. *We call* symmetrizable dioid *a dioid* $(E, \oplus, \otimes)$ *for which the operation* $\oplus$ *is cancellative, that is to say such that* $(E, \oplus)$ *is a* hemi-group. *(see Sect. 3.5).*

Example 6.3.2. The set $\mathbb{N}$ of natural numbers endowed with the ordinary operations $+$ and $\times$ is a symmetrizable dioid. Indeed, $(\mathbb{N}, +)$ is a hemi-group (see Sect. 3.5). Similarly, the set $\mathbb{R}_+$ endowed with operations $+$ and $\times$ is a symmetrizable dioid. On the other hand $(\hat{\mathbb{R}}_+, +, \text{Min})$ is not a dioid because Min is not distributive with respect to $+$: $\text{Min}\{2, 1 + 5\} \neq \text{Min}\{2, 1\} + \text{Min}\{2, 5\}$. ||

The symmetrization of a symmetrizable dioid produces a ring. A symmetrizable dioid could therefore be referred to as a *hemi-ring*.

Remark 6.3.3. In the literature on the subject, a different type of symmetrization of a dioid has also been investigated; it is called *weak symmetrization* (see Gaubert 1992). As in Remark 2.3.6, from the equivalence relation $\overline{\mathbb{R}}$ on the ordered pairs of elements of E^2 defined as:

$$(a_1, a_2)\ \mathcal{R}\ (b_1,\ b_2) \Leftrightarrow \begin{cases} a_1 \neq b_1, a_2 \neq b_2 \quad \text{and} \quad a_1 \oplus b_2 = b_1 \oplus a_2 \\ (a_1, a_2) = (b_1, b_2) \quad \text{otherwise} \end{cases}$$

weak symmetrization consists in defining three types of elements: "positive" elements isomorphic to the elements of E and corresponding to the classes $\overline{(a, \varepsilon)}$, the "negative" elements corresponding to the classes $\overline{(\varepsilon, a)}$, and the "balanced" elements corresponding to the classes $\overline{(a, a)}$.

These *weakly symmetrizable dioids* can be useful for instance to express in algebraic form combinatorial properties of dioids (see Chap. 2); they can also be used in the framework of studying solutions of linear equations of the form: $Ax \oplus b = Cx \oplus d$ (see Chap. 4).

Refer to Gaubert (1992) for a detailed study of *weak symmetrization*. ||

6.4. Idempotent and Selective Dioids

Definition 6.4.1. *(idempotent dioid)*

We call idempotent dioid *a dioid in which the addition* $\oplus$ *is commutative and idempotent.*

A frequently encountered special case is one where addition $\oplus$ is not only idempotent, but *selective* (i.e.: $\forall a, b \in E$: $a \oplus b = a$ or b).

Definition 6.4.2. *(selective dioid)*

We call selective dioid *a dioid in which the addition $\oplus$ is commutative and selective.*

Idempotent dioids form a particularly rich class of dioids which contains many sub-classes, in particular:

- Doubly-idempotent dioids and distributive lattices (see Sect. 6.5);
- Doubly selective dioids (see Sect. 6.5);
- Idempotent-cancellative dioids and selective-cancellative dioids (see Sect. 6.6);
- Idempotent-invertible dioids and selective-invertible dioids (see Sect. 6.7).

6.5. Doubly-Idempotent Dioids and Distributive Lattices. Doubly-Selective Dioids

Definition 6.5.1. *We call* doubly-idempotent dioid *a dioid which has a commutative idempotent monoid structure for $\oplus$ and an idempotent monoid structure for $\otimes$.*

Definition 6.5.2. *We call* doubly selective *dioid a dioid which has commutative and selective monoid structure for $\oplus$ and for $\otimes$.*

Example 6.5.3. Let us take for E the set of reals $\overline{\mathbb{R}} = \mathbb{R} \cup \{+\infty\} \cup \{-\infty\}$ and let us define the operations $\oplus$ and $\otimes$ as:

$$\forall a, b \in E: \quad a \oplus b = \text{Min}\{a, b\}$$
$$\forall a, b \in E: \quad a \otimes b = \text{Max}\{a, b\}$$

$(E, \oplus)$ and $(E, \otimes)$ are commutative and selective monoids having neutral elements $\varepsilon = \{+\infty\}$ and $e = \{-\infty\}$ respectively.

$(E, \oplus, \otimes)$ is therefore a doubly-selective dioid. ||

As we are going to show, doubly-idempotent dioids are algebraic structures which are very close to *distributive lattices*.

Definition 6.5.4. *We call* lattice a *set* E *ordered by an order relation $\propto$ and which, for this relation, is at the same time a sup-semi-lattice and an inf-semi-lattice (see Sect.* 3.6*).*

Consequently, in a lattice E, to every pair of elements $a, b \in E$, we can let correspond:

- an upper bound $a \vee b$
- a lower bound $a \wedge b$.

A lattice is said to be *complete* if every subset of E (of finite or infinite cardinality) has an upper bound and a lower bound.

A lattice is said to be *distributive* if and only if the operation $\wedge$ is right and left distributive with respect to the operation $\vee$, that is to say:

$$\forall x, y, z \in E: x \wedge (y \vee z) = (x \wedge y) \vee (x \wedge z)$$
$$(x \vee y) \wedge z = (x \wedge z) \vee (y \wedge z)$$

(*N.B.*: it can be shown that the distributivity of $\vee$ with respect to $\wedge$ is a consequence of the above, see e.g. Dubreil and Dubreil-Jacotin 1964, p. 288).

Example 6.5.5. $\mathbb{R}$ endowed with the usual order relation is a distributive lattice with, $\forall a, b \in \mathbb{R}$: $a \vee b = \text{Max}\{a, b\}$; $a \wedge b = \text{Min}\{a, b\}$. ||

Example 6.5.6. If S is a set, $\mathcal{P}(S)$, the power set of S ordered by the inclusion is a distributive lattice with, $\forall A \subset S$, $\forall B \subset S$: $A \vee B = A \cup B$ and $A \wedge B = A \cap B$. ||

Distributive lattices form a particular family of interesting dioids as the following proposition shows.

Proposition 6.5.7. *If* E *is a distributive lattice, then* $(E, \vee, \wedge)$ *is a doubly-idempotent dioid, the order relation (canonical) of the dioid being defined as:*

$$a \le b \Leftrightarrow a \vee b = b.$$

Conversely, let $(E, \oplus, \otimes)$ *be a doubly-idempotent dioid for which* $\le$*, the canonical order relation relative to the law* $\oplus$ *is also a canonical order relation for* $\otimes$*:*

$$x \le y \Leftrightarrow x \otimes y = x.$$

Then E *is a distributive lattice.*

Proof. The if part of the proposition is easy to verify. In particular, we observe that ε, the neutral element of $(E, \vee)$ is the smallest element of E (in the sense of the canonical order relation) which implies the property of absorption: $\forall x \in E$: $x \wedge \varepsilon = \varepsilon$. Let us now prove the converse.

Following Proposition 3.6.2 in Sect. 3, $(E, \oplus)$ is a sup-semi-lattice for the canonical order relation relative to the law $\oplus$ and which is defined as:

$$a \le b \Leftrightarrow a \oplus b = b.$$

Similarly $(E, \otimes)$ is a sup. semi-lattice for the order relation $\le'$ defined as:

$$a \le' b \Leftrightarrow a \otimes b = b.$$

It is therefore an inf-semi-lattice relative to the order relation $\le''$ below:

$$a \le'' b \Leftrightarrow a \otimes b = a.$$

It is thus seen that the order relations $\le$ and $\le''$ coincide and consequently E is a lattice with:

$$\sup(a, b) = a \oplus b$$
$$\inf(a, b) = a \otimes b.$$

Finally, this lattice is distributive because of the property of distributivity of the dioid $(E, \oplus, \otimes)$. $\square$

Example 6.5.8. $(\mathbb{N}, \text{lcm}, \text{gcd})$ where $\forall a, b \in N, a \oplus b = \text{lcm}(a, b)$ and $a \otimes b = \text{gcd}(a, b)$ is a doubly idempotent dioid. According to the canonical order relation, $a \leq b$ if and only a divides b. In this case, we clearly have $a \otimes b = \text{gcd}(a, b) = a$, which proves, following Proposition 6.5.7 that $(\mathbb{N}, \text{lcm}, \text{gcd})$ is a distributive lattice (see also Chap. 8 Sect. 4.6.2). ||

Let us also observe that a dioid defined from two idempotent laws is not necessarily a lattice as seen in the following example:

Example 6.5.9. Let us take for E the set of reals endowed with the $\oplus$ law defined as:

$$\forall a, b \in \mathbb{R}: \quad a \oplus b = \text{Min}\{a, b\}$$

and multiplication $\otimes$ defined as:

$$\forall a, b \in \mathbb{R}: \quad a \otimes b = a \text{ (the result is always the first operand)}$$

(we easily check the idempotency and associativity of $\otimes$, as well as right and left distributivity of $\otimes$ with respect to $\oplus$).

$(E, \oplus, \otimes)$ is therefore clearly a dioid. To be convinced it is not a lattice just observe that $\otimes$ is not commutative. ||

A lattice $(E, \oplus, \otimes)$ is said to be *complemented* if, $\forall a \in E$: $e \leq a \leq \varepsilon$ and, $\forall a \in E$, there exists $\bar{a} \in E$ such that: $a \oplus \bar{a} = e$ and $a \otimes \bar{a} = \varepsilon$.

A distributive and complemented lattice is called a *Boolean lattice*.

Examples 6.5.5 and 6.5.6 correspond to Boolean lattices. Example 6.5.8 is not a complemented lattice.

Lattices are fundamental structures which have been extensively studied in the literature, see e.g. Birkhoff (1979), Mc Lane and Birkhoff (1970), and Dubreil and Dubreil-Jacotin (1964). See also Exercises 4, 6–9 at the end of the chapter.

6.6. Idempotent-Cancellative Dioids. Selective-Cancellative Dioids

Definition 6.6.1. *We call* idempotent-cancellative dioid *a dioid which has a commutative idempotent monoid structure for* $\oplus$ *and a cancellative monoid structure for* $\otimes$.

Example 6.6.2. Let us return to Example 2.1.13 in Sect. 2.

A being a set of letters ("alphabet"), the set of words on A (the so-called free monoid on A) is denoted A^*. Every subset (whether finite or infinite) of A^*, $L \in \mathcal{P}(A^*)$, is called a *language* on A. We denote $\mathcal{L}$ the set of all the languages on A. The sum of two languages $L_1 \oplus L_2$ is defined as the set union of the words of L_1 and the words of L_2.

The product of two languages $L_1 \otimes L_2$ is the set of the words formed by the concatenation of a word m_1 of L_1 and a word m_2 of L_2 (in this order).

We easily verify:

- that $(\mathcal{L}, \oplus)$ is a commutative idempotent monoid for $\oplus$ with neutral element $\emptyset$ the empty language (i.e. not containing any word of A^*);
- that $(\mathcal{L}, \otimes)$ is a (non commutative) cancellative monoid with neutral element $L_\emptyset$, the language formed by the empty word;
- that $\otimes$ is right and left distributive with respect to $\oplus$.

$(\mathcal{L}, \oplus, \otimes)$ defined above is therefore an *idempotent-cancellative dioid*: this is the algebraic structure underlying the theory of *regular languages* (see e.g. Salomaa 1969, Eilenberg, 1974). (Let us observe however that the axioms of regular languages include, in addition to the above, the closure operation denoted *). ||

An interesting special case of an idempotent-cancellative dioid is one where the operation $\oplus$ is not only idempotent but also *selective*.

Definition 6.6.3. *We call* selective-cancellative dioid *a dioid which has a selective monoid structure for* $\oplus$ *and a cancellative monoid structure for* $\otimes$.

Example 6.6.4. Let us take for E the set of nonnegative reals $\mathbb{R}_+ \cup \{+\infty\}$ and let us define the operations $\oplus$ and $\otimes$ as:

$$\forall a, b \in E: \quad a \oplus b = \text{Min}\{a, b\}$$
$$\forall a, b \in E: \quad a \otimes b = a + b \quad \text{(addition of reals)}$$

$(E, \oplus)$ is a selective monoid with neutral element $\varepsilon = +\infty$, and $(E, \otimes)$ is a cancellative monoid with neutral element $e = 0$.

The structure $(E, \oplus, \otimes)$ above is therefore a selective-cancellative dioid. ||

Remark. A special case of idempotent-cancellative dioids is one where $(E, \otimes)$ is, not only a cancellative monoid but a hemi-group (see Sect. 3.5). This is the situation encountered in Example 6.6.4 above where $(\mathbb{R}_+ \cup \{+\infty\}, +)$ is a cancellative monoid canonically ordered by $+$, and therefore a hemi-group. ||

6.7. Idempotent-Invertible Dioids. Selective-Invertible Dioids

Definition 6.7.1. *We call* idempotent-invertible dioid *a dioid* $(E, \oplus, \otimes)$ *which has a commutative idempotent monoid structure for* $\oplus$ *and a group structure for* $\otimes$ *(every element of* $E\backslash\{\varepsilon\}$ *having an inverse for* $\otimes$*).*

To insist on the fact that the set has a group structure relative to the second law, we can also refer to such a dioid as an idempotent-group dioid. This terminology will be preferred to that of semi-field (see Sect. 5.2).

An important special case for applications is when the $\oplus$ law is *selective*.

Definition 6.7.2. *We call selective-invertible-dioid a dioid* $(E, \oplus, \otimes)$ *which has a selective monoid structure for* $\oplus$ *and a group structure for* $\otimes$ *(every element of* $E\backslash\{\varepsilon\}$ *having an inverse for* $\otimes$*).*

Example 6.7.3. "Min-Plus" Dioid

Let us take for E the set of reals $\hat{\mathbb{R}} = \mathbb{R} \cup \{+\infty\}$ and let us define the operations $\oplus$ and $\otimes$ as:

$$\forall a, b \in E: \quad a \oplus b = \text{Min}\{a, b\}$$
$$\forall a, b \in E: \quad a \otimes b = a + b \quad \text{(addition of reals)}$$

$(E, \oplus)$ is a selective monoid with neutral element $\varepsilon = +\infty$, and $(E, \otimes)$ is a group with neutral element $e = 0$.

The structure $(E, \oplus, \otimes)$ above is therefore a selective-invertible-dioid. ||

Note that, in the terminology of the theory of languages and automata, selective-invertible dioids such as Min-Plus or Max-Plus dioids are sometimes referred to as *Tropical* semirings (see for example Simon 1994).

6.8. Product of Dioids

Given p dioids $(E_i, \oplus_i, \otimes_i)$, the *product dioid* is defined as the set $E = E_1 \times E_2 x \cdots x\ E_p$ endowed with the "product" laws $\oplus$ and $\otimes$ defined as:

$$\forall x = \begin{pmatrix} x_1 \\ \vdots \\ x_p \end{pmatrix} \in E, \quad \forall y = \begin{pmatrix} y_1 \\ \vdots \\ y_p \end{pmatrix} \in E: \quad x \oplus y = \begin{pmatrix} x_1 \oplus_1 y_1 \\ x_2 \oplus_2 y_2 \\ \vdots \\ x_p \oplus_p y_p \end{pmatrix}$$

and:

$$x \otimes y = \begin{pmatrix} x_1 \otimes_1 y_1 \\ x_2 \otimes_2 y_2 \\ \vdots \\ x_p \otimes_p y_p \end{pmatrix}$$

We already know (see Sect. 5.4) that $(E, \oplus, \otimes)$ is a semiring. It is moreover canonically ordered by $\oplus$ in view of the fact that the monoids $(E_i, \oplus_i)$ are canonically ordered.

We easily check the following:

Proposition 6.8.1. *The product of p dioids is a dioid.*

Example 6.8.2. Dioids of signed non standard numbers

Let us consider the pair $(a, s) \in \mathbb{R}_+ \times S$ where $S = \{+, -, 0, ?\}$ is the set of signs of qualitative algebra (see Example 6.1.3). With every real number x, we thus associate four non standard numbers $x^+, x^-, x^\circ, x^?$ corresponding respectively to: x obtained as the limit of a sequence of numbers $>x(x^+)$; of a sequence of numbers $<x(x^-)$; of a sequence of numbers all equal to $x(x^\circ)$; of a sequence of numbers convergent towards $x(x^?)$.

We define the addition $\oplus$ of two signed non standard numbers (a, s) and (b, σ) as

$$(a, s) \oplus (b, \sigma) = (a + b, s \dot{+} \sigma)$$

and the multiplication $\otimes$ as

$$(a, s) \otimes (b, \sigma) = (ab, s \dot{\times} \sigma)$$

where $\dot{+}$ and $\dot{\times}$ are the addition and multiplication of qualitative algebra (see Example 6.1.3).

$(\mathbb{R}_+ \times S, \oplus, \otimes)$ is then a dioid, as a product of the dioids $(\mathbb{R}_+, +, \times)$ and $(S, \dot{+}, \dot{\times})$.

In the case where we consider the non standard numbers on $\mathbb{R} \times S$, we no longer obtain a dioid, but a semiring, see Chap. 8 Sect. 3.1.2. ||

Observe that, as a general rule, the product of a dioid and a ring is neither a dioid nor a ring but a semiring (see Chap. 8 Sect. 3.1).

6.9. Dioid Canonically Associated with a Semiring

Proposition 6.9.1. *Let* $(E, \oplus, \otimes)$ *be a semiring in which the canonical preorder relation* $\leq$ *is not an order relation. Let* $\mathcal{R}$ *be the equivalence relation defined on* E *as:* $\forall a, b \in E$*:*

$$a\mathcal{R}b \Leftrightarrow a \leq b \quad \text{and} \quad b \leq a.$$

Then the set $E' = E/\mathcal{R}$*, endowed with the laws induced by* $\oplus$ *and* $\otimes$ *is a dioid, which we call* dioid canonically associated with $(E, \oplus, \otimes)$.

Proof. The relation $\mathcal{R}$, defined above is clearly reflexive, transitive and symmetric. It is therefore an *equivalence relation*. The elements of E′ are the equivalence classes relative to $\mathcal{R}$ on E and we still denote $\oplus$ and $\otimes$ the operations induced on E′ by the operations $\oplus$ and $\otimes$ on E. The neutral elements are the equivalence classes corresponding to the neutral elements ε and e. Clearly $(E', \oplus, \otimes)$ is a semiring.

Furthermore, the preorder relation $\leq$ on E induces on E′ an antisymmetric preorder relation, that is to say an order relation.

Finally, as $(E, \oplus, \otimes)$ is a semiring, ε is absorbing by $\otimes$. It follows that in $E' = E/\mathcal{R}$, the class of the element ε is absorbing for the law induced by $\otimes$. $(E', \oplus, \otimes)$ is therefore a dioid. □

Example 6.9.2. Let E be the semiring product of the dioids $E_1 = (\mathbb{N} \cup \{+\infty\}, \text{Min}, +)$, and $E_2 = (\mathbb{N}, +, \times)$ and of the ring: $E_3 = (\mathbb{Z}, +, \times)$

The elements of E are therefore ordered triples $\begin{pmatrix} x_1 \\ x_2 \\ x_3 \end{pmatrix}$.

We easily verify that the dioid canonically associated with $(E, \oplus, \otimes)$ is, in this example, isomorphic to the product dioid $E_1 \times E_2$ (see Sect. 6.8). ||

6.10. Classification of Dioids

Table 3 sums up the main properties of the various types of dioids. The first line indicates the basic properties common to all dioids, the following lines only show the additional properties corresponding to the sub-classes under consideration.

Figure 3 provides a graphic representation of the classification.

Table 3 The main types of dioids and their basic properties

	Properties of $(E, \oplus)$	Properties of $(E, \otimes)$	Relation $\leq$	Additional properties comments
Dioid	Commutative monoid, neutral elem. ε	Monoid neutral elem. e	Order	$\otimes$ right and left distributive with respect to $\oplus$ ε absorbing for $\otimes$
Symmetrizable dioid	Hemi-group			
Idempotent dioid	Idempotent monoid			
Doubly – idempotent dioid	Idempotent monoid	Idempotent monoid		
Distributive lattice	Idempotent monoid	Idempotent monoid		Doubly idempotent - dioid with the additional property: $x \leq y \Leftrightarrow x \otimes y = x$ where $\leq$ is the canonical order relation with respect to $\oplus$
Idempotent-cancellative dioid	Idempotent monoid	Cancellative monoid		Idempotent dioid and $\otimes$ cancellative
Idempotent-invertible dioid	Idempotent Monoid	Group		Idempotent dioid and $\otimes$ invertible
Selective dioid	Selective monoid		Total order	
Selective-cancellative dioid	Selective monoid	Cancellative monoid	Total order	Selective dioid and $\otimes$ cancellative
Selective-invertible dioid	Selective monoid	Group	Total order	Selective dioid and $\otimes$ invertible
Doubly selective dioid	Selective monoid	Selective monoid	Total order	

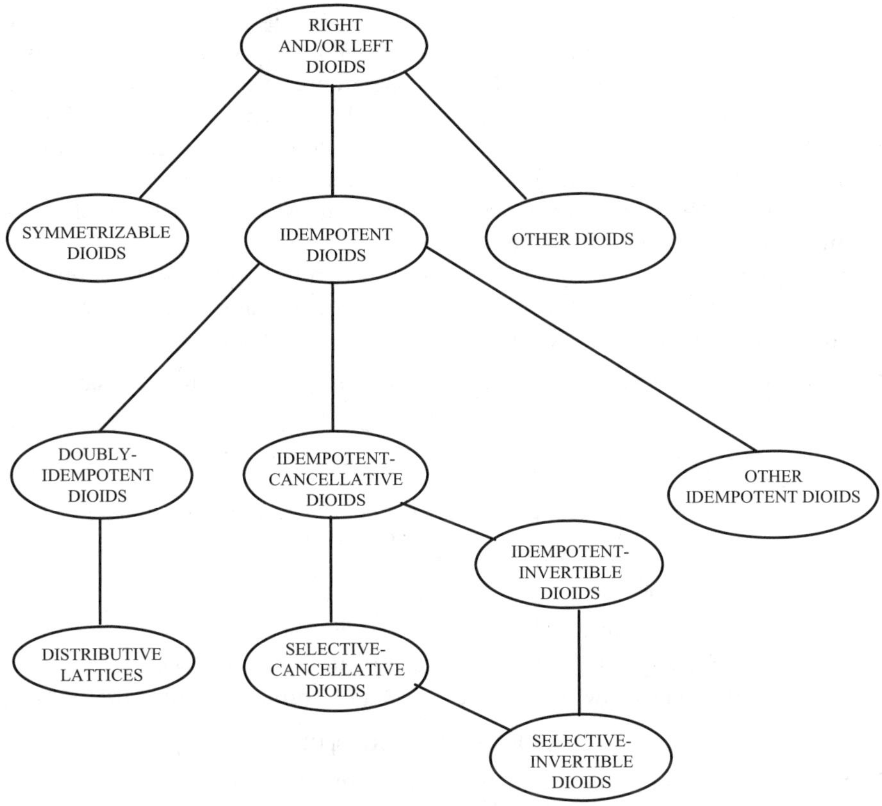

Fig. 3 Classification of dioids

Exercises

Exercise 1. We consider $\mathbb{R}_+$ endowed with the law $\oplus$ defined as:

$$a \oplus b = a(1 + b^2)^{1/2} + b(1 + a^2)^{1/2}$$

(1) Show that $(\mathbb{R}_+, \oplus)$ is a semigroup by establishing the formula of associativity:

$$(a \oplus b) \oplus c = a(1 + b^2)^{1/2}(1 + c^2)^{1/2} + b(1 + a^2)^{1/2}(1 + c^2)^{1/2} \\ + c(1 + a^2)^{1/2}(1 + b^2)^{1/2} + abc$$

(2) Find the same result by carrying out the change of variable $a = \mathrm{sh}(\alpha)$ and $b = \mathrm{sh}(\beta)$ and observing that $a \oplus b = \mathrm{sh}(\alpha + \beta)$.

(3) Show that $[0, 1]$ endowed with the law $\oplus$ defined as $a \oplus b = a(1 - b^2)^{1/2} + b(1 - a^2)^{1/2}$ is also a semigroup.

Exercise 2. Study of t-norms and t-conorms

Boolean algebra provides a very natural model of binary logic. To generalize the logical AND and OR, a great variety of operations on the interval [0, 1] have been proposed in the literature, in particular in the context of so-called "fuzzy" logic. The most classical ones are the operations referred to as *triangular norms* (or t-norms) which generalize logical AND and *triangular conorms* (or t-conorms) which generalize logical OR. They were introduced by Menger in 1942, to define a triangular inequality in stochastic geometry.

A binary law $*$ on [0, 1] is a triangular norm (t-norm) if ([0, 1], $*$) is an ordered commutative monoid having 1 as neutral element and 0 as absorbing element. It is Archimedean if and only if $*$ is continuous and $a * a < a \; \forall a \in]0, 1[$.

(1) Consider a function φ: $[0, 1] \to [0, +\infty[$ continuous and decreasing, satisfying $\varphi(1) = 0$. If $\varphi(0) = +\infty$, we set:

$$a * b = \varphi^{-1}[\varphi(a) + \varphi(b)].$$

If $\varphi(0) < +\infty$, we call pseudo-inverse of φ the function $\varphi^{(-1)}$ defined as:

$$\varphi^{(1)}(x) = \begin{cases} \varphi^{-1}(x) & \text{if} \quad x \in [0, \varphi(0)] \\ 0 & \text{if} \quad x \in]\varphi(0), +\infty[\end{cases}$$

and we set $a * b = \varphi^{(-1)}[\varphi(a) + \varphi(b)]$ (see illustration in Fig. 4).

Show that the binary law $*$ is an Archimedean t-norm. φ is called the additive generator of the t-norm. Possible examples are: $\varphi(x) = -\ln x$, $\varphi(x) = \dfrac{1-x}{x}$.

(2) A binary law $\oplus$ on [0, 1] is a triangular conorm (t-conorm) if ([0, 1], $\oplus$) is an ordered commutative monoid having 0 as neutral element and 1 as absorbing element. It is Archimedean if and only if $\oplus$ is continuous and $a \oplus a > a \; \forall a \in]0, 1[$.

Consider a function g: $[0, 1] \to [0, +\infty[$ continuous and increasing satisfying $g(0) = 0$. If $g(1) = +\infty$, we set:

$$a \oplus b = g^{-1}[g(a) + g(b)]$$

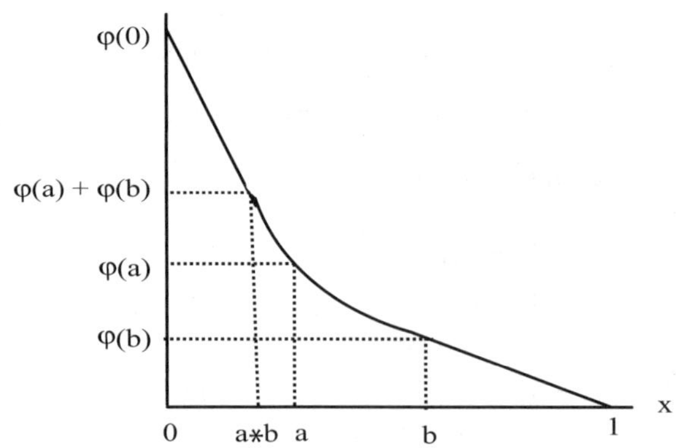

Fig. 4

If $g(1) < +\infty$, we call pseudo-inverse of g the function $g^{(-1)}$ defined as:

$$g^{(-1)}(x) = \begin{cases} g^{-1}(x) & \text{if} \quad x \in [0, g(1)] \\ 0 & \text{if} \quad x \in]g(1), +\infty[\end{cases}$$

and we set

$$a \oplus b = g^{(-1)}\,[g(a) + g(b)]$$

Show that the binary law $\oplus$ is an Archimedean t-conorm. g is called the additive generator of the t-conorm. Possible examples are: $g(x) = x^{\alpha}$, $g(x) = -\ln(1-x)$, $g(x) = e^{x/h} - 1 (h > 0)$.

Show that, through duality, every t-norm $*$ generates a t-conorm $\oplus$ and conversely according to:

$$a \oplus b = 1 - [(1-a) * (1-b)]$$
$$a * b = 1 - [(1-a) \oplus (1-b)].$$

In the literature on fuzzy sets, the triangular conorms $\oplus$ are considered as generalizations of the set union and the triangular norms $*$ are considered as generalizations of set intersection.

(3) Show that all the continuous triangular norms are constructed through isomorphism (or ordinal sum) from one of the three fundamental norms:

$$* = \text{product} \quad \text{(strictly monotone triangular norms)}$$
$$a * b = \text{Max}(0, a + b - 1) \quad \text{(nilpotent norms)}$$
$$a * b = \text{Min}(a, b) \quad \text{(this is the largest of the triangular norms, and the only one that is idempotent).}$$

[*Indications*: (1) and (2) see Dubois (1987), (3) Schweizer and Sklar (1983). For the study of pseudo-inverses, refer to Chap. 3, Sect. 8 where *residuable* functions are discussed.]

Exercise 3. Passing to the limit in the monoid $(\mathbb{R}, \oplus_p)$

(1) Show that the operation $\oplus_p$ (p integer) defined on $\mathbb{R}_+$ as

$$a \oplus_p b = \left(a^p + b^p\right)^{1/p} \text{ endows } \mathbb{R} \text{ with a regular monoïd structure.}$$

Show that $\oplus_p$ "tends" towards the max operation when $p \to +\infty$.

(2) Show that the operation $\oplus_p$ (p odd integer) endows $\mathbb{R}$ with a group structure.

$$\text{Show that} \quad \lim_{p \to +\infty} a \oplus_p b = \begin{cases} b & \text{if} \quad |a| < |b| \\ b & \text{if} \quad a = b \\ 0 & \text{if} \quad a = -b \end{cases}$$

Show that this limit operation is not associative.

[*Answers*: (2°) Counter-example: $(a \oplus_p b) \oplus_p c \neq a \oplus_p (b \oplus_p c)$ with $b = -a$, $|a| > |c|$].

Exercise 4. Left semiring and lattice

We consider a left semiring $S = (E, \oplus, \otimes)$, therefore only satisfying left distributivity $a \otimes (b \oplus c) = a \otimes b \oplus a \otimes c$ (see Sect. 5.1).

(1) We assume that S satisfies the following relation:

(i) $a \otimes b \oplus c = (a \oplus c) \otimes (b \oplus c) \quad \forall a, b, c \in S$
Show that $\oplus$ and $\otimes$ are idempotent.
Deduce that the semiring is a distributive lattice.

(2) We assume that S satisfies the relation (ii):

(ii) $a \oplus e = e \quad \forall a \in S$
Show that if S satisfies (ii) and $\otimes$ is idempotent, (i) holds.

(3) Show that (i) implies (ii).

[*Indications*:

(1) $a = b = \varepsilon$ in (i) implies $c = c^2$ and $a = \varepsilon$, $b = \varepsilon$ implies $c = c \oplus c$.
According to Proposition 6.5.7, it suffices to show that $a \leq b \Rightarrow a \otimes b = a$.
From (i), with $a = \varepsilon$, we derive $c = c \otimes b \oplus c$.]

Exercise 5. Positive semiring and positive dioid

A semiring is said to be *positive* (see Eilenberg) if and only if it satisfies:

(i) $a \oplus b = \varepsilon \Rightarrow a = \varepsilon$ and $b = \varepsilon$
(ii) $a \otimes b = \varepsilon \Rightarrow a = \varepsilon$ or $b = \varepsilon$

(1) Show that a dioid always satisfies (i).
A dioid is therefore positive if and only if it satisfies (ii) (such dioids are also referred to as "entire dioids").
(2) Show that a dioid which is a group for the second law is a positive dioid. Positive semirings (and consequently positive dioids) are often referred to as *semi-fields* (see Exercise 8).

[*Answer*: (1) see Proposition 3.4.8].

Exercise 6. Show that a complete sup-semi-lattice having a smallest element is a complete lattice.
[*Indications*: see (Dubreil and Dubreil-Jacotin 1964, pp. 175–176).

For a subset A of the sup-semi-lattice $(E, \leq)$, consider the set T of lower bounds of A and show that the upper bound of T belongs to T].

Exercise 7. Ideals and filters

(1) Show that the descending sets of an ordered set $(E, \leq)$, endowed with set union and set intersection, form a complete lattice.
(We recall, see Sect. 3.1, that a set $S \subset E$ is called *descending* if $x \in S$ and $y \leq x$ imply $y \in S$).

(2) Show that the set of ideals of an ordered set, endowed with the two laws $I \wedge J = I \cap J$, $I \vee J = \bigcap_{K \text{ ideal}, K \subset I \cup J} K$ is a complete lattice. (We recall, see Sect. 3.1, that an ideal S is a descending set such that for all $a, b \in S$, $a \vee b \in S$).

Exercise 8. Idempotent semi-field and inf-dioid

We call *inf-dioid* (Gaubert 1992) an idempotent dioid $(E, \oplus, \otimes)$ such that every pair of elements $(a, b) \in E^2$ has a lower-bound (denoted $a \wedge b$) with respect to the canonical order.

(1) Show that an idempotent-cancellative dioid (idempotent semi-field) is an inf-dioid and that $a \wedge b = b\,(a \oplus b)^{-1}\,a$.

Show in addition that the lower bound distributes with respect to the product: $(a \wedge b) \otimes c = (a \otimes c) \wedge (b \otimes c)$, $c \otimes (a \wedge b) = (c \otimes a) \wedge (c \otimes b)$ (the group $(E\backslash\{\varepsilon\}, \otimes)$ is therefore reticulated).

(2) Show that a complete idempotent dioid is a inf-dioid and that the law $\wedge$, defined as: $a \wedge b = \oplus\{x | x \leq a \text{ and } x \leq b\}$ makes $(E, \oplus, \wedge)$ a complete lattice.

(3) Show that the dioid B [X] of the polynomials with Boolean coefficients in an indeterminate X is an example of an inf-dioid which is neither complete, nor an idempotent semi-field.

(4) Show that a non-trivial idempotent semi-field (i.e. non reducible to $\{\varepsilon, e\}$) does not have a largest element, and in particular that it is not complete.

[*Answers*: (1) see Gaubert (1992). Complete counter-example $\left(\bigoplus_{k \in N} X^k \notin B[X]\right)$, idempotent $((e \oplus X) \otimes (e \oplus X^2) = (e \oplus X) \otimes (e \oplus X \oplus X^2)$, therefore $\otimes$ is not cancellative)].

Exercise 9. Interval algebras

We consider the set Int($\mathbb{R}$) of the intervals of $\mathbb{R}$ of the form $a = [\underline{a}, \bar{a}]$ with $\underline{a} \leq \bar{a}$.

(1) On Int($\mathbb{R}$), let us consider the addition of intervals $\oplus$ defined as:

$$a \oplus b = [\underline{a} + \underline{b}, \bar{a} + \bar{b}].$$

Show that (Int($\mathbb{R}$), $\oplus$) is a commutative monoid with neutral element [0, 0] non canonically ordered. Let Int($\mathbb{R}_+$), the set of the intervals on $\mathbb{R}_+$ $(0 \leq \underline{a} \leq \bar{a})$. Show that (Int($\mathbb{R}_+$), $\oplus$) is canonically ordered and that each interval is a cancellative element of Int($\mathbb{R}_+$).

On Int($\mathbb{R}$), we now consider the multiplication $\otimes$ defined as:

$$a \otimes b = \{x | x = \alpha \cdot \beta, \alpha \in a, \beta \in b\}.$$

Show that $a \otimes b = [\min(\underline{a} \cdot \underline{b}, \underline{a} \cdot \bar{b}, \bar{a} \cdot \underline{b}, \bar{a} \cdot \bar{b}), \max(\underline{a} \cdot \underline{b}, \underline{a} \cdot \bar{b}, \bar{a} \cdot \underline{b}, \bar{a} \cdot \bar{b})]$ and that $\oplus$ is not distributive with respect to $\otimes$.

Show on the other hand that (Int ($\mathbb{R}_+$), $\oplus$, $\otimes$) is a *symmetrizable dioid.*

(2) On Int($\mathbb{R}$), consider the operations $\oplus$ and $\otimes$ defined as

$$a \oplus b = [\min(\underline{a}, \underline{b}), \max(\bar{a}, \bar{b})],$$
$$a \otimes b = [\underline{a} + \underline{b}, \bar{a} + \bar{b}].$$

Show that (Int($\mathbb{R}$), $\oplus$, $\otimes$) is an idempotent *dioid*.

(3) On Int($\mathbb{R}$), consider the operations $\oplus$ and $\otimes$ defined as

$$a \oplus b = [\min(\underline{a}, \underline{b}), \max(\bar{a}, \bar{b})],$$
$$a \otimes b = a \cap b.$$

Show that (Int ($\mathbb{R}$), $\oplus$, $\otimes$) is a *nondistributive lattice*.

(4) On Int($\mathbb{R}$), consider the operations $\oplus$ and $\otimes$ defined as

$$a \oplus b = a \cap b,$$
$$a \otimes b = [\underline{a} + \underline{b}, \bar{a} + \bar{b}].$$

Show that $\otimes$ is not distributive with respect to $\oplus$.

[*Indications*:
(1) Chapter 8, Sects. 1.1.1, 1.4.6, 4.2.2. (2) Chap. 8, Sect. 4.3.4. (3) nondistributive: $a = [-3, -2]$, $b = [2, 3]$, $c = [-1, 1]$; $(a \oplus b) \otimes c = c$, $(a \otimes c) \oplus (b \otimes c) = d$].

Exercise 10. Newton polygon of an algebraic equation

(1) Let us consider the algebraic equation on the reals:

$$\varepsilon^2 + \varepsilon y - y^3 = 0$$

When $\varepsilon \to 0^+$, I. Newton seeks solutions in the form

$$y(\varepsilon) = b_1\varepsilon^{\mu_1} + b_2\varepsilon^{\mu_2} + \cdots$$

with $0 < \mu_1 < \mu_2 < \cdots$

Show that the μ_i are the nondifferentiability points (corners) of the concave function

$$f(\mu) = \min\{2, 1 + \mu, 3\,\mu\}.$$

Deduce the solutions.

(2) In a more general way, consider the algebraic equation

$$P(\varepsilon, y) = \sum_{j=0}^{n} \left(\sum_{i=0}^{N_j} a_{ij}\, \varepsilon^{\nu_{ij}} \right) y^j = 0.$$

When $\varepsilon \to 0^+$, we seek the n solutions in the form:

$$y(\varepsilon) = b_1\varepsilon^{\mu_1} + b_2\varepsilon^{\mu_2} + \cdots$$

with $0 < \mu_1 < \mu_2 < \cdots$

Show that the μ_i are the nondifferentiability points (corners) of the concave function

$$f(\mu) = \underset{i,j}{\mathrm{Min}}\{\nu_{ij} + i\mu\}.$$

[*Indications*:

(1) $y(\varepsilon) = -\sqrt{\varepsilon} - \varepsilon$, $y(\varepsilon) = +\sqrt{\varepsilon} - \varepsilon$.
(2) Newton–Puiseux theorem (after Gaubert 1998).

See also Dieudonné 1980. pp. 106–112]

Exercise 11. Pap's g-calculus

Let us consider a strictly monotone function defined on the finite interval $[a, b] \subset \mathbb{R}$ with values in $[0, +\infty]$, such that either $g(a) = 0$ and $g(b) = +\infty$, or $g(b) = 0$ and $g(a) = +\infty$.

We set

$$u \oplus v = g^{-1}(g(u) + g(v))$$
$$u \otimes v = g^{-1}(g(u) \cdot g(v))$$

(1) Show that $([a, b], \oplus, \otimes)$ is a dioid with $\varepsilon = g^{-1}(0)$ and $e = g^{-1}(1)$.
(2) Now, we assume g to be continuously differentiable on [a, b].
Let f: $[c, d] \to [a, b]$. If the function f is differentiable on [c, d] and has the same monotony as the function g, then we define the *g-derivative* of f at the point $x \in [c, d]$ as:

$$\frac{d^{\oplus} f(x)}{dx} = g^{-1}\left(\frac{d}{dx} g(f(x))\right).$$

Let f_1 and f_2 be two g-differentiable functions on [c, d] with values in [a, b]. Show that for every $\lambda \in [a, b]$ we have

(i) $\dfrac{d^{\oplus}(f_1 \oplus f_2)}{dx} = \dfrac{d^{\oplus} f_1}{dx} \oplus \dfrac{d^{\oplus} f_2}{dx}$

(ii) $\dfrac{d^{\oplus}(\lambda \otimes f)}{dx} = \lambda \otimes \dfrac{d^{\oplus} f}{dx}$

(iii) $\dfrac{d^{\oplus} x}{dx} = g^{-1}(1) = e$

(iv) $\dfrac{d^{\oplus}(f_1 \otimes f)}{dx} = \left(\dfrac{d^{\oplus} f_1}{dx} \otimes f_2\right) \oplus \left(f_1 \otimes \dfrac{d^{\oplus} f_2}{dx}\right)$

Calculate $\dfrac{d^{\oplus} f(x)}{dx}$ for $g(u) = e^{-u}$ and $g(u) = \ln \dfrac{1+u}{1-u}$.

(3) We define recursively the n-g-derivative of f: $[c, d] \to [a, b]$ (if it exists) from the $(n-1)$-g-derivative as

$$\frac{d^{(n)\oplus} f}{dx^n} = \frac{d^{\oplus}}{dx}\left(\frac{d^{(n-1)\oplus} f}{dx^{n-1}}\right).$$

(3a) If the n-g-derivative of f exists, show that we have:

$$\frac{d^{(n)\oplus}f}{dx} = g^{-1}\left(\frac{d^n}{dx^n}g(f(x))\right)$$

(3b) Let h be a function defined on $[a, b] \subset [0, \infty]$ and $f: [c, d] \to [a, b]$ and $F(x) = h(f(x))$. We assume that f is derivable at $x_0 \in [c, d]$ and that h has a derivative at $f(x_0)$.
Show that $\dfrac{d^{\oplus}F(x_0)}{dx}$ exists and that

$$\frac{d^{\oplus}F(x_0)}{dx} = \frac{d^{\oplus}[h(f(x_0))]}{dx} \otimes \frac{d^{\oplus}}{dx}g^{-1}(h(f(x_0)) \otimes \frac{d^{\oplus}g^{-1}(f(x_0))}{dx}.$$

(4) For a measurable function $f: [c, d] \to [a, b]$, we define the g-integral

$$\int\limits_{[c,d]}^{\oplus} f dx = g^{-1}\left(\int\limits_{c}^{d} g(f)dx\right).$$

(4a) Show that the g-integral is linear with respect to $(\oplus, \otimes)$.
(4b) Show that

$$f_1 \le f_2 \Rightarrow \int\limits_{[c,d]}^{\oplus} f_1 dx \le \int\limits_{[c,d]}^{\oplus} f_2 dx, \quad \int\limits_{[c,d]\cup[e,f]} f dx = \int\limits_{[c,d]}^{\oplus} f\,dx \oplus \int\limits_{[e,f]}^{\oplus} f dx.$$

(4c) If f is continuous on [c, d], then

$$\frac{d^{\otimes}}{dx}\int\limits_{[c,x]}^{\oplus} f dx = f(x)$$

(4d) If f has a continuous g-derivative on [c, d], then

$$\int\limits_{[c,x]}^{\oplus} \frac{d^{\oplus}f}{dx} \oplus f(c) = f(x) \quad \text{for all } x \in [c, d].$$

(4e) Calculate $\int\limits_{[x,+\infty]}^{\oplus} x dx$ for $g(u) = e^{-u}$.

(5) Let us assume that $f: [c, d] \times [a, b] \to [a, b]$ is continuous, that ψ is defined and continuous on $J = \{x: x_0 - h < x < x_0 + h\} \subset [c, d]$ with values in [a, b] and that $(x_0, y_0) \in [c, d] \times [a, b]$ with $\psi(x_0) = y_0$.

(5a) Show that a necessary and sufficient condition for ψ to be the solution of:

$$\frac{d^{\oplus}\psi}{dx} = f(x, \psi(x))$$

on J is that ψ satisfies the g-integral equation

$$\psi(x) = y_0 \oplus \int_{(x_0,x)}^{\oplus} f(t, \psi(t))\, dt \quad \text{for } x \in J.$$

(5b) We consider the second order ordinary differential equation for $p > 0$ and $n \in R^+$:

$$y'' + (p-1)y^{-1}(y')^2 + (2npx^{-1} - x^{-np}y^p)\, y' + n(np-1)\, x^{-2}\, y' = 0 \quad (1)$$

Show that this equation can be expressed successively in the following forms:

$$\begin{aligned} y \cdot (y^p)' &= (x^{np}y^p)'' \\ y \otimes \frac{d^{\oplus}}{dx}(y) &= \frac{d^{(2)\oplus}}{dx}(x^n \otimes y) \end{aligned} \quad (2)$$

where the generator g(x) is equal to x^p.

(5c) Show that the inverse x^* of an element $x \in [a, b]$ is equal to $x^* = g^{-1}\left(\frac{1}{g(x)}\right)$ and that, for every $n \in N$,

$$\underbrace{x \oplus x \oplus \cdots \oplus x}_{n \text{ times}} = g^{-1}(n) \otimes x.$$

(5d) Show that for a g-derivable function f(x), we have

$$f \otimes \frac{d^{\oplus}f}{dx} = g^{-1}\left(\frac{1}{2}\right) \otimes \frac{d^{\otimes}}{dx}(f \otimes f).$$

(5e) Show that equation (2) can be solved and yields

$$\frac{1}{2}g(y)^2 + c = \frac{d}{dx}g(x^n\, y) \quad (3)$$

where c is a constant.

By setting $t = y^p$, show that (t, x) satisfies the Riccati equation:

$$t' = \frac{1}{2}x^{-np}t^2 - np\, x^{-1}t + c\, x^{-np}.$$

Investigate the case where $np = 1$.

[*Indications*:
see Pap (1995), Sects. 8.3 and 8.4.

(2) For $g(u) = e^{-u}$, $u \oplus v = -\ln(e^{-u} + e^{-v})$, $u \otimes v = u + v$ and

$$\frac{d^{\otimes} f(x)}{dx} = f(x) - \ln(-f'(x)) \quad \text{for} \quad f'(x) < 0.$$

For $g(u) = \ln \dfrac{1+u}{1-u}$, $g^{-1}(u) = \dfrac{e^u - 1}{e^u + 1}$ and $\dfrac{d^{\otimes} f(x)}{dx} = \dfrac{\exp\left(\frac{2f'(x)}{1-f^2(x)}\right) - 1}{\exp\left(\frac{2f'(x)}{1-f^2(x)}\right) + 1}$.

(4) $\displaystyle\int_{[x,+\infty]}^{\otimes} x\, dx = x.$

(5b) We multiply (1) by $p\, x^{np}\, y^{p-1}$ and we integrate.
If $g(x) = x^p$ for $p > 0$, $u \oplus v = (u^p + v^p)^{1/p}$ and $u \otimes v = u \cdot v$.

(5e) We apply 5d to (2) to obtain

$$g^{-1}\left(\frac{1}{2}\right) \otimes y \otimes y \oplus c_1 = \frac{d^{\otimes}}{dx}(x^n \otimes y)$$

where c_1 is a constant.
For $n\, p = 1$, we have $t' = x^{-1}\left(\dfrac{1}{2}t^2 - t + c\right)$.
This corresponds to the equation initially considered:

$$y'' + (p-1)\, y^{-1}\, (y')^2 + x^{-1}(2 - y^p)\, y' = 0. \,]$$

Chapter 2
Combinatorial Properties of (Pre)-Semirings

1. Introduction

Many results of classical linear algebra, such as the well-known Cayley–Hamilton theorem, first established in the context of vector spaces on fields, do not actually require all the properties of these structures. We show in this chapter that many known results of this type are deduced from purely combinatorial properties which are valid in more elementary algebraic structures such as semirings and pre-semirings. We will not even require the dioid structure since there is no need to assume the presence of a canonical order relation.

In the present chapter we will thus consider matrices, polynomials and formal series with elements or coefficients in a pre-semiring or in a semiring.

The basic definitions concerning matrices, polynomials and formal series are introduced in Sects. 2 and 3.

Definitions and basic properties for permutations are recalled in Sect. 4.1, and the concepts of a *bideterminant* and of the *characteristic bipolynomial* of a matrix are introduced in Sects. 4.2 and 4.3.

Section 5 presents a combinatorial proof of the extended version of the classical identity for the determinant of the product of two matrices. Section 6 provides a combinatorial proof of the Cayley–Hamilton theorem generalized to commutative pre-semirings.

In Sect. 7, we focus on the links between the bideterminant of a matrix and the arborescences of the associated directed graph. An extension to semirings of the classical "Matrix Tree Theorem" is first established in Sects. 7.1 and 7.2. A more general form of this result is then studied in Sect. 7.4, which may be considered as an extension to semirings, of the so-called "All Minors Matrix Tree Theorem".

Finally, a version of the well-known Mac Mahon identity, generalized to commutative pre-semirings, is presented in Sect. 8.

In order to derive each of the identities discussed in this chapter, a superficial analysis might lead one to believe that it is enough to start from the corresponding classical result (usually stated in the field of real numbers) and to simply rewrite it by moving all the negative terms to the other side to make them appear positively.

The result of Sect. 5 (about the bideterminant of the product of two matrices), as well as the generalization of the classical "All-Minors Matrix Tree Theorem", which is studied in Sect. 7.4, provide concrete examples where such an approach would lead to a wrong result; this indeed confirms the necessity of new direct proofs, different from those previously known for the standard case.

2. Polynomials and Formal Series with Coefficients in a (Pre-) Semiring

2.1. Polynomials

Let $(E, \oplus, \otimes)$ be a pre-semiring or a semiring with neutral elements ε and e (for $\oplus$ and $\otimes$ respectively).

Definition 2.1.1. *A polynomial* P *of degree* n *in the variable* x *is defined by specifying a mapping* $f: \{0, 1, \ldots. n\} \to E$ *where,* $\forall k, 0 \le k \le n, f(k) \in E$ *is called the coefficient of* x^k *in the polynomial* P. P *can thus be represented by the sum:*

$$P(x) = \sum_{k=0}^{n} f(k) \otimes x^k$$

where the sum is to be understood in the sense of the operation $\oplus$ *(by convention* $x^0 = e$ *and,* $\forall k$: $\varepsilon \otimes x^k = \varepsilon$).

In accordance with classical notation, we denote E[x] the set of polynomials in x with coefficients in E.

Let P and Q be two polynomials of E [x] defined as:

$$P(x) = \sum_{k=0}^{p} f(k) \otimes x^k$$

$$Q(x) = \sum_{k=0}^{q} g(k) \otimes x^k$$

The sum of P and Q, denoted $S = P \oplus Q$, is the polynomial of degree at most $s = \text{Max}\{p, q\}$ defined as:

$$S(x) = \sum_{k=0}^{s} (f(k) \oplus g(k)) \otimes x^k$$

(we agree to set $f(j) = \varepsilon$ for $j > p$ and $g(j) = \varepsilon$ for $j > q$).

The product of P and Q, denoted $T = P \otimes Q$ is the polynomial of degree $r = p + q$ defined as:

$$T(x) = \sum_{k=0}^{r} t(k) \otimes x^k$$

with, $\forall k = 0 \ldots . r$:

$$t(k) = \sum_{\substack{0 \le i \le p \\ 0 \le j \le q \\ i+j=k}} f(i) \otimes g(j)$$

ε being the neutral element of $\oplus$, E[x] has, as neutral element for $\oplus$, the polynomial denoted $\varepsilon(x)$, of degree 0, defined as: $\varepsilon(x) = \varepsilon \otimes x^0 = \varepsilon$. Likewise, e being the neutral element of $\otimes$, E[x] has as neutral element for $\otimes$, the polynomial denoted e(x) of degree 0 defined as: $e(x) = e \otimes x^0 = e$.

Proposition 2.1.2. *(i) If* $(E, \oplus, \otimes)$ *is a pre-semiring, then* $(E[x], \oplus, \otimes)$ *is a pre-semiring*
(ii) If $(E, \oplus, \otimes)$ *is a semiring, then* $(E[x], \oplus, \otimes)$ *is a semiring*
(iii) If $(E, \oplus, \otimes)$ *is a dioid, then* $(E[x], \oplus, \otimes)$ *is a dioid.*

Proof. It follows from the fact that the elementary properties of $\oplus$ and $\otimes$ on E induce the same properties on E[x]. Let us just show that, in case (iii), the canonical preorder relation on E[x] defined as:

$$P \le Q \Leftrightarrow \exists R \in E[x] \quad \text{such that:} \quad Q = P \oplus R$$

is an order relation.

$$\text{If } P(x) = \sum_{k=0}^{p} f(k) \otimes x^k$$

$$Q(x) = \sum_{k=0}^{q} g(k) \otimes x^k$$

then $P \le Q \Rightarrow \exists R$ with: $R(x) = \sum\limits_{k=0}^{r} h(k) \otimes x^k$, such that: $Q = P \oplus R$

Similarly $Q \le P \Rightarrow \exists R'$ with: $R'(x) = \sum\limits_{k=0}^{r'} h'(k) \otimes x^k$ such that: $P = Q \oplus R'$

Set $K = \text{Max}\{p, q, r, r'\}$ and let us agree that:

$$\begin{array}{llll} \text{If} \quad K > p, & f(j) = \varepsilon & \text{for every} & j \in [p+1, K] \\ \text{If} \quad K > q, & g(j) = \varepsilon & \text{for every} & j \in [q+1, K] \\ \text{If} \quad K > r, & h(j) = \varepsilon & \text{for every} & j \in [r+1, K] \\ \text{If} \quad K > r', & h'(j) = \varepsilon & \text{for every} & j \in [r'+1, K] \end{array}$$

We deduce $\forall k = 0, \ldots. K$:

$$\exists\, r(k): \quad g(k) = f(k) \oplus r(k)$$
$$\exists\, r'(k): \quad f(k) = g(k) \oplus r'(k)$$

in other words:

$$f(k) \le g(k), \quad \text{and} \quad g(k) \le f(k)$$

Since $(E, \oplus, \otimes)$ is a dioid, we deduce $\forall k$: $f(k) = g(k)$ and therefore $P = Q$. $(E[x], \oplus, \otimes)$ is thus clearly a dioid in this case. □

The above is easily generalized to *multivariate* polynomials in several commutative indeterminates $x_1, x_2, \ldots. x_m$, the set of these polynomials being denoted $E[x_1, x_2, \ldots. x_m]$.

2.2. Formal Series

Let $(E, \oplus, \otimes)$ be a pre-semiring or a semiring with neutral elements ε and e (for $\oplus$ and $\otimes$, respectively).

Definition 2.2.1. *A formal series* F *in* m *commutative indeterminates* $x_1, x_2, \ldots x_m$ *is defined by specifying a mapping* $f\colon N^m \to E$*, where:* $\forall (k_1, k_2, \ldots. k_m) \in N^m$, $f(k_1, k_2, \ldots. k_m)$ *is the coefficient of the term* $x_1^{k_1} \otimes x_2^{k_2} \otimes \cdots \otimes x_m^{k_m}$

Formally, we represent F *by the (infinite) sum:*

$$F = \sum_{\substack{(k_1, k_2, \ldots, k_m)\\ \in N^m}} f(k_1, k_2, \ldots k_m) \otimes x_1^{k_1} \otimes \cdots \otimes x_m^{k_m}$$

Let us consider two formal series with coefficients $f(k_1, k_2 \cdots k_m)$ and $g(k_1, \ldots, k_m)$. The sum is the formal series of coefficients $s(k_1, \ldots k_m)$ defined as:

$$\forall (k_1, k_2 \cdots k_m) \in N^m\colon s(k_1 \cdots. k_m) = f(k_1 \cdots k_m) \oplus g(k_1 \cdots. k_m).$$

The product is the formal series of coefficients $t(k_1 \cdots k_m)$ defined as: $\forall (k_1 \ldots, k_m) \in N^m$: $t(k_1, k_2 \ldots, k_m) = \Sigma f(i_1, i_2 \ldots i_m) \otimes g(j_1, \ldots j_m)$ where the sum extends to all the pairs of m-tuples $(i_1, \ldots i_m) \in N^m$, $(j_1, j_2, \ldots j_m) \in N^m$ such that:

$$i_1 + j_1 = k_1, i_2 + j_2 = k_2, \ldots, i_m + j_m = k_m.$$

Proposition 2.1.2 of Sect. 2.1 easily extends to formal series as defined above.

3. Square Matrices with Coefficients in a (Pre)-Semiring

Let $(E, \oplus, \otimes)$ be a pre-semiring or a semiring. We denote $M_n(E)$ the set of square $n \times n$ matrices with elements in E.

Given two matrices $A = (a_{ij})$ and $B = (b_{ij})$ of $M_n(E)$

- The sum, denoted $A \oplus B$, is the matrix $S = (s_{ij})$ defined as:

$$\forall i, j: s_{ij} = a_{ij} \oplus b_{ij}$$

- The product, denoted $A \otimes B$, is the matrix $T = (t_{ij})$ defined as:

$$\forall i, j: t_{ij} = \sum_{k=1}^{n} a_{ik} \otimes b_{kj} \quad \text{(sum in the sense of } \oplus\text{)}.$$

If E has a neutral element ε for ⊕, the matrix:

$$\Sigma = \begin{bmatrix} \varepsilon, \varepsilon, \ldots, \varepsilon \\ \vdots \\ \varepsilon, \varepsilon, \ldots, \varepsilon \end{bmatrix}$$

is the neutral element of $M_n(E)$ for ⊕.

If, moreover, E has unit element e, and ε is absorbing for ⊗, then the matrix:

$$I = \begin{bmatrix} e & & & \\ & e & & \varepsilon \\ & \varepsilon & \ddots & \\ & & & e \end{bmatrix}$$

is the unit element of $M_n(E)$ for ⊗.

It is then easy to prove the following:

Proposition 3.1. *(i) If* (E, ⊕, ⊗) *is a pre-semiring then* ($M_n(E)$, ⊕, ⊗) *is a pre-semiring*

(ii) If (E, ⊕, ⊗) *is a semiring, then* ($M_n(E)$, ⊕, ⊗) *is a semiring (in general noncommutative)*

(iii) If (E, ⊕, ⊗) *is a dioid, then* ($M_n(E)$, ⊕, ⊗) *is a dioid (in general noncommutative)*

In the subsequent sections, we study properties of square $n \times n$ matrices with elements in a commutative pre-semiring (E, ⊕, ⊗). For some of the properties considered, we will have to assume that (E, ⊕, ⊗) has a semiring structure.

4. Bideterminant of a Square Matrix. Characteristic Bipolynomial

In this section we introduce the concept of *bideterminant* for matrices with coefficients in a pre-semiring.

4.1. Reminder About Permutations

Let π be a permutation of $X = \{1, 2, \ldots, n\}$ where, $\forall i \in X$, $\pi(i) \in X$ denotes the element corresponding to i through π. The *graph associated with* π is the directed graph G_π having X as set of vertices and n arcs of the form $(i, \pi(i))$. This graph can contain loops (when $\pi(i) = i$).

It is well-known that the permutation graph decomposes into *disjoint elementary circuits* (each connected component is an elementary circuit). If a connected component is reduced to a single vertex i, the corresponding circuit is the loop (i, i).

Figure 1 below represents the permutation graph of $\{1, \ldots 7\}$ defined as:

$$\pi(1) = 7,\ \pi(2) = 4,\ \pi(3) = 5,\ \pi(4) = 2,\ \pi(5) = 1,\ \pi(6) = 6,\ \pi(7) = 3.$$

The *parity* of a permutation π, is defined as the parity of the number of transpositions necessary to transform the permutation π into the identity permutation.

Thus, in the above example, a possible sequence of transpositions would be:

$$\begin{pmatrix}7\\4\\5\\2\\1\\6\\3\end{pmatrix} \to \begin{pmatrix}1\\4\\5\\2\\7\\6\\3\end{pmatrix} \to \begin{pmatrix}1\\2\\5\\4\\7\\6\\3\end{pmatrix} \to \begin{pmatrix}1\\2\\3\\4\\7\\6\\5\end{pmatrix} \to \begin{pmatrix}1\\2\\3\\4\\5\\6\\7\end{pmatrix}$$

The permutation of Fig. 1 is therefore *even*.

More generally, we can prove:

Property 4.1.1. The parity of a permutation π is equal to the parity of the number of circuits of even length of the graph G_π associated with the permutation.

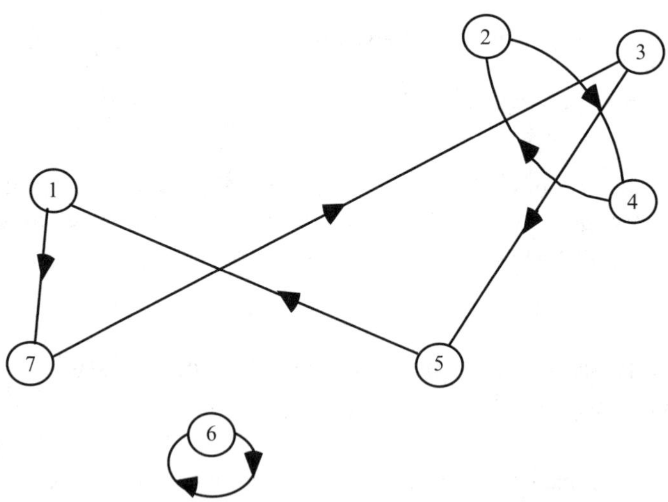

Fig. 1 Permutation graph

Example. The graph of Fig. 1 contains two circuits of even length (1, 7, 3, 5) and (2, 4), the corresponding permutation is therefore *even*. ||

We call *signature* of a permutation π, the quantity sign (π) defined as:

$$\begin{aligned} \text{sign}(\pi) &= +1 \quad \text{if } \pi \text{ is even} \\ \text{sign}(\pi) &= -1 \quad \text{if } \pi \text{ is odd} \end{aligned}$$

It is easy to see that the signature of a permutation π can be calculated as:

$$\text{sign}(\pi) = \prod_{\text{C circuit of } G_\pi} (-1)^{|C|-1}$$

(where $|C|$ is the cardinality of the circuit, and where the product extends to the set of the circuits of G_π).

In the example of Fig. 1 we have three circuits: $C_1 = (6)$ of odd length and $C_2 = (2, 4)$; $C_3 = (1, 3, 5, 7)$ of even length. We clearly have:

$$\begin{aligned} \text{sign}(\pi) &= (-)^{|C_1|-1} \times (-1)^{|C_2|-1} \times (-1)^{|C_3|-1} \\ &= +1 \end{aligned}$$

Hereafter we denote:

Per(n) the set of all the permutations of $\{1, 2, \ldots, n\}$
$\text{Per}^+(n)$ the set of all the even permutations of $\{1, 2, \ldots, n\}$ (the set of the permutations of signature +1)
$\text{Per}^-(n)$ the set of *odd* permutations of $\{1, 2 \ldots, n\}$ (of signature − 1)

We will also make use of the concept of *partial permutation*: a *partial permutation* of $X = \{1, \ldots, n\}$ is simply a permutation of a subset S of X.

Example. If $X = \{1, \ldots, 7\}$ $S = \{2, 3, 5, 7\}$ then σ defined as:

$$\sigma(2) = 3; \quad \sigma(3) = 7; \quad \sigma(5) = 5; \quad \sigma(7) = 2.$$

is a permutation of S and a partial permutation of X. The domain of definition of σ, denoted dom (σ), is $S = \{2, 3, 5, 7\}$

With every partial permutation σ of $X = \{1, \ldots, n\}$ we can associate the permutation $\hat{\sigma}$ of $\{1, \ldots, n\}$ defined as:

$$\begin{cases} \hat{\sigma}(i) = \sigma(i) & \text{if} \quad i \in \text{dom}(\sigma) \\ \hat{\sigma}(i) = i & \text{if} \quad i \in X \backslash (\text{dom}(\sigma)) \end{cases}$$

$\hat{\sigma}$ will be referred to as the *extension* of σ.

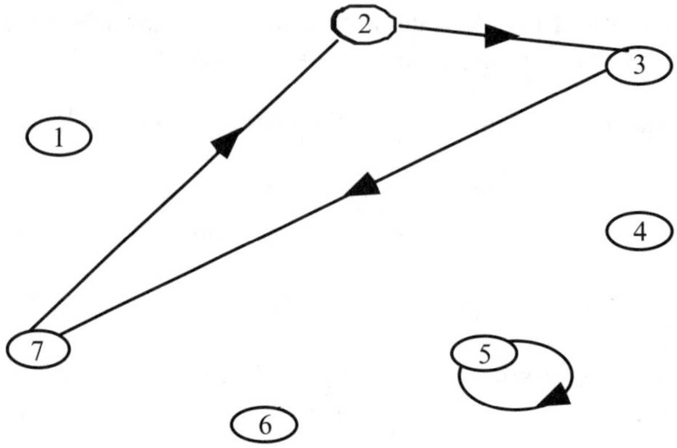

Fig. 2 Graph associated with a partial permutation σ of characteristic +1: $\sigma \in \text{Part}^+(7)$

The *parity* (resp. *signature*) of a partial permutation σ is the *parity* (resp. *signature*) of its extension $\hat{\sigma}$.

The *characteristic* of a partial permutation σ, denoted char (σ), is defined as:

$$\text{char}(\sigma) = \text{sign}(\sigma) \times (-1)^{|\sigma|}$$

$|\sigma|$ denoting the cardinality of dom(σ).

We observe that, if σ is a partial permutation of order k (i.e. $|\sigma| = |\text{dom}(\sigma)| = k$) and *cyclic* (i.e. such that the associated graph contains a single circuit covering all the vertices of dom (σ)) then: $\text{sign}(\sigma) = \text{sign}(\hat{\sigma}) = (-1)^{k-1}$, hence:

$$\text{char}(\sigma) = (-1)^{2k+1} = -1.$$

From the above, we deduce:

Property 4.1.2. For every partial permutation σ, $\text{char}(\sigma) = (-1)^r$ where r is the number of circuits in the graph associated with σ.

Example. For the partial permutation of {1, ..., 7} defined as:

$$\sigma(2) = 3; \quad \sigma(3) = 7; \quad \sigma(5) = 5; \quad \sigma(7) = 2.$$

the associated graph (see Fig. 2) contains two circuits, therefore: $\text{char}(\sigma) = +1$. ||

Hereafter, we denote Part(n) the set of all the partial permutations of {1, ..., n} (Observe that $\text{Per}(n) \subset \text{Part}(n)$).

The set of partial permutations of characteristic +1, (resp. of characteristic −1), will be denoted $\text{Part}^+(n)$ (resp. $\text{Part}^-(n)$).

4.2. Bideterminant of a Matrix

For a square matrix of order n, $A = (a_{ij})$ with elements in $\mathbb{R}$ endowed with the standard operations, the determinant det (A) is classically defined as:

$$\det(A) = \sum_{\pi \in \mathrm{Per}(n)} \mathrm{sign}(\pi) \left(\prod_{i=1}^{n} a_{i,\pi(i)} \right) \tag{1}$$

or equivalently, with the notation of Sect. 4.1., as:

$$\det(A) = \sum_{\pi \in \mathrm{Per}^{+}(n)} \left(\prod_{i=1}^{n} a_{i,\pi(i)} \right) - \sum_{\pi \in \mathrm{Per}^{-}(n)} \left(\prod_{i=1}^{n} a_{i,\pi(i)} \right) \tag{2}$$

(the above sums should be understood in the sense of the addition of reals). This notation is possible given that $(\mathbb{R}, +)$ is a *group*.

If one wishes to generalize the concept of determinant to algebraic structures featuring fewer properties, where addition does not induce a group structure, one must introduce the concept of *bideterminant*.

Definition 4.2.1. *(Bideterminant)*

Let $A = (a_{ij})$ *be a square* $n \times n$ *matrix with elements in a commutative pre-semiring* $(E, \oplus, \otimes)$. *We call bideterminant of* A *the pair* $(\det^{+}(A), \det^{-}(A))$ *where the values* $\det^{+}(A) \in E$ *and* $\det^{-}(A) \in E$ *are defined as:*

$$\det^{+}(A) = \sum_{\pi \in \mathrm{Per}^{+}(n)} \left(\prod_{i=1}^{n} a_{i,\pi(i)} \right) \tag{3}$$

$$\det^{-}(A) = \sum_{\pi \in \mathrm{Per}^{-}(n)} \left(\prod_{i=1}^{n} a_{i,\pi(i)} \right) \tag{4}$$

(the above sums and products should be understood in the sense of the operations $\oplus$ and $\otimes$ of the pre-semiring).

4.3. Characteristic Bipolynomial

In the case of a real $n \times n$ matrix A, the characteristic polynomial is defined as the polynomial in the variable λ equal to the determinant of the matrix $\lambda I - A$ where I is the $n \times n$ unit matrix:

$$\begin{aligned} P_A(\lambda) &= \det(\lambda I - A) \\ &= \sum_{\pi \in \mathrm{Per}(n)} \mathrm{sign}(\pi) \left(\prod_{i=1}^{n} b_{i,\pi(i)} \right) \end{aligned}$$

$$\text{where, } \forall i, j: \begin{cases} b_{ij} = -a_{ij} & \text{if } i \neq j \\ b_{ij} = \lambda - a_{ij} & \text{if } i = j \end{cases}$$

We observe that, for every q, $1 \le q \le n$, the coefficient of the term involving λ^{n-q} in the above expression can be expressed as:

$$\begin{aligned}
&\sum_{\substack{\sigma\in \text{Part}(n)\\ |\sigma|=q}} \text{sign}(\sigma)\left(\prod_{i\in \text{dom}(\sigma)} (-a_{i,\sigma(i)})\right)\\
&= \sum_{\substack{\sigma\in \text{Part}\ (n)\\ |\sigma|=q}} (-1)^{|\sigma|}\cdot \text{sign}(\sigma)\left(\prod_{i\in \text{dom}(\sigma)}^{n} (a_{i,\sigma(i)})\right)
\end{aligned} \tag{5}$$

For $q = 0$, the λ^n term has coefficient equal to 1. Observing that $(-1)^{|\sigma|}$ sign(σ) is none other than the characteristic char(σ) (see Sect. 4.1), (5) is rewritten:

$$\sum_{\substack{\sigma\in \text{Part}(n)\\ |\sigma|=q}} \text{car}(\sigma)\left(\prod_{i\in \text{dom}(\sigma)}^{n} (a_{i,\sigma(i)})\right) \tag{6}$$

By denoting (see Sect. 4.1) $\text{Part}^+(n)$ (resp. $\text{Part}^-(n)$) the set of partial permutations of $\{1, \dots, n\}$ with characteristic $+1$ (resp. with characteristic -1) then the above sum becomes:

$$\sum_{\substack{\sigma\in \text{Part}^+(n)\\ |\sigma|=q}} \left(\prod_{i\in \text{dom}(\sigma)}^{n} (a_{i,\sigma(i)})\right) - \sum_{\substack{\sigma\in \text{Part}^-(n)\\ |\sigma|=q}} \left(\prod_{i\in \text{dom}(\sigma)}^{n} (a_{i,\sigma(i)})\right) \tag{7}$$

Now, when $A = (a_{ij})$ is a matrix with coefficients in a pre-semiring $(E, \oplus, \otimes)$, one is then naturally lead to define the *characteristic bipolynomial* as follows.

Definition 4.3.1. ***(characteristic bipolynomial)*** *Let $A = (a_{ij})$ be a square $n \times n$ matrix with elements in a commutative pre-semiring $(E, \oplus, \otimes)$. We call characteristic bipolynomial the pair $(P_A^+(\lambda), P_A^-(\lambda))$ where $P_A^+(\lambda)$ and $P_A^-(\lambda)$ are two polynomials of degree n in the variable λ, defined as:*

$$P_A^+(\lambda) = \sum_{q=1}^{n}\left(\sum_{\substack{\sigma\in \text{Part}^+(n)\\ |\sigma|=q}} \left(\prod_{i\in \text{dom}(\sigma)}^{n} (a_{i,\sigma(i)})\right)\right) \otimes \lambda^{n-q} \oplus \lambda^n \tag{8}$$

and:

$$P_A^-(\lambda) = \sum_{q=1}^{n}\left(\sum_{\substack{\sigma\in \text{Part}^-(n)\\ |\sigma|=q}} \left(\prod_{i\in \text{dom}(\sigma)}^{n} (a_{i,\sigma(i)})\right)\right) \otimes \lambda^{n-q} \tag{9}$$

(the sums and the products above are to be understood in the sense of the addition $\oplus$ and the multiplication $\otimes$ of the pre-semiring $(E, \oplus, \otimes)$).

We observe that, in the case where (E, ⊕, ⊗) is a semiring, ε the neutral element of ⊕, is absorbing and the formulae (8)–(9) give:

$$P_A^+(\varepsilon) = \sum_{\substack{\sigma\in \mathrm{Part}^+(n) \\ |\sigma|=n}} \left(\prod_i a_{i,\sigma(i)}\right)$$

$$P_A^-(\varepsilon) = \sum_{\substack{\sigma\in \mathrm{Part}^-(n) \\ |\sigma|=n}} \left(\prod_i a_{i,\sigma(i)}\right)$$

Since, for $|\sigma| = n$, $\mathrm{char}(\sigma) = (-1)^n \,\mathrm{sign}(\sigma)$, we see that for *even n*, $\mathrm{Part}^+(n) = \mathrm{Per}^+(n)$ and consequently:

$$P_A^+(\varepsilon) = \det{}^+(A),\ P_A^-(\varepsilon) = \det{}^-(A)$$

For odd n, we have $\mathrm{Part}^+(n) = \mathrm{Per}^-(n)$ and consequently:

$$P_A^+(\varepsilon) = \det{}^-(A),\ P_A^-(\varepsilon) = \det{}^+(A).$$

We thus find again the analogue of the classical property for the characteristic polynomial:

$$P_A(0) = \det(-A) = (-1)^n \det(A).$$

5. Bideterminant of a Matrix Product as a Combinatorial Property of Pre-Semirings

Given two square n × n real matrices, a classical result of linear algebra is the identity:

$$\det(A \times B) = \det(A) \times \det(B)$$

In the present section we study the generalization of this result to square matrices with elements in a commutative *pre-semiring* (E, ⊕, ⊗).

$$\text{If}\quad A = (a_{ij})\quad B = (b_{ij})\quad \text{and}\quad C = A \otimes B = (c_{ij})$$

with:

$$c_{ij} = \sum_{k=1}^{n} a_{ik} \otimes b_{kj} \quad \text{(sum in the sense of the operation } \oplus)$$

Then, by definition (see Sect. 4.2):

$$\det{}^+(A \otimes B) = \sum_{\pi\in \mathrm{Per}^+(n)} \left(\prod_{i=1}^{n} c_{i,\pi(i)}\right) \tag{10}$$

For $\pi \in \text{Per}^+(n)$ fixed, we can write:

$$\prod_{i=1}^{n} c_{i,\pi(i)} = \prod_{i=1}^{n} \left(\sum_{k=1}^{n} a_{ik} \otimes b_{k,\pi(i)} \right) \tag{11}$$

By using distributivity, each term in the expansion of expression (11) is obtained by choosing, for each value of $i (1 \leq i \leq n)$, a value of $k \in \{1, \ldots, n\}$. In other words, each term in the expanded expression is associated with a mapping $f : \{1, \ldots, n\} \to \{1, \ldots, n\}$, and the value of the corresponding term in (11) is:

$$\prod_{i=1}^{n} (a_{i,f(i)} \otimes b_{f(i),\pi(i)})$$

By denoting F(n) the set of mappings: $\{1, \ldots, n\} \to \{1, \ldots, n\}$, (10) can therefore be rewritten:

$$\det{}^+(A \otimes B) = \sum_{f \in F(n)} \sum_{\pi \in \text{Per}^+(n)} \prod_{i=1}^{n} (a_{i,f(i)} \otimes b_{f(i),\pi(i)}) \tag{12}$$

We would similarly obtain:

$$\det{}^-(A \otimes B) = \sum_{f \in F(n)} \sum_{\pi \in \text{Per}^-(n)} \prod_{i=1}^{n} (a_{i,f(i)} \otimes b_{f(i),\pi(i)}) \tag{13}$$

Among the mappings of F(n), we find (even and odd) permutations, i.e.:

$$F(n) = \text{Per}^+(n) \cup \text{Per}^-(n) \cup F'(n)$$

where $F'(n)$ denotes the set of all the mappings of F(n) which are not permutations.

Expression (12) therefore decomposes into the sum of three sub-expressions:

$$\alpha^+ = \sum_{f \in \text{Per}^+(n)} \sum_{\pi \in \text{Per}^+(n)} \prod_{i=1}^{n} (a_{i,f(i)} \otimes b_{f(i),\pi(i)}) \tag{14}$$

$$\beta^+ = \sum_{f \in \text{Per}^-(n)} \sum_{\pi \in \text{Per}^+(n)} \prod_{i=1}^{n} (a_{i,f(i)} \otimes b_{f(i),\pi(i)}) \tag{15}$$

$$\gamma^+ = \sum_{f \in F'(n)} \sum_{\pi \in \text{Per}^+(n)} \prod_{i=1}^{n} (a_{i,f(i)} \otimes b_{f(i),\pi(i)}) \tag{16}$$

In cases where f is a permutation, let g be the permutation $\pi \circ f^{-1}$. In the expressions (14) and (15) above, we can rewrite the term:

$$\left(\prod_{i=1}^{n} a_{i,f(i)} \right) \otimes \left(\prod_{i=1}^{n} b_{f(i),\pi(i)} \right) \text{ as: } \left(\prod_{i=1}^{n} a_{i,f(i)} \right) \otimes \left(\prod_{i=1}^{n} b_{i,g(i)} \right)$$

Let us then consider the expression α^+.

f being an even permutation, f^{-1} is even and g, as the product of two even permutations is even. Then α^+ can be rewritten:

$$\alpha^+ = \left(\sum_{f \in Per^+(n)} \prod_{i=1}^{n} a_{i,f(i)} \right) \otimes \left(\sum_{g \in Per^+(n)} \prod_{i=1}^{n} b_{i,g(i)} \right) \\ = \det{}^+(A) \otimes \det{}^+(B) \tag{17}$$

Let us now consider the expression β^+.

f being odd, f^{-1} is odd and g, as the product of an even permutation and an odd permutation, is odd. Then β^+ can be rewritten:

$$\beta^+ = \left(\sum_{f \in Per^-(n)} \prod_{i=1}^{n} a_{i,f(i)} \right) \otimes \left(\sum_{g \in Per^-(n)} \prod_{i=1}^{n} b_{i,g(i)} \right) \\ = \det{}^-(A) \otimes \det{}^-(B) \tag{18}$$

From the above, we deduce:

$$\det{}^+(A \otimes B) = \det{}^+(A) \otimes \det{}^+(B) \oplus \det{}^-(A) \otimes \det{}^-(B) \oplus \gamma^+ \tag{19}$$

Through similar reasoning, we would prove that:

$$\det{}^-(A \otimes B) = \det{}^+(A) \otimes \det{}^-(B) \oplus \det{}^-(A) \otimes \det{}^+(B) \oplus \gamma^- \tag{20}$$

with:

$$\gamma^- = \sum_{f \in F'(n)} \sum_{\pi \in Per^-(n)} \prod_{i=1}^{n} (a_{i,f(i)} \otimes b_{f(i),\pi(i)}) \tag{21}$$

Now we prove:

Lemma 5.1. *The two expressions γ^+, given by* (16)*, and γ^-, given by* (21)*, take the same value.*

Proof. Let us consider an arbitrary term of the sum (16) whose value is:

$$\theta = \prod_{i=1}^{n} a_{i,f(i)} \otimes b_{f(i),\pi(i)}$$

with $f \in F'(n)$ and $\pi \in Per^+(n)$.

We are going to show that we associate it with a term θ' of expression (21) such that $\theta' = \theta$.

Since $f \in F'(n)$, f is not a permutation of $X = \{1, \ldots, n\}$, which therefore implies that there exists $i_0 \in X$, $i'_0 \in X$, $i'_0 \neq i_0$, $k \in X$ such that:

$$f(i_0) = k = f(i'_0) \tag{22}$$

If there exist several ordered triples (i_0, i'_0, k) satisfying (22) we choose the smallest possible value of k and, for this value of k, the two smallest possible values for i_0 and i'_0.

From the permutation π, let us define the following permutation π':

$$\begin{cases} \pi'(j) = \pi(j) \forall j \in X \backslash \{i_0, i'_0\}, \\ \pi'(i_0) = \pi(i'_0), \\ \pi'(i'_0) = \pi(i_0) \end{cases}$$

We observe that π' is deduced from π by transposition of the elements i_0 and i'_0, consequently $\pi' \in \mathrm{Per}^-(n)$. Furthermore, we observe that the same construction that obtains (f, π') from (f, π) enables one to obtain (f, π) from (f, π').

Finally, we have:

$$\begin{aligned} \theta' &= \prod_{i=1}^{n} \left(a_{i,f(i)} \otimes b_{f(i),\pi'(i)}\right) \\ &= \left(\prod_{\substack{i=1 \\ i \neq i_0 \\ i \neq i'_0}}^{n} a_{i,f(i)} \otimes b_{f(i),\pi'(i)} \right) \otimes a_{i_0,k} \otimes b_{k,\pi'(i_0)} \otimes a_{i'_0,k} \otimes b_{k,\pi'(i'_0)} \\ &= \left(\prod_{\substack{i=1 \\ i \neq i_0 \\ i \neq i'_0}}^{n} a_{i,f(i)} \otimes b_{f(i),\pi(i)} \right) \otimes a_{i_0,k} \otimes b_{k,\pi(i_0)} \otimes a_{i'_0,k} \otimes b_{k,\pi(i'_0)} \\ &= \theta \end{aligned}$$

which completes the proof. □

We have therefore obtained:

Theorem 1. *Let* A *and* B *be two square* $n \times n$ *matrices with coefficients in a commutative pre-semiring* $(E, \oplus, \otimes)$.

Then:

$$\det{}^+(A \otimes B) = \det{}^+(A) \otimes \det{}^+(B) \oplus \det{}^-(A) \otimes \det{}^-(B) \oplus \gamma$$

and:

$$\det{}^-(A \otimes g) = \det{}^+(A) \otimes \det{}^-(B) \oplus \det{}^-(A) \otimes \det{}^+(B) \oplus \gamma$$

where:

$$\gamma = \sum_{f\in F'(n)} \sum_{\pi\in Per^+(n)} \left(\prod_{i=1}^{n} a_{i,f(i)} \otimes b_{f(i),\pi(i)}\right)$$
$$= \sum_{f\in F'(n)} \sum_{\pi\in Per^-(n)} \left(\prod_{i=1}^{n} a_{i,f(i)} \otimes b_{f(i),\pi(i)}\right)$$

$F'(n)$, *in the above expressions, denoting the set of the mappings* $f\colon \{1, \ldots n\} \to \{1, \ldots n\}$ *which are not permutations.* □

As an immediate consequence of the above, we find again the well-known result:

Corollary 5.2. *If* $(E, \oplus)$ *is a group, then:*

$$\det(A \otimes B) = \det(A) \otimes \det(B)$$

As already pointed out in the introduction, Theorem 1 above clearly *does not* directly follow from the classical result (on the real field). Indeed a different proof is needed for the case of pre-semirings to get the exact expression of the additional term γ arising in both expressions of $\det^+(A \otimes B)$ and $\det^-(A \otimes B)$.

6. Cayley–Hamilton Theorem in Pre-Semirings

The Cayley–Hamilton theorem is a classical result of linear algebra (on the field of real numbers) according to which a matrix satisfies its own characteristic equation.

Combinatorial proofs of this theorem have been provided by Straubing (1983) and previously by Rutherford (1964). Rutherford's result constituted, moreover, a generalization of the theorem to the case of *semirings*.

Below we give a combinatorial proof inspired from Straubing (1983) and Zeilberger (1985), but which further generalizes the theorem to the case of *commutative pre-semirings* (indeed, it does not need to assume that ε, the neutral element of $\oplus$, is absorbing for $\otimes$).

Theorem 2. *Let* $(E, \oplus, \otimes)$ *be a commutative pre-semiring with neutral elements* ε *and* e.

Let A *be a square* $n \times n$ *matrix with coefficients in* $(E, \oplus, \otimes)$, *and let* $(P_A^+(\lambda), P_A^-(\lambda))$ *be the characteristic bipolynomial of* A.

$$\text{Then we have: } P_A^+(A) = P_A^-(A) \tag{23}$$

where:

$P_A^+(A)$ *and* $P_A^-(A)$ *are matrices obtained by replacing* λ^{n-q} *by the matrix* A^{n-q} *in the expression of* $P_A^+(\lambda)$ *and* $P_A^-(\lambda)$, *and where the following conventional notation is used:* A^o *denotes the matrix with diagonal terms equal to e and nondiagonal terms equal to* ε*; for every* $\alpha \in E$, $\alpha \otimes A^o$ *denotes the matrix with diagonal terms equal to* α *and nondiagonal terms equal to* ε.

Proof. We show that each entry (i, j) of the matrix $P_A^+(A)$ is equal to the entry (i, j) of the matrix $P_A^-(A)$.

Let us therefore consider i and j as fixed.

For $q = 0, 1, \ldots, n-1$, the value of term (i, j) of the matrix A^{n-q} is:

$$(A^{n-q})_{ij} = \sum_{\substack{p \in P_{ij} \\ |p| = n-q}} \left(\prod_{(k,l) \in p} a_{k,l} \right)$$

where P_{ij} is the set of (nonnecessarily elementary) paths joining i to j in the complete directed graph on the set of vertices $\{1, \ldots, n\}$, and where $|p|$ denotes the cardinality (number of arcs) of the path $p \in P_{ij}$.

For $q = n$, consistently with the adopted notational convention, $(A^{n-q})_{i,j} = (A^0)_{ij}$ is equal to ε for $i \neq j$, and to e for $i = j$.

Furthermore, the coefficient of A^{n-q} in $P_A^+(A)$ is:

$$\sum_{\substack{\sigma \in \mathrm{Part}^+(n) \\ |\sigma| = q}} \left(\prod_{i \in \mathrm{dom}(\sigma)} a_{i,\sigma(i)} \right)$$

and, consequently, the term (i, j) of the matrix $P_A^+(A)$ (by using the distributivity of ⊗ with respect to ⊕) is given by the following formulae. For $i \neq j$:

$$\sum_{q=1}^{n-1} \left[\left(\sum_{\substack{p \in P_{ij} \\ |p| = n-q}} \prod_{(k,l) \in p} a_{k,l} \right) \otimes \left(\sum_{\substack{\sigma \in \mathrm{Part}^+(n) \\ |\sigma| = q}} \prod_{i \in \mathrm{dom}(\sigma)} a_{i,\sigma(i)} \right) \right] \oplus \left[\sum_{\substack{p \in P_{ij} \\ |p| = n}} \prod_{(k,l) \in p} a_{k,l} \right] \tag{24}$$

For $i = j$, we must add to expression (24) the extra term:

$$\sum_{\substack{\sigma \in \mathrm{Part}^+(n) \\ |\sigma| = n}} \left(\prod_{i \in \mathrm{dom}(\sigma)} a_{i,\sigma(i)} \right)$$

(which may be viewed as corresponding to the value $q = n$).

Let us denote $\mathcal{F}_{ij}^+$ (resp. $\mathcal{F}_{ij}^-$) the family of graphs having $X = \{1, 2, \ldots, n\}$ as vertex set and whose set of arcs U decomposes into: $U = P \cup C$ where:

- P is a set of arcs forming a path from i to j;
- C is a set of arcs such that the graph $G = [X, C]$ is the graph associated with a partial permutation σ of X with $\sigma \in \mathrm{Part}^+(n)$ (resp. $\sigma \in \mathrm{Part}^-(n)$).
  In other words, [X, C] is a union of an even (resp. odd) number of disjoint circuits (loops are allowed) not necessarily covering all the vertices.
- $|U| = |P| + |C| = n$.

The *weight* w(G) of a graph G = [X, U] belonging to $\mathcal{F}_{ij}^{+}$ or to $\mathcal{F}_{ij}^{-}$ is defined as:

$$w(G) = \prod_{(k,l)\in U} a_{k,l}$$

In the case where $i \neq j$, by expanding (24) (distributivity) we then observe that entry (i, j) of $P_A^{+}(A)$ is:

$$\sum_{G\in\mathcal{F}_{ij}^{+}} w(G) \tag{25}$$

In the case where i = j, by considering that the path P can be empty in the decomposition $U = P \cup C$, the additional term corresponding to q = n is clearly taken into account in expression (25).

Similarly, it is easy to see that the entry (i, j) of $P_A^{-}(A)$ is, in all cases, (i = j and $i \neq j$), equal to:

$$\sum_{G\in\mathcal{F}_{ij}^{-}} w(G) \tag{26}$$

It therefore remains to show that the two expressions (25) and (26) are equal. To do so, let us show that, with any graph G of $\mathcal{F}_{ij}^{+}$, we can associate a graph G′ of $\mathcal{F}_{ij}^{-}$ of the same weight, w(G′) = w(G), the correspondence thus exhibited between $\mathcal{F}_{ij}^{+}$ and $\mathcal{F}_{ij}^{-}$ being one-to-one.

Let us therefore consider $G = [X, P \cup C] \in \mathcal{F}_{ij}^{+}$. [X, C] is a union of an even number (possibly zero) of vertex-disjoint circuits (Fig. 3 shows an example where $n = 8, i = 1, j = 4$).

Since $|P| + |C| = n$, we observe that the sets of vertices covered by P and C necessarily have at least one common element. Furthermore, the path P not necessarily being elementary, P can contain one (or several) circuit(s).

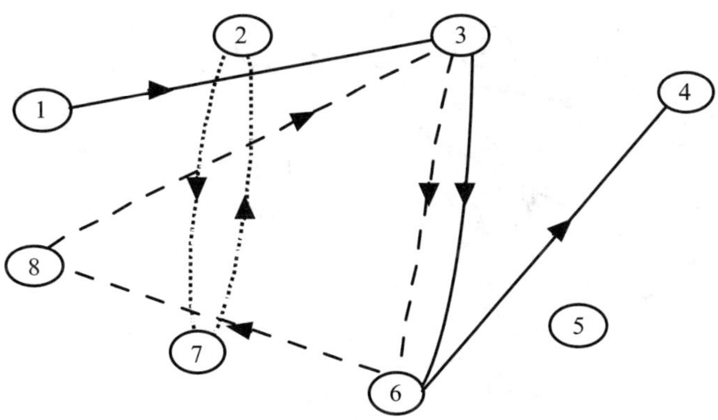

Fig. 3 Example illustrating the proof of the Cayley–Hamilton theorem. A graph $G \in \mathcal{F}_{ij}^{+}$ for $n = 8$, with $i = 1$ and $j = 4$. The path P is indicated in full lines and the partial permutation σ of characteristic +1 (as it contains two vertex-disjoint circuits) is indicated with dotted lines

Let us follow the path P starting from i until one of the following two situations occurs:

Case 1. We arrive at a vertex of P already traversed without meeting a vertex covered by C;
Case 2. We arrive at a vertex k covered by C.

In *case 1* we have identified a circuit Γ of P which does not contain any vertex covered by C. In this case, we construct $G' = [X, P' \cup C']$ where:

- P′ is deduced from P by eliminating the circuit Γ;
- C′ is deduced from C by adding the circuit Γ.

We observe that C′ now contains an odd number of disjoint circuits, therefore $G' \in \mathcal{F}_{ij}^{-}$.

In *case 2*, let Γ be the circuit of C containing the vertex k. We construct $G' = [X, P' \cup C']$ where:

- P′ is deduced from P by adding the circuit Γ;
- C′ is deduced from C by eliminating the circuit Γ.

Here again, C′ contains an odd number of disjoint circuits, therefore $G' \in \mathcal{F}_{ij}^{-}$.

Furthermore, we observe that in the two cases, G and G′ have the same set of arcs, therefore $w(G') = w(G)$.

Finally, it is easy to see that, the same construction by which G is transformed into G′ can be used to transform G′ back into G: there is therefore a one-to-one correspondence between $\mathcal{F}_{ij}^{+}$ and $\mathcal{F}_{ij}^{-}$. (see illustration in Fig. 4)

From the above we deduce:

$$\sum_{G \in \mathcal{F}_{ij}^{+}} w(G) = \sum_{G \in \mathcal{F}_{ij}^{-}} w(G)$$

which completes the proof of Theorem 2. □

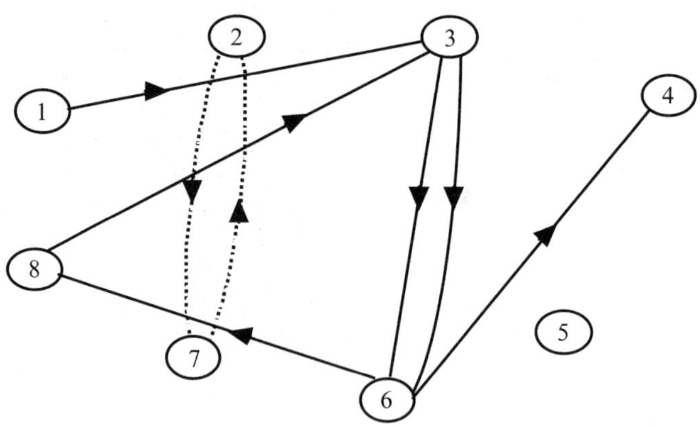

Fig. 4 The graph G′ obtained by including the circuit (3, 6, 8) in P is an element of $\mathcal{F}_{ij}^{-}$ and it has the same weight as G

7. Semirings, Bideterminants and Arborescences

In the present section we consider a square $n \times n$ matrix, $A = (a_{ij})$ with elements in a commutative semiring $(E, \oplus, \otimes)$. We assume therefore:

- That $\oplus$ has a neutral element ε
- That $\otimes$ has a neutral element e.
- That ε is *absorbing* for $\otimes$ that is to say,

$$\forall x \in E: \qquad \varepsilon \otimes x = x \otimes \varepsilon = \varepsilon$$

For $r \in [1, n]$ we denote $\bar{A}$ the $(n-1) \times (n-1)$ matrix deduced from A by deleting row r and column r.

We denote I the $(n-1) \times (n-1)$ identity matrix of $M_{n-1}(E)$ with all diagonal terms equal to e and all other terms equal to ε.

7.1. An Extension to Semirings of the Matrix-Tree Theorem

Let us begin by stating below the result which will be proved in Sect. 7.2, and which may be viewed as a generalization, to semirings, of the classical "Matrix-Tree-Theorem" by Borchardt (1860) and Tutte (1948).

Theorem 3. *(Minoux, 1997)*

Let A *be a square* $n \times n$ *matrix with coefficients in a commutative semiring* $(E, \oplus, \otimes)$. *Let* $\bar{A}$ *be the matrix deduced from* A *by deleting row* r *and column* $r (r \in [1, n])$ *and let* B *be the* $(2n-2) \times (2n-2)$ *matrix of the form:*

$$B = \begin{bmatrix} \overline{D} & \vdots & \bar{A} \\ \cdots & \cdot & \cdots \\ I & \vdots & I \end{bmatrix}$$

where I is the identity matrix of $M_{n-1}(E)$ *and* $\overline{D}$ *the diagonal matrix whose diagonal terms are:*

$$d_{ii} = \sum_{j=1}^{n} a_{ij} \quad \forall i \in \{1, \dots, n\} \backslash \{r\}$$

(sum in the sense of $\oplus$*).*

Let us denote by $\mathcal{G}$ *the complete directed 1-graph on the vertex set* $X = \{1, 2, \dots, n\}$ *and by* $\mathcal{T}_r$ *the set of the arborescences rooted at* r *in* $\mathcal{G}$. *For an arbitrary partial graph* G *of* $\mathcal{G}$, *the weight of* G, *denoted* w (G), *is the product (in the sense of* $\otimes$*) of the values* a_{ij} *for all the arcs* (i, j) *of* G.

Then we have the identity:

$$\det{}^{+}(B) = \det{}^{-}(B) \oplus \sum_{\mathcal{G} \in \mathcal{T}_r} w(G) \qquad \square$$

7.2. Proof of Extended Theorem

To prove Theorem 3, let us consider the following $(2n-2) \times (2n-2)$ square matrix:

$$B' = \begin{bmatrix} \bar{A} & \vdots & \overline{D} \\ \cdots & \cdots & \cdots \\ I & \vdots & I \end{bmatrix}$$

We observe that the permutation applied to the columns of B to obtain B' is even if $n-1$ is even, and odd if $n-1$ is odd. Consequently, if $n-1$ is even we have $\det^+(B) = \det^+(B')$ and $\det^-(B) = \det^-(B')$. If $n-1$ is odd, we have: $\det^+(B) = \det^-(B')$ and $\det^-(B) = \det^+(B')$.

Let us begin by studying the properties of the bideterminant of $B' = (b'_{ij})$. We have:

$$\det{}^+(B') = \sum_{\pi \in Per^+(2n-2)} \left(\prod_{i=1}^{2n-2} b'_{i,\pi(i)} \right) \tag{27}$$

In the above expression, all the terms corresponding to permutations π of $\{1, \ldots, 2n-2\}$ such that $b'_{i,\pi(i)} = \varepsilon$ for some $i \in [1, 2n-2]$ disappear because of the absorption property.

Consequently, in (27), we only have to take into account the permutations π of $Per^+(2n-2)$ such that, for $1 \le i \le n-1$:

$$\pi(i+n-1) = i \quad \text{or} \quad \pi(i+n-1) = i+n-1$$

Each admissible permutation π can therefore be associated with a partition of $\overline{X} = \{1, \ldots, n-1\}$ in two subsets U and V where:

$$U = \{i/i \in \overline{X}; \quad \pi(i+n-1) = i\}$$
$$V = \{i/i \in \overline{X}; \quad \pi(i+n-1) = i+n-1\}$$

Furthermore, we observe that the columns of B' indexed $i+n-1$ with $i \in U$ can only be covered by rows with index $i \in U$. Given that $\overline{D}$ is diagonal, we must therefore have:

$$\forall i \in U: \quad \pi(i) = i+n-1$$

Each admissible permutation π can therefore be considered as derived from a permutation σ of V (a partial permutation of $X = \{1, \ldots, n\}$) as follows:

$$\begin{cases} \forall i \in V: \begin{cases} \pi(i) = \sigma(i) \\ \pi(i+n-1) = i+n-1 \end{cases} \\ \forall i \in U: \begin{cases} \pi(i) = i+n-1 \\ \pi(i+n-1) = i \end{cases} \end{cases}$$

The graph representing π on the set of vertices $\{1, \ldots, 2n-2\}$ therefore consists of:

- Elementary circuits representing the partial permutation σ;
- $|V|$ loops on the vertices $i+n-1$ $(i \in V)$;
- $|U|$ circuits of length 2 (therefore even) of the form $(i, i+n-1), i \in U$.

The signature of π is therefore equal to

$$\text{sign}(\pi) = \text{sign}(\sigma) \times (-1)^{|U|}$$

hence:

$$\begin{aligned}\text{sign}(\pi) &= \text{sign}(\pi) \times (-1)^{2\times|V|} \\ &= \text{sign}(\sigma) \times (-1)^{|V|} \times (-1)^{|U|+|V|} \\ &= \text{char}(\sigma) \times (-1)^{n-1}\end{aligned}$$

(since $V = \text{dom}(\sigma)$).

Let us first assume that $n-1$ is even. In this case, $\text{sign}(\pi)$ is none other than the characteristic of σ as a partial permutation of $\overline{X}$, and $\pi \in \text{Per}^+(2n-2)$ if and only if $\sigma \in \text{Part}^+(n-1)$. Then, (27) can be rewritten:

$$\begin{aligned}\det{}^+(B') &= \sum_{\sigma\in\text{Part}^+(n-1)} \left(\prod_{i\in V} a_{i,\sigma(i)}\right) \otimes \left(\prod_{i\in U} d_{ii}\right) \\ &= \det{}^+(B)\end{aligned} \tag{28}$$

We would obtain a similar expression for $\det^-(B') = \det^-(B)$ simply by replacing $\sigma \in \text{Part}^+(n-1)$ in (28) with $\sigma \in \text{Part}^-(n-1)$. (Fig. 5)

Let us now consider the case where $n-1$ is odd. We then have $\text{sign}(\pi) = -\text{char}(\sigma)$, and, consequently, we have:

$$\begin{aligned}\det{}^+(B') &= \sum_{\sigma\in\text{Part}^-(n-1)} \left(\prod_{i\in V} a_{i,\sigma(i)}\right) \otimes \left(\prod_{i\in U} d_{ii}\right) \\ &= \det{}^-(B)\end{aligned} \tag{29}$$

(we obtain the expression of $\det^-(B') = \det^+(B)$ by replacing $\sigma \in \text{Part}^-(n-1)$ in (29) with $\sigma \in \text{Part}^+(n-1)$).

Thus it is seen that, in both cases ($n-1$ even or odd), the expression giving $\det^+(B)$ is:

$$\det{}^+(B) = \sum_{\sigma\in\text{Part}^+(n-1)} \left(\prod_{i\in V} a_{i,\sigma(i)}\right) \otimes \left(\prod_{i\in U} d_{ii}\right) \tag{30}$$

(where $V = \text{dom}(\sigma)$ and $U = \overline{X}\backslash V$). The expression giving $\det^-(B)$ is simply deduced from the above by replacing $\sigma \in \text{Part}^+(n-1)$ with $\sigma \in \text{Part}^-(n-1)$.

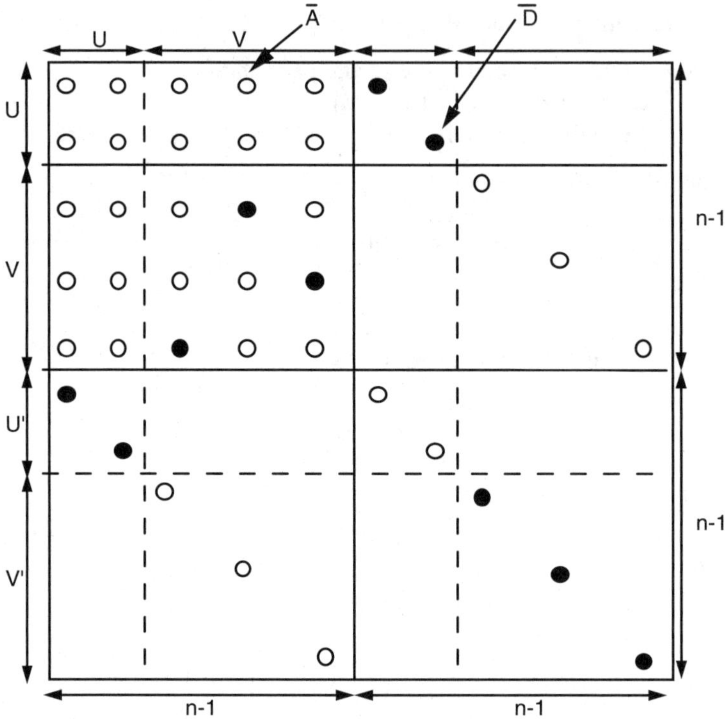

Fig. 5 The matrix B′ and a partition of $\overline{X} = \{1, \ldots, n-1\}$ into two subsets U and V corresponding to an admissible permutation π of $\{1, \ldots, 2n-2\}$. Only the terms distinct from ε (neutral element of ⊕) are represented (by circles). The terms indicated in black are those corresponding to the permutation π. The partial permutation σ is the one induced by π on the sub-matrix of $\bar{A}$ restricted to the rows and columns of V

Let us denote $\mathcal{F}^+$ (resp. $\mathcal{F}^-$) the family of all directed graphs constructed on the vertex set $X = \{1, 2, \ldots, n\}$, of the form $G = [X, C \cup Y]$ where:

- C is a set of arcs constituting vertex-disjoint circuits and containing an even (resp. odd) number of circuits;
- Y is a set of arcs such that, for every $i \in X\backslash\{r\}$ not covered by C, Y contains a single arc of the form (i, j) (the possibility j = i being authorized, as well as the possibility j = r).

By expanding expression (30), that is to say by replacing each term d_{ii} by $\sum_{j=1}^{n} a_{ij}$ and by using distributivity, we then observe that $\det^+(B)$ can be expressed in the form:

$$\det^+(B) = \sum_{G \in \mathcal{F}^+} w(G) \quad (31)$$

where the "weight" w(G) of the graph $G = [X, C \cup Y]$ is:

$$w(G) = \prod_{(k,l)\in C\cup Y} a_{k,l}$$

We would prove similarly that:

$$\det{}^{-}(B) = \sum_{G\in\mathcal{F}^-} w(G) \tag{32}$$

Among the graphs of $\mathcal{F}^+ \cup \mathcal{F}^-$, those *which do not contain a cycle* play a special role. Indeed, in this case, $C = \emptyset$, and the set Y does not contain a cycle and is composed of $n - 1$ arcs (an arc originating at each vertex $i \in X\backslash\{r\}$). Y therefore forms an *arborescence rooted at* r.

Since $C = \emptyset$, the subclass $\mathcal{T}_r$ (the set of arborescences rooted at r) is necessarily included in $\mathcal{F}^+$.

If we denote $$\mathcal{F}^+ = \mathcal{T}_r \cup \mathcal{F}_c^+$$

we can therefore write:

$$\det{}^{+}(B) = \sum_{G\in\mathcal{T}_r} w(G) \oplus \sum_{G\in\mathcal{F}_c^+} w(G) \tag{33}$$

The end of the proof uses the following result (Zeilberger, 1985):

Lemma 7.2.1.

$$\sum_{G\in\mathcal{F}_c^+} w(G) = \sum_{G\in\mathcal{F}^-} w(G) \tag{34}$$

Proof. It proceeds by showing that, with each graph $G \in \mathcal{F}_c^+$ we can associate a graph G' of $\mathcal{F}^-$ with $w(G') = w(G)$, and that the correspondence is one-to-one.

Let us therefore consider a graph G of $\mathcal{F}_c^+$ of the form $G = [X, C \cup Y]$.

This graph contains at least one circuit and [X, C] contains an even number (possibly zero) of circuits. Among all the circuits of G, let us consider the one which meets the vertex with the smallest index number and let Γ be the set of its arcs.

If $\Gamma \subset Y$ then let us define $G' = [X, C' \cup Y']$ with

$$C' = C \cup \Gamma$$
$$Y' = Y\backslash\Gamma$$

If $\Gamma \subset C$ then let us define C' and Y' as:

$$C' = C\backslash\Gamma$$
$$Y' = Y \cup \Gamma$$

In both cases, C′ contains an odd number of circuits, therefore $G' \in \mathcal{F}^-$, and as G and G′ have the same sets of arcs:

$$w(G') = w(G).$$

Furthermore, we observe that the same construction which transforms G to G′ enables one to transform G′ back to G.

We would prove in the same way that, with every $G \in \mathcal{F}^-$ we can associate $G' \in \mathcal{F}_c^+$ such that $w(G') = w(G)$.

This completes the proof of Lemma 7.2.1. □

By using Lemma 7.2.1, (33) is then rewritten:

$$\det{}^+(B) = \sum_{G \in \mathcal{T}_r} w(G) \oplus \det{}^-(B), \text{ which establishes Theorem 3.} \quad \square$$

7.3. The Classical Matrix-Tree Theorem as a Special Case

In the special case where A is a real matrix on the field of real numbers, we see that

$$\sum_{G \in \mathcal{T}_r} w(G) = \det{}^+(B) - \det{}^-(B) = \det(B)$$

where det(B) is the determinant of B in the usual sense and:

$$\begin{aligned} \det(B) &= \det \begin{bmatrix} \overline{D} & \vdots & \bar{A} \\ \cdots & \cdot & \cdots \\ I & \vdots & I \end{bmatrix} \\ &= \det \begin{bmatrix} \overline{D} - \bar{A} & \vdots & \bar{A} \\ \cdots & \cdot & \cdots \\ 0 & \vdots & I \end{bmatrix} \\ &= \det(\overline{D} - \bar{A}) \end{aligned}$$

From the above, we deduce the following corollary, known as the "Matrix Tree Theorem", due independently to Borchardt (1860) and Tutte (1948):

Corollary 7.3.1. *Let* $A = (a_{ij})$ *be a square* $n \times n$ *matrix with real coefficients;* D *the diagonal matrix whose ith diagonal term is* $d_{ii} = \sum_{j=1}^{n} a_{ij}$; $\bar{A}$ *and* $\overline{D}$ *the matrices deduced from* A *and* D *by eliminating the rth row and the rth column (for any fixed* r, $1 \le r \le n$*). Then det* $(\overline{D} - \bar{A})$ *is equal to the sum of the weights of the arborescences rooted at* r *in the graph associated with matrix* A.

Theorem 3 can thus be considered as an extension to semirings of the "Matrix-Tree Theorem".

7.4. A Still More General Version of the Theorem

A more general version of the "Matrix Tree Theorem", known as the "All Minors Matrix Tree Theorem" (see Chen (1976), Chaiken (1982)) can also be extended to semirings. We present this extension below (Theorem 4).

Let $A = (a_{ij})$ be a square $n \times n$ matrix with coefficients in a commutative semi-ring $(E, \oplus, \otimes)$, such that $\forall i = 1, \ldots, n$: $a_{ii} = \varepsilon$ (the neutral element of $\oplus$ in E).

For every $i \in X = \{1, 2, \ldots, n\}$ set:

$$d_{ii} = \sum_{\substack{k=1 \\ k \neq i}}^{n} a_{ik}$$

Let $L \subset X$ be a subset of rows of A and $K \subset X$ a subset of columns of A with $|L| = |K|$.

Let $\bar{A}$ be the sub-matrix of A obtained by eliminating the rows of L and the columns of K. The rows and the columns of $\bar{A}$ are therefore indexed by $\overline{L} = X \backslash L$ and $\overline{K} = X \backslash K$.

By setting $m = |\overline{L}| = |\overline{K}|$ and $p = |\overline{L} \cap \overline{K}|$ let us consider the $(m + p) \times (m + p)$ square matrix B having the block structure:

$$B = \begin{bmatrix} \bar{A} & \vdots & Q \\ \cdots & \vdots & \cdots \\ R & \vdots & I_p \end{bmatrix}$$

where:

I_p is the $p \times p$ identity matrix of the semiring $(E, \oplus, \otimes)$.

Q is a $m \times p$ matrix whose rows are indexed by $\overline{L}$ and whose columns are indexed by $\overline{L} \cap \overline{K}$; all its terms are equal to ε except those indexed (i, i) with $i \in \overline{L} \cap \overline{K}$ which are equal to d_{ii}.

R is a $p \times m$ matrix whose lines are indexed by $\overline{L} \cap \overline{K}$ and whose columns are indexed by $\overline{K}$; all its terms are equal to ε except those indexed (i, i) with $i \in \overline{L} \cap \overline{K}$ which are equal to e (the neutral element of $\otimes$ in E).

For every subset $Y \subset X = \{1, 2, \ldots, n\}$ let us denote $\text{sign}(Y, X) = (-1)^{\nu(Y,X)}$ where:

$$\nu(Y, X) = |\{(i, j)/i \in X \backslash Y, j \in Y, i < j\}|$$

and $s(L, K) = \text{sign}(L, X) \times \text{sign}(K, X) \times (-1)^m$.

Let us also consider the set $\mathcal{T} = \mathcal{T}^+ \cup \mathcal{T}^-$ of all the directed forests H on the vertex set X satisfying the following three properties:

(i) H contains exactly $|L| = |K|$ trees;
(ii) Each tree of H contains exactly a vertex of L and a vertex of K;
(iii) Each tree of H is an arborescence, the root of which is the unique vertex of K which it contains.

The subsets $\mathcal{T}^+$ and $\mathcal{T}^-$ are then defined as follows.

With each $H \in \mathcal{T}$ we can associate a one-to-one correspondence $\pi^*: L \to K$ defined as: $\pi^*(j) = i$ if and only if $i \in K$ and $j \in L$ belong to the same tree of H.

Then $\mathcal{T}^+$ (resp. $\mathcal{T}^-$) is the set of the directed forests of $\mathcal{T}$ such that sign $(\pi^*) = +1$ (resp. sign $(\pi^*) = -1$).
We can then state:

Theorem 4. *(Minoux, 1998a)*
If $s(L, K) = +1$ *then there exists* $\Delta \in E$ *such that:*

$$\begin{cases} \det^+(B) = \sum\limits_{H \in \mathcal{T}^+} w(H) \oplus \Delta \\ \det^-(B) = \sum\limits_{H \in \mathcal{T}^-} w(H) \oplus \Delta \end{cases}$$

If $s(L, K) = -1$ *then there exists* $\Delta \in E$ *such that:*

$$\begin{cases} \det^+(B) = \sum\limits_{H \in \mathcal{T}^-} w(H) \oplus \Delta \\ \det^-(B) = \sum\limits_{H \in \mathcal{T}^+} w(H) \oplus \Delta \end{cases}$$

Proof. Refer to Exercise 1 at the end of the chapter where the exact expression of Δ is specified. □

The above result suggests, once again, an essential remark concerning the general approach followed in the present chapter. In fact, suppose that we apply the simple trick which consists in formally deducing the generalized result from the classical result. The reader will easily be convinced that we can reformulate the classical "All-Minors Matrix-Tree Theorem" as:

$$\det(B) = \sum_{H \in \mathcal{T}^+} w(H) - \sum_{H \in \mathcal{T}^-} w(H)$$

If one thinks that it then suffices to rewrite the classical result by switching each term appearing negatively to the other side of the equation, one is led to propose a generalized version of the form:

$$\det^+(B) \oplus \sum_{H \in \mathcal{T}^-} w(H) = \det^-(B) \oplus \sum_{H \in \mathcal{T}^+} w(H)$$

which is not correct. Indeed, the above formula does not take into account the additional term Δ which cancels itself in the classical result.

Only a direct proof, specialized to the semiring structure, can exhibit this term and provide the exact expression (see Exercise 1 at the end of the chapter).

8. A Generalization of the Mac Mahon Identity to Commutative Pre-Semirings

Let us consider a square $n \times n$ matrix, $A = (a_{ij})$ with coefficients in a commutative pre-semiring $(E, \oplus, \otimes)$.

$x_1, x_2, \dots, x_n$ being indeterminates and $m_1, m_2, \dots, m_n$ natural integers, we consider the expression:

$$
\begin{aligned}
&(a_{11} \otimes x_1 \oplus a_{12} \otimes x_2 \oplus \cdots a_{1n} \otimes x_n)^{m_1} \\
&\otimes (a_{21} \otimes x_1 \oplus \cdots \oplus a_{2n} \otimes x_n)^{m_2} \\
&\quad \vdots \\
&\otimes (a_{n1} \otimes x_1 \oplus \cdots \oplus a_{nn} \otimes x_n)^{m_n}
\end{aligned}
\tag{35}
$$

and we denote $K(m_1, m_2, \dots, m_n)$ the coefficient of the term involving $x_1^{m_1} \otimes x_2^{m_2} \otimes \cdots \otimes x_n^{m_n}$ in the expansion of expression (35).

The Mac Mahon identity (1915) (recalled in Sect. 8.2 below) establishes a link between the formal series S in $x_1, x_2, \dots, x_n$, with coefficients $K(m_1, m_2, \dots, m_n)$, and the expansion of the inverse of the determinant of the matrix $I - A\,D_x$, where D_x is the diagonal matrix whose diagonal terms are the indeterminates $x_1, x_2, \dots, x_n$.

In Sect. 8.1, we establish a more general version of this result for commutative pre-semirings by giving a combinatorial proof generalizing that of Foata (1965), Cartier and Foata (1969) (see also Zeilberger, 1985). In Sect. 8.2 we show that the classical identity can be found again as a special case.

8.1. The Generalized Mac Mahon Identity

Theorem 5. *(Minoux 1998b, 2001)*

Let $(E, \oplus, \otimes)$ *be a commutative pre-semiring and* $A = (a_{ij}) \in M_n(E)$.

Let S denote the formal series:

$$S = \sum_{(m_1,\dots m_n)} K(m_1, \dots, m_n) \otimes x_1^{m_1} \otimes x_2^{m_2} \otimes \cdots \otimes x_n^{m_n} \tag{36}$$

where the sum extends to all distinct n*-tuples of natural integers.*

Then we have the following generalized Mac Mahon identity:

$$
\begin{aligned}
&S \otimes \left(\sum_{\sigma \in \mathrm{Part}^+(n)} \prod_{i \in \mathrm{dom}(\sigma)} a_{i,\sigma(i)} \otimes x_{\sigma(i)} \right) \\
&= e \oplus S \otimes \left(\sum_{\sigma \in \mathrm{Part}^-(n)} \prod_{i \in \mathrm{dom}(\sigma)} a_{i,\sigma(i)} \otimes x_{\sigma(i)} \right)
\end{aligned}
\tag{37}
$$

Proof. Let us consider the family $\mathcal{G}(m_1, \dots, m_n)$ of all the directed multigraphs of the form $G = [X, Y]$ where $X = \{1, 2, \dots, n\}$ is the vertex set and where the set of arcs Y satisfies the two conditions:

(1) $\forall i \in X$, Y contains exactly m_i arcs origining at i
(2) $\forall i \in X$, Y contains exactly m_i arcs terminating at i

(observe that the graphs of the family $\mathcal{G}(m_1, \dots, m_n)$ can obviously contain loops).

The weight of $G = [X, Y]$ is defined as the formal expression:

$$w(G) = \prod_{(k,l)\in Y} (a_{kl} \otimes x_l)$$

(product in the sense of $\otimes$) with the convention $w(G) = e$ if $Y = \emptyset$.

We then verify that:

$$K(m_1, \ldots, m_n) x_1^{m_1} \otimes x_2^{m_2} \otimes \cdots \otimes x_n^{m_n} = \sum_{G\in\mathcal{G}(m_1,\ldots,m_n)} w(G)$$

Consequently, the expression S given by (36) can be rewritten:

$$S = \sum_{(m_1,\ldots,m_n)} \sum_{G\in\mathcal{G}(m_1,\ldots,m_n)} w(G) = \sum_{G\in\mathcal{G}} w(G)$$

$$\text{with} \quad \mathcal{G} = \bigcup_{(m_1,\ldots,m_n)} \mathcal{G}(m_1, \ldots, m_n)$$

(union extended to all distinct n-tuples of natural integers).

Let us now consider the family $\mathcal{F}^+$ (resp. $\mathcal{F}^-$) of all the graphs of the form $G = [X, Y \cup C]$ where:

- $[X, Y] \in \mathcal{G}$
- $[X, C]$ is the graph representative of a partial permutation $\sigma \in \text{Part}^+(n)$ (resp. $\sigma \in \text{Part}^-(n)$). It is therefore a set of arcs forming an even number (resp. odd number) of elementary vertex-disjoint circuits (some of these circuits may be loops).

We then observe that the left-hand side of (37) is equal to: $\sum_{G\in\mathcal{F}^+} w(G)$ and the right-hand side of (37) is equal to: $e \oplus \sum_{G\in\mathcal{F}^-} w(G)$.

Among all the graphs of the family $\mathcal{F}^+ \cup \mathcal{F}^-$, let us consider $G_0 = [X, Y \cup C]$ with $Y = \emptyset$ and $C = \emptyset$. In this case, the graph $[X, Y]$ corresponds to $m_1 = 0, m_2 = 0, \ldots m_n = 0$, it is therefore the unique element of the family $\mathcal{G}(0, 0, \ldots 0)$. Furthermore, $G_0 \in \mathcal{F}^+$ since $C = \emptyset$ corresponds to an even number of circuits, and $w(G_0) = e$.

Consequently, it suffices to establish that:

$$\sum_{G\in\mathcal{F}^+\backslash G_0} w(G) = \sum_{G\in\mathcal{F}^-} w(G) \tag{38}$$

To do so, we are going to exhibit a one-to-one correspondence between $\mathcal{F}^+\backslash G_0$ and $\mathcal{F}^-$ such that, if $G \in \mathcal{F}^+\backslash G_0$ and $G' \in \mathcal{F}^-$ are images through this one-to-one correspondence, then $w(G') = w(G)$.

All the graphs of the form $[X, Y \cup C]$ in $(\mathcal{F}^+\backslash G_o) \cup \mathcal{F}^-$ are assumed to be represented by adjacency lists with the following convention: for every $i \in X$, if i belongs to a circuit in $[X, C]$, then the arc of origin i in C is placed in the first position of the list of the arcs of origin i.

Now, let us consider $G = [X, Y \cup C] \in \mathcal{F}^+ \backslash G_0$. Since $G \neq G_0$, there exists at least one vertex of nonzero degree in G. Among these, let i_0 be the vertex having minimum index number.

Observe that C consists of an even number of vertex-disjoint circuits (this number may possibly be zero).

Let us traverse the partial graph [X, Y] starting from vertex i_0 by using the arcs of Y as follows: from every intermediate vertex i encountered that is not covered by C, we take the arc (i, j) which appears first in the adjacency list of vertex i. The traversal stops when one of the two following situations arises:

Case 1. We arrive at a vertex already encountered in the pathway before having encountered a vertex covered by C;
Case 2. We arrive at a vertex k covered by C.

In the first case, we have exhibited a circuit of the partial graph [X, Y], which does not have a common vertex with C. Let $\Gamma \subset Y$ be the set of its arcs.

$$\begin{aligned} \text{We then form} \quad & G' = [X, Y' \cup C'] \\ \text{with} \quad & Y' = Y \backslash \Gamma \\ & C' = C \cup \Gamma \end{aligned}$$

In the second case, C contains a circuit passing through k and let Γ be the set of its arcs. Then we form $G' = [X, Y' \cup C']$ with:

$$\begin{aligned} Y' &= Y \cup \Gamma \\ C' &= C \backslash \Gamma \end{aligned}$$

Moreover, the adjacency list of each node i covered by the circuit Γ is modified in such a way that the arc of Γ which originates at i becomes the first in the adjacency list for i.

In both cases, C' contains an odd number of vertex-disjoint circuits. Furthermore, the sets of arcs of G and G' being the same, we have $w(G') = w(G)$.

Finally, we observe that, thanks to the convention established concerning the order of arcs in the adjacency lists, the same construction which transforms G into G' enables one to transform G into G'. This is therefore a one-to-one correspondence between $\mathcal{F}^+ \backslash G_0$ and $\mathcal{F}^-$, which completes the proof of Theorem 5. □

8.2. The Classical Mac Mahon Identity as a Special Case

It is interesting to verify that the generalized form (37) of the Mac Mahon identity includes, as a special case, the usual form on the field of real numbers, which is expressed by the following corollary:

Corollary 8.2.1. S *being defined as in expression (*36*), and* B *denoting the matrix* $B = (b_{ij})_{\substack{i=1,\dots n \\ j=1,\dots n}} = (a_{ij} x_j)_{\substack{i=1,\dots n \\ j=1,\dots n}}$, *we have:*

$$S \times \det(I - B) = 1 \tag{39}$$

Proof. See Exercise 2 at the end of the chapter and Minoux (1998b, 2001). □

Exercises

Exercise 1. We consider the real matrix:

$$A = \begin{bmatrix} 4 & 2 & 0 & 1 \\ 0 & 3 & -1 & 2 \\ 2 & 0 & 5 & -3 \\ -2 & 1 & 6 & 0 \end{bmatrix}$$

on a dioid $(\overline{\mathbb{R}}, \oplus, \otimes)$.

(1) Give the formal expression of the bideterminant of A, by formally stating $\det^+(A)$ and $\det^-(A)$.
(2) Compute the value of the bideterminant when the dioid under consideration is $(\overline{\mathbb{R}}, \text{Max}, \text{Min})$. Check that $\text{Max}\{\det^+(A);\ \det^-(A)\}$ is indeed equal to the optimal value of the « bottleneck » (Max-Min) assignment problem.
(3) Compute the value of the bideterminant when the dioid under consideration is $(\overline{\mathbb{R}}, \text{Max}, +)$. Check that $\text{Max}\{\det^+(A);\ \det^-(A)\}$ is indeed equal to the optimal value of the assignment problem (where the objective is to maximize the sum of the selected entries).
(4) Check the Cayley–Hamilton theorem for A in both cases ($(\overline{\mathbb{R}}, \text{Max}, \text{Min})$ and $(\overline{\mathbb{R}}, \text{Max}, +)$).

Exercise 2. We consider the real 4×4 matrix with entries in the dioid $(\overline{\mathbb{R}}, \text{Min}, +)$:

$$A = \begin{bmatrix} \infty & 4 & 0 & 1 \\ 0 & \infty & -1 & 2 \\ 3 & 5 & \infty & -3 \\ -2 & 1 & 6 & \infty \end{bmatrix}$$

which is a generalized adjacency matrix corresponding to the complete oriented graph.

(1) Set up the list of all arborescences with root $r = 1$ in the above graph, and calculate the sum S (in the sense of $\oplus = \text{Min}$) of the weights of these arborescences. We recall that, in the Matrix-Tree Theorem (see Theorem 3, Sect. 7.1), the arborescences involved are those having arcs oriented from the pending vertices to the root. The vertex $r = 1$ has thus zero out-degree.
(2) Check the generalized version of the « matrix tree theorem » on this example, in other words that $\det^+(B) = \text{Min}\ \{\det^-(B);\ S\}$

where B is the 6×6 matrix: $\begin{bmatrix} \overline{D} & \vdots & \bar{A} \\ \cdots & \cdots & \cdots \\ I & \vdots & I \end{bmatrix}$

where:

$\bar{A}$ is deduced from A by deleting the first row and the first column of A; $\overline{D}$ is the diagonal matrix with diagonal entries:

$$d_{ii} = \underset{j=1,\ldots,n}{\text{Min}} \{a_{ij}\} \quad \forall i = 2, 3, 4.$$

[*Answers*:

(1) There are 16 distinct arborescences rooted at $r = 1$ in this example. For instance the arborescence composed of the arcs (2,1) (3,1) (4,1) with weight 1 ($= 0 + 3 - 2$); the arborescence composed of the arcs (2, 1) (3, 1) (4, 2) with weight 4, etc. The minimum of the weights of these 16 arborescences is $S = -6$, and corresponds to the arborescence $(4, 1)(2, 3)(3, 4)$.

(2) We have $\bar{A} = \begin{bmatrix} \infty & -1 & 2 \\ 5 & \infty & -3 \\ 1 & 6 & \infty \end{bmatrix}$ and $\overline{D} = \begin{bmatrix} -1 & \infty & \infty \\ \infty & -3 & \infty \\ \infty & \infty & -2 \end{bmatrix}$

and it can be checked that:

$$\det{}^{+}(B) = -6, \quad \det{}^{-}(B) = -3$$

and that the extended Matrix-Tree Theorem holds since:

$$\det{}^{+}(B) = \text{Min}\{\det{}^{-}(B), S\} = \text{Min}\{-3, -6\}.]$$

Exercise 3. (Proof of Theorem 4: generalized "All Minors Matrix Tree Theorem")

In this exercise, we refer to the concepts and notation used in Sect. 7.4.

Given two subsets U and V of X of equal cardinality ($|U| = |V|$), we refer to as *matching* every one-to-one correspondence $\pi: U \to V$. The *signature* of a matching $\pi: U \to V$, denoted $\text{sign}(\pi)$, is defined as follows. A pair (i, j) of elements of U is said to be in *inversion* relatively to π if $i < j$ and $\pi(i) > \pi(j)$. By denoting $\nu(\pi)$ the number of pairs (i, j) $i \in U, j \in U$, which are in inversion relatively to π, then sign $(\pi) = (-1)^{\nu(\pi)}$. We observe that, in the special case where $U = V = X$, a matching is none other than a permutation of X, and we verify that in this case the definition of the matching signature is consistent with that of the permutation signature.

The *characteristic* of a matching $\pi: U \to V$ is defined as:

$$\text{char}(\pi) = \text{sign}(\pi) \times (-1)^{|W|}$$
$$\text{where} \quad W = \{i / i \in U, \pi(i) = i\}$$

We now denote by $\mathcal{F}^{+}$ (resp. $\mathcal{F}^{-}$) the set of all the directed graphs on X having as set of arcs $S \cup T$ where:

- S is the set of arcs of the form $(i, \pi(i))$ for every $i \in L$ such that $i \neq \pi(i)$, where $\pi: \overline{L} \to \overline{K}$ is a matching of characteristic $+1$ (resp. of characteristic -1).
- T is a set of arcs such that, for every $i \in \overline{L}$ satisfying $\pi(i) = i$, there is exactly one arc in T of the form (k, i) with $k \in X$, $k \neq i$ (note that $\pi(i) = i$ implies $i \in \overline{L} \cap \overline{K}$).

Among the graphs H of the family $\mathcal{F}^{+}$ (resp. $\mathcal{F}^{-}$) those which are circuitless are exactly those of $\mathcal{T}^{+}$ (resp. $\mathcal{T}^{-}$) (see Sect. 7.4). We can therefore write:

$\mathcal{F}^{+} = \mathcal{T}^{+} \cup \mathcal{F}_c^{+}$ and $\mathcal{F}^{-} = \mathcal{T}^{-} \cup \mathcal{F}_c^{-}$ where $\mathcal{F}_c^{+}$ (resp. $\mathcal{F}_c^{-}$) denotes the family of sub-graphs $H \in \mathcal{F}^{+}$ (resp. $H \in \mathcal{F}^{-}$) which contain nontrivial circuits (i.e. circuits which are not loops).

(1) Prove that we have:

$$\det{}^{+}(B) = \sum_{H \in \mathcal{F}^{+}} w(H) \quad \text{and} \quad \det{}^{-}(B) = \sum_{H \in \mathcal{F}^{-}} w(H)$$

where B is the matrix $\begin{bmatrix} \bar{A} & \vdots & Q \\ \cdots & \vdots & \cdots \\ R & \vdots & I_p \end{bmatrix}$ defined in Sect. 7.4.

(2) Show, by using an argument similar to the one used by Chaiken (1982), that

$$\sum_{H \in \mathcal{F}_c^{+}} w(H) = \sum_{H \in \mathcal{F}_c^{-}} w(H)$$

(3) Then show that Theorem 4 is deduced from the above by taking:

$$\Delta = \sum_{H \in \mathcal{F}_c^{+}} w(H) = \sum_{H \in \mathcal{F}_c^{-}} w(H)$$

[*Answers*: refer to Minoux (1998a)].

Exercise 4. Where we recover the classical Mac Mahon identity

Here we take the field of real numbers as the basic algebraic structure.

(1) Let B be a $n \times n$ matrix with coefficients in $\mathbb{R}$, and I the identity matrix of $M_n(\mathbb{R})$. Prove that:

$$\det(I - B) = \sum_{\sigma \in \mathrm{Part}^{+}(n)} \left(\prod_{j \in \mathrm{dom}(\sigma)} b_{i,\sigma(i)} \right) - \sum_{\sigma \in \mathrm{Part}^{-}(n)} \left(\prod_{i \in \mathrm{dom}(\sigma)} b_{i,\sigma(i)} \right).$$

(2) By using the above relation, deduce from Theorem 5 (see Sect. 8.1) the classical Mac Mahon identity:

$$S \times \det(I - B) = 1$$
$$\text{with} \quad B = (b_{ij})_{\substack{i=1\cdots n \\ j=1\cdots n}} = (a_{ij} x_j)_{\substack{i=1\cdots n \\ j=1\cdots n}}.$$

[*Answers*: refer to Minoux (1998b, 2001)]

Chapter 3
Topology on Ordered Sets: Topological Dioids

1. Introduction

This chapter is devoted to the study of topological properties, first in general ordered sets, then in dioids (this will eventually lead to the concept of *topological dioids*) and to the solution of equations of the fixed-point type.

Various types of topologies may be introduced, depending on the nature of the ordered sets considered. The simplest cases correspond to a totally ordered set, or to a product of totally ordered sets (e.g. $\mathbb{R}^n$ with the partial order induced by the usual order on $\mathbb{R}$). The relevant topologies on such sets are extensions of usual topologies. We will concentrate here on the more general case of partially ordered sets (or "posets"). In relation to these sets, we introduce in Sect. 2 two basic topologies: the *sup-topology* and the *inf-topology*.

Then we show in Sect. 3 that the sup-topology may be interpreted in terms of *limit sup of increasing sequences*; and likewise that the inf-topology may be interpreted in terms of the *limit inf of decreasing sequences*. The notions of continuity and semi-continuity for functions on partially ordered sets are introduced in Sect. 4.

We then discuss the fixed-point theorem, first in the context of general ordered sets (Sect. 5), and next in the context of topological dioids, in view of solving linear equations of the fixed-point type. Section 7 is devoted to the concept of p-stable element in a dioid which guarantees the existence of a quasi-inverse, and which turns out to be useful in the solution of various types of equations, whether linear (Sect. 7.2) or nonlinear (Sect. 7.3).

Finally, Sect. 8 introduces and discusses the concepts of residuation and of generalized solutions.

2. Sup-Topology and Inf-Topology in Partially Ordered Sets

Let (E, $\leq$) be an ordered set, where $\leq$ is a reflexive, transitive and antisymmetric binary relation. To define a *topology on* E, it is known that it suffices to provide a *fundamental system of neighborhoods*. This can be achieved in various ways.

For example, if we choose as the system of fundamental neighborhoods the set of subsets of E (*ideals*) of the form $\downarrow a = \{x/x \le a\}$ for a running through E, we obtain a so-called (*left*) *Alexandrov topology*. We could similarly, choose as fundamental neighborhoods *filters* of the form $\uparrow a = \{x/a \le x\}$. We would then obtain the *right Alexandrov topology*.

The Alexandrov topologies *are not separated* (see Example 2.1.3. below). We recall that a topology is *separated* if and only if, given two arbitrary distinct elements $a \in E$, $a' \in E$, we can find a neighborhood of a and a neighborhood of a' that are *disjoint*. It is an essential property to guarantee *uniqueness* for the limit of a sequence.

In this chapter, we study separated topologies finer than the Alexandrov topologies: the *Sup-topology* and the *Inf-topology*. We will see that the Sup-topology corresponds to the concept of *upper limit of* sequences and that the Inf-topology corresponds to the concept of *lower limit* of sequences.

2.1. The Sup-Topology

For every $a \in E$ introduce the *ideal*:

$\downarrow a = \{x \in E/x \le a\}$ and the *anti-ideal*:

$\ddagger a = \{x \in E/x \not\le a\}$.

Definition 2.1.1. E *being an ordered set with respect to the relation* $\le$, *we call Sup-topology on* E *the topology for which the system of fundamental neighborhoods for any element* $a \in E$ *consists of all the sets of the form:*

$$V = \downarrow a \cap \ddagger b_1 \cap \ddagger b_2 \ldots \cap \ddagger b_k \tag{1}$$

where $(b_1, b_2, \ldots b_k)$ *is a finite (possibly empty) family of elements of* E *such that,* $\forall i = 1, 2, \ldots k$: $a \not\le b_i$ (*that is to say,* b_i *belongs to* $\ddagger$ a) (*see Betrema 1982*).

A neighborhood of $a \in E$, *in the sense of the Sup-topology, is therefore formed by every subset of* E *containing a subset of the form* V *defined as* (1).

The set of the neighborhoods of $a \in E$ for the Sup-topology will be denoted $\mathcal{V}_s(a)$.

With the above definitions, we easily check the properties:

(i) If $V \in \mathcal{V}_s(a)$ and $V' \in \mathcal{V}_s(a)$ then $V \cap V' \in \mathcal{V}_s(a)$
(ii) If $V \in \mathcal{V}_s(a)$ and $U \supset V$ then $U \in \mathcal{V}_s(a)$
(iii) If $V_i \in \mathcal{V}_s(a)$ for $i \in I \subset \mathbb{N}$, then $\bigcup_{i \in I} V_i \in \mathcal{V}_s(a)$.

Since every neighborhood in the (left) Alexandrov topology is a neighborhood in the Sup-topology (but not the converse), we see that the Sup-topology is *finer* than the left Alexandrov topology.

We can state:

Property 2.1.2. *The Sup-topology is separated.*

Proof. Let $a \in E$, $a' \in E$, $a \neq a'$. We distinguish two cases.

Case 1: $a' \leq a$
In this case: $a \nleq a'$ (since $a \neq a'$).
Then $V(a) = \downarrow a \cap \not\downarrow a'$ and $V(a') = \downarrow a'$
are two neighborhoods of a and a′ respectively, and they are disjoint.
Case 2: $a' \nleq a$
Then $V(a) = \downarrow a$ and $V(a') = \downarrow a' \cap \not\downarrow a$
are two neighborhoods of a and a′ respectively, and they are disjoint. □

Contrary to the Sup-topology, the left Alexandrov topology *is not separated*, as seen in the following example.

Example 2.1.3. We consider $E = \mathbb{R}^2$ endowed with the order relation $\begin{pmatrix} a \\ a' \end{pmatrix} \propto \begin{pmatrix} b \\ b' \end{pmatrix} \Leftrightarrow a \leq b$ and $a' \leq b'$ (where $\leq$ denotes the standard order relation on $\mathbb{R}$). The two elements $\begin{pmatrix} 1 \\ 1 \end{pmatrix}$ and $\begin{pmatrix} 3 \\ 2 \end{pmatrix}$ are distinct, but every neighborhood of $\begin{pmatrix} 3 \\ 2 \end{pmatrix}$ in the sense of the left Alexandrov topology contains $\downarrow \begin{pmatrix} 1 \\ 1 \end{pmatrix}$. Therefore, in the left Alexandrov topology, every neighborhood of $\begin{pmatrix} 3 \\ 2 \end{pmatrix}$ intersects an arbitrary neighborhood of $\begin{pmatrix} 1 \\ 1 \end{pmatrix}$. ||

2.2. The Inf-Topology

The Inf-topology can be defined similarly to the above.

For arbitrary $a \in E$, let us introduce the *filter*:
$\uparrow a = \{x \in E / a \leq x)$ and the *anti-filter*:
$\not\uparrow a = \{x \in E / a \nleq x\}$.

Definition 2.2.1. *We call Inf-topology on* E *the topology for which the system of fundamental neighborhoods of an arbitrary element* $a \in E$ *is composed of all the sets of the form:*

$$V = \uparrow a \cap \not\uparrow c_1 \cap \not\uparrow c_2 \ldots \cap \not\uparrow c_\ell \tag{2}$$

where $c_1, c_2, \ldots c_\ell$ *is a finite (possibly empty) family of elements of* E *such that:* $\forall i = 1, \ldots \ell$: $c_i \nleq a$, *that is to say:* $c_i \in \not\uparrow a$.

We denote $\mathcal{V}_i(a)$ *the set of neighborhoods for the inf-topology.*

Remark 2.2.2. In the case of a totally ordered set or of a set defined as the product of totally ordered sets, we can choose the system of fundamental neighborhoods of an element $a \in E$ as the family of sets of the form:

$$V = \bigcap_{i \in I} (\not\downarrow b_i) \bigcap_{j \in J} (\not\uparrow c_j)$$

where $(b_i)_{i \in I}$ and $(c_j)_{j \in J}$ are finite (possibly empty) families of elements of E such that, $\forall i \in I: a \not\leq b_i$ (that is to say: $b_i \in \not\uparrow a$) and, $\forall j \in J$: $c_j \not\leq a$ (that is to say: $c_j \in \not\downarrow a$). ||

The various topologies introduced above (Sup-topology, Inf-topology), endow E with a structure of *topological space*. Let us briefly recall some definitions and basic properties valid in every topological space.

- A subset A of E is called an *open neighborhood* if and only if A is a neighborhood of each of its elements.
- Let $A \subset E$ and $x \in A$. We say that x is *interior* to A if there exists a neighborhood V of x such that $V \subset A$. The interior of A, denoted $\mathring{A}$, is the set of the elements interior to A.
- A subset $A \subset E$ is *open*, if and only $A = \mathring{A}$.
- A subset $B \subset E$ is *closed* if B is the complement in E of an open subset A.
- Let $A \subset E$ and $x \in E$. We say that x is *adherent* to A if every neighborhood of x in E intersects A.
- The set of adherent points of A is called the *closure* of A and is denoted $\bar{A}$.
- The closure $\bar{A}$ of $A \subset E$ is the smallest closed subset of E containing A.
- A subset A of E such that $\bar{A} = E$ is said to be *dense* in E. A is dense in E if and only if every non empty open subset of E intersects A.

3. Convergence in the Sup-Topology and Upper Bound

Let $(E, \leq)$ be an ordered set endowed with the Sup-topology.

3.1. Definition (Sup-Convergence)

An infinite sequence $\{x_n\}_{n \in N}$ *of elements of* E *is said to be convergent in the sense of the Sup-topology, with limit* $\bar{x} \in E$, *if and only if,* $\forall V \in \mathcal{V}_s(\bar{x})$, *there exists* $K \in \mathbb{N}$ *such that:* $k \geq K \Rightarrow x_k \in V$. *This convergence will be denoted* $x_n \rightharpoonup \bar{x}$.

Observe that, as the Sup-topology is *separated*, the limit of a sequence, when it exists, is *unique*.

The following result (Betrema 1982) establishes the equivalence between the limits (for the Sup-topology) of nondecreasing sequences and their least upper bounds.

Theorem 1. *Let* (E, $\leq$) *be an ordered set endowed with the Sup-topology.*

(i) Let $\{x_n\}_{n\in\mathbb{N}}$ *be a nondecreasing sequence bounded from above and having a least upper bound* $\bar{x}$*. Then the sequence* $\{x_n\}$ *is convergent for the Sup-topology and has limit* $\bar{x}$*.*

(ii) Let $\{x_n\}_{n\in\mathbb{N}}$ *be a convergent sequence in the sense of the Sup-topology (but not necessarily nondecreasing) and with limit* $\bar{x}(x_n \rightharpoonup \bar{x})$*. Then there exists an integer* $p \in \mathbb{N}$ *such that:* $\bar{x} = \sup\{x_n \colon n \geq p\}$

Proof. (i) By assumption, the set $\{x_n \colon n \in \mathbb{N}\}$ has a least upper bound $\bar{x} = \sup\{x_n \colon n \in \mathbb{N}\}$

Let $V_s(\bar{x})$ be an arbitrary neighborhood of $\bar{x}$ in the sense of the Sup-topology. It is of the form:

$$V_s(\bar{x}) = \downarrow \bar{x} \cap \not\downarrow y_1 \cap \ldots \cap \not\downarrow y_k$$

where, $\forall i = 1, \ldots k$: $\bar{x} \not\leq y_i$, that is to say $y_i \in \not\uparrow \bar{x}$. Consider any i, $1 \leq i \leq k$; y_i is not an upper bound of the sequence $\{x_n\}$. Consequently, $\forall i = 1, \ldots, k$, there exists an integer p_i such that $x_{p_i} \not\leq y_i$.

Since the sequence $\{x_n\}$ is nondecreasing, $n \geq p_i \Rightarrow x_n \geq x_{p_i}$, therefore necessarily $x_n \not\leq y_i$. In other words:

$$n \geq p_i \Rightarrow x_n \in \not\downarrow y_i$$

Thus:

$$n \geq p = \underset{i=1,\ldots,k}{\mathrm{Max}} \{p_i\} \Rightarrow \forall i = 1, \ldots, k\colon x_n \in \not\downarrow y_i.$$

Since $x_n \leq \bar{x}$, we have, $\forall n \geq p$: $x_n \in \downarrow \bar{x} \cap \not\downarrow y_1 \cap \ldots \cap \not\downarrow y_k$.

This proves that the sequence $\{x_n\}$ is convergent in the sense of the Sup-topology and that $\bar{x}$ is its limit.

(ii) The sequence $\{x_n\}$ being convergent, let us choose $\downarrow \bar{x}$ as a neighborhood of $\bar{x}$: there exists an integer $p \in \mathbb{N}$ such that $n \geq p \Rightarrow x_n \in \downarrow \bar{x} \Rightarrow x_n \leq \bar{x}$. Consequently $\bar{x}$ is an upper bound of $X_p = \bigcup_{n\geq p} \{x_n\}$. Let us assume that X_p has another upper bound $\bar{y} \neq \bar{x}$ such that $\bar{y} \leq \bar{x}$. We necessarily have $\bar{x} \not\leq \bar{y}$ (otherwise we would have $\bar{x} = \bar{y}$, which would give rise to a contradiction).

Then $\downarrow \bar{x} \cap \not\downarrow \bar{y}$ is a neighborhood of $\bar{x}$ which contains none of the $x_n (n \geq p)$, thus contradicting the convergence of $\{x_n\}$ towards $\bar{x}$ (in the sense of the Sup-topology). This clearly proves that:

$$\bar{x} = \sup\{x_n : n \geq p\} \qquad \square$$

For the Inf-topology, we would obtain results similar to Theorem 1 by replacing "nondecreasing sequences" with "nonincreasing sequences" and "least upper bound" with "greatest lower bound." The Inf-topology can thus be interpreted as the topology corresponding to the *lower limit of sequences*.

Thus, for a sequence $\{x_n\}_{n\in N}$ that converges in the sense of the Inf-topology and with limit $\bar{x}$ (this convergence will be denoted $x_n \rightharpoondown \bar{x}$) there exists an integer $p \in \mathbb{N}$ such that:

$$\bar{x} = \inf\{x_n : n \geq p\}. \tag{3}$$

Remark 3.1.1. Sup-convergence, inf-convergence and finite convergence

If a sequence $\{x_n\}$ (not necessarily monotone) is convergent both in the sense of the Sup-topology and in the sense of the Inf-topology *towards the same limit*, then *it converges finitely*.

Indeed, if $\{x_n\}$ is convergent in the sense of the Sup-topology with limit $\bar{x}$, then there exists $P \in \mathbb{N}$ such that:

$$n \geq P \Rightarrow x_n \in \downarrow \bar{x}.$$

If, furthermore, $\{x_n\}$ is convergent in the sense of the Inf-topology with limit $\bar{x}$, then there exists $Q \in \mathbb{N}$ such that:

$$n \geq Q \Rightarrow x_n \in \uparrow \bar{x}.$$

We then deduce that for $n \geq \text{Max}\{P, Q\}$ we have: $x_n \in \downarrow \bar{x} \cap \uparrow \bar{x} = \{\bar{x}\}$. ||

3.2. Concepts of Limit-sup and Limit-inf

Let us now discuss the case where the order relation $\leq$ induces on the set E a *complete lattice* structure. In other words, every subset of E with finite or infinite cardinality bounded from above (resp. bounded from below) has a least upper bound (resp. a greatest lower bound).

We can then introduce the concepts of *upper limit* (resp. *lower limit*) for arbitrary (not necessarily monotone) sequences bounded from above (resp. bounded from below).

Let us consider an arbitrary sequence $\{x_n\}_{n \in \mathbb{N}}$ assumed to be bounded, that is to say such that there exists $m \in E$ and $M \in E$ such that: $\forall n \in \mathbb{N}$: $m \leq x_n$ and $x_n \leq M$.

For an arbitrary p, $p \in \mathbb{N}$, let us consider the set $X_p = \{x_n : n \geq p\}$

This set is bounded from above. It has therefore a least upper bound σ_p: $\sigma_p = \sup\{x_n : n \geq p\}$. The sequence $\left\{\sigma_p\right\}_{p \in \mathbb{N}}$ is monotone nonincreasing and bounded from below by m. It is therefore convergent in the sense of the Inf-topology, and its limit $\bar{\sigma}$ will be referred to as the *upper limit* of the sequence $\{x_n\}$. Thus we have: $\bar{\sigma} = \text{lim-sup}\{x_n\} = \inf_{p \in \mathbb{N}} \sup\{x_n : n \geq p\}$

Similarly, we can define the sequence $\left\{\theta_p\right\}_{p \in \mathbb{N}}$ as:

$$\forall p \in \mathbb{N}: \theta_p = \inf\{x_n : n \geq p\}.$$

This sequence is monotone nondecreasing, therefore convergent in the sense of the Sup-topology, and its limit $\bar{\theta}$ will be referred to as the *lower limit* of the sequence $\{x_n\}$. Thus:

$$\bar{\theta} = \text{lim-inf}\{x_n\} = \sup_{p \in \mathbb{N}} \inf\{x_n : n \geq p\}.$$

4. Continuity of Functions, Semi-Continuity

Let $(E, \overset{E}{\leq})$ and $(F, \overset{F}{\leq})$ be two ordered sets each of them endowed with the Sup-topology.

A function f: $E \to F$ is *nondecreasing* if and only if:

$$\forall x, y \in E \quad x \overset{E}{\leq} y \Rightarrow f(x) \overset{F}{\leq} f(y).$$

Definition 4.1. *A function f:* $E \to F$ *is said to be* continuous *in the sense of the Sup-topology (resp. of the Inf-topology) if and only if, for every sequence* $\{x_n\}$ *convergent in* E *of limit* $\bar{x}$*, the sequence* $f(x_n)$ *is convergent in* F *in the sense of the Sup-topology (resp. of the Inf-topology), its limit being* $\bar{y} = f(\bar{x})$.

Definition 4.2. *A function* f: $E \to F$ *is said to be* upper semi-continuous *(u.s.c.) in the sense of the Sup-topology (resp.* lower semi-continuous *(l.s.c.) in the sense of the Inf-topology) if and only if, for every sequence* $\{x_n\}$ *convergent in* E *with limit* $\bar{x}$*, the sequence* $f(x_n)$ *is convergent in* F *in the sense of the Sup-topology (resp. in the sense of the Inf-topology) and has limit* $\bar{y} \overset{F}{\leq} f(\bar{x})$ *(resp.* $\bar{y} \overset{F}{\geq} f(\bar{x})$*).*

The following result establishes upper semi-continuity in the sense of the Sup-topology for nondecreasing functions.

Theorem 2. *Let* $(E, \overset{E}{\leq})$ *and* $(F, \overset{F}{\leq})$ *be two ordered sets endowed with the Sup-topology. We assume moreover that in* F*, every nondecreasing sequence bounded from above has a least upper bound. Then every increasing function* f: $E \to F$ *is upper semi-continuous in the sense of the Sup-topology.*

Proof. Let us consider a nondecreasing sequence $\{x_n\}$ convergent in E in the sense of the Sup-topology and let $\bar{x}$ be its limit.

Since f is nondecreasing, the sequence $\{f(x_n)\}$ is nondecreasing in F. Let us show that it is bounded from above by $f(\bar{x})$.

According to Theorem 1: $\bar{x} = \sup\{x_n : n \in \mathbb{N}\}$, therefore $\forall n$: $x_n \overset{E}{\leq} \bar{x}$,

hence: $f(x_n) \overset{F}{\leq} f(\bar{x})$.

Following from the assumptions of the theorem, the sequence $\{f(x_n)\}$ therefore has a least upper bound $\bar{y} \in F$ which is its limit in the sense of the Sup-topology:

$$f(x_n) \rightharpoonup \bar{y} = \sup\{f(x_n) : n \in \mathbb{N}\}.$$

As $f(\bar{x})$ is an upper bound of $\{f(x_n) : n \in \mathbb{N}\}$, we deduce: $\bar{y} \leq f(\bar{x})$, which proves the upper semi-continuity of f. □

Let us observe here that we could establish a similar result concerning the lower semi-continuity (in the sense of the Inf-topology), of nonincreasing functions: $E \to F$.

5. The Fixed-Point Theorem in an Ordered Set

Definition 5.1. *(fixed point and lower fixed point)*

Let $(E, \leq)$ *be an ordered set and* $f \colon E \to E$ *a nondecreasing function. We call* lower fixed point of f *every* $x \in E$ *satisfying:* $x \leq f(x)$.

We call fixed point of f *every* $x \in E$ *satisfying* $x = f(x)$.

For every $x \in E$, we denote:

$$f^{(2)}(x) = f(f(x)) \quad \text{and,} \quad \forall k \in N\colon \; f^{(k)}(x) = f(f^{(k-1)}(x)).$$

Theorem 3. *Let* $(E, \leq)$ *be an ordered set having a smaller element* ε*, and in which every nondecreasing sequence bounded from above has a least upper bound. Let* $f \colon E \to E$ *be a nondecreasing function. Then if the sequence* $\{f^{(n)}(\varepsilon)\}$ *is bounded from above, the function* f *has a lower fixed point* x_f *satisfying*

$$x_f = \sup\{f^{(n)}(\varepsilon) : n \in \mathbb{N}\}. \tag{4}$$

If, moreover, the function f *is continuous for the Sup-topology, then:*

- x_f *is a fixed point of* f,
- x_f *is the smallest element of* $X_f = \{y : y = f(y)\}$ *(smallest fixed point).*

Proof. ε being the smallest element of E, we have: $\varepsilon \leq f(\varepsilon)$, and since f is nondecreasing, we deduce:

$$\varepsilon \leq f(\varepsilon) \leq f^{(2)}(\varepsilon) \leq \ldots \leq f^{(n)}(\varepsilon) \leq \ldots$$

The sequence $\{f^{(n)}(\varepsilon)\}$ being nondecreasing and bounded from above, it converges (in the sense of the Sup-topology) towards a limit x_f and:

$$x_f = \sup\{f^{(n)}(\varepsilon) : n \in \mathbb{N}\}$$

Thus $\forall n\colon f^{(n)}(\varepsilon) \leq x_f$, and consequently:

$\forall n\colon f^{(n+1)}(\varepsilon) \leq f(x_f)$, which shows that $f(x_f)$ is an upper bound of the sequence $\{f^{(n)}(\varepsilon)\}$.

Thus $x_f \leq f(x_f)$, which shows that x_f is a lower fixed point of f.

If we assume now that f is *continuous* for the Sup-topology, we have: $f(x_f) = x_f$ therefore x_f is a fixed point of f.

Let us then show that x_f defined by (4) is the smallest element of X_f.

Let y be an arbitrary element of $X_f = \{y\colon f(y) = y\}$.

We have: $\varepsilon \leq y$ and, since f is nondecreasing: $f(\varepsilon) \leq f(y) = y$.

By induction we obtain: $\forall n\colon f^{(n)}(\varepsilon) \leq y$ which shows that y is an upper bound of the sequence $\{f^{(n)}(\varepsilon)\}$. Since x_f is, by definition, the least upper bound of this sequence, we deduce that $x_f \leq y$. This shows that x_f is the smallest fixed point of f on E. □

6. Topological Dioids

Let (E, ⊕, ⊗) be a dioid. E being an ordered set for the canonical order relation ≤ induced by ⊕, we can endow E with one of the topologies defined in Sect. 2. We will consider in particular the Sup-topology. Moreover, in this section, we restrict our attention to the topologies constructed directly from the canonical order relation ≤ relative to ⊕. Many other topologies could be constructed differently, even if compatibility ("continuity") with the ⊕ and ⊗ operations is required.

6.1. Definition

We call topological dioid *relative to the Sup-topology, a dioid* (E, ⊕, ⊗) *endowed with the Sup-topology associated with the canonical order ≤, and having the following additional properties:*

(i) Every nondecreasing sequence bounded from above has a least upper bound (every nondecreasing sequence bounded from above is therefore convergent in the sense of the Sup-topology, its limit being equal to its least upper bound).
(ii) Taking the limit is compatible with the two laws ⊕ and ⊗ of the dioid (in other words the operations ⊕ and ⊗ are continuous *w.r.t. the Sup-topology).*

Example 6.1.1. The dioid (E, ⊕, ⊗), where $E = (R \cup \{-\infty\})^2$ is endowed with the operations ⊕ and ⊗ defined as:

$$\begin{pmatrix} x \\ x' \end{pmatrix} \oplus \begin{pmatrix} y \\ y' \end{pmatrix} = \begin{pmatrix} \text{Max}\{x, y\} \\ \text{Max}\{x', y'\} \end{pmatrix}$$

$$\begin{pmatrix} x \\ x' \end{pmatrix} \otimes \begin{pmatrix} y \\ y' \end{pmatrix} = \begin{pmatrix} x + x' \\ x' + y' \end{pmatrix}$$

is a *topological dioid.*

Indeed, the canonical order relation ≤ is defined as:

$$\begin{pmatrix} x \\ x' \end{pmatrix} \le \begin{pmatrix} y \\ y' \end{pmatrix} \Leftrightarrow \begin{cases} x \le y \\ \text{and} \\ x' \le y' \end{cases}$$

and every nondecreasing sequence bounded from above is formed by a sequence of pairs $\begin{pmatrix} x_n \\ x'_n \end{pmatrix}$ ($n \in \mathbb{N}$) where $\{x_n\}$ and $\{x'_n\}$ are nondecreasing sequences of reals bounded from above. Such a sequence is therefore convergent in the sense of the Sup-topology. Furthermore, it is easily seen that taking the limit is compatible with the laws ⊕ and ⊗. ||

The following result shows that assumption (ii) for the ⊕ law is automatically satisfied in selective dioids.

Property 6.1.2. *If* $(E, \oplus, \otimes)$ *is a selective dioid* ($\oplus$ *selective), then taking the limit for nondecreasing convergent sequences is compatible with the* $\oplus$ *law.*

Proof. Let $\{x_n\}$ and $\{y_n\}$ be two nondecreasing sequences bounded from above (therefore convergent for the Sup-topology), their limits being respectively $\bar{x}$ and $\bar{y}$.

We have:

$$\bar{x} = \lim\{x_n\} = \sup\{x_n : n \in \mathbb{N}\}$$
$$\bar{y} = \lim\{y_n\} = \sup\{y_n : n \in \mathbb{N}\}$$

Since $\leq$ is compatible with $\oplus$, $\{x_n \oplus y_n\}$ is also a nondecreasing sequence bounded from above by $\bar{x} \oplus \bar{y}$. It is therefore convergent and its limit is $\bar{z} \leq \bar{x} \oplus \bar{y}$.

Let us show that $\bar{z} = \bar{x} \oplus \bar{y}$.

Since $\oplus$ is selective, we have:

$$\forall n\colon x_n \oplus y_n = \sup\{x_n, y_n\}$$

and:

$$\begin{aligned}
\bar{x} \oplus \bar{y} &= \sup\{\bar{x}, \bar{y}\} \\
&= \sup\{\sup\{x_n : n \in \mathbb{N}\}; \sup\{y_n : n \in \mathbb{N}\}\} = \sup\left\{\bigcup_n \{x_n, y_n\}\right\} \\
&= \sup\left\{\bigcup_n \{\sup\{x_n, y_n\}\}\right\} = \sup\{x_n \oplus y_n : n \in \mathbb{N}\}
\end{aligned}$$

We therefore deduce:

$$\bar{x} \oplus \bar{y} = \lim\{x_n \oplus y_n\} \quad \square$$

Another direct consequence of the definition of a topological dioid is the existence of a quasi-inverse (see Definition 6.2.1 below):

Property 6.1.3. *Let* $(E, \oplus, \otimes)$ *be a topological dioid where* e, *the neutral element for* $\otimes$, *satisfies:* $e \oplus e = e$. *Then every element* $a \leq e$ *has a quasi-inverse* a^*.

Proof. We have:

$$\begin{aligned}
& a \leq e \\
\text{hence, } \forall k\colon\ & a^k \leq e \\
\text{and }\ & a^{(k)} = e \oplus a \oplus a^2 \oplus \ldots \oplus a^k \leq e.
\end{aligned}$$

The sequence $a^{(k)}$ is thus nondecreasing and bounded from above. It therefore has a limit a^* referred to as the *quasi-inverse of* a. $\square$

Observe that the condition $e \oplus e = e$ will be satisfied, in particular,

- when $\oplus$ is idempotent,
- when e is the largest element of E.

6.2. Fixed-Point Type Linear Equations in a Topological Dioid: Quasi-Inverse

In a *topological dioid* (E, ⊕, ⊗), let us consider solving equations of the type:

$$x = a \otimes x \oplus e \quad \text{and} \quad x = x \otimes a \oplus e \tag{5}$$

where $a \in E$ is a given element.

For $k \in N$ set:

$$a^{(k)} = e \oplus a \oplus a^2 \oplus \ldots \oplus a^k$$

where $a^k = a \otimes a \otimes \ldots \otimes a$ (k times)

Since: $a^{(k+1)} = a \otimes a^{(k)} \oplus e$, we have: $a^{(k+1)} = a^{(k)} \oplus a^{k+1}$, hence: $a^{(k+1)} \geq a^{(k)}$.

The sequence $a^{(k)}$ is therefore nondecreasing. If, moreover, it is bounded from above, it is convergent; in this case, its limit $\sup_{k \geq 0}\{a^{(k)}\}$ will be denoted a^*. This leads to the following definition.

Definition 6.2.1. *(quasi-inverse)*

We call quasi-inverse *of the element* $a \in E$, *denoted* a^*, *the limit, when it exists, of the sequence* $a^{(k)}$ *where, for every* $k \in \mathbb{N}$.

$$a^{(k)} = e \oplus a \oplus a^2 \oplus \ldots \oplus a^k.$$

Proposition 6.2.2. *If* a^* *(quasi-inverse of a) exists, then it is the minimal solution of the equations*

$$x = a \otimes x \oplus e$$
$$x = x \otimes a \oplus e \tag{5}$$

in other words, a^* *is the smallest element of the set of solutions.*

Proof. We have: $a^{(k+1)} = a \otimes a^{(k)} \oplus e$.

As a consequence of the compatibility of ⊕ and ⊗ with taking the limit, $a \otimes a^{(k)}$ is convergent and has limit $a \otimes a^*$; similarly $a \otimes a^{(k)} \oplus e$ is convergent and has the limit $a \otimes a^* \oplus e$. We deduce the relation:

$$a^* = a \otimes a^* \oplus e$$

Similarly, we establish that: $a^* = a^* \otimes a \oplus e$

Furthermore, from equation $x = a \otimes x \oplus e$
we easily deduce by induction: $x = a^{k+1} \otimes x \oplus a^{(k)}$
which proves that every solution of (5) is an upper bound of the sequence $a^{(k)}$.
a^*, least upper bound of the sequence $a^{(k)}$, is therefore the minimal solution to (5). □

Remark. The above proposition could also be deduced directly from the fixed point theorem (Theorem 3 Sect. 5) observing that, as a result of the properties of the topological dioid $(E, \oplus, \otimes)$, the function $f: E \to E$ defined as:

$$f(x) = a \otimes x \oplus e$$

is *continuous* in the sense of the Sup-topology. ||

Proposition 6.2.3. *If* $a \in E$ *has a quasi-inverse* a^*, *then* $\forall b \in E$, $a^* \otimes b$ *(resp.* $b \otimes a^*$*) is the minimal solution to the equation:*

$$y = a \otimes y \oplus b \quad (\text{resp. } y = y \otimes a \oplus b).$$

Proof. $a^* \otimes b$ is obviously a solution because, by using the above proposition,

$$a \otimes a^* \otimes b \oplus b = (a \otimes a^* \oplus e) \otimes b = a^* \otimes b$$

Let us show that this is the minimal solution. Let y be an arbitrary solution:

$$\begin{aligned} y &= a \otimes y \oplus b \\ &= a(a \otimes y \oplus b) \oplus b \\ &= a^2 \otimes y \oplus a \otimes b \oplus b \\ &= a^3 \otimes y \oplus a^2 \otimes b \oplus a \otimes b \oplus b \end{aligned}$$

and in a general way: $y = a^{k+1} \otimes y \oplus a^{(k)} \otimes b$

Thus for every $k \in N$: $y \geq a^{(k)} \otimes b$.

y is therefore an upper bound of the sequence $\{a^{(k)} \otimes b\}$. This sequence is nondecreasing, convergent towards $a^* \otimes b$, the least upper bound of the sequence. Therefore we clearly have: $y \geq a^* \otimes b$. □

Proposition 6.2.4. *Let* $(E, \oplus, \otimes)$ *be a topological dioid and two elements* $a \in E$, $b \in E$ *such that the quasi-inverses* a^* *and* $(b \otimes a^*)^*$ *exist. Then* $(a \oplus b)^*$ *exists and we have the identity:*

$$(a \oplus b)^* = a^* \otimes (b \otimes a^*)^*.$$

Similarly, if a^* *and* $(a^* \otimes b)^*$ *exist then* $(a \oplus b)^*$ *exists and we have the identity*

$$(a \oplus b)^* = (a^* \otimes b)^* \otimes a^*.$$

Proof. Let us first show that $a^* \otimes (b \otimes a^*)^*$ is a solution to the equation

$$x = (a \oplus b) \otimes x \oplus e \tag{6}$$

$u = a^*$ is the minimal solution to: $u = a \otimes u \oplus e$.

Likewise $v = (b \otimes a^*)^*$ is the minimal solution to:

$$v = (b \otimes a^*) \otimes v \oplus e = (b \otimes u \otimes v) \oplus e.$$

We deduce:

$$\begin{aligned} u \otimes v &= a \otimes u \otimes v \oplus v \\ &= (a \otimes u \otimes v) \oplus (b \otimes u \otimes v) \oplus e \end{aligned}$$

hence we deduce that $u \otimes v = a^* \otimes (b \otimes a^*)^*$ is a solution to (6).

In the same way as in the proof of 6.2.3, we have for every solution x of (6), ∀k:

$$x \geq (a \oplus b)^{(k)}$$

From this we deduce that the nondecreasing sequence $(a \oplus b)^{(k)}$ is bounded from above. E being a topological dioid, it therefore has a limit in E, which proves the existence of $(a \oplus b)^* = \bar{x}$ minimal solution to (6).

We can then deduce from the above the inequality $\bar{x} = (a \oplus b)^* \leq a^* \otimes (b \otimes a^*)^*$.

Let us now prove the reverse inequality. To do so, it suffices to observe that $\bar{x}$ is the minimal solution in x of the equation:

$$x = x \otimes a \oplus \bar{x} \otimes b \oplus e$$

(indeed let $x' \leq \bar{x}$ satisfying $x' = x' \otimes a \oplus \bar{x} \otimes b \oplus e$. By setting $\bar{x} = x' \oplus h$ we would have:

$x' = x' \otimes a \oplus x' \otimes b \oplus h \otimes b \oplus e$ which shows that:

$$x' \geq (h \otimes b \oplus e) \otimes (a \oplus b)^* \geq \bar{x}$$

hence we deduce $x' = \bar{x}$).

Since a^* exists we therefore have $\bar{x} = (\bar{x} \otimes b \oplus e) \otimes a^*$ which shows that $\bar{x}$ is a solution to the equation:

$$\bar{x} = \bar{x} \otimes (b \otimes a^*) \oplus a^*$$

Since $(b \otimes a^*)^*$ exists we therefore have

$$\bar{x} \geq a^* \otimes (b \otimes a^*)^*.$$

The second part of the proposition is proved in a similar way, starting from the equation $x = x \otimes (a \oplus b) \oplus e$. □

Proposition 6.2.5. *Let* (E, ⊕, ⊗) *be a topological dioid and two elements* $a \in E$, $b \in E$ *such that the quasi-inverses* a^* *and* $(a \oplus b)^*$ *exist. Then* $(b \otimes a^*)^*$ *and* $(a^* \otimes b)^*$ *exist, and we have the identities:*

$$(a \oplus b)^* = a^* \otimes (b \otimes a^*)^* = (a^* \otimes b)^* \otimes a^*.$$

Proof. $\bar{x} = (a \oplus b)^*$ is the minimal solution to the equation:

$$x = x \otimes a \oplus x \otimes b \oplus e$$

therefore (see proof of 6.2.4) $\bar{x}$ is the minimal solution in x to the equation:

$$x = x \otimes a \oplus \bar{x} \otimes b \oplus e$$

and since a^* exists, we have:

$$\bar{x} = (\bar{x} \otimes b \oplus e) \otimes a^*$$

$\bar{x}$ is therefore a solution in x to the equation

$$x = x \otimes (b \otimes a^*) \oplus a^*$$

For every integer $k \geq 1$ we can therefore write:

$$\bar{x} = \bar{x} \otimes (b \otimes a^*)^{k+1} \oplus a^* \otimes (b \otimes a^*)^{(k)}$$

and consequently, $\forall k \geq 1$:

$$a^* \otimes (b \otimes a^*)^{(k)} \leq \bar{x}$$

The sequence $a^* \otimes (b \otimes a^*)^{(k)}$ is therefore nondecreasing and bounded from above. Since E is a topological dioid, it therefore has a limit which is necessarily $a^* \otimes (b \otimes a^*)^*$.

One would similarly prove the existence of $(a^* \otimes b)^*$. The claimed identities are then deduced from Proposition 6.2.4. □

The above result will be used, in particular, in Chap. 4 Sect. 4.2 to prove the convergence of the generalized Gauss-Seidel algorithm.

Proposition 6.2.6. *Let* $A = (a_{ij}) \in M_n(E)$ a $n \times n$ *matrix with entries in a topological dioid* $(E, \oplus, \otimes)$. *Assume the existence of the quasi-inverse* A^* *of* A *defined as the limit of* $A^{(k)} = I \oplus A \oplus A^2 \oplus \ldots \oplus A^k$ *as* $k \to \infty$. *Then each diagonal entry* a_{ii} *of* A *has a quasi-inverse* $(a_{ii})^*$.

More generally, each square principal submatrix of A *has a quasi-inverse.*

Proof. In Chap. 4 Sect. 3.2 it will be shown that $(A^k)_{ii}$ can be expressed as the sum of the weights of all cardinality−k circuits through node i in the graph G(A) associated with the A matrix. One of the circuits involved in this sum is the loop (i, i) taken k times, the weight of which is $(a_{ii})^k$. We can therefore write:

$$\left(A^k\right)_{ii} = (a_{ii})^k \oplus \delta(i, k)$$

(where the term $\delta(i, k)$ accounts for the sum of the weights of all other cardinality-k circuits).

From this we can deduce that:

$$(A^{(k)})_{ii} = (a_{ii})^{(k)} \oplus \Delta(i, k)$$

where $\Delta(i, k) = \delta(i, 1) \oplus \delta(i, 2) \oplus \ldots \oplus \delta(i, k)$.

Since $A^{(k)}$ is a nondecreasing sequence of matrices with limit A^* we have, $\forall k$: $\left(A^{(k)}\right)_{ii} \le (A^*)_{ii}$, and thus we can write:

$$(a_{ii})^{(k)} \oplus \Delta(i, k) \le (A^*)_{ii}$$

which clearly implies:

$$(a_{ii})^{(k)} \le (A^*)_{ii}.$$

From this we conclude that $(a_{ii})^{(k)}$ is a nondecreasing sequence of elements of E, bounded from above. Since $(E, \oplus, \otimes)$ is a topological dioid, we deduce that this sequence is convergent and has a limit $(a_{ii})^*$, the quasi-inverse of a_{ii}.

Consider now the more general case of a principal submatrix $A_{[S]}$ deduced from A by considering only the rows and columns in a given subset of indices $S \subset \{1, 2, \ldots, n\}$. Denote $(A_{[S]})^k$ the kth power of $A_{[S]}$. Then $(A_{[S]})^k_{ij}$ can be expressed as the sum of the weights of all cardinality-k i-j paths in G(A) with the restriction of using only nodes in S.

We therefore have, $\forall k$:

$$\left(A_{[S]}\right)^k \le (A^k)_{[S]}$$

and thus:

$$\left(A_{[S]}\right)^{(k)} \le (A^*)_{[S]}$$

From this it is seen that the nondecreasing sequence of matrices $(A_{[S]})^{(k)}$ is bounded from above. Since $M_n(E)$ is a topological dioid, the existence of $(A_{[S]})^*$ is deduced. □

7. P-Stable Elements in a Dioid

In many applications involving dioids, it is not necessary to resort to Sup-topology to study the convergence properties of sequences. It is enough to guarantee *finite convergence* of some nondecreasing sequences (discrete topology).

Let $(E, \oplus, \otimes)$ be a dioid. For $a \in E$ we recall the notation introduced in Sect. 6.2:

$$a^{(k)} = e \oplus a \oplus a^2 \oplus \ldots \oplus a^k.$$

Definition 7.1. (p *stable element)*

For $p \ge 0$ *integer, an element* a *is said to be* p-*stable if and only if:*

$$a^{(p+1)} = a^{(p)}$$

We then have $a^{(p+2)} = e \oplus a \otimes a^{(p+1)} = e \oplus a \otimes a^{(p)} = a^{(p+1)}$, hence by induction

$$a^{(p+r)} = a^{(p)} \quad \forall r \ge 0, \quad \text{integer.}$$

For each p-stable element $a \in E$, we therefore deduce the existence of a^*, *quasi-inverse* of a, defined as:

$$a^* = \lim_{k \to +\infty} a^{(k)} = a^{(p)}$$

which satisfies the equations

$$a^* = a \otimes a^* \oplus e = a^* \otimes a \oplus e. \tag{7}$$

Let us study some examples of dioids with p-stable elements.

7.1. Examples

Example 7.1.1. Consider the dioid $(\mathbb{R}_+ \cup \{+\infty\}, \max, \min)$, with neutral elements $\varepsilon = 0$ and $e = +\infty$. Since $e \oplus a = \max(+\infty, a) = \{+\infty\} = e$, $\forall a \in E$, we deduce that every element a is 0-stable and has as quasi-inverse $a^* = e$.

This example is encountered, e.g. in the study of the maximum capacity path of a graph (see Chap. 4 Sect. 6.3). ||

Example 7.1.2. In the case of the dioid $(\mathbb{R} \cup \{+\infty\}, \min, +)$, with neutral elements $\varepsilon = +\infty$ and $e = 0$, if the element a is nonnegative, $e \oplus a = \min(0, a) = 0 = e$ and it is 0-stable. Every nonnegative element therefore has a quasi-inverse $a^* = e$.

If the element a is negative, then $e \oplus a = \min(0, a) = a$ and $a^{(k)} = \min(0, a, 2a, \ldots, ka) = ka$. $a^{(k)}$ tends towards $-\infty$ and a *does not have a quasi-inverse* in $R \cup \{+\infty\}$.

This example can be extended to the case of a matrix, see Example 7.1.8 and Chap. 4. It is encountered in the study of the shortest path problem in a graph (see Chap. 4 Sects. 2 and 6.5). ||

Example 7.1.3. In the case of the dioid $([0, 1], \max, \times)$, with neutral elements $\varepsilon = 0$ and $e = 1$, $e \oplus a = \max(1, a) = 1$ $\forall a \in E$. We deduce that every element of a is 0-stable and has as quasi-inverse $a^* = e$.

This example arises in the study of the maximum reliability path of a graph (see Chap. 4 Sect. 6.6). ||

Example 7.1.4. Let E be the cone of $\overline{\mathbb{R}}^2$ defined as follows: $a = (a_1, a_2) \in E$ if and only if $a_1 \in \overline{\mathbb{R}}$, $a_2 \in \overline{\mathbb{R}}$ and $a_1 \le a_2$.

We then define two operations $\oplus$ and $\otimes$ on E as follows:

If $a \in E$ and $b \in E$, then

$$c = a \oplus b = (c_1, c_2) \quad \text{with} \quad c_1 = \min{}_1(a_1, b_1),\ c_2 = \min{}_2(a_1, b_1, a_2, b_2)$$
$$c = a \otimes b = (c_1, c_2) \quad \text{with} \quad c_1 = \min{}_1(a_i + b_j)_{\substack{i=1,2\\ j=1,2}},\ c_2 = \min{}_2(a_i + b_j)_{\substack{i=1,2\\ j=1,2}}$$

where $\min_1$ and $\min_2$ correspond respectively to the first minimum and second minimum among the set of values under consideration.

We observe that $\oplus$ is not idempotent, for instance:

$$(2, 3) \oplus (2, 3) = (2, 2) \neq (2, 3).$$

We easily check that these two laws endow E with a dioid structure where $\varepsilon = (+\infty, +\infty)$ and $e = (0, +\infty)$.

Let us show that any $a = (a_1, a_2) \in E$ such that $a_1 \geq 0$ is 1-stable.

We have $e \oplus a = (0, a_1)$, and since $a^2 = (2\, a_1, a_1 + a_2)$, $e \oplus a \oplus a^2 = e \oplus a$

Thus, every element a with nonnegative components is 1-stable and has a quasi-inverse $a^* = e \oplus a$.

This example can be extended to the case where E is the cone of $\overline{\mathbb{R}}^k$ defined as $a = (a_1, a_2, \ldots, a_k)$ with $a_i \in \overline{\mathbb{R}}$ and $a_1 \leq a_2 \leq \ldots \leq a_k$. The operations $\oplus$ and $\otimes$ on E are then defined as

$$a \oplus b = a\ \mathrm{Min}_{(k)} b$$

$$a \otimes b = a \overset{(k)}{+} b$$

(where the laws $\mathrm{Min}_{(k)}$ and $\overset{(k)}{+}$ are defined in Chap. 8 Sects. 1.3.1 and 1.1.5).

We verify that every element a with nonnegative components is (k − 1) stable and has a quasi-inverse $a^* = a^{(k-1)}$.

This example arises in the study of the kth shortest path problem in a graph (see Chap. 4 Sect. 6.8). ||

Example 7.1.5. Let us consider a dioid for which $\oplus$ and $\otimes$ are idempotent. Then every element is 1-stable. Indeed $e \oplus a \oplus a^2 = e \oplus a \oplus a = e \oplus a$.

This example is linked to an important class of applications where the property of 1-stability is always satisfied: namely *distributive lattices* (see Chap. 8 Sect. 4.6). ||

Example 7.1.6. Let E be the set of the sequences $a = (a_1 \leq a_2 \leq \ldots \leq a_q)$ with $a_i \in \overline{\mathbb{R}}$ and $a_q \leq a_1 + \eta$ where η is a given positive real. The operations $\oplus$ and $\otimes$ on E are then defined as:

$$a \oplus b = a\ \mathrm{Min}_{(\leq\eta)} b$$

$$a \otimes b = a \overset{(\leq\eta)}{+} b$$

(where the laws $\mathrm{Min}_{(\eta)}$ and $\overset{(\leq\eta)}{+}$ are defined in Chap. 8 Sects. 1.3.2 and 1.1.6).

We verify that these two laws endow E with a dioid structure where $\varepsilon = (+\infty)$ (the sequence formed by a single element equal to $+\infty$) and $e = (0)$ (the sequence formed by a single element equal to 0).

Moreover, it can be shown that every a with strictly positive components is p-stable with $p = \lceil \eta / a_1 \rceil$ (where $\lceil\ \rceil$ denotes the smallest integer greater than or equal to).

This example arises in the study of η-optimal paths of a graph (see Chap. 4 Sect. 6.10). ||

Example 7.1.7. Let us take for E the set of polynomials in several variables with coefficients in $\mathbb{Z}$, *idempotent* for ordinary multiplication (therefore in particular with Boolean variables $x_i^2 = x_i$). We take for addition $\oplus$ the symmetric difference $(a \oplus b = a \oplus b - ab)$, for multiplication $\otimes$ ordinary multiplication. Thus $\varepsilon = 0$ and $e = 1$.

Let us show that $\oplus$ is an internal law: if $a \in E$ and $b \in E$, we have $a^2 \in E$ and $b^2 \in E$; then:

$$\begin{aligned}(a \oplus b)^2 &= (a + b - ab)^2 = a^2 + b^2 + a^2b^2 + 2ab - 2ab^2 - 2a^2b \\ &= a + b + ab + 2ab - 2ab = a + b - ab = a \oplus b.\end{aligned}$$

$(E, \oplus, \otimes)$ is a dioid and it can be shown that every element is 0-stable.

This example arises in the study of the reliability of a network (see Chap. 4 Sect. 6.9). ||

Example 7.1.8. We consider here the dioid $(M_n(E), \oplus, \otimes)$ of square matrices of order n with elements in a given dioid $(E, \oplus, \otimes)$.

In Chap. 4 (Theorems 1–3), a number of sufficient conditions for a matrix A to be p-stable will be studied. It will be seen, in particular, that if the elements of E are 0-stable, then the elements of $M_n(E)$ are n-1-stable (see Theorem 1 of Chap. 4). ||

Example 7.1.9. On the dioid $(E, \oplus, \otimes)$, consider the set F of functions: $\mathbb{N} \to E$. Given two functions f(x) and g(x) of F we define the sum $f \oplus g \in F$ as: $\forall x \in \mathbb{N}$: $f \oplus g(x) = f(x) \oplus g(x)$. We also define the *convolution product* of two functions f(x) and g(x) of F by:

$$(f * g)(x) = \sum_{k+l=x} f(k) \otimes g(\ell)$$

where Σ corresponds to the addition $\oplus$ on the dioid E and + to the addition in $\mathbb{N}$.

We can check that $(F, \oplus, *)$ is a dioid. Moreover, if all the elements in E are p-stable, then all the elements in F are p-stable. ||

In the following section, we show how the concept of p-stable element can be used to solve fixed-point type equations in dioids.

7.2. Solving Linear Equations

Let us consider in $(E, \oplus, \otimes)$ the linear equation:

$$x = a \otimes x \oplus b \tag{8}$$

Proposition 7.2.1. *If the element* a *is* p*-stable, then* $a^* \otimes b$ *is the minimum solution to (8).*

Proof. It follows directly from Proposition 6.2.3 of Sect. 6. □

Observe that it is the canonical order relation $\leq$ of the dioid which guarantees the unicity of the minimum solution to (8).

In the same way, if a is p-stable, $b \otimes a^*$ is the minimum solution to

$$x = x \otimes a \oplus b. \tag{9}$$

We deduce from the above that a^* is the minimum solution to each of the equations

$$x = a \otimes x \oplus e \quad \text{and} \quad x = x \otimes a \oplus e \tag{10}$$

Example 7.2.2. In the case of Example 7.1.1 where $(E, \oplus, \otimes) \equiv (\overline{\mathbb{R}}_+, \max, \min)$, (9) leads to:

$$x = \max\{\min\{a; x\}; b\} \tag{11}$$

If $a \leq b$, $\max\{\min\{a;\ x\};\ b\} = b(\forall x)$ and $x = b$ is the only solution to (11).

If $a > b$, the set of the solutions to (11) is the interval [b, a], and b is the smallest of these solutions; this is consistent with Proposition 7.2.1. ||

Remark 7.2.3. Since $M_n(E)$ is a dioid if E is a dioid, we will see in Chap. 4 Sect. 3 that Proposition 7.2.1 can also be applied to matrix equations of the form:

$$X = A \otimes X \oplus B \tag{12}$$

for A a p-stable matrix in $M_n(E)$. ||

Definition 7.2.4. *(stable element)*

We say that an element is stable *if there exists an integer* p *for which it is* p-*stable.*

Thus, every stable element has a quasi-inverse which corresponds to the formal expansion, in the sense of standard algebra, of $(e - a)^{-1}$ if it is finite (finite convergence).

This formal correspondence then enables one to obtain the following expressions when the elements involved are stable (see Backhouse and Carré 1975, p. 165):

$$(ab)^* = e \oplus a(ba)^*b \tag{13}$$

$$(a \oplus b)^* = a^*(ba^*)^* = (a^*b)^*a^* \tag{14}$$

$$(ba^*c)^* = e \oplus b(a \oplus cb)^*c \tag{15}$$

(we have omitted the $\otimes$ signs for the sake of notational simplicity).

Thus (13) can be obtained from the formal correspondence with the identity:

$$(e - ab)^{-1} = e + a(e - ba)^{-1}b$$

Similarly (14) can be obtained from the identity:

$$(e - (a + b))^{-1} = [e - (e - a)^{-1}b]^{-1}(e - a)^{-1}$$

and (15) from the identity:

$$(e - b(e - a)^{-1}c)^{-1} = e + b(e - (a + cb))^{-1}c$$

We deduce (again omitting the $\otimes$ signs for the sake of notational simplicity):

Proposition 7.2.5.

(i) If ba *is stable,* ab *is stable and its quasi-inverse is given by* (13)
(ii) if a^* *exists and if* ba^* *is stable,* $a \oplus b$ *is stable and its quasi-inverse satisfies* (14)
(iii) if a^* *exists and if* $a \oplus b$ *is stable, then* ba^* *(resp.* a^* b*) is stable and its quasi-inverse satisfies*

$$(ba^*)^* = e \oplus b(a \oplus b)^* \ (\text{resp. } (a^*b)^* = e \oplus (a \oplus b)^*b) \tag{16}$$

Proposition 7.2.6. *(Gondran 1979)*
If a *and* b *are stable and if* $\otimes$ *is commutative,* $a \oplus b$ *and* $a^* b$ *are stable.*

Proof. According to Proposition 7.2.5, we only have to show that $a^* b$ is stable. Let us assume that a is p-stable and that b is q-stable.

To show that $a^* b$ is $(q+1)$-stable, it suffices to show that $(a^* b)^{q+2}$ is absorbed by $(a^* b)^{(q+1)}$. To do so, it is enough to show that, for any k, $a^k b^{q+2}$ is absorbed by $(a^* b)^{(q+1)}$.

But, since $\otimes$ is commutative, from the expression of $(a^* b)^{(q+1)} = e \oplus (a^* b) \oplus \ldots \oplus (a^* b)^{q+1}$, we can extract $a^k b \oplus a^k b^2 \oplus \ldots \oplus a^k b^{q+1} = a^k b (e \oplus b \oplus \ldots \oplus b^q)$ which clearly absorbs $a^k b^{q+2}$ since b is q-stable. □

The commutativity of $\otimes$ is essential for Proposition 7.2.6 except if a is 0-stable $(a^* = e)$ or if $b = e$. Let us consider the latter case.

Proposition 7.2.7. *(Gondran 1979)*
If e *is stable we have:*

$$e^* = e^* \oplus e^* = e^* e^* \tag{17}$$

and for every stable a, a^* *is stable and:*

$$(a^*)^* = e^* a^* = a^* e^*. \tag{18}$$

Proof. If e is q-stable

$$e^* = e \oplus e \oplus \ldots \oplus e = qe$$

and (17) is obvious.

We observe that e^* commutes with all the elements for the operation $\otimes$; let us then consider $(a^*)^{(k)}$. It is a polynomial in the variable a:

$$(a^*)^{(k)} = \sum_{i=0}^{kq} \alpha_\ell a^\ell.$$

Let us show that for $\ell \le q$, its coefficients α_ℓ are larger than q as soon as $\frac{1}{2} k(k-1) \ge q$. To do so, let us consider

$$(a^*)^r = \left(\sum_{i=0}^{q} a^i \right)^r = \sum_{s=0}^{s=qr} \beta_{s,r} a^s.$$

For $s \le q$, we have $\beta_{s,r} \ge r$. Then, since $\ell \le q$, we have $\alpha_\ell = \sum_{r=1}^{k} \beta_{\ell,r}$; we deduce

$$\alpha_\ell \ge \sum_{r=1}^{k} r = \frac{1}{2} k(k-1) \ge q.$$

This inequality therefore implies for $\ell \le q$, $\alpha_\ell\, a^\ell = e^*\, a^\ell$. We deduce $\sum_{\ell=0}^{q} \alpha_\ell\, a^\ell = e^*\, a^*$.

For $\ell > q$, the a^ℓ from $(a^*)^{(k)}$ are absorbed by $\sum_{\ell=0}^{q} \alpha_\ell\, a^\ell = e^*\, a^*$.

Thus $(a^*)^{(k)} = e^*\, a^*$ as soon as $\frac{1}{2}\, k(k-1) \ge q$, and (18) follows. □

Set $\overline{E} = \{c/c = e^*\, a, a \in E) = e^*\, E$

Then (17) imply:

$$c \oplus c = c \quad \forall c \in \overline{E}$$
$$e^* c = c \quad \forall c \in \overline{E}.$$

7.3. Solving "Nonlinear" Equations

Assuming p-stability of elements we now turn to investigate sufficient conditions for the existence of solutions to some nonlinear equations.

7.3.1. Quasi-Square Root

For $a \in E$, let

$$a^{(k)_2} = e \oplus a \oplus 2a^2 \oplus 5a^3 \oplus \cdots \oplus \alpha_k a^k \tag{19}$$

with

$$\alpha_n = \frac{1}{n} \binom{2n}{n-1} \in \mathbb{N} \text{ (the so-called Catalan numbers)}$$

Then, if a is p-stable, we have:

$$a^{(p)_2} = a^{(p+1)_2} = a^{(p+2)_2} = \cdots$$

since the terms involving a^{p+1} are absorbed by $a^{(p)}$.

For each p-stable element $a \in E$, we then deduce the existence of $a^{*/2}$, *quasi-square root* of a, defined as:

$$a^{*/2} = \lim_{k \to +\infty} a^{(k)_2} = a^{(p)_2} = a^{(p+1)_2} = \cdots \tag{20}$$

Observe that the quasi-square root corresponds to the formal expansion of:

$$\frac{e - \sqrt{e - 4a}}{2a}$$

and (omitting the ⊗ signs for the sake of notational simplicity) satisfies the equations:

$$a^{*/2} = a(a^{*/2})^2 \oplus e = (a^{*/2})^2 a \oplus e = a^{*/2} a\, a^{*/2} \oplus e$$

$a^{*/2}$ is therefore a solution to the equations

$$y = ay^2 \oplus e = y^2 a \oplus e = y\, a\, y \oplus e \tag{21}$$

Still omitting the ⊗ sign for the sake of notational simplicity, let us now consider for integer n >0 the polynomial:

$$f(y) = ay^n \oplus e. \tag{22}$$

Thus: $f^2(y) = f(f(y)) = a\,(ay^n \oplus e)^n \oplus e$.

Lemma 7.3.1.1. *For every* $m \geq 1, f^m(y)$ *can be expressed as:*

$$f^m(y) = f^m(\varepsilon) \oplus a^m y^n g_m(y) \tag{23}$$

where the $g_m\,(y)$ *are polynomials in* y *and in a such that* $g_m\,(e) \geq e$.

Proof. Let us prove (23) by induction. The assumption is true for $m = 1$ ($f\,(\varepsilon) = e$, $g_1\,(y) = e$). Let us therefore assume it to be true for m. We then have:

$$f^{m+1}(y) = f^m[f(y)] = f^m(\varepsilon) \oplus a^m[f(y)]^n g_m[f(y)].$$
$$f^{m+1}(y) = f^m(\varepsilon) \oplus a^m(e \oplus ay^n)^n g_m(e \oplus ay^n).$$

Since g_m is a polynomial, $g_m(e \oplus ay^n) = g_m(e) \oplus ay^n h_m(y)$ where $h_m\,(y)$ is a polynomial in y and in a. Similarly, $(e \oplus ay^n)^n = e \oplus ay^n \ell_n\,(y)$ where $\ell_n\,(y)$ is a polynomial in y and in a.

Observe, moreover, that $\ell_n(e) \geq e$. We therefore deduce:

$$f^{m+1}(y) = f^m(\varepsilon) \oplus a^m g_m(e) \oplus a^{m+1}y^n[h_m(y) \oplus \ell_n(y)g_m(e) \oplus \ell_n(y)ay^n h_m(y)],$$

which yields (23) by setting:

$$f^{m+1}(\varepsilon) = f^m(\varepsilon) \oplus a^m g_m(e) \tag{24}$$

$g_{m+1}(y) = h_m(y) \oplus \ell_n(y)g_m(e) \oplus \ell_n(y)a\,y^n h_m(y)$. Since $\ell_n(e) \geq e$ and $g_m\,(e) \geq e$, we have $\ell_n(e)g_m(e) \geq e$ and we deduce that $g_{m-1}(e) \geq e$; this shows that $f^m(\varepsilon)$ is a polynomial in a with degree at least $m - 1$. □

Let us then consider the equation:

$$y = ay^n \oplus e. \tag{25}$$

Proposition 7.3.1.2. *(Gondran 1979)*
If a *is* p*-stable,* $f^{p+1}\,(\varepsilon)$ *is the minimal solution to* (25).

Proof. According to Lemma 7.3.1.1, $f^{p+1}(\varepsilon)$ is a polynomial in a where all the coefficients indexed from 0 to p are ≥ e. We deduce that the subsequent terms are absorbed by the first p + 1 ones, since a is p-stable.

Thus:

$$f^{p+1}(\varepsilon) = \sum_{k=0}^{p} \beta_k a^P.$$

$g_m(e)$ being a polynomial in a, and a being p-stable, (24) therefore implies:

$$f^{p+2}(\varepsilon) = f^{p+1}(\varepsilon).$$

This shows that $f^{p+1}(\varepsilon)$ is a solution to (25). On the other hand, according to (23), every solution y to (25) is expressed as:

$$y = f^m(\varepsilon) \oplus a^m y^n g_m(y)$$

which shows that $f^{p+1}(\varepsilon)$ is indeed the minimal solution to (25). □

Let us consider now the case $n = 2$, that is to say the equation:

$$y = ay^2 \oplus e. \tag{26}$$

Corollary 7.3.1.3. *If a is* p*-stable,* $a^{*/2}$ *is the minimal solution to* (26).

Proof. We proceed by induction by showing that $f^m(\varepsilon)$ is expressed as $a^{(m-1)_2} \oplus a^m g_m(a)$ where $g_m(a)$ is a polynomial in a.

The property is true for $m = 1$, with $g_1(a) = \varepsilon$.

Let us therefore assume it to be true for m. We then have:

$$f^{m+1}(\varepsilon) = f[f^m(\varepsilon)] = a[f^m(\varepsilon)]^2 \oplus e$$
$$[f^m(\varepsilon)]^2 = [a^{(m-1)_2}] \oplus a^m[2a^{(m-1)_2} g_m(a) \oplus g_m(a) a^m (g_m(a))^2]$$

Let us then consider the sum of the first m terms of $(a^{(m-1)_2})^2$; it is equal to:

$$\sum_{r=0}^{r=m-1} \left(\sum_{k=0}^{r} \alpha_k \alpha_{r-k} \right) a^r.$$

However, as the first Catalan numbers satisfy the equality

$$\sum_{k=0}^{r} \alpha_k \alpha_{r-k} = \alpha_{r+1}$$

the sum of the first m terms of $(a^{(m-1)_2})^2$ is therefore equal to $\sum_{r=0}^{m-1} \alpha_{r+1}\, a^r$.

We deduce that:

$$[f^m(\varepsilon)]^2 = \sum_{k=1}^{m} \alpha_k a^{k-1} \oplus a^{m+1} g_{m+1}(a)$$

where $g_{m+1}(a)$ is a polynomial in a.

Finally, we obtain:

$$\begin{aligned} f^{m+1}(\varepsilon) &= e \oplus \sum_{k=1}^{m} \alpha_k a^k \oplus a^{m+1} g_{m+1}(a) \\ &= a^{(m)_2} \oplus a^{m+1} g_{m+1}(a) \end{aligned}$$

which proves correctness of the induction.

The corollary then immediately follows by observing that if a is p-stable, the recurrence implies:

$$f^{p+1}(\varepsilon) = a^{(p)_2}. \quad \square$$

Corollary 7.3.1.4. *If* ac *is stable, and if* $\otimes$ *is commutative,* c $(ac)^{*/2}$ *is a solution to*

$$y = ay^2 \oplus c \tag{27}$$

Proof. Since $(ac)^{*/2} = ac\,[(ac)^{*/2}]^2 \oplus e$, we deduce: $c\,(ac)^{*/2} = a\,[c\,(ac)^{*/2}]^2 \oplus c$, and consequently $c(ac)^{*/2}$ is a solution to (27). $\square$

Corollary 7.3.1.5. *If* b *and* ac *are stable and if* $\otimes$ *is commutative,* $b^*c(b^{*2}ac)^{*/2}$ *exists and is a solution to:*

$$y = ay^2 \oplus by \oplus c \tag{28}$$

Proof. b and ac being stable, b^*ac is stable according to Proposition 7.2.5. b and b^*ac being stable, we furthermore have that $b^{*2}ac$ is stable, therefore $b^*c(b^{*2}ac)^{*/2}$ exists.

$$\text{For} \quad y_0 = b^*c(b^{*2}ac)^{*/2}, \quad \text{let us calculate} \quad f(y_0) = a\,y_0^2 \oplus b\,y_0 \oplus c:$$

$$\begin{aligned} f(y_0) &= ab^{*2}c^2[(b^{*2}ac)^{*/2}]^2 \oplus b\,b^*c(b^{*2}ac)^{*/2} \oplus c \\ &= c(b^{*2}ac[(b^{*2}ac)^{*/2}]^2 \oplus e \oplus b\,b^*(b^{*2}ac)^{*/2}) \\ &= c\{(b^{*2}ac)^{*/2} \oplus b\,b^*(b^{*2}ac)^{*/2}\} = c(e \oplus bb^*)(b^{*2}ac)^{*/2} \\ &= cb^*(b^{*2}ac)^{*/2} = y_0. \quad \square \end{aligned}$$

Remark 7.3.1.6. If e is stable, $e^{*/2} = e^*$. Moreover, if $e^* = e$ then: $a^{*/2} = a^*$. ||

7.3.2. Quasi-nth-root

The above results are easily generalized to solving equations of the form:

$$y = P(y) \tag{29}$$

where P (y) is a degree-n polynomial in y, thanks to the introduction of a *quasi-nth-root* of an element a, denoted $a^{*/n}$, minimal solution to (25).

To define in a simple way the expansion of $a^{*/n}$, we use the Lagrange theorem on generating series, recalled below:

Lagrange Theorem

If $y = x\varphi(y)$ *where* $\varphi(y) = f_0 + f_1\,y + f_2\,y^2 + \ldots$, *then the coefficient of* x^k *in* y (x) *is equal to* $\frac{1}{k}$ *times the coefficient of* y^{k-1} *in* φ^k (y).

Thus, for the equation $y = x\,(y+1)^n$, the coefficient of x^k in y(x) is equal to $\frac{1}{k}\binom{nk}{k-1}$.

For $a \in E$, let

$$a^{(k)_n} = e \oplus a \oplus na^2 \oplus \frac{n(3n-1)}{2}a^3 \oplus \ldots \oplus \frac{1}{k}\binom{nk}{k-1}a^k \tag{30}$$

Then, if a is p-stable, we have:

$$a^{(p)_n} = a^{(p+1)_n} = a^{(p+2)_n} = \cdots$$

since the terms involving a^{p+1} are absorbed by $a^{(p)}$.

For each p-stable element, we then deduce the existence of $a^{*/n}$, quasi-nth-root of a, defined as

$$a^{*/n} = \lim_{k \to +\infty} a^{(k)_n} = a^{(p)_n} = a^{(p+1)_n} = \cdots \tag{31}$$

Examples making use of the quasi-nth-root will be found in Exercises 2 and 3 at the end of this chapter.

8. Residuation and Generalized Solutions

In the two previous sections, we studied solutions for fixed-point type equations, $x = f(x)$. The theory of residuation, which we review here, enables one to introduce the concept of *generalized solutions* for equations of the form:

$$f(x) = b \tag{32}$$

in cases where f can be non-surjective (problem of existence) and/or non-injective (problem of unicity).

The work by Blyth and Janowitz (1972) is a basic reference on the subject; see also Gaubert (1992) and Baccelli et al. (1992).

The generalized solutions correspond to *residuable mappings* often referred to as *Galois correspondences.*

Definition 8.1. *(lower-solution, upper-solution)*

Let E *and* F *be two ordered sets,* $b \in F$, *and* f *a mapping:* $E \to F$. *We say that* x *is a* lower-solution *to* (32) *if we have* $f(x) \le b$.

If the set $\{x \in E \mid f(x) \le b\}$ *has a least upper bound, then, if* f *is lsc,*

$$f^{\uparrow}(b) = \sup\{x \in E \mid f(x) \le b\}$$

is the largest lower-solution *to* (32).

We say that y *is an* upper-solution *to* (32) *if we have*

$$f(y) \ge b.$$

If the set $\{x \in E \mid f(x) \ge b\}$ *has a greatest lower bound, then, if* f *is usc,*

$$f^{\downarrow}(b) = \inf\{x \in E \mid f(x) \ge b\}$$

is the smallest upper-solution *to* (32).

To ensure the consistency of this definition, we must verify that $f^{\uparrow}(b)$ (resp. $f^{\downarrow}(b)$) is clearly a lower-solution (resp. an upper-solution) to (32). According to the assumption, since $\{x \in E \mid f(x) \leq b\}$ has an upper bound, then for every sequence x_n of lower-solutions converging towards $f^{\uparrow}(b)$, we have $f(x_n) \leq b$ and since f is l.s.c, $f \circ f^{\uparrow}(b) \leq b$.

$f^{\uparrow}(b)$ is therefore clearly the largest lower-solution to (32).

$f^{\uparrow}$ will be referred to as the *sup-pseudo-inverse* of f and $f^{\downarrow}$ as the *inf-pseudo-inverse* of f.

Thus $f \circ f^{\uparrow}(b) \leq b$ and $f \circ f^{\downarrow}(b) \geq b$.

We moreover check that:

$$f^{\uparrow} \circ f(b) = \sup\{y \in E \mid f(x) \leq f(b)\} \geq b$$

$$f^{\downarrow} \circ f(b) = \inf\{x \in E \mid f(x) \geq f(b)\} \leq b$$

and that the functions $f^{\uparrow}$ and $f^{\downarrow}$, if they exist, are monotonic.

Proposition 8.2. *Let* E *and* F *be two ordered sets, and* f *a mapping:* $E \to F$. *Let us denote* Id_E *(resp.* Id_F*) the identity mapping of* E *(resp. of* F*). The following statements are equivalent:*

(i) there exists g: $F \to E$ *nondecreasing such that* $f \circ g \leq Id_F$ *and* $g \circ f \geq Id_E$*,*
(ii) for every y *in* F*, the set* $\{x \in E \mid f(x) \leq y\}$ *has a largest element.*

Proof. Let us assume (i). If $f(x) \leq y$, $x \leq g \circ f(x) \leq g(y)$.

Moreover, $f \circ g(y) \leq y$ which shows that g(y) is the largest element of $\{x \in E \mid f(x) \leq y\}$. Conversely, under assumption (ii), the mapping $g\colon y \to \sup\{x \in E \mid f(x) \leq y\}$ satisfies (i). □

Thus, $f^{\uparrow}$ (resp. $f^{\downarrow}$), if it exists, is the unique nondecreasing function (resp. nonincreasing) satisfying:

$$f \circ f^{\uparrow} \leq Id_F \quad \text{and} \quad f^{\uparrow} \circ f \geq Id_E \tag{33}$$

(resp. $f^{\downarrow} \circ f \leq Id_E$ and $f \circ f^{\downarrow} \geq Id_F$)

Definition 8.1 is classically applied to the case of a nondecreasing function f. In this case, if $f^{\uparrow}(b)$ (resp. $f^{\downarrow}(b)$) exists for every b, we say that the function f is *residuable* (resp. *dually residuable*) and $f^{\uparrow}$ is called the *residue mapping* of f (resp. $f^{\downarrow}$, the *dual residue mapping* of f).

We denote $Res^{\uparrow}(E, F)$ the set of residuable mappings: $E \to F$.

We provide below a few examples of residuable mappings.

Example 1. (reciprocal image)

Given a mapping $f\colon A \to B$, the associated mapping $\varphi_f\colon (\mathcal{P}(A), \subset) \to (\mathcal{P}(B), \subset)$, defined, for every $X \in \mathcal{P}(A)$ as: $\varphi_f(X) = f(X)$ is residuable and we have $\varphi_f^{\uparrow}(Y) = f^{-1}(Y)$ (reciprocal image of Y by f). ||

Example 2. (integer part)

The canonical injection $(\mathbb{Z}, \leq) \to (\mathbb{R}, \leq)$ is residuable and the residue mapping is the "integer part" mapping. ||

Example 3. (orthogonal subspace)

Let us endow $\mathbb{R}^d$ with the standard scalar product.

The mapping

$$\varphi: (\mathcal{P}(\mathbb{R}^d), \subset) \to (\mathcal{P}(\mathbb{R}^d), \supset)$$

which, with every set X associates its orthogonal $X^\perp = \{y \in \mathbb{R}^d \mid \forall x \in X, x . y = 0\}$ is residuable (see Gaubert 1992). ||

Example 4. (convex conjugate)

The Fenchel transform:

$\mathcal{F} : (\overline{\mathbb{R}}^{\mathbb{R}^n}, \geq) \to (\overline{\mathbb{R}}^{\mathbb{R}^n}, \leq)$, defined as $\mathcal{F}\, f(p) = \sup\limits_{x \in \mathbb{R}^n} \{px - f(x)\}$, is residuable (observe the reversal of the order between the start and end sets), and we have $\mathcal{F}^\uparrow = \mathcal{F}$. ||

Proposition 8.3. *For every residuable function f, we have:*

(i) $f \circ f^\uparrow \circ f = f$
(ii) $f^\uparrow \circ f \circ f^\uparrow = f^\uparrow$

Proof. (i) $f = f \circ I_{d_E} \leq f \circ (f^\uparrow \circ f) = (f \circ f^\uparrow) \circ f \leq I_{d_F} \circ f = f$.
(ii) Similar proof. □

The residue mapping $f^\uparrow$, or "sup-pseudo-inverse," therefore plays a role formally analogous to the classical pseudo-inverse $A^\#$ for a matrix A in standard linear algebra, which, is known to satisfy:

$$AA^\#A = A \quad \text{and} \quad A^\#AA^\# = A^\#.$$

Proposition 8.4. *Let* $f: E \to F$ *be a residuated function (thus isotone) with residual* $f^\uparrow$. *Then the equation* $f(x) = y$ *for* $y \in F$ *has a solution if and only if* $f(f^\uparrow(y)) = y$. *Moreover in this case,* $f^\uparrow(y)$ *is the maximum solution.*

Proof. Suppose that $f(x_0) = y$ for some $x_0 \in E$. Then from Definition 8.1 we deduce:

$$f^\uparrow(y) = f^\uparrow(f(x_0)) \geq x_0.$$

Now, using the monotonicity of f, we get:

$$f(f^\uparrow(y)) \geq f(x_0) = y$$

and, from Definition 8.1 again:

$f(f^\uparrow(y)) \leq y$ which proves that $f^\uparrow(y)$ is indeed a solution. Moreover, $f^\uparrow(y) \geq x_0$ proves that this solution is indeed maximal. □

Proposition 8.5. *For any* $f, f_i, g \in Res^\uparrow(E, F)$, *we have:*

(i) $(f \circ g)^\uparrow = g^\uparrow \circ f^\uparrow$
(ii) $(f \vee g)^\uparrow = f^\uparrow \wedge g^\uparrow$
(iii) If E *and* F *are complete lattices, we have for any finite subset of mappings* f_i: $(\vee_i f_i)^\uparrow = \wedge_i f_i^\uparrow$.

Proof. (i) $(f \circ g) \circ (g^{\uparrow} \circ f^{\uparrow}) = f \circ (g \circ g^{\uparrow}) \circ f^{\uparrow} \le f \circ \mathrm{Id} \circ f^{\uparrow} \le \mathrm{Id}$.
The other inequality is proved in the same way by using the other inequality of Proposition 8.2.

(ii) We decompose $t \to f(t) \vee g(t)$ as the product of the following mappings:

$$\begin{array}{ccc} E \xrightarrow{\varphi_1} E^2; & E^2 \xrightarrow{\varphi_2} F^2; & F^2 \xrightarrow{\varphi_3} F \\ t \to (t, t) & (u, v) \to (f(u), g(v)) & (x, y) \to x \vee y \end{array}$$

and we apply (i) observing that the associated residue mappings are given by:

$$\varphi_3^{\uparrow}((x', y')) = x' \wedge y', \quad \varphi_2^{\uparrow}(u', v') = (f^{\uparrow}(u'), g^{\uparrow}(v'),), \quad \varphi_1^{\uparrow}(t') = (t', t').$$

We thus have: $(f \vee g)^{\uparrow} = (\varphi_3 \circ \varphi_2 \circ \varphi_1)^{\uparrow} = \varphi_1^{\uparrow} \circ \varphi_2^{\uparrow} \circ \varphi_3^{\uparrow} = f^{\uparrow} \wedge g^{\uparrow}$.
The proof of (iii) is similar to that of (ii). □

Proposition 8.6. *Let* $(E, \le)$ *and* $(F, \le)$ *be two complete ordered sets with smallest elements* ε_E *and* ε_F *respectively. A nondecreasing mapping* $f: E \to F$ *is residuable if and only if* f *is continuous and* $f(\varepsilon_E) = \varepsilon_F$.

Proof. If f is residuable, the set $\{x \in E \mid f(x) \le \varepsilon_F\}$ has a largest element x_0 and since f is nondecreasing, $f(\varepsilon_E) \le f(x_0) \le \varepsilon_F$. Since, moreover, $f(\varepsilon_E) \ge \varepsilon_F$ we have $f(\varepsilon_E) = \varepsilon_F$. Let us then show that f is continuous. E and F being complete sets, let us denote $\bigvee_{x \in X} x$ the upper bounds of every set $X \subset E$ or $X \subset F$. Since f is nondecreasing, we have for every $X \subset E$: $f(\bigvee_{x \in X} x) \ge \bigvee_{x \in X} f(x)$. Let $f^{\uparrow}$ be the residue mapping associated with f. By using the inequalities (33) and the fact that $f^{\uparrow}$ is nondecreasing:

$$f(\bigvee_{x \in X} x) \le f(\bigvee_{x \in X} f^{\uparrow} \circ f(x)) \le f \circ f^{\uparrow}(\bigvee_{x \in X} f(x)) \le \bigvee_{x \in X} f(x)$$

which proves continuity.

Conversely, if $f(\varepsilon_E) = \varepsilon_F$, for every $y \in F$, the set $X = \{x \mid f(x) \le y\}$ is non empty. Furthermore, by continuity of f, $f(\bigvee_{x \in X} x) = \bigvee_{x \in X} f(x)$ and therefore X has a largest element. □

Definition 8.7. *We call* closure *a nondecreasing mapping* $\varphi: E \to E$, *such that* $\varphi \circ \varphi = \varphi$ *and* $\varphi \ge I_d$.

Proposition 8.8. *A residuable closure* φ *satisfies*

$$\varphi = \varphi^{\uparrow} \circ \varphi = \varphi \circ \varphi^{\uparrow}.$$

Proof.

$$\varphi = \mathrm{Id} \circ \varphi \ge \varphi \circ \varphi^{\uparrow} \circ \varphi \ge \varphi^{\uparrow} \circ \varphi = \varphi^{\uparrow} \circ \varphi \circ \varphi \ge \mathrm{Id} \circ \varphi. \quad \square$$

Definition 8.9. *(closed elements)*
If f *is residuable,* $f^{\uparrow} \circ f$ *is a closure, and we refer to as* closed *the elements of the form* $f^{\uparrow} \circ f(x)$.

In the case of Examples 2–4, the closed elements are respectively the integers (Example 2), the vector sub-spaces (Example 3), the convex functions (Example 4).

We can similarly define *dual closure*. We observe that f is a one-to-one correspondence between the set of closed elements and the set of dually closed elements.

Proposition 8.10. *(projection lemma)*

Let E *be an complete ordered set and* F *a complete subset of* E *containing* ε*, the minimal element of* E*. The canonical injection* i: F → E *is residuable. The residue mapping* $\mathrm{pr_F} = \mathrm{i}^{\uparrow}$ *satisfies:*

(i) $\mathrm{pr_F} \circ \mathrm{pr_F} = \mathrm{pr_F}$
(ii) $\mathrm{pr_F} \leq \mathrm{Id_E}$
(iii) $\mathrm{x} \in \mathrm{F} \Leftrightarrow \mathrm{pr_F}\,(\mathrm{x}) = \mathrm{x}$.

Proof. The residuability of i results from Proposition 8.5.

(i): $\mathrm{i}^{\uparrow} \circ \mathrm{i}^{\uparrow} = (\mathrm{i} \circ \mathrm{i})^{\uparrow} = \mathrm{i}^{\uparrow}$.
(ii): $\mathrm{pr_F} = \mathrm{i} \circ \mathrm{pr_F} = \mathrm{i} \circ \mathrm{i}^{\uparrow} \leq \mathrm{Id}$.
(iii): if $\mathrm{x} \in \mathrm{F}$, then $\mathrm{x} = \mathrm{i}\,(\mathrm{x})$, therefore $\mathrm{pr_F}(\mathrm{x}) = \mathrm{pr_F} \circ \mathrm{i}\,(\mathrm{x}) \geq \mathrm{x}$.

The other inequality is given by (ii). The converse is straightforward. □

Properties (i) and (ii) assert that $\mathrm{pr_F}$ is a dual closure, property (iii) asserts that the dually closed elements are elements of F.

Example. Let $\mathrm{E} = \mathbb{R}^{\overline{\mathbb{R}}}$, $\mathrm{F} = \mathrm{Inc}(\mathbb{R}, \overline{\mathbb{R}})$ the complete subset of nondecreasing functions: $\overline{\mathbb{R}} \to \mathbb{R}$. The projection lemma shows that for every $\mathrm{u} \in \mathbb{R}^{\overline{\mathbb{R}}}$, there exists *a largest nondecreasing function* $\bar{\mathrm{u}}$ *smaller than or equal* to u. One can then show that:

$$\bar{\mathrm{u}}(\mathrm{t}) = \inf_{\tau \geq \mathrm{t}} \{\mathrm{u}(\tau)\}$$

In a dual manner, there exists a smallest *nondecreasing function* $\underline{\mathrm{u}}$ *larger than or equal to* u, given by

$$\underline{\mathrm{u}}(\mathrm{t}) = \sup_{\tau \leq \mathrm{t}} \{\mathrm{u}(\tau)\}.$$

Exercises

Exercise 1. Let us consider a dioid (E, ⊕, ⊗)

If the element a is p-stable, show that $\mathrm{a}^{*/\mathrm{n}}$ given by (31) in Sect. 7 is the minimal solution to:

$$\mathrm{y} = \mathrm{a}\mathrm{y}^{\mathrm{n}} \oplus \mathrm{e}.$$

Exercise 2. Let us consider a dioid (E, ⊕, ⊗), and two elements a and c in E.

Show that if $\mathrm{a\,c^{n-1}}$ is stable, and if ⊗ is commutative, $\mathrm{c}\big(\mathrm{a\,c^{n-1}}\big)^{*/\mathrm{n}}$ is a solution to:

$$\mathrm{y} = \mathrm{a}\mathrm{y}^{\mathrm{n}} \oplus \mathrm{c}.$$

Exercise 3. Let us consider a dioid $(E, \oplus, \otimes)$ and three elements a, b, c in E.

Show that if b and a c^{n-1} are stable, and if $\otimes$ is commutative, $b^* c \left((b^*)^n \, a \, c^{n-1}\right)^{*/n}$ exists and is a solution to $y = a\, y^n \oplus by \oplus c$.

Exercise 4. Let E denote the closed interval [0, 1] and $\otimes$ the operation: $a \otimes b = \text{Min}\{a, b\}$.

Also consider the operation $\otimes'$ defined as:

$$\begin{aligned} a \otimes' b &= 1 \quad \text{if } a \leq b, \\ &= b \quad \text{otherwise.} \end{aligned}$$

For any $\rho \in E$, we now define the function r: $E \to E$ as: $r(x) = \rho \otimes x \ (\forall x \in E)$.

Show that r is residuable and that $r^{\uparrow}$ is expressed as: $r^{\uparrow}(x) = \rho \otimes' x$.

Check that $r^{\uparrow}$ is l.s.c.

Exercise 5. E being a given partially ordered set, we denote E^n the set of n-vectors with components in E. For any a, b $\in$ E, $a \vee b$ (resp. $a \wedge b$) denotes the sup (resp. inf) in E.

Let $r_{11}, r_{12}, \ldots, r_{mn}$ be residuated functions with respective residuals $r_{11}^{\uparrow}$, $r_{12}^{\uparrow}, \ldots, r_{mn}^{\uparrow}$. Define $\mathcal{R}$: $E^n \to E^m$ and $\tilde{\mathcal{R}}$: $E^m \to E^n$ as:

$$[\mathcal{R}(x)]_i = \bigoplus_{j=1}^{n} r_{ij}(x_j) \quad (\forall i = 1, \ldots, m)$$

$$\left[\tilde{\mathcal{R}}(x)\right]_j = \bigoplus_{i=1}^{m} r_{ij}^{\uparrow}(y_i) \quad (\forall j = 1, \ldots, n)$$

Taking $\oplus = \vee$, show that $\mathcal{R}$ is residuated with residual $\mathcal{R}^{\uparrow} = \tilde{\mathcal{R}}$.

Show that the same property holds when considering $\oplus = \wedge$.

Exercise 6. Consider the dioid $(E, \oplus, \otimes)$ where $E = [0, 1], \oplus = \text{Max}$, $\otimes = \text{Min}$.

Let $\oplus'$ and $\otimes'$ be the two dual operations defined as:

$$a \oplus' b = \text{Min}\{a, b\}$$

$$a \otimes' b = \begin{cases} 1 & \text{if } a \leq b \\ b & \text{otherwise} \end{cases}$$

For any two n-vectors $u \in E^n$, $v \in E^n$, we denote $u \leq v$ iff $u_i \leq v_i$ for all $i = 1, \ldots, n$.

Given two matrices $P \in M_{m,\ell}(E)$ and $Q \in M_{\ell,n}(E)$ we define two "dual" products:

$$P \otimes Q = R = (r_{ij}) \quad \text{with} \quad r_{ij} = \bigoplus_{k=1}^{\ell} p_{ik} \otimes q_{kj}$$

$$P \otimes' Q = T = (t_{ij}) \quad \text{with} \quad t_{ij} = \bigoplus_{k=1}^{\ell}{}' \, p_{ik} \otimes' q_{kj}$$

(1) By using the results of Exercises 4 and 5, show that for a given matrix $R \in M_{m,n}(E)$ the function: $\mathcal{R}(x)$: $E^n \to E^m$ defined as:
$\mathcal{R}(x) = \mathcal{R} \otimes x$, is residuable, with residual $\mathcal{R}^*(y)$: $E^m \to E^n$ defined as:

$$\mathcal{R}^*(y) = R^T \otimes' y.$$

(2) Let $R \in M_{m,n}(E)$ and $b \in E^m$ be given.
Show that the equation $R \otimes x = b$ has a solution if and only if $b = R \otimes (R^T \otimes' b)$ and that, in this case, $x^*\ (R, b) \triangleq R^T \otimes' b$ is the maximum solution.

(3) Show that $d^* = R \otimes (R^T \otimes' b)$ is the maximum d such that the equation $R \otimes x = d$ has a solution and $d \le b$.

(4) Consider the following numerical example:

$$R = \begin{pmatrix} 1 & 0.2 & 0.4 \\ 0 & 0.2 & 0.2 \\ 0 & 0.6 & 0.6 \\ 0.5 & 0.3 & 0.4 \end{pmatrix} \qquad b = \begin{pmatrix} 1 \\ 0.7 \\ 0.3 \\ 0.5 \end{pmatrix}$$

Show that $x^*\ (R, b) = (1, 0.3, 0.3)^T$ and that $d^* = (1, 0.2, 0.3, 0.5)^T$.
Show that the system $R \otimes x = b$ has no solution for all $d = (1, \alpha, 0.3, 0.5)^T$ with $0.2 < \alpha \le 0.7$.

(5) Show that the results of questions 1, 2, 3 also apply to the various dioids (E, Max, Min) with either $E = \overline{\mathbb{R}}$, $E = \overline{\mathbb{R}}_+$, or $E = [\alpha, \beta]$. Indicate, in each of these cases, which is the dual law $\otimes'$ to be considered.

[*Answers*: refer to Cuninghame-Green and Cechlárová (1995)]

Chapter 4
Solving Linear Systems in Dioids

1. Introduction

How can we expect to solve a linear equality system in algebraic structures consisting of a set with two internal laws $\oplus$ and $\otimes$ which are not a priori invertible, i.e. where one cannot always solve $a \oplus x = b$ and $a \otimes x = b$?

The key idea in the present chapter is to observe that the solution of a "fixed point" type equation such as $x = a \otimes x \oplus b$ only requires the existence of the *quasi-inverse* a^* of the element a, defined in the previous chapter as the "limit" of the series: $e \oplus a \oplus a^2 \oplus \cdots$

It is indeed remarkable that neither the additive inverse nor the multiplicative inverse are needed to compute "$(e - a)^{-1}$"!

However, in order to guarantee some form of uniqueness, it will be necessary to work in *canonically ordered* semirings, i.e. in *dioids*.

The purpose of this chapter is thus to discuss how to solve linear systems of the fixed point type, which will lead to generalizations of the main known algorithms for solving linear systems in classical linear algebra.

This chapter does not address the problem of solving linear systems of the form $A\,x = b$ in dioids. This actually relates to residuation theory introduced in Chap. 3, Sect. 8, and involves a concept of generalized pseudo-inverse.

Given a $n \times n$ matrix: $A = (a_{ij})$ with entries in a dioid $(E, \oplus, \otimes)$ and $b \in E^n$ a given n-vector, we will focus here on the solution of linear systems of the form:

$$y = y \otimes A \oplus b^T \tag{1}$$

or

$$z = A \otimes z \oplus b \tag{2}$$

where we denote $y \otimes A$ the row-vector, the jth component of which is:

$$\sum_{i=1}^{n} y_i \otimes a_{ij}$$

and A ⊗ z the column-vector the ith component of which is:

$$\sum_{j=1}^{n} a_{ij} \otimes z_j$$

(the above sums are to be understood in the sense of ⊕)

Section 2 illustrates the general perspective of the chapter by considering as an introductory example the case of linear systems on the dioid $(\hat{\mathbb{R}}, \text{Min}, +)$ which corresponds to the shortest path problem in a graph.

By viewing the matrix A as the generalized incidence matrix of an associated graph G(A), it is shown in Sect. 3 that the successive powers of A and the quasi-inverse A^* of A may be interpreted in terms of path weights of G(A). It will be shown how the quasi-inverse A^* can be used to compute the *minimal solution* to each of the equations (1) or (2), and how minimal solutions can be interpreted as solutions to associated path-finding problems in G(A).

The general algorithms for solving linear systems (1) or (2) are studied in Sect. 4 and 5: iterative methods in Sect. 4, together with an extension to algebras of endomorphisms in Sect. 4.4; then, the "direct" methods in Sect. 5. A broad overview of applications to modeling and solving a huge variety of path-finding problems in graphs is finally presented in Sect. 6.

2. The Shortest Path Problem as a Solution to a Linear System in a Dioid

A typical example of a path-finding problem in graphs which can be formulated in terms of solution to a linear system of type (1) or (2) is the determination of the *shortest paths of fixed origin in a valued directed graph.*

Let us consider a directed graph G = [X, U] where the vertices are numbered $1, 2, \ldots n$, and in which each arc (i, j) is assigned a length $a_{ij} \in \mathbb{R}$. Given a vertex $i_o \in X$ chosen as origin, we seek the lengths $y_j (j = 1 \cdots n)$ of the shortest paths between the vertex i_o and the other vertices j of the graph.

It is not restrictive to assume that G is a 1-graph, i.e. that, for any ordered pair of vertices (i, j), there exists at most one arc of the form (i, j). (Indeed, if U contains several arcs $u_1, u_2, \ldots u_p$ of the form (i, j) and of lengths $\ell(u_1), \ell(u_2), \ldots \ell(u_p)$, to solve the shortest path problem, it is enough to consider that there exists a single arc (i, j) of length $a_{ij} = \underset{k=1,\ldots,p}{\text{Min}} \{\ell(u_k)\}$ and to ignore all the other arcs).

2.1. The Linear System Associated with the Shortest Path Problem

According to the well known principle of dynamic programming (Bellman 1954, 1958), it is known that the values y_i satisfy the following equations (referred to as "optimality" conditions):

$$\begin{cases} y_{i_o} = 0 \\ \forall j \neq i_o: \quad y_j = \underset{i \in \Gamma_j^{-1}}{\text{Min}} \{y_i + a_{ij}\} \end{cases}$$

where $\Gamma_j^{-1} = \{i/(i, j) \in U\}$ is the set of direct predecessors of vertex j in G.

By agreeing to set $a_{ij} = +\infty$ if arc (i, j) does not exist and $a_{ii} = 0$ ($\forall i = 1, \ldots n$) the previous relations can be rewritten as:

$$\begin{cases} y_{i_o} = 0 \\ \forall j \neq i_o: \quad y_j = \underset{i=1,\ldots,n}{\text{Min}} \{y_i + a_{ij}\} \end{cases}$$

Assuming, for the sake of simplification, that the lengths a_{ij} are all nonnegative, we must have $y_j \geq 0$ ($\forall j = 1 \cdots n$) and consequently the previous relations can be further rewritten:

$$\begin{cases} y_{i_o} = \text{Min}\left\{ \underset{i=1,\ldots,n}{\text{Min}} \{y_i + a_{i,i_o}\};\, 0\right\} \\ \text{and, } \forall j \neq i_o \\ y_j = \text{Min}\left\{ \underset{i=1,\ldots,n}{\text{Min}} \{y_i + a_{ij}\}; +\infty\right\} \end{cases}$$

Let us then consider the dioid $(E, \oplus, \otimes)$ where $E = \mathbb{R} \cup \{+\infty\}$ and where the internal laws $\oplus$ and $\otimes$ are defined as:

$$\forall a \in E, \forall b \in E:$$
$$a \oplus b = \text{Min}\{a, b\}$$
$$a \otimes b = a + b$$

$\varepsilon = +\infty$ is the neutral element of $\oplus$ and $e = 0$ is the neutral element of $\otimes$

(see Chap. 8, Sect. 4.7.1). It is then observed that the above relations can be written in the form of the linear system:

$$\begin{cases} y_{i_o} = \sum_{i=1}^{n} y_i \otimes a_{i,i_o} \oplus e \\ y_j = \sum_{i=1}^{n} y_i \otimes a_{i,j} \oplus \varepsilon \quad \text{for any } j \neq i_o \end{cases}$$

that is to say, in matrix notation:

$$y = y \otimes A \oplus b^T \tag{1}$$

where A denotes the matrix $(a_{ij})_{\substack{i=1\ldots,n \\ j=1,\ldots,n}}$ and b the n-vector defined as:

$$b_{i_o} = e \quad \text{and,} \quad \forall j \neq i_o: b_j = \varepsilon.$$

It is thus seen that determining the vector y where the components are the lengths of the shortest paths of origin i_o amounts to solving the linear system (1) in the dioid $(E, \oplus, \otimes)$.

2.2. Bellman's Algorithm and Connection with Jacobi's Method

As will be seen in Sects. 4 and 5 of the present chapter, most algorithms for solving the shortest path problem can be interpreted as variants of known methods in classical linear algebra.

This is the case, for example, of Bellman's algorithm (1958) which consists in starting from y^o defined as:

$$y^o_{i_o} = 0, y^o_j = +\infty \quad (\forall j \neq i_o)$$

and then in performing the following computations iteratively:

$$\begin{cases} y^{t+1}_{i_o} = 0 \\ y^{t+1}_j = \underset{i \in \Gamma^{-1}_j}{\text{Min}} \{y^t_i + a_{ij}\} \end{cases}$$

By using system (1) this algorithm can be expressed in the equivalent form:

$$\begin{cases} y^o = b^T & (3) \\ y^{t+1} = y^t \otimes A \oplus b^T & (4) \end{cases}$$

where one can recognize the analog to *Jacobi's method* in classical linear algebra.

From (3) and (4) we deduce, for an arbitrary integer $t > 0$:

$$y^t = b \otimes (I \oplus A \oplus A^2 \oplus \cdots \oplus A^t)$$

where $I = \begin{bmatrix} e & & & \\ & e & & \varepsilon \\ & \varepsilon & \cdot & \\ & & & \cdot \\ & & & & e \end{bmatrix}$ is the identity matrix of $M_n(E)$.

The convergence of Bellman's algorithm is therefore intimately related to the convergence of the series $I \oplus A \oplus A^2 \oplus \cdots \oplus A^t \oplus \cdots$

2.3. Quasi-Inverse of a Matrix with Elements in a Semiring

It is well known that Bellman's algorithm converges in at most $n - 1$ iterations, in other words that y^{n-1} is the vector of shortest path lengths, if and only if the graph does not contain a circuit of negative length. In these conditions, we will see (see Theorem 1 Sect. 3.3) that the series $A^{(k)} = I \oplus A \oplus A^2 \oplus \cdots \oplus A^k$ converges in at most $n - 1$ steps, i.e. that $A^{(n-1)} = A^{(n)} = A^{(n+1)} = \cdots$ and the limit of this series, denoted A^* will be referred to as the *quasi-inverse of* the matrix A.

Bellman's algorithm then converges towards $y = b^T \otimes A^*$.

We verify that $b \otimes A^*$ clearly corresponds to a solution of (1) because:

$$b^T \otimes A^* \otimes A \oplus b^T = b^T \otimes (A^* \otimes A \oplus I)$$

Now, if $A^* = A^{(n-1)}$, this yields:

$$A^* \otimes A \oplus I = A^{(n)} = A^{(n-1)} = A^*$$

We therefore clearly deduce:

$$b^T \otimes A^* = (b \otimes A^*) \otimes A \oplus b^T$$

We now turn to show that the solution $b \otimes A^*$ provided by Bellman–Jacobi's Algorithm is by no means an arbitrary solution to (1).

2.4. Minimality of Bellman–Jacobi Solution

By using the fact that $(\hat{\mathbb{R}}, \text{Min}, +)$ is a *dioid* (i.e. a canonically ordered semiring) let us show that $b^T \otimes A^*$ is the *minimal solution* to linear system (1).

Let y then be an arbitrary solution of $y = y \otimes A \oplus b^T$

We can write:

$$\begin{aligned} y &= y \otimes A \oplus b^T \\ &= (y \otimes A \oplus b^T) \otimes A \oplus b^T \\ &= y \otimes A^2 \oplus b^T \otimes (I \oplus A) \end{aligned}$$

By transferring the above expression of y into (1) we similarly obtain:

$$y = y \otimes A^3 \oplus b^T \otimes (I \oplus A \oplus A^2)$$

By reiterating the argument, we obtain for any $k \geq 2$:

$$y = y \otimes A^k \oplus b^T \otimes A^{(k-1)}$$

Thus for $k \geq n$, this yields:

$$y = y \otimes A^k \oplus b^T \otimes A^*$$

By denoting $\leq$ the canonical order relation of the dioid $(\hat{\mathbb{R}}, \text{Min}, +)$ this result shows that $b^T \otimes A^* \leq y$, and this holds for any solution y to (1). $b^T \otimes A^*$ is therefore clearly *the minimal solution to* (1).

For the shortest path problem (as well as for many other path-finding problems in graphs, see Sect. 6 below), the problem to be solved is not only a matter of finding an arbitrary solution of (1) or (2) but the *minimal solution*.

The example in Fig. 1 illustrates the fact that a non minimal solution to (1) is not relevant with respect to the shortest path problem (by convention the non represented arcs of the complete graph have a length $+\infty$). The vector $y = (0, 1, 1, 1)$ is a solution to system (1). Indeed, we verify that we clearly have:

$$\begin{aligned} y_1 &= \text{Min}\{y_1 + a_{11}, y_2 + a_{21}, y_3 + a_{31}, y_4 + a_{41}, 0\} \\ y_2 &= \text{Min}\{y_1 + a_{12}, y_2 + a_{22}, y_3 + a_{32}, y_4 + a_{42}, +\infty\} \end{aligned}$$

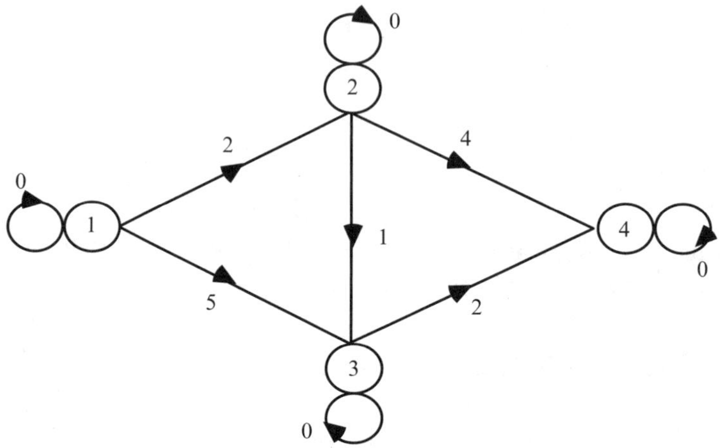

Fig. 1 Example of an oriented graph for which we want to determine the lengths of the shortest paths originating at vertex 1. The vector $y = (0, 1, 1, 1)$ is a solution to system (1), but is not the solution to the shortest path problem of origin 1 because it *is not the minimal solution to (1)*

$$y_3 = \mathrm{Min}\{y_1 + a_{13}, y_2 + a_{23}, y_3 + a_{33}, y_4 + a_{43}, +\infty\}$$
$$y_4 = \mathrm{Min}\{y_1 + a_{14}, y_2 + a_{24}, y_3 + a_{34}, y_4 + a_{44}, +\infty\}$$

However, it can be observed that the components of y have nothing to do with the lengths of the shortest paths originating at vertex 1 in the graph of Fig. 1. Only the minimal solution (0, 2, 3, 5) to system (1) has components equal to the desired shortest path lengths. (Remember that $2 \leq 1, 3 \leq 1, 5 \leq 1$, in the sense of the canonical order relation of the dioid $(\hat{\mathbb{R}}, \mathrm{Min}, +)$).

3. Quasi-Inverse of a Matrix with Elements in a Semiring Existence and Properties

3.1. Definitions

The concept of quasi-inverse of an element of E was introduced and studied in Chap. 3, Sects. 6 and 7. Here, we generalize this concept to the case of matrices $A \in M_n(E)$, then we show that the minimal solutions to systems such as (1) or (2) can be easily deduced from the quasi-inverse A^* of A (when the latter exists).

Let $A \in M_n(E)$ be a square n x n matrix with elements in a semiring $(E, \oplus, \otimes)$.

For any $k \in N$, denote by A^k the kth power of A, i.e. $A \otimes A \otimes \cdots \otimes A$ (k times) and define the matrices $A^{(k)}$ by:

$$A^{(k)} = I \oplus A \otimes A^2 \oplus \cdots \oplus A^k$$

where $I = \begin{bmatrix} e & & & & \\ & e & & \varepsilon & \\ & & \cdot & & \\ & \varepsilon & & \cdot & \\ & & & & e \end{bmatrix}$ is the identity matrix of $M_n(E)$.

We observe that, in the special case where the operation $\oplus$ is *idempotent*, the following result yields an alternative expression of $A^{(k)}$ involving the matrix $A' = I \oplus A$

Proposition 3.1.1. *If the* $\oplus$ *laws is idempotent, then* $A^{(k)} = (I \oplus A)^k$

Proof. The expansion of $(I \oplus A)^k$ gives $(I \oplus A)^k = I \oplus \sum_{r=1}^{k} C_k^r A^r$ where the sum is to be understood in the sense of $\oplus$ and where $C_k^r\ A^r$ denotes the sum $A^r \oplus A^r \oplus \cdots \oplus A^r C_k^r$ times $\left(C_k^r = \dfrac{k!}{r!\,(k-r)!}\right)$. Since $\oplus$ is idempotent we therefore have $C_k^r A^r = A^r$ and the proposition is deduced. □

Definition 3.1.2. *(Quasi-inverse of a matrix)*

We call quasi-inverse of $A \in M_n(E)$, *denoted* A^*, *the limit, when it exists, of the sequence of matrices* $A^{(k)}$ *as* $k \to \infty$*:*

$$A^* = \lim_{k\to\infty} A^{(k)}$$

We will now study sufficient conditions of existence for the matrix A^*. These conditions involve the interpretation of the matrices A^k and $A^{(k)}$ in terms of paths of the graph G(A) associated with the matrix A.

We distinguish between two cases:

- The case where $(E, \oplus, \otimes)$ is a semiring without being a dioid, i.e. the case where there is no canonical order relation on E. This means it will not be possible to use the topologies on ordered sets introduced in Chap. 3 to define the convergence of the sequence of the matrices $A^{(k)}$. In this situation, we will therefore have to limit ourselves to cases of finite convergence of $A^{(k)}$ towards A^*.
- The case where $(E, \oplus, \otimes)$ is a *topological dioid* i.e. where E is a (canonically) ordered set endowed with sup-topology (topology of the upper limit of nondecreasing sequences, see Chap. 3 Sect. 3).

3.2. Graph Associated with a Matrix. Generalized Adjacency Matrix and Associated Properties

Let $A \in M_n(E)$ be a square $n \times n$ matrix with entries in E. We define the graph G(A) associated with A as follows:

- The set of vertices of G(A) is $\{1, 2, \ldots n\}$ the set of indices of the rows (or columns) of A;

– The set of arcs of G(A) is the set of ordered pairs (i, j) corresponding to the terms a_{ij} of A distinct from ε (the neutral element of ⊕). If A contains a diagonal term $a_{ii} \neq \varepsilon$, then G(A) contains an arc (i, i) also referred to as a *loop*.

We observe that the graph G(A) thus defined is a *valued* graph: each arc (i, j) is endowed with the value $a_{ij} \in E$ ($a_{ij} \neq \varepsilon$) of the corresponding term of the matrix A.

Conversely, the matrix A can be considered as the (generalized) *adjacency matrix* of the valued graph G(A).

Property 3.2.1 below is the basis for the developments to follow, enabling us to interpret the coefficients of the matrices A^k and $A^{(k)}$ in terms of *paths* and *circuits* of the associated graph G(A).

For any $k \in N$, let us denote:

- P_{ij}^k the set of paths of G(A) (not necessarily elementary) joining vertex i to vertex j and containing *exactly* k arcs;
- $P_{ij}^{(k)}$ the set of paths of G(A) (not necessarily elementary) joining vertex i to vertex j and containing *at most* k arcs.

Moreover, with any path $\mu \in P_{ij}^k$, composed of the sequence of vertices $i_o, i_1, i_2, \ldots i_{k-1}, i_k$ (with $i_0 = i$ and $i_k = j$), we associate its *weight* $w(\mu) = a_{i_o,i_1} \otimes a_{i_1,i_2} \otimes \cdots . \otimes a_{i_{k-1},i_k}$

One can then state:

Property 3.2.1. (i) Each term (i, j) of matrix A^k is equal to the sum of the weights of the paths of P_{ij}^k:

$$\left(A^k\right)_{i,j} = \sum_{\mu \in P_{ij}^k} w(\mu) \tag{5}$$

(ii) Each term (i, j) of matrix $A^{(k)}$ is equal to the sum of the weights of the paths of $P_{ij}^{(k)}$:

$$\left(A^{(k)}\right)_{i,j} = \sum_{\mu \in P_{ij}^{(k)}} w(\mu) \tag{6}$$

Proof. This is easily proved by induction on k, taking into account the fact that ε is absorbing for multiplication ⊗, and that the latter is distributive with respect to addition. □

The paths belonging to P_{ij}^k or $P_{ij}^{(k)}$ are not necessarily elementary, i.e. can contain circuits. In the general case where the multiplication ⊗ is not commutative, the weight of a circuit γ passing successively through the vertices $i_1, i_2, \ldots i_k, i_1$ depends on the way it is traversed. Thus, if the starting vertex is i_1 the weight of the circuit is defined as:

$$a_{i_1 i_2} \otimes a_{i_2 i_3} \otimes \cdots \otimes a_{i_k,i_1}$$

if the starting vertex is i_2, the weight is defined as:

$$a_{i_2 i_3} \otimes a_{i_3 i_4} \otimes \cdots \otimes a_{i_k,i_1} \otimes a_{i_1 i_2}$$

and these two quantities can be different. One is thus lead to introduce the concept of *pointed circuit*.

Definition 3.2.2. *(concept of pointed circuit)*

We say that we have a pointed circuit *of* G(A) *when we are given a circuit of* G(A) *together with a special vertex of this circuit taken as the origin of the circuit. The weight of the pointed circuit* $\gamma = \{i_1 i_2, \dots i_k, i_1\}$ *of origin* i_1 *is:*

$$w(\gamma) = a_{i_1 i_2} \otimes a_{i_2 i_3} \otimes \cdots \otimes a_{i_k, i_1}$$

Thus to a circuit of G composed of k vertices, we let correspond k pointed circuits. Each of these pointed circuits can therefore have a different weight. In the case where multiplication is commutative, all the pointed circuits corresponding to a given circuit have the same weight and the concept of pointed circuit is not necessary.

We say that a graph G has no p-absorbing circuit *if the weight of each pointed circuit of the graph* G *is* a p-stable element of E. *(see Chap.* 3 *Sect.* 7*:* $a \in E$ *is said to be p-stable if and only if* $a^{(p)} = a^{(p+1)} = \dots$*, where,* $\forall k \in \mathbb{N}$: $a^{(k)} = e \oplus a \oplus a^2 \oplus \cdots \oplus a^k$).

Property 3.2.1 then becomes:

Property 3.2.3. If G(A) has no p-absorbing circuit, then:

$$\left(A^{(k)}\right)_{ij} = \sum_{\mu} w(\mu) \tag{7}$$

where the sum extends to the set of paths from i to j containing at most k arcs and traversing no more than p times successively each pointed circuit of G(A).

Proof. To prove the proposition, it is enough to show that any path traversing more than p times successively a pointed circuit does not need to be taken into account in (6). Let us therefore consider a path μ from i to j successively taking $p+q$ times $(q \geq 1)$ some pointed circuit. Let $\gamma_{\ell\ell}$ be this pointed circuit of origin ℓ. The path μ may therefore be decomposed into a path $\mu_{i\ell}$ from i to ℓ, in $p+q$ times the circuit $\gamma_{i\ell}$, and then a path $\mu_{\ell j}$ from ℓ to j; we will denote:

$$\mu = \mu_{i\ell}(\gamma_{\ell\ell})^{p+q}\mu_{\ell j}$$

If $\mu \in P_{ij}^{(k)}$ then each path:

$$\mu^r = \mu_{i\ell}\left(\gamma_{\ell\ell}\right)^{r+q-1}\mu_{\ell j} \text{ with } r = 0, 1, \dots, p$$

also belongs to $P_{ij}^{(k)}$.

Let us show that the path μ can be absorbed in (6) by the set of paths μ^r (for $r = 0, 1, \ldots, p$). Indeed:

$$\mathrm{w}(\mu) \oplus \sum_{0 \leq r \leq p} \mathrm{w}(\mu^r) = \mathrm{w}(\mu_{i\ell}) \otimes \mathrm{w}(\gamma_{\ell\ell})^{p+q} \otimes \mathrm{w}(\mu_{\ell_j})$$
$$\oplus \sum_{0 \leq r \leq p} \mathrm{w}(\mu_{i\ell}) \oplus \mathrm{w}(\gamma_{\ell\ell})^{r+q-1} \oplus \mathrm{w}(\mu_{\ell j})$$

hence, taking into account the right and left distributivity of $\otimes$:

$$\mathrm{w}(\mu) \oplus \sum_{0 \leq r \leq p} \mathrm{w}(\mu^r)$$
$$= \mathrm{w}(\mu_{i\ell}) \otimes \mathrm{w}(\gamma_{\ell\ell})^{q-1} \otimes \left[e \oplus \mathrm{w}(\gamma_{\ell\ell}) \oplus \cdots \oplus \mathrm{w}(\gamma_{\ell\ell})^{p} \oplus \mathrm{w}(\gamma_{\ell\ell})^{p+1} \right] \otimes \mathrm{w}(\mu_{\ell j})$$

Then, using the fact that G(A) has no p-absorbing circuit, this yields: $\mathrm{w}(\gamma_{\ell\ell})^{(p+1)} = \mathrm{w}(\gamma_{\ell\ell})^{(p)}$ and the previous equation becomes:

$$\mathrm{w}(\mu) \oplus \sum_{o \leq r \leq p} \mathrm{w}(\mu^r) = \sum_{o \leq r \leq p} \mathrm{w}(\mu^t)$$

We deduce the proposition by applying this property of absorption to all the paths successively taking more than p times a pointed circuit of G(A). □

A special case, frequently encountered in the examples (see Sect. 6 below), is when $p = 0$, i.e. the case where, for every pointed circuit γ, we have:

$$\mathrm{w}(\gamma) \oplus e = e.$$

Let us denote:

- $P_{ij}^{(k)}(o)$ the set of *elementary* paths (not traversing the same vertex twice) in $P_{ij}^{(k)}$.
- $P_{ij}(o)$ the set of all elementary paths from i to j.

In this case, we have the following corollary:

Corollary 3.2.4. *If* G(A) *has no* 0*-absorbing circuit, then:*

$$\left(A^{(k)}\right)_{ij} = \sum_{\mu \in P_{ij}^{(k)}(0)} \mathrm{w}(\mu) \tag{8}$$

$$\left(A^{(n-1)}\right)_{ij} = \sum_{\mu \in P_{ij}(0)} \mathrm{w}(\mu) \tag{9}$$

Proof. (8) is an immediate consequence of (7) taking into account the fact that $p = 0$. (9) is then deduced from (8) observing that an elementary path contains at most $n - 1$ arcs. □

3.3. Conditions for Existence of the Quasi-Inverse A^*

In the case where G(A) has no 0-absorbing circuit, one can immediately deduce from Corollary 3.2.4 of Sect. 3.2:

Theorem 1. *(Carré et al. 1971; Gondran 1973)*

If G(A) *has no* 0*-absorbing circuit, then the sequence of matrices* $A^{(k)}$ *has a limit* A^* *when* $k \to \infty$*, and this limit is reached for* $k \le n - 1$:

$$A^* = \lim_{k \to \infty} A^{(k)} = A^{(n-1)} = A^{(n)} = \cdots \tag{10}$$

Furthermore, A^* *(quasi-inverse of* A*) satisfies the matrix equations:*

$$A^* = I \oplus A \otimes A^* = I \oplus A^* \otimes A \tag{11}$$

Proof. The fact that the sequence $A^{(k)}$ has a limit which is certainly reached for $k = n - 1$ follows directly from (8) and (9).

Moreover we have:

$$\begin{aligned} I \oplus A \otimes A^* &= I + A \otimes (I \oplus A \oplus \cdots \oplus A^{n-1}) \\ &= A^{(n)} = A^{(n-1)} = A^* \end{aligned}$$

which proves (11). □

The previous result shows that if the weight of all the circuits of G(A) are 0-stable elements of E, then the matrix A is $(n - 1)$ stable in $M_n(E)$.

Remark. Theorem 1 can be generalized to the case of right dioids (resp. left dioids) where $\otimes$ is only right-distributive (resp. left-distributive) with respect to $\oplus$. In this case, it must be assumed that G has no 0-absorbing circuit on the right (resp. on the left), i.e. that, for any circuit γ, we must have, $\forall a \in E$: $a \oplus w(\gamma) \otimes a = a$ (resp. $a \oplus a \otimes w(\gamma) = a$).

Refer to Exercise 4 at the end of the chapter. ||

Let us now study the case where G(A) has no p-absorbing circuit with $p \ge 1$. In this situation, the assumptions of Property 3.2.3 alone are not sufficient to prove the finite convergence of $A^{(k)}$, as the cardinality of the paths involved in (7) cannot be bounded. We will therefore successively examine two types of additional assumptions:

- The commutativity of multiplication
- The p-nilpotency of the set of entries of matrix A.

We first consider the case where the *multiplication* is supposed to be *commutative*.

Let us denote:

- $P_{ij}^{(k)}(p)$ the set of paths of $P_{ij}^{(k)}$ traversing no more than p times each elementary circuit of G(A).
- $P_{ij}(p)$ the set of paths from i to j traversing no more than p times each elementary circuit of G(A).

We observe that the cardinality (number of arcs) of the paths of $P_{ij}(p)$ is bounded from above by $n - 1 + pnt$, where t denotes the total number of elementary circuits of G(A).

We then have the following theorem:

Theorem 2. *(Gondran, 1973)*

If G *has no* p*-absorbing circuit and if the multiplication is commutative, then:*

$$\left(A^{(k)}\right)_{ij} = \sum_{\mu \in P_{ij}^{(k)}(p)} w(\mu) \tag{12}$$

$$\left(A^{(n_p)}\right)_{ij} = \sum_{\mu \in P_{ij}(p)} w(\mu) \tag{13}$$

where n_p *is the maximum number of arcs of the paths of* P_{ij} (p) *(*$n_p \le n - 1 + pnt$ *where* t *is the total number of elementary circuits of* G(A)*).*

Furthermore, the sequence of matrices $A^{(k)}$ *has a limit* A^* *when* $k \to \infty$*, and this limit is reached for* $k \le n_p$*:*

$$A^* = \lim_{k \to +\infty} A^{(k)} = A^{(n_p)} = A^{(n_p+1)} = \cdots \tag{14}$$

A^* *(quasi inverse of* A*) then satisfies the matrix equations* (11).

Proof. (12) is deduced directly from Property 3.2.3 since, from the commutativity of ⊗, it is possible to consider any circuit as a product of elementary circuits, without taking into account the order in which these circuits are traversed.

The rest of the theorem is proved in the same way as for Theorem 1. □

The previous result shows that, if the multiplication is commutative and if the weights of the circuits of G(A) are p-stable elements of E, then the matrix A is n_p-stable in $M_n(E)$ (with $n_p \le n - 1 + pnt$ where t is the number of elementary circuits of G(A)).

Let us consider now the case where ⊗ is not commutative, but where the set of entries of matrix A is p-*nilpotent*.

Let F be a subset of E and p an integer > 0. We will say that F is p-*nilpotent* if and only if, for any sequence of p + 1 elements (not necessarily all distinct) $a_0, a_1, a_2, \ldots a_p$ taken in F, we have:

$$a_0 \otimes a_1 \otimes \cdots \otimes a_p = \varepsilon \quad \text{(the neutral element of } \oplus\text{)}$$

The following result shows then that, due to p-nilpotency, the commutativity of multiplication is not required to establish the convergence of the sequence $A^{(k)}$.

Theorem 3. *(Minoux, 1976)*

If the set F *of entries* a_{ij} *of the matrix* A *is p-nilpotent, then the matrix* $A^{(k)}$ *has a limit* A^* *which is reached for* $k \leq p$*:*

$$A^* = \lim_{k \to +\infty} A^{(k)} = A^{(p)} = A^{(p+1)} = \cdots$$

A^* *then satisfies the matrix equations* (11).

Proof. It suffices to observe that all the paths of more than p arcs do not need to be taken into account in (6). (11) is proved as for Theorem 1. □

Theorem 3 thus shows that, if the set of terms of A is p-nilpotent, then the A matrix is p-stable. We observe that all the elements of a p-nilpotent set are p-stable: the p-nilpotency assumption is therefore stronger than that of p-stability. An important class of problems for which we will have to assume p-nilpotency concerns generalized path algebras involving endomorphisms (see Sect. 4.4 below). Theorem 3 will be used there to generalize iterative algorithms to compute quasi-inverses of matrices of endomorphisms.

3.4. Quasi-Inverse and Solutions of Linear Systems. Minimality for Dioids

Let us assume that the matrix A has a quasi-inverse A^* which satisfies the relations (11).

The following result shows that A^* determines solutions for systems (1) and (2) of Sect. 1.

Property 3.4.1. Let $A \in M_n(E)$ and let us assume that its quasi-inverse A^* exists and satisfies:

$$A^* = I \oplus A \otimes A^* = I \oplus A^* \otimes A \tag{11}$$

Then:

(i) For any matrix $B \in E^{m \times n}$
(m integer $1 \leq m \leq n$)
$Y = B \otimes A^* \in E^{m \times n}$ is a solution to the linear system:

$$Y = Y \otimes A \oplus B \tag{15}$$

(ii) For any matrix $B \in E^{n \times m}$ (m integer $1 \leq m \leq n$) $Z = A^* \otimes B \in E^{n \times m}$ is a solution to the linear system:

$$Z = A \otimes Z \oplus B \tag{16}$$

Proof. We obtain (15) with $Y = B \otimes A^*$ by left multiplying (11) by B and (16) with $Z = A^* \otimes B$, by right multiplying (11) by B. □

In the special case where $m = 1$, the above property shows how to construct, using A^*, solutions to systems of type (1) and (2) introduced in Sect. 1.

Let us consider now the case where $(E, \oplus, \otimes)$ is a dioid, i.e. where the preorder canonical relation $\leq$ is an order relation.

We have seen in Chap. 2, Sect. 3 that this order relation can be extended naturally to the vectors of E^n and to the matrices of $E^{m\times n}$.

The following property then establishes *minimality* in the sense of this order relation of the solutions constructed by means of the quasi-inverse.

Property 3.4.2. Let us assume that there exists $K \in N$ such that $A^* = A^{(K)} = A^{(K+1)} = \cdots$.

If $(E, \oplus, \otimes)$ is a dioid endowed with the canonical order relation $\leq$, then $Y = B \otimes A^*$ (resp. $Z = A^* \otimes B$) is the *minimal solution* in the set of solutions to (15) (resp. in the set of solutions to (16)) ordered by the order relation induced by $\leq$.

Proof. Let Y be an arbitrary solution of (15)

We can write:

$$\begin{aligned} Y &= B \oplus (B \oplus Y \otimes A) \otimes A \\ &= B \otimes (I \oplus A) \oplus Y \otimes A^2 \end{aligned}$$

and, generally, for any $k \in N$:

$$Y = B \otimes (I \oplus A \oplus \cdots \oplus A^{k-1}) \oplus Y \otimes A^k$$

We therefore have for $k \geq K$:

$$Y = B \otimes A^* \oplus Y \otimes A^k$$

which shows that $B \otimes A^* \leq Y$

Consequently $B \otimes A^*$ is the minimal solution in the set of solutions to (15).

We would prove, similarly, that $A^* \otimes B$ is the minimal solution to (16). □

One thus again finds, in a generalized form, the property of minimality already encountered in Sect. 2.4 in relation to the shortest path problem in a graph.

For example, taking for B the row-vector b^T (viewed as a $1 \times n$ matrix) with all components ε except the component of index i_o equal to e, it can be observed that $b^T \otimes A^*$, which is none other than the row of index i_o of A^*, is the minimal solution to the equation:

$$y = y \otimes A \oplus b^T$$

Thus, when E is a dioid, it is equivalent to compute A^* or to determine minimal solutions for each of the n linear system of type (1) obtained successively taking b equal to the n unitary vectors:

$$\begin{gathered} (e, \varepsilon, \varepsilon, \ldots \varepsilon)^T \\ (\varepsilon, e, \varepsilon \ldots .\varepsilon)^T \\ \vdots \\ (\varepsilon, \varepsilon, \ldots, e)^T \end{gathered}$$

4. Iterative Algorithms for Solving Linear Systems

For a given $A \in M_n(E)$, we have shown in Sect. 3 how solving linear systems of the type:

$$y = y \otimes A \oplus b^T \tag{1}$$

or

$$z = A \otimes z \oplus b \tag{2}$$

(and more generally linear matrix systems of type (15) or (16)) reduces to determining the quasi-inverse A^* of A.

The computation of A^* can be performed, depending on the case, rowwise, columnwise or globally.

4.1. Generalized Jacobi Algorithm

Let us assume, to set the ideas, that we wish to determine the first row of the A^* matrix. In other words, we are seeking $y = b^T \otimes A^*$ with $b^T = (e, \varepsilon, \varepsilon, \dots, \varepsilon)$.

From Sect. 3, we know that y solves the equation (of the "fixed point" type):

$$y = y \otimes A \oplus b^T \tag{1}$$

Algorithm 1 below can then be seen as the direct analogue to Jacobi's method, in classical linear algebra, to solve a system of type (1).

Algorithm 1 *(Generalized Jacobi)*
Determination of the first row of A^*, *or proof that* $A^{(K)} \neq A^*$

(a) set $y^o = b^T = (e, \varepsilon, \dots \varepsilon)$; $t \leftarrow 0$;
(b) At iteration t, let y^t *be the current solution. Compute*

$$y^{t+1} = y^t \otimes A \oplus b^T$$

If $y^{t+1} = y^t$, *the algorithm terminates and* $y^t = b^T \otimes A^*$ *(the first row of* A^**)*
If $y^{t+1} \neq y^t$ *and* $t \leq K$ *then set* $t \leftarrow t + 1$ *and return to (b).*
If $y^{t+1} \neq y^t$ *and* $t = K$ *then interrupt the computation:* $A^* \neq A^{(K)}$.

Theorem 4. *(i) If there exists an integer* K *such that* $A^* = A^{(K)}$, *then the generalized Jacobi algorithm constructs, in at most* K *iterations, the first row of* A^*. *In the opposite case, the algorithm proves in at most* K *iterations that* $A^* \neq A^{(K)}$.
(ii) If A^* *exists and if* $(E, \oplus, \otimes)$ *is a topological dioid, then the generalized Jacobi algorithm generates a nondecreasing and convergent sequence (in the sense of the sup-topology) towards the first row of* A^*.

Proof. Let us first prove (i), i.e. the case when finite convergence occurs.

We have $y^o = b^T$ hence $y^1 = b^T \otimes (I \oplus A)$

$y^2 = b^T \otimes (I \oplus A \oplus A^2)$ and, by induction, $\forall t \in N$:

$$y^t = b^T \otimes (I \oplus A \oplus A^2 \oplus \cdots \oplus A^t) = b^t \otimes A^{(t)}$$

which shows that the sequence of the solutions generated by the algorithm is monotone nondecreasing.

If there exists $K \in \mathbb{N}$ such that $A^{(K)} = A^*$ then this yields: $y^K = b^T \otimes A^* = y^{K+1} = \cdots$.

Consequently, if $y^{K+1} \neq y^K$, it is because $A^* \neq A^{(K)}$.

Let us now prove (ii), i.e. convergence in topological dioids.

According to the above we have, $\forall t$: $y^t = b^T \otimes A^{(t)}$. Since A^* exists, the sequence $A^{(t)}$ is nondecreasing and convergent towards A^*. Since taking the limit is compatible with the laws $\oplus$ and $\otimes$ (since we are dealing with a topological dioid, see Chap. 3 Sect. 6), we deduce that y^t converges to $b^T \otimes A^*$, the first row of A^*. □

In practice, it is often possible to obtain an upper bound on the number K such that $A^{(K)} = A^*$ with additional assumptions. For example:

- In the absence of a 0-absorbing circuit (see Sect. 3, Theorem 1) then $K \leq n - 1$;
- If the multiplication $\otimes$ is commutative and in the absence of a p-absorbing circuit (see Sect. 3, Theorem 2) then $K \leq n - 1 + pn\,t$, where t is the total number of elementary circuits of G(A);
- If the set F of the entries of A is p-nilpotent (see Sect. 3, Theorem 3) then $K \leq p$.

In the case where the algorithm terminates at iteration $t = K$ with $y^{K+1} \neq y^K$, we end up at a situation which is inconsistent with the hypotheses. The Jacobi algorithm can therefore also be considered as a means of algorithmically checking the relevance of the assumptions used.

The following result states the complexity of Algorithm 1.

Proposition 4.1.1. *Assuming that the complexity of each of the operations $\oplus$ and $\otimes$ is $\mathcal{O}(1)$ the generalized Jacobi algorithm has complexity $\mathcal{O}(Kn^2)$ if all the entries of the matrix* A *are distinct from* ε, *and $\mathcal{O}(KM)$ if* M *is the number of entries of* A *different from* ε.

Proof. Each iteration requires a matrix-vector product (n^2 operations $\oplus$ and $\otimes$) and a sum of two vectors (n operations $\oplus$). The complexity is therefore $\mathcal{O}(Kn^2)$. If the number of entries of A distinct from ε is $M \ll n^2$, this complexity can be reduced to $\mathcal{O}(K.M)$ by using a compact representation of A. □

In the case of the shortest path problem ($E = \mathbb{R} \cup \{+\infty\}, \oplus = \text{Min}, \otimes = +$), Algorithm 1 is recognized as Bellman's algorithm (1958).

4.2. Generalized Gauss–Seidel Algorithm

By analogy with the classical version of the Gauss–Seidel algorithm, the idea is to decompose matrix A by expressing it as the sum of a lower triangular matrix L and an upper triangular matrix U. To simplify the presentation, we will first assume that the diagonal elements of A are all equal to ε. We therefore have:

$$A = U \oplus L$$

The basic iteration of the generalized Jacobi algorithm (see Sect. 4.1) then takes the form:

$$y^{t+1} = b^T \oplus y^t \otimes U \oplus y^t \otimes L$$

Thus, for any component $j = 1, \dots n$, the computation of y_j^{t+1} is achieved via the relation:

$$y_j^{t+1} = b_j \oplus \sum_{i=1}^{j-1} y_i^t \otimes a_{ij} \oplus \sum_{i=j+1}^{n} y_i^t \otimes a_{ij}$$

During the computation of $y_j{}^{t+1}$, all the terms $y_1{}^{t+1}, y_2{}^{t+1}, \dots, y_{j-1}{}^{t+1}$ have already been determined. The idea of the Gauss–Seidel algorithm is then to replace the values $y_j{}^t$ in the first summation by the new values $y_j{}^{t+1}$ already determined, which leads to the recurrence:

$$y_j^{t+1} = b_j \oplus \sum_{i=1}^{j-1} y_i^{t+1} \otimes a_{ij} \oplus \sum_{i=j+1}^{n} y_i^t \otimes a_{ij}$$

or equivalently, in matrix notation:

$$y^{t+1} = b^T \oplus y^{t+1} \otimes U \oplus y^t \otimes L \tag{17}$$

This gives rise to the following algorithm:

Algorithm 2 *(Generalized Gauss–Seidel)*
Determination of the first row of A^* *or proof that* $A^{(K)} \neq A^*$

(a) Set $b^T = (e, \varepsilon, \cdots \varepsilon)^T$ *and* $y^o = (\varepsilon, \varepsilon, \varepsilon, \cdots \varepsilon)$; $t \leftarrow 0$;
(b) At iteration t, let y^t *be the current solution. Compute:* y^{t+1} *satisfying:*

$$y^{t+1} = b^T \oplus y^{t+1} \otimes U \oplus y^t \otimes L$$

If $y^{t+1} = y^t$, *the algorithm terminates and* $y^t = b^T \otimes A^*$ *(the first row of* A^**)*
If $y^{t+1} \neq y^t$ *and* $t \leq K$ *then, set* $t \leftarrow t + 1$ *and return to (b)*
If $y^{t+1} \neq y^t$ *and* $t = K$ *then terminate the computations:* $A^* \neq A^{(K)}$.

Theorem 5. *(i) If there exists* $K \in \mathbb{N}$ *such that* $A^{(K)} = A^*$, *then the generalized Gauss–Seidel algorithm constructs, in at most* K *iterations, the first row of* A^*. *In the opposite case, the algorithm proves, in at most* K *iterations, that* $A^* \neq A^{(K)}$.
(ii) If A^* *exists and if* $(E, \oplus, \otimes)$ *is a topological dioid, then the generalized Gauss–Seidel algorithm generates a nondecreasing and convergent (in the sense of the Sup-Topology) sequence towards the first row of* A^*.

Proof. Let us first prove (i), i.e. the case where finite convergence occurs.

In order to do so, let us consider $\hat{y}^t$ to be the solution obtained from $\hat{y}^0 = (\varepsilon, \varepsilon, \dots \varepsilon)$ at the tth iteration of Jacobi's method, i.e. by:

$$\hat{y}_j^{t+1} = b_j \oplus \sum_{i=1}^{j-1} \hat{y}_i^t \otimes a_{ij} \oplus \sum_{i=j+1}^{n} \hat{y}_i^t \otimes a_{ij}$$

Let us recall that, by construction $\hat{y}^{t+1} \geq \hat{y}^t$.

Let us prove then by induction that, $\forall t$: and $\forall j = 1, \dots, n$: $\hat{y}_j^t \leq y^t$

This relation is true for $t = 0$. Let us therefore assume it to be true for an arbitrary $t \in \mathbb{N}$ and assume that for an arbitrary given $j (2 \leq j \leq n)$, for any $i \leq j - 1$, we have ${y_i}^{t+1} \geq \hat{y}_i^{t+1}$.

Then we can write, $\forall j = 1, \dots, n$:

$$\begin{aligned}
y_j^{t+1} &= b_j \oplus \sum_{i=1}^{j-1} y_i^{t+1} \otimes a_{ij} \oplus \sum_{i=j+1}^{n} y_i^t \otimes a_{ij} \\
&\geq b_j \oplus \sum_{i=1}^{j-1} \hat{y}_i^{t+1} \otimes a_{ij} \oplus \sum_{i=j+1}^{n} \hat{y}_i^t \otimes a_{ij} \\
&\geq b_j \oplus \sum_{i=1}^{j-1} \hat{y}_i^t \otimes a_{ij} \oplus \sum_{i=j+1}^{n} \hat{y}_i^t \otimes a_{ij}
\end{aligned}$$

hence the we deduce ${y_j}^{t+1} \geq \hat{y}_j^{t+1}$. By induction on j we therefore deduce $y^{t+1} \geq \hat{y}^{t+1}$ which proves the property for $t + 1$.

Therefore, in the case where the Jacobi algorithm converges finitely, the generalized Gauss–Seidel algorithm cannot require more iterations. We deduce (i).

Let us now prove (ii), i.e. convergence in topological dioids. The matrix U being upper triangular, it is n-nilpotent, i.e. $U^n = U^{n+1} = U^{n+2} = \cdots = \Sigma$ ($n \times n$ matrix where all the entries are equal to ε). Then, the quasi-inverse U^* of U exists, and is equal to $U^{(n-1)}$.

Moreover, because of the form of the iteration (17), we observe that y^{t+1} is a minimal solution in y of the equation:

$$y = y \otimes U \oplus b^T \oplus y^t \otimes L$$

We therefore have:

$$y^{t+1} = (b^T \oplus y^t \otimes L) \otimes U^* \tag{18}$$

which provides the expression of y^{t+1} in terms of y^t only.

One can then interpret (18) as an iteration of the Jacobi type applied to the equation:

$$y = b^T \otimes U^* \oplus y \otimes (L \otimes U^*) \tag{19}$$

By using the proof of Theorem 4, and since $y^{\circ} = (\varepsilon, \varepsilon, \ldots, \varepsilon)$, we can write, $\forall t$:

$$y^t = (b^T \otimes U^*) \otimes [I \oplus (L \otimes U^*) \oplus (L \otimes U^*)^2 \oplus \cdots \oplus (L \otimes U^*)^t]$$

Since $A^* = (U \oplus L)^*$ exists, according to Proposition 6.2.5 of Chap. 3, $(L \otimes U^*)^*$ exists and this yields:

$$(U \oplus L)^* = U^* \otimes (L \otimes U^*)^*.$$

The sequence y^t generated by the generalized Gauss–Seidel algorithm is therefore nondecreasing and bounded from above with limit:

$$\begin{aligned} &b^T \otimes U^* \otimes (L \otimes U^*)^* \\ &= b^T \otimes (U \oplus L)^* = b^T \otimes A^* \end{aligned}$$

which is recognized as the minimal solution to the equation $y = y \otimes A \oplus b^T$. □

Applied to the shortest path problem in a graph, Algorithm 2 is none other than Ford's algorithm (1956). In this case, we have finite convergence.

Remark. In the case where the diagonal elements of A are not all equal to ε but are quasi-invertible, the computation of y^{t+1} from y^t at each iteration must be modified, and be carried out according to the following procedure:

for $j = 1, 2, \ldots, n$

$$y_j^{t+1} = \left(b_j \oplus \sum_{i=1}^{j-1} y_i^{t+1} \otimes a_{ij} \oplus \sum_{i=j+1}^{n} y_i^t \otimes a_{ij} \right) \otimes a_{jj}^*$$

(where a_{jj}^* denotes the quasi inverse of the diagonal term a_{jj}). ||

4.3. Generalized Dijkstra Algorithm ("Greedy Algorithm") in Some Selective Dioids

We are going to show now that one can obtain an algorithm generally more efficient than those described in the previous paragraphs by restricting to a special class of dioids.

We will thus assume, throughout this section, that $(E, \oplus, \otimes)$ is a *selective dioid* in which e (the neutral element of $\otimes$) is the largest element (in the sense of the order relation of the dioid), in other words: $\forall a \in E$: $e \oplus a = e$. The order relation being compatible with multiplication, we therefore have, in such a dioid:

$$\forall a \in E, b \geq c \Rightarrow b \otimes e \geq c \otimes a \Rightarrow b \geq c \otimes a \quad (20)$$

$(\mathbb{R}_+ \cup \{+\infty\}, \text{Min}, +), ([0, 1], \text{Max}, \times) (\{0, 1\}, \text{Max}, \text{Min}), (\mathbb{R}_+ \cup \{+\infty\}, \text{Max}, \text{Min})$ are examples of selective dioids in which e is the largest element.

Let us recall that, in a dioid where e is the largest element, any element of E is 0-stable. Any matrix $A \in M_n(E)$ is therefore quasi-invertible since G(A) then has no 0-absorbing circuit (see Theorem 1, Sect. 3.3).

Algorithm 3, described hereafter, uses the following result.

Theorem 6. *(Gondran, 1975b)*

Let $(E, \oplus, \otimes)$ *be a selective dioid in which e is the largest element.*

Given $A \in M_n(E)$*, let us consider the linear system:*

$$z = z \otimes A \oplus b \tag{21}$$

where b *is a given row-vector with n components and let* $\bar{z} = b \otimes A^*$ *be the minimal solution to this system. Then there exists* $i_o \in [1, n]$ *such that* $\bar{z}_{i_o} = b_{i_o}$ *and this index* i_o *satisfies:*

$$b_{i_o} = \sum_{i=1}^{n} b_i \tag{22}$$

Proof. (a) Let i_o be defined as satisfying (22). Let us show first of all that there exists a solution $\hat{z}$ to (21) such that $\hat{z}_{i_o} = b_{i_o}$

By setting $z_{i_o} = b_{i_o}$ the equations (21) are written, for $i \neq i_o$:

$$z_i = \sum_{\substack{j=1 \\ j \neq i_o}}^{n} z_j \otimes a_{ji} \oplus \left(b_i \oplus b_{io} \otimes a_{i_o i}\right) \tag{23}$$

and for $i = i_o$:

$$z_{i_o} = \sum_{j=1}^{n} z_j \otimes a_{ji_o} \oplus b_{i_o} \tag{24}$$

By denoting $\tilde{A}$ the matrix deduced from A by deleting the row i_o and the column i_o, $\tilde{z}$ the vector deduced from z by deleting the component i_o, the relations (23) are written in the form of the system:

$$\tilde{z} = \tilde{z} \otimes \tilde{A} \oplus \tilde{b} \tag{25}$$

where $\tilde{b}$ is the vector with components $(b_i \oplus b_{i_o} \otimes a_{i_o i})$ for $i \neq i_o$. This system has $\tilde{z} = \tilde{b} \otimes \tilde{A}^*$ as minimal solution. Then let $\hat{z}$ be the n- vector such that $\hat{z}_{i_o} = b_{i_o}$ and where the components $\hat{z}_i (i \neq i_o)$ are those of $\tilde{z} = \tilde{b} \otimes \tilde{A}^*$.

By construction, $\hat{z}$ satisfies (23). Let us show that it also satisfies (24).

In order to do so, and due to the idempotency of $\oplus$, it is enough to show that:

$$\forall j \neq i_o: \quad b_{i_o} \geq \hat{z}_j \otimes a_{ji_o} \tag{26}$$

According to (22): $b_{i_o} \geq b_i$, hence, it follows from (20): $b_{i_o} \geq b_i \otimes \left(\tilde{A}^*\right)_{ij}$.

Still using (20), one can also write:

$$b_{i_o} \geq b_{i_o} \otimes a_{i_o i} \otimes \left(\tilde{A}^*\right)_{ij}$$

Then, $\hat{z}_j$ being defined, $\forall j \neq i_o$, by the relation:

$$\hat{z}_j = \sum_{\substack{i=1 \\ i \neq i_o}}^{n} (b_i \oplus b_{i_o} \otimes a_{i_o i}) \otimes \left(\tilde{A}^*\right)_{ij}$$

we deduce from the above: $b_{i_o} \geq \hat{z}_j$
which, by (20), implies (26). Thus, $\hat{z}$ defined above solves (21).

(b) It remains to show that $\hat{z}$ is the minimal solution to (21).
For any solution z to (21) we have (24) therefore $z_{i_o} \geq b_{i_o} = \hat{z}_{i_o}$.
Moreover, $\tilde{\tilde{z}}$ (deduced from z by deleting the component i_o) solves

$$\tilde{\tilde{z}} = \tilde{\tilde{z}} \otimes \tilde{A} \oplus \tilde{\tilde{b}} \tag{27}$$

where $\tilde{\tilde{b}}$ is the vector with components $(b_i \oplus z_{i_o} \otimes a_{i_o i})$ for $i \neq i_o$.
Since $\tilde{\tilde{b}} \geq \tilde{z}$, and since $\tilde{\tilde{b}} \otimes \tilde{A}^*$ is the minimal solution to (27) this yields:

$$\tilde{\tilde{z}} \geq \tilde{\tilde{b}} \otimes \tilde{z}^* \geq \tilde{b} \otimes \tilde{A}^* = \tilde{z}$$

We deduce $z \geq \hat{z}$, i.e. the minimality of $\hat{z}$. □

The previous result can be used to determine the matrix A^* row by row. For example, to determine the first row of A^*, we will choose $b = (e, \varepsilon, \varepsilon, \ldots, \varepsilon)$. The computation of $\bar{z} = \left(A^*_{1,1}, A^*_{1,2} \cdots .A^*_{1,n}\right)$ can then be carried out step by step, as follows.

We determine an index $i_o \in [1, n]$ such that:

$$b_{i_o} = \sum_{i=1}^{n} b_i \quad \text{(sum to be understood in the sense of } \oplus\text{)}.$$

By Theorem 6, we deduce the component i_o of the desired solution: $\bar{z}_{i_o} = b_{i_o}$.

With this information, one can then reduce the problem to the determination of the minimal solution to the reduced linear system:

$$\tilde{z} = \tilde{z} \otimes \tilde{A} \oplus \tilde{b} \tag{25}$$

with, $\forall i \neq i_o$: $\tilde{b}_i = b_i \oplus b_{i_o} \otimes a_{i_o i}$, and where $\tilde{A}$ is the matrix deduced from A by deletion of row i_o and column i_o.

By again applying Theorem 6 to the latter problem, we determine an index $i_1 \in [1, n] \backslash \{i_o\}$ satisfying:

$$\tilde{b}_{i_1} = \sum_{i \in [1,n] \backslash \{i_o\}} \tilde{b}_i$$

and we obtain $\bar{z}_{i_1} = \tilde{b}_{i_1}$, and so on.

At each step, we obtain the value of a new component of the solution $\bar{z}$ and the dimension of the problem is decreased by one unit. After n steps, the solution vector is therefore obtained.

One is thus lead to the following algorithm:

Algorithm 3 *(generalized Dijkstra)*
Determination of a row of A^* *corresponding a chosen index* $r (r \in [1, n])$

(a) Initialization:

$$\text{Set:} \begin{cases} \pi(r) \leftarrow e \\ \pi(i) \leftarrow \varepsilon \text{ for } i \in [1, n]\backslash\{r\} \end{cases}$$
$$T = \{1, 2, \ldots, n\}$$

(b) Current step:

(b_1) *Determine an index* $i \in T$ *satisfying:*

$$\pi(i) = \sum_{j \in T} \pi(j) \quad \textit{(sum to be understood in the sense of } \oplus)$$

then set: $T \leftarrow T\backslash\{i\}$
If $T = \emptyset$*, end of algorithm: the vector* $(\pi(1), \pi(2), \ldots \pi(n))$ *is the row* r *of* A^*.
If $T \neq \emptyset$ go to (b_2).

(b_2) *For all* $j \in T$ *set:*
$$\pi(j) \leftarrow \pi(j) \oplus \pi(i) \otimes a_{ij}$$
and return to (b).

The following result states the complexity of Algorithm 3.

Proposition 4.3.1. *Algorithm 3 requires* $\mathcal{O}(n^2)$ *operations* $\oplus$ *and* $\mathcal{O}(n^2)$ *operations* $\otimes$.

Proof. For iteration k where $|T| = n - k + 1$, the computation of the index i in step (b_1) requires $n - k + 1$ operations $\oplus$ and the updating of the $\pi(j)$ values in step (b_2) requires $n - k$ operations $\oplus$ and $n - k$ operations $\otimes$. The result is deduced by summing up for k from 1 to n. □

In the special case of the shortest path problem in a graph with all nonnegative lengths, we find again DIJKSTRA's classical algorithm (1959).

4.4. Extensions of Iterative Algorithms to Algebras of Endomorphisms

We are going to show in this section that the iterative algorithms of Sects. 4.1 and 4.2 as well as the generalized Dijkstra Algorithm of Sect. 4.3 ("greedy" algorithm) can (under appropriate assumptions to be specified) be generalized for the "point wise" computation of the quasi-inverse Φ^* of a matrix $\Phi = (\varphi_{ij})_{\substack{i=1,\ldots,n \\ j=1,\ldots,n}}$ where the entries are endomorphisms of a monoid (S, ⊕). As will be seen, this extension opens up the way to solving new problems, among which a huge variety of "non classical" path-finding problems in graphs. A typical example concerns the shortest path problem *with time dependent lengths on the arcs*, which can be stated as follows. With each arc (i, j) of a graph G we associate a function φ_{ij} giving the time $t_j = \varphi_{ij}(t_i)$ of arrival in j when starting at i at the instant t_i.

Starting from vertex 1 at a given instant t_1, we want to determine a path reaching vertex i in minimum time. Without special assumptions on the form of the functions φ_{ij}, such a problem does not fit into the framework of the classical model introduced in Sect. 2. If one analyzes the causes of the difficulty, it can be observed that, in the classical shortest path problem, the information attached to the vertices and the information attached to the arcs of the graph are of the same nature: these are elements of a same basic set, the dioid ($\hat{\mathbb{R}}$, Min, +). On the contrary, in the shortest path with time dependent lengths on the arcs, the information attached to the vertices $i \in X$ are times, i.e. real numbers, whereas the information attached to the arcs are *functions*: $\mathbb{R} \to \mathbb{R}$.

Starting then from the idea that a good algebraic model for this type of problem must take this distinction into account, Minoux (1976, 1977) introduced very general algebraic structures which will be called *algebras of endomorphisms*.

Clearly, when such an algebra has the properties of a semiring (resp. of a dioid), we will refer to of a *semiring of endomorphisms* (resp. to a dioid of endomorphisms).

4.4.1. Endomorphism Algebra – Definition

An algebra of endomorphisms (of a monoid) is defined as a given quadruple: (S, H, ⊕, ∗), where:

- S is the *ground set*, endowed with an internal law ⊕ which induces a commutative monoid structure with neutral element ε (in the case of a path-finding problem, its elements correspond to the information attached to the vertices of the graph).
- H is the set of mappings: S → S, satisfying:

$$h(a \oplus b) = h(a) \oplus h(b) \quad \forall h \in H, a \in S, b \in S.$$
$$h(\varepsilon) = \varepsilon \quad \forall h \in H.$$

  H is therefore the set of endomorphisms of (S, ⊕) satisfying $h(\varepsilon) = \varepsilon$. The unit endomorphism will be denoted e: $e(a) = a, \forall a \in S$.
- The ⊕ law on S induces on H an operation also denoted ⊕, defined as:

$$(h \oplus g)(a) = h(a) \oplus g(a) \quad \forall h \in H, g \in H, a \in S.$$

  Observe that ⊕ is an internal law on H. The neutral element of ⊕ in H is the endomorphism denoted h^ε which, with any $a \in S$, associates $\varepsilon \in S$; we clearly have:

$$h^\varepsilon \oplus h = h \oplus h^\varepsilon = h \quad \forall h \in H.$$

- H is moreover endowed with a second law, denoted ∗ defined as:

$$h \in H, g \in H, h * g = g \circ h$$

  where ∘ is the classical product mapping.

$*$ is therefore an internal, *associative* law on H which has the unit endomorphism e as neutral element.

Furthermore, $*$ is right and left *distributive* with respect to the $\oplus$ law. It has h^{ε} as absorbing element because:

$$(g * h^{\varepsilon})(a) = h^{\varepsilon}(g(a)) = \varepsilon \qquad \forall a \in S, \quad \forall g \in H.$$
$$(h^{\varepsilon} * g)(a) = g[h^{\varepsilon}(a)] = g(\varepsilon) = \varepsilon \quad \forall a \in S, \quad \forall g \in H.$$

We deduce that $(H, \oplus, *)$ is a semiring.

4.4.2. Quasi-Inverse of a Matrix of Endomorphisms

Given a semiring of endomorphisms $(S, H, \oplus, *)$, let us consider now a $n \times n$ matrix: $\Phi = (\varphi_{ij})_{\substack{i=1,\dots,n \\ j=1,\dots,n}}$ with elements in H.

We now focus on the existence and computation of the *quasi-inverse* Φ^*. Since Φ is a matrix of endomorphisms, we will have to make it clear in what sense one can speak of the computation of Φ^*.

Let us first of all address the question of *existence*.

If it is assumed, for example, that all the endomorphisms φ_{ij} corresponding to the entries of Φ satisfy:

$$\forall a \in S: \quad a \oplus \varphi_{ij}(a) = a$$

in other words that, $\forall i, j$:

$$e \oplus \varphi_{ij} = e$$

it is then easy to see that the graph G associated with Φ has no 0-absorbing circuit, which implies the existence of the matrix $\Phi^* = (\varphi^*_{ij})_{\substack{i=1,\dots,n \\ j=1,\dots,n}}$ *quasi-inverse of* Φ (see Theorem 1 Sect. 3.3).

Let us observe, however, that, as opposed to the classical case dealt with above, the fact that Φ^* exists does not necessarily guarantee that the computation of Φ^* can be performed explicitly and efficiently. Indeed, the entries φ^*_{ij} are endomorphisms, therefore mappings: $S \to S$, which, apart from special cases, can only be exactly known through the set of images of *all the* elements of S. Thus, from the point of view of *computation* and information storage this poses a problem whenever (and this is the most frequent case) $|S| = +\infty$.

Moreover, in many applications:

- Either each endomorphism φ_{ij} (each entry of the Φ matrix) is only known by its effect on each element of S;
- Or the product of two endomorphisms $\varphi_{ij} * \varphi_{jk}$ is known only by its effect on each element of S.

Indeed, the algorithms to be described in Sect. 4.4.4 will be limited to efficiently computing each endomorphism φ^*_{ij} *point-wise*, i.e. they will compute $\varphi^*_{ij}(a)$ for a given arbitrary $a \in S$.

We first provide below a few typical examples of problems that can be modeled as the point-wise computation of the elements of Φ^*, quasi-inverse of a matrix of endomorphisms Φ.

4.4.3. Some Examples

Example 1. Shortest path with time dependent lengths on the arcs (see Chap. 1, Sect. 6.2.)

We will take here $S = \hat{\mathbb{R}}, \oplus = \text{Min}, \varepsilon = +\infty$. The set H will be the set of endomorphisms h of S satisfying $h(+\infty) = +\infty$. Since here $\oplus = \text{Min}$, the condition of endomorphism amounts to:

$$h(\text{Min}\{t, t'\}) = \text{Min}\{h(t), h(t')\}$$

which is equivalent to requiring that the functions h: $\mathbb{R} \to \mathbb{R}$ be nondecreasing.

With each arc (i, j) of the graph we associate $\varphi_{ij} \in H$ having the following meaning:

$t_j = \varphi_{ij}(t_i)$ is the instant of arrival at the vertex j when starting from i at the instant t_i through arc (i, j).

If one starts from vertex i_0 chosen as the origin at the instant t_0, one wishes to determine, for every vertex $j \neq i_0$, an optimum path in order to arrive in j as early as possible. The aim therefore is to determine all the values $\varphi^*_{i_o,j}(t_0)$ which correspond to the row i_0 of the quasi-inverse Φ^*.

Example 2. Shortest path with discounting (see Chap. 1, Sect. 6.2.)

With each arc (i, j) of a graph G, we associate a length which depends, in a path, on the number of arcs previously taken. If we interpret, for example, the transversal of the arc (i, j) as corresponding to an annual investment program, the cost of the arc (i, j) is $c_{ij}/(1+\tau)^t$ if t is the number of arcs previously taken by the path, that is to say, the year of the expenditure c_{ij} (τ is a given parameter referred to as the *discount rate*). For every arc (i, j) the value C_{ij}, assumed to be given, therefore represents the amount of the investment corresponding to the arc (i, j) assuming that this is the first arc of the path to be determined.

We seek the shortest path with discounting from a given vertex 1 to each of the other vertices.

If T is the number of time intervals under consideration, we will take for S the set of (T + 1)-vectors with components in $\mathbb{R}_+ \cup \{+\infty\}$. If $a = (a_0, a_1, \ldots, a_T)$ and $b = (b_0, b_1, \ldots, b_T)$, we define $d = a \oplus b = (d_0, d_1, \ldots, d_T)$ by setting $d_t = \min\,(a_t, b_t)$, for all t from 0 to T. $\varepsilon = (+\infty, \ldots, +\infty)$. Then we define the endomorphism φ_{ij} as:

$$\varphi_{ij}(a) = b$$

$$\text{with} \quad \begin{cases} b_o = +\infty \\ b_t = a_{t-1} + \dfrac{c_{ij}}{(1+\tau)^{t-1}} \end{cases} \quad (t = 1, \ldots, T)$$

We observe that such endomorphisms are T-nilpotent (see Sect. 3.3) and the matrix Φ^* therefore exists (as a consequence of Theorem 3, Sect. 3.3).

The initial state of the vertex 1 being taken as equal to $\alpha = (0, +\infty, +\infty, \dots, +\infty)$, the problem is reduced to the determination of $\varphi^*_{1j}(\alpha)$ for $j = 2, 3, \dots, n$.

The shortest path with discounting between 1 and j has the value

$$\underset{t=0,\dots,T}{\text{Min}} \left\{ \left[\varphi^*_{1,j}(\alpha) \right]_t \right\}$$

Example 3. Shortest path with time constraints

With each arc (i, j) of a graph G, we associate:

- a duration $d_{ij} > 0$ measuring the transit time on arc (i, j),
- a set of intervals $V_{ij} \subset [0, +\infty[$ representing the set of instants of possible departure from vertex i towards vertex j using arc (i, j).

With each vertex i, we associate a set of intervals $W_i \subset [0, +\infty[$ representing the set of instants where parking is authorized in vertex i.

The problem is to determine the shortest path joining two specified vertices x and y (in the sense of the transit time) compatible with the time constraints specified by the V_{ij} (on the arcs) and the W_i (on the vertices).

The information associated with a vertex i is the set E_i of the instants of possible arrival in i from the origin x. An element of S will therefore be a set of intervals $\subset [0, +\infty[$. We define on S the operation $\oplus$ (union of two sets of intervals) by:

$$a \oplus b = \{t/t \in a \text{ or } t \in b\} \quad \forall a, b \in S.$$

The empty set Ø is the neutral element of $\oplus$. To define the endomorphisms φ_{ij}, we define the *transition* between i and j in several steps:

- If E_i corresponds to the set of instants of possible arrival in i, then the set D_i of the instants of possible departure from i will be:

$$D_i = E_i \perp W_i$$

where the operation $\perp$ is defined as follows:

$$\text{If } E_i = \left\{ [\alpha_1, \alpha'_1], [\alpha_2, \alpha'_2], \dots, [\alpha_p, \alpha'_p] \right\}$$
$$W_i = \left\{ [\beta_1, \beta'_1], [\beta_2, \beta'_2], \dots, [\beta_q, \beta'_q] \right\}$$

then:

$$D_i = [\gamma_1, \gamma'_1] \oplus [\gamma_2, \gamma'_2] \oplus [\gamma_p, \gamma'_p]$$

where, for k from 1 to p:

$$[\gamma_k, \gamma'_k] = \begin{cases} [\alpha_k, \alpha'_k] \text{ if } \alpha'_k \notin W_i \\ [\alpha_k, \beta'_j] \text{ if } \alpha'_k \in [\beta_j, \beta'_j] \quad \text{for some } j; \end{cases}$$

– The set of instants of possible departure from i towards j using arc (i, j) will be then:

$$D_i \cap V_{ij}$$

– Let us define on S an external operation $\top$ (translation) as:

$$a \in S, \quad \tau \in \mathbb{R}_+, \quad \tau \top a = \{t + \tau / t \in a\}.$$

The set of instants of possible arrival in j from i through arc (i, j) will therefore be:

$$d_{ij} \top (D_i \cap V_{ij}).$$

Finally we define φ_{ij} as:

$$\varphi_{ij}(E_i) = d_{ij} \top [(E_i \perp W_i) \cap V_{ij}].$$

It is easily checked that φ_{ij} is an endomorphism of $(S, \oplus)$.

We observe that the endomorphism φ_{ij} is entirely determined by the triple (W_i, V_{ij}, d_{ij}) but the product of two such endomorphisms will not generally correspond to such a triple.

One can then show that for some $p \in \mathbb{N}$ deduced from the problem data, the set of endomorphisms φ_{ij} is a p-nilpotent set (see Exercise 1 at the end of the chapter), which implies the existence of A^* (Theorem 3, Sect. 3.3).

The minimum time to reach y will be then the minimum element of $E_y = \varphi^*_{x,y}(E_x)$.

4.4.4. A Few Solution Algorithms (*Minoux, 1976*)

We now turn to show how the classical Jacobi, Gauss–Seidel (see Sects. 4.1 and 4.2) and Greedy algorithms (Sect. 4.3) can be generalized to the problem of (point-wise) computation of the quasi-inverse Φ^* of a matrix of endomorphisms Φ. Here we will focus on paths of origin 1 for example, i.e. on the (point-wise) computation of the first row of Φ^*.

The generalization of the Jacobi algorithm readily leads to the following.

Algorithm 1′ *(Generalized Jacobi algorithm)*

(a) Initialization of the states of the various vertices. E_1^0 *is the initial state of vertex* 1.

$$E_j^0 \leftarrow \varepsilon, \forall j = 2, \ldots, n. \quad k \leftarrow 0.$$

(b) Repeat (current iteration)

$$k \leftarrow k + 1;$$

$$\begin{cases} E_1^k = \sum_{i=1}^{n} \varphi_{i1}\left(E_i^{k-1}\right) \oplus E_1^0 \\ E_j^k = \sum_{i=1}^{n} \varphi_{ij}(E_i^{k-1}) \end{cases}$$

While ($\exists$ *i such that:* $E_i^k \neq E_i^{k-1}$)

(c) *When the iterations terminate, the current state of each vertex* i *is the desired result:* $E_i = \varphi^*_{1,i}(E^0_1)$

If we have $\Phi^{(K)} = \Phi^*$ (Theorem 1, 2 or 3), then the previous algorithm converges in at most K steps.

When $\oplus$ is idempotent, the previous algorithm can be improved by transforming it into an algorithm of the Gauss–Seidel type to obtain Algorithm 2′ below.

Algorithm 2′ *(Generalized Gauss–Seidel algorithm)*
(Case where $\oplus$ is idempotent)

(a) *Initialization of the states of the various vertices:* $E_1 = E^0_1$ *initial state of vertex* 1.

$$E_i \leftarrow \varepsilon \quad \text{for} \quad i = 2, \dots, n; \quad k \leftarrow 0.$$

Test ← FALSE.

(b) *Repeat (current iteration)*
  k ← k + 1; Test ← FALSE;
  For (any arc $u = (i, j) \in U$ *such that* $E_i \neq \varepsilon$*) proceed as follows:*
    Compute $E_j' = E_j \oplus \varphi_{ij}(E_i)$*;*
    If $(E_j' \neq E_j)$ *then*
      Test ← TRUE;
    Endif
    $E_j \leftarrow E_j'$;
  Endfor
While (Test = TRUE)

(c) *When the iterations terminate, the state of each vertex i is the desired result* $E_i = \varphi^*_{1,i}(E^0_1)$.

Just as in algorithm 1′, if $A^{(K)} = A^*$, then algorithm 2′ converges in at most K iterations.

We are now going to make an improvement to algorithm 2′, still in the case where $\oplus$ is idempotent. A vertex i will be referred to as labelled at iteration k if its state E_i was modified by this iteration. A vertex i will be referred to as examined at iteration k if for all the vertices $j \in \Gamma_i$, we have carried out the transformations:

$$\begin{cases} E_j' \leftarrow E_j \oplus \varphi_{ij}(E_i) \\ E_j \leftarrow E_j' \end{cases}$$

We then observe that it is unnecessary in algorithm 2′ to examine, at iteration k, a vertex i examined at some iteration $\ell \leq k$ and not labelled since then.

Indeed, at iteration ℓ, we calculated:

$$E_j^2 = E_j^1 \oplus \varphi_{ij}(E_i^1)$$

for all $j \in \Gamma_i$, and thus at iteration k, the state of j can be written in full generality:

$$E_j^3 = E_j^1 \oplus \varphi_{ij}(E_i^1) \oplus F_i$$

(F_i is the contribution of the new labellings which have taken place between iteration l and iteration k from vertices other than i).

Since, in iteration k, the state of i is still $E_i{}^1$, the labelling of j will give:

$$E_j^4 = E_j^3 \oplus \varphi_{ij}(E_i^1) = E_j^1 \oplus \varphi_{ij}(E_i^1) \oplus F_j \oplus \varphi_{ij}(E_i^1) = E_j^3$$

since $\oplus$ is idempotent. This is true $\forall j \in \Gamma_i$ and the property thus follows.

Taking this remark into account, and denoting at the current step, the set of labelled vertices by X_2, we obtain the generalized Moore algorithm.

Algorithm 2″ *(Generalized Moore algorithm)*
(Case where $\oplus$ is idempotent)

(a) Initialization: $E_1 \leftarrow E_1^0$
$E_i \leftarrow \varepsilon$ for $i = 2, \dots n$
$X_2 \leftarrow \{1\}$; $k \leftarrow 0$;
(b) Repeat (current iteration)
$k \leftarrow k + 1$;
Select an arbitrary vertex $i \in X_2$.
Set: $X_2 \leftarrow X_2 \backslash \{i\}$;
For (j *running through* Γ_i*) do:*
compute $E_j' = E_j \oplus \varphi_{ij}(E_i)$
If ($E_j' \neq E_j$), *then:*
$X_2 \leftarrow X_2 \cup \{j\}$
Endif
$E_j \leftarrow E_j'$
Endfor
While ($X_2 \neq \emptyset$)
(c) When the iterations terminate, the state of each vertex i *is the desired result:* $E_i = \varphi_{1,i}^* (E_1^0)$.

Let us now assume that there exists on S a total preorder relation[1] denoted $\propto$, compatible with $\oplus$, and such that:

$$\varphi_{ij}(a) \propto a \quad \forall i, j \tag{26}$$

We observe that a relation of this kind is not necessarily antisymmetrical: $a \propto b$ and $b \propto a$ does not necessarily imply $a = b$.

In the case of the shortest path with discounting (see Sect. 4.4.3 above), the preorder relation will be defined as:

$a = (a_0, a_1, \dots, a_T) \propto b = (b_0, b_1, \dots, b_T)$ if and only if: $\underset{i=0,\dots,T}{\operatorname{Min}} \{a_i\} \leq \underset{i=0,\dots,T}{\operatorname{Min}} \{b_i\}$

[1] Let us recall that a total preorder relation is a binary relation $\propto$ reflexive ($a \propto a$), transitive ($a \propto b, b \propto c \Rightarrow a \propto c$) and such that: $\forall a, b \in S \Rightarrow a \propto b$ or $b \propto a$. It is compatible with $\oplus$ if, $\forall a, b, c \in S$, we have: $a \propto b \Rightarrow a \oplus c \propto b \oplus c$.

In the case of the shortest path with time constraints (see Sect. 4.4.3 above), the preorder relation will be:

$$\mathrm{E} \propto \mathrm{F} \Leftrightarrow \min\{t/t \in \mathrm{E}\} \leq \min\{t/t \in \mathrm{F}\}.$$

It is clear that these two relations are not antisymmetrical.

In the case of the shortest path with time dependent lengths on the arcs, the preorder relation will be the usual order relation defined on $\mathbb{R}$.

Now, the total preorder relation together with (26) will be used to remove the indetermination existing with regard to which vertex of X_2 should be selected in algorithm 2″. This leads to the following algorithm which generalizes Algorithm 3 of Sect. 4.3 ("greedy" algorithm).

Algorithm 3′ *(Generalized Dijkstra Algorithm)*

(Case where $\oplus$ is idempotent and where there exists a total preorder relation $\propto$ on S).

(a) *Initialization:* $E_1 \leftarrow E_1^0$,
  $E_i \leftarrow \varepsilon$ for $i = 2, \ldots, n$.
  $X_1 \leftarrow \emptyset$; $X_2 \leftarrow \{1\}$; $k \leftarrow 0$.
(b) *Repeat (current iteration)*
    $k \leftarrow k + 1$;
    Select $r \in X_2$ *such that:*
    $E_r \propto E_i, \forall i \in X_2$.
    If $(r \notin X_1)$ *then*
      $X_1 \leftarrow X_1 \cup \{r\}$
      $X_2 \leftarrow X_2 \backslash \{r\}$
    Endif
    If $(X_1 \neq X)$ *then*
      For (j running through Γ_r*) do:*
        Compute $E_j' \neq E_j \oplus \varphi_{rj}\ (E_r)$
        If $(E_j' \neq E_j)$ *then:*
          $X_2 \leftarrow X_2 \cup \{j\}$
        Endif
        $E_j \leftarrow E_j'$
      Endfor
    Endif
  While $((X_2 \neq \emptyset)$ *or* $(X_1 \neq X))$
(c) *When the iterations terminate, the state of each vertex i is the desired result:*
  $E_i = \varphi_{1,i}^*(E_1^0)$.

The finite convergence of Algorithm 3′ follows from that of Algorithm 2″. Indeed, it only differs from the previous one by:

- A selection rule for the node to be examined;
- An additional stopping criterion.

Clearly, the selection rule of r does not influence the convergence. Moreover, an additional stopping criterion can only reduce the number of iterations; hence Algorithm 3′ will be preferred whenever the required properties are present.

To show that Algorithm 3′ indeed achieves a minimal label E_j for every vertex j, it is enough to check that E_r is minimal in S when r is transferred for the first time to X_1 (see proof in Exercise 2).

In the special case where $\propto$ is a total order relation (this is the case for example for the shortest path problem with time dependent lengths on the arcs), it will always hold true that $X_1 \cap X_2 = \emptyset$ (see proof in Exercise 2).

Each vertex is then examined once at step (b) and the maximum time taken by Algorithm 3′ is then $\mathcal{O}$ (N^2) assuming that each computation of $E_j \oplus \varphi_{ij}$ (E_i) takes place in time $\mathcal{O}$ (1). Algorithm 3′ then appears as a generalization of Dijkstra's Algorithm or of algorithm 3 of Sect. 4.3 (greedy algorithm).

5. Direct Algorithms: Generalized Gauss–Jordan Method and Variations

In this section, we generalize the classical computation of the inverse of a matrix via Gaussian elimination to the computation of the quasi-inverse of a matrix $A \in M_n(E)$ in a semiring or in a dioid, when this quasi-inverse exists.

We know then (see Sect. 3.4) that the quasi-inverse A^* satisfies the linear system (11) and that, moreover, if (E, ⊕, ⊗) is a dioid, A^* is the minimal solution to the matrix systems:

$$Y = Y \otimes A \oplus I \tag{28}$$

and

$$Z = A \otimes Z \oplus I \tag{29}$$

where $Y = (y_{ij}) \in M_n(E)$ and $Z = (z_{ij}) \in M_n(E)$ denote, in each case, the (unknown) matrix to be determined. I denotes the identity matrix of $M_n(E)$.

5.1. Generalized Gauss–Jordan Method: Principle

We will describe the generalized Gauss–Jordan method by considering system (28)′:

$$Y = Y \otimes A \oplus B \tag*{(28)′}$$

where $B = (b_{ij}) \in M_n(E)$ is a given arbitrary matrix.

We will assume that A^* exists, therefore, implying that (28)′ has a minimal solution $B \otimes A^*$

Note that we will be able to obtain similar formulas to solve the system:

$$Z = A \otimes Z \oplus B \tag*{(29)′}$$

In the sequel, it will be assumed that (E, ⊕, ⊗) is a topological dioid; we recall that in such a dioid, any nondecreasing sequence bounded from above has a limit (see Chap. 3, Sect. 6).

As will be seen, this assumption will enable to show that at each step, the method calculates expressions leading to minimal solutions. In the case where (E, ⊕, ⊗) is not a dioid, but only a semiring, the method will possibly yield a solution to (28)′ which does not necessarily have the property of minimality.

In the special case where B = I (the identity matrix of $M_n(E)$) and where (E, ⊕, ⊗) is a dioid, the generalized Gauss–Jordan method therefore yields the minimal solution to (28) i.e. the quasi-inverse A^* of A.

The first equation of (28)′ is written:

$$y_{11} = y_{11} \otimes a_{11} \oplus \sum_{j=2}^{n} y_{1j} \otimes a_{j1} \oplus b_{11} \tag{30}$$

Since (E, ⊕, ⊗) is a topological dioid we deduce that the entry a_{11} has a quasi-inverse a_{11}^* (see Chap. 3, Proposition 6.2.6) and consequently from (30) we obtain the expression of y_{11} in terms of the other variables y_{1j} (j = 2, ... n):

$$y_{11} = \sum_{j=2}^{n} y_{1j} \otimes a_{j1} \otimes a_{11}^* \oplus b_{11} \otimes a_{11}^*$$

We proceed similarly with all the other equations of (28)′ corresponding to the first column of Y. We therefore have, for any i = 1, ..., n:

$$y_{i1} = y_{i1} \otimes a_{11} \oplus \sum_{j=2}^{n} y_{ij} \otimes a_{j1} \oplus b_{i1}$$

which, by using a_{11}^*, enables one to express y_{i1} in terms of the other variables y_{ij} (j = 2, ... n):

$$y_{i1} = \sum_{j=2}^{n} y_{ij} \otimes a_{j1} \otimes a_{11}^* \oplus b_{i1} \otimes a_{11}^* \tag{31}$$

Once the expressions of y_{i1} (i = 1 ... n) given by (31) are obtained, one can substitute them in the other equations of system (28)′, which gives, for i arbitrary and k ≥ 2:

$$y_{ik} = \sum_{j=1}^{n} y_{ij} \otimes a_{jk} \oplus b_{ik}$$
$$= y_{i1} \otimes a_{1k} \oplus \sum_{j=2}^{n} y_{ij} \otimes a_{jk} \oplus b_{ik}$$

By using (31), the above expression takes the form:

$$y_{ik} = \sum_{j=2}^{n} y_{ij} \otimes (a_{j1} \otimes a_{11}^{*} \otimes a_{1k} \oplus a_{jk}) \oplus b_{i1} \otimes a_{11}^{*} \otimes a_{lk} \oplus b_{ik} \tag{32}$$

Let us denote $Y^{[1]}$ the matrix deduced from Y by replacing all the entries in the first column by ε and let us introduce the square n × n matrices

$$A^{[1]} = \left(a_{ij}^{[1]}\right) \quad \text{and} \quad B^{[1]} = \left(b_{ij}^{[1]}\right)$$

defined for any $j = 1 \cdots n$ and $k = 1 \cdots n$ as:

$$a_{jk}^{[1]} = a_{jk} \oplus a_{j1} \otimes a_{11}^{*} \otimes a_{1k} \tag{33}$$

and

$$b_{jk}^{[1]} = b_{jk} \oplus b_{j1} \otimes a_{11}^{*} \otimes a_{1k} \tag{34}$$

It can be observed that the equations (31), which define the first column of Y can also be written:

$$\begin{bmatrix} y_{11} \\ y_{21} \\ \cdot \\ \cdot \\ \cdot \\ y_{n1} \end{bmatrix} = Y^{[1]} \otimes \begin{bmatrix} a_{11}^{[1]} \\ a_{21}^{[1]} \\ \cdot \\ \cdot \\ a_{n1}^{[1]} \end{bmatrix} \oplus \begin{bmatrix} b_{11}^{[1]} \\ b_{21}^{[1]} \\ \cdot \\ \cdot \\ b_{n1}^{[1]} \end{bmatrix} \tag{35}$$

Indeed, for i = 1, ... n, the expression of y_{i1} given by (35) reads:

$$\begin{aligned} y_{i1} &= \sum_{j=2}^{n} y_{ij} \otimes a_{j1}^{[1]} \oplus b_{i1}^{[1]} \\ &= \sum_{j=2}^{n} y_{ij} \otimes (a_{j1} \oplus a_{j1} \otimes a_{11}^{*} \otimes a_{11}) \oplus b_{i1} \oplus b_{i1} \otimes a_{11}^{*} \otimes a_{11} \end{aligned}$$

Now, since a_{11}^{*} is the quasi-inverse of a_{11}, we have that:

$$a_{j1} \oplus a_{j1} \otimes a_{11}^{*} \otimes a_{11} = a_{j1} \otimes (e \oplus a_{11}^{*} \otimes a_{11}) = a_{j1} \otimes a_{11}^{*}$$

and similarly:

$$b_{i1} \oplus b_{i1} \otimes a_{11}^{*} \otimes a_{11} = b_{i1} \otimes a_{11}^{*}$$

from which we can write:

$$y_{i1} = \sum_{j=2}^{n} y_{ij} \otimes a_{j1} \otimes a_{11}^{*} \oplus b_{i1} \otimes a_{11}^{*}$$

which is exactly expression (31).

The relations (32) and (35) therefore show that after elimination of the variables y_{i1} $(i = 1 \cdots n)$, system (28)′ takes the form:

$$Y = Y^{[1]} \otimes A^{[1]} \oplus B^{[1]} \tag{36}$$

Moreover (33) and (34) can be written in matrix terms:

$$A^{[1]} = A \otimes M^{[1]} \tag{37}$$

and

$$B^{[1]} = B \otimes M^{[1]} \tag{38}$$

where the matrix $M^{[1]} \in M_n(E)$ is the transformation matrix defined as:

$$M^{[1]} = \begin{bmatrix} a_{11}^* & a_{11}^* \otimes a_{12} & a_{11}^* \otimes a_{13} & \cdot & \cdot & \cdot & a_{11}^* \otimes a_{1n} \\ \varepsilon & e & \varepsilon & \cdot & \cdot & \cdot & \varepsilon \\ \varepsilon & \varepsilon & e & \cdot & \cdot & \cdot & \cdot \\ \cdot & \cdot & & \cdot & \cdot & \cdot & \cdot \\ \cdot & \cdot & & \cdot & \cdot & \cdot & \cdot \\ \varepsilon & \varepsilon & \cdot & \cdot & \cdot & \cdot & e \end{bmatrix}$$

i.e. the matrix deduced from the identity matrix of $M_n(E)$ by replacing the entries in the first row by:

$$\begin{cases} m_{11}^{[1]} = a_{11}^* \\ m_{1j}^{[1]} = a_{11}^* \otimes a_{1j} \ (j = 2, \ldots, n) \end{cases}$$

(for the entries in the first column of $A^{[1]}$, (33) yields: $a_{j1}^{[1]} = a_{j1} \otimes (e \oplus a_{11}^* \otimes a_{11}) = a_{j1} \otimes a_{11}^*$; in the same way (34) yields $b_{j1}^{[1]} = b_{j1} \otimes a_{11}^*$)

By analogy with the classical Gauss–Jordan method, we will say that (36) is deduced from (28)′ via a pivot operation on the first row and the first column (the entry a_{11} is referred to as the *pivot element*).

It is now easy to see that the elimination technique explained above can be iterated: since A^* exists, $a_{22}^{[1]}$ is quasi-invertible (see remark below) and the element $a_{22}^{[1]}$ can be used as a pivot element to eliminate all the variables of the second column of Y. By denoting $Y^{[2]}$ the matrix deduced from $Y^{[1]}$ by replacing all the terms of the second column with ε, the system obtained at the second iteration reads:

$$Y = Y^{[2]} \otimes A^{[2]} \oplus B^{[2]}$$

where

$$A^{[2]} = A^{[1]} \otimes M^{[2]}$$
$$B^{[2]} = B^{[1]} \otimes M^{[2]}$$

$M^{[2]}$ being the matrix deduced from the identity matrix I by replacing the second row by:

$$\begin{cases} m_{22}^{[2]} = (a_{22}^{[1]})^* \\ m_{2j}^{[2]} = (a_{22}^{[1]})^* \otimes a_{2j}^{[1]} \quad (j \neq 2) \end{cases}$$

Remark. The fact that the existence of A^* implies that $a_{22}^{[1]}$ has a quasi-inverse can be derived as follows.

Consider $\widetilde{A} = \begin{bmatrix} a_{11} & a_{12} \\ a_{21} & a_{22} \end{bmatrix}$ corresponding to the first two rows and columns. Since A^* exists, we know (from Proposition 6.2.6 in Chapter 3) that $\widetilde{A}$ has a quasi-inverse $\widetilde{A}^*$ which solves the 2×2 system:

$$\begin{bmatrix} u_{11} & u_{12} \\ u_{21} & u_{22} \end{bmatrix} = \begin{bmatrix} u_{11} & u_{12} \\ u_{21} & u_{22} \end{bmatrix} \otimes \widetilde{A} \oplus \begin{bmatrix} e & \varepsilon \\ \varepsilon & e \end{bmatrix}$$

carrying out on the above system a pivot operation with pivot element a_{11} yields the relation

$$\begin{aligned} u_{22} &= u_{22} \otimes \left(a_{21} \otimes a_{11}^* \otimes a_{12} \oplus a_{22}\right) \oplus e \\ &= u_{22} \otimes a_{22}^{[1]} \oplus e \end{aligned}$$

For the solution $u_{22} = \left(\widetilde{A}^*\right)_{22}$, the above implies, $\forall k \in \mathbb{N}$:

$$u_{22} = u_{22} \otimes \left(a_{22}^{[1]}\right)^k \oplus \left(a_{22}^{[1]}\right)^{(k)}$$

hence:

$$\left(a_{22}^{[1]}\right)^{(k)} \leq u_{22}.$$

Thus the sequence $\left(a_{22}^{[1]}\right)^{(k)}$ is nondecreasing and bounded from above in a topological dioid, hence the existence of $\left(a_{22}^{[1]}\right)^*$ follows. ||

In a general way, the matrices $A^{[k]}$ and $B^{[k]}$ are defined recursively as:

$$A^{[0]} = A \quad B^{[0]} = B \tag{39}$$

and, $\forall k = 1, 2, \ldots. n$:

$$A^{[k]} = A^{[k-1]} \otimes M^{[k]} \tag{40}$$

$$B^{[k]} = B^{[k-1]} \otimes M^{[k]} \tag{41}$$

with $M^{[k]}$ deduced from I by replacing the row k by:

$$\begin{cases} m_{kk}^{[k]} = (a_{kk}^{[k-1]})^* & (42) \\ m_{kj}^{[k]} = (a_{kk}^{[k-1]})^* \otimes a_{kj}^{[k-1]} \quad (j \neq k) & (43) \end{cases}$$

(The fact that $\left(a_{kk}^{[k-1]}\right)^*$ exists is obtained along the same line as in the above remark, considering the principal submatrix induced by the first k rows and columns and carrying out k−1 pivot operations).
We then have:

Theorem 7. *Let* $A \in M_n(E)$ *where* $(E, \oplus, \otimes)$ *is a topological dioid. It is assumed that* A *has a quasi-inverse* A^*.

Then $Y = B^{[n]}$ *obtained by* (39)–(43) *is the minimal solution to the system*

$$Y = Y \otimes A \oplus B \tag{28'}$$

In the special case where $B = B^{[0]} = I$, *then* $B^{[n]}$ *is the quasi-inverse* A^* *of* A.

Proof. All the intermediate operations performed in applying the recurrence relations (39)–(43) comply with each equation of ((28)′). The matrix $Y = B^{[n]}$ obtained at the nth iteration is therefore clearly a solution.

The fact that it is indeed a minimal solution can be easily proved by induction as follows.

Let $\overline{Y} = (\bar{y}_{ij})$ be an arbitrary solution to (28′). At iteration 1, $\overline{Y}$ therefore satisfies: $\forall i = 1 \dots n$:

$$y_{i1} = y_{i1} \otimes a_{11} \oplus \sum_{j=2}^{n} y_{ij} \otimes a_{j1} \oplus b_{i1} \tag{44}$$

$\overline{Y}$ is therefore a solution to (44)

Thus $\bar{y}_{i1}$ solves:

$$y_{i1} = y_{i1} \otimes a_{11} \oplus \sum_{j=2}^{n} \bar{y}_{ij} \otimes a_{j1} \oplus b_{i1}$$

and consequently:

$$\bar{y}_{i1} \geq \left(\sum_{j=2}^{n} \bar{y}_{ij} \otimes a_{j1} \oplus b_{i1}\right) \otimes a_{11}^*$$

This shows that $\overline{Y}$ satisfies system (36) obtained at the end of the first iteration with the inequality, in other words that:

$$\overline{Y} \geq \overline{Y}^{[1]} \otimes A^{[1]} \oplus B^{[1]} = \overline{Y}^{[1]} \otimes A^{[1]} \oplus B \otimes M^{[1]}$$

There therefore exists a matrix $H^{[1]} \in M_n(E)$ such that $\overline{Y}$ satisfies

$$\overline{Y} = \overline{Y}^{[1]} \otimes A^{[1]} \oplus B \otimes M^{[1]} \oplus H^{[1]} \tag{45}$$

The second iteration, corresponding to the elimination of all the variables y_{i2}, performed on system (45) would lead, similarly, to showing that $\overline{Y}$ satisfies:

$$\overline{Y} \geq \overline{Y}^{[2]} \otimes A^{[2]} \oplus \left(B \otimes M^{[1]} \oplus H^{[1]}\right) \otimes M^{[2]}$$

There therefore exists a matrix $H^{[2]} \in M_n(E)$ such that:

$$\overline{Y} = \overline{Y}^{[2]} \otimes A^{[2]} \oplus \left(B \otimes M^{[1]} \oplus H^{[1]}\right) \otimes M^{[2]} \oplus H^{[2]}$$

One can therefore deduce, by induction, the existence of matrices $H^{[1]}\ H^{[2]} \dots H^{[n]}$ such that: $\overline{Y} = Y^{[n]} \otimes A^{[n]} \oplus (((\cdots(B \otimes M^{[1]} \oplus H^{[1]}) \cdots \otimes M^{[2]} \oplus H^{[2]}) \otimes \cdots)$ $M^{[n]} \oplus H^{[n]})$

Since $Y^{[n]} = \begin{bmatrix} \varepsilon\ \varepsilon & & \varepsilon \\ \varepsilon & & \\ & & \\ \varepsilon & & \varepsilon \end{bmatrix}$ we deduce

$$\overline{Y} \geq B \otimes M^{[1]} \otimes M^{[2]} \otimes \cdots \otimes M^{[n]} = B^{[n]}$$

This proves that $B^{[n]}$ is a minimal solution to (28)′. □

5.2. Generalized Gauss–Jordan Method: Algorithms

Let us consider here the case where $B = I$. Applying the induction formulae (39)–(43) we then construct the quasi-inverse A^* of A.

We have, in this case:

$$B^{[n]} = A^* = M^{[1]} \otimes M^{[2]} \otimes \cdots \otimes M^{[n]}$$

and:

$$\begin{aligned} A^{[n]} &= A \otimes M^{[1]} \otimes M^{[2]} \otimes \cdots \otimes M^{[n]} \\ &= A \otimes B^{[n]} \\ &= A \otimes A^* \\ &= A^+ \end{aligned}$$

It is seen that to compute $A^{[n]} = A^+$ we just have to use the induction formula (40). A^* is then directly deduced from A^+ by: $A^* = I \oplus A^+$.

Since $M^{[k]}$ only depends on $A^{[k-1]}$, this remark shows that one can compute A^* by only working on each iteration with a single matrix $A^{[k]}$.

We then have:

Algorithm 4 *(Generalized Gauss–Jordan)*
Computation of the matrices A^+ *and* A^*

(a) *Set* $A^{[0]} = A$
(b) *For* $k = 1, 2, \dots n$ *successively do:*
$a_{kk}^{[k]} = \left(a_{kk}^{[k-1]}\right)^*$;
For $i = 1 \dots n$

For $j = 1 \ldots n$

If $(i \neq k$ *or* $j \neq k)$ *do:*

$a_{ij}^{[k]} \leftarrow a_{ij}^{[k-1]} \oplus a_{ik}^{[k-1]} \otimes a_{kk}^{[k]} \otimes a_{kj}^{[k-1]}$

Endfor

Endfor

Endfor

(c) At the end of step (b) we obtain $A^+ = A^{[n]}$. A^* *is then deduced by:*

$A^* = A^{[n]} \oplus I.$

The following result states the complexity of *Algorithm 4.*

Proposition 5.2.1. *Algorithm 4 requires n computations of the quasi-inverse of an element, and* $\mathcal{O}(n^3)$ *operations* $\oplus$ *and* $\otimes$.

Proof. Each iteration requires the computation of $\left(a_{kk}^{[k-1]}\right)^*$ (quasi-inverse of the pivot element) and $\mathcal{O}(n^2)$ operations $\oplus$ and $\otimes$. From this the result follows. □

A special case of interest is the one where the graph G(A) does not contain a 0-absorbing circuit. In this case, $a_{kk}^{[k-1]}$ is a 0-stable element (see chap. 3 Sect. 7) therefore

$$a_{kk}^{[k]} = \left(a_{kk}^{[k-1]}\right)^* = e$$

Algorithm 4 then specializes as follows:

Algorithm 4′ *Computation of* A^+ *and* A^* *in the case where* G(A) *does not contain a 0-absorbing circuit*

(a) $A^{[0]} = A$

(b) For $k = 1 \ldots n$

For $i = 1 \ldots n$

For $j = 1 \ldots n$

$a_{ij}^{[k]} \leftarrow a_{ij}^{[k-1]} \oplus a_{ik}^{[k-1]} \otimes a_{kj}^{[k-1]}$;

Endfor

Endfor

Endfor

(c) $A^+ = A^{[n]}$ *and* $A^* = I \oplus A^{[n]}$

In the case of the shortest path problem in a graph without negative length circuits ($\oplus = \text{Min}, \otimes = +$) algorithm 4′ is none other than Floyd's algorithm (1962).

5.3. Generalized "Escalator" Method

A being a $n \times n$ matrix with elements in a dioid E, for any $k \in [1, \ldots, n]$ let us denote $A_{\{k\}}$ the sub-matrix of A formed by the first k rows and the first k columns of A. With this notation this yields: $A = A_{\{n\}}$

The principle of the "escalator" method is to determine $(A_{\{2\}})^*$ from $(A_{\{1\}})^* = a_{11}^*$; then $(A_{\{3\}})^*$ from $(A_{\{2\}})^*$, and so on, until one obtains $A^* = (A_{\{n\}})^*$ from $(A_{\{n-1\}})^*$.

To implement this recursion in the form of an algorithm, we will use the formula of quasi-inversion by blocks given by the following result.

Proposition 5.3.1. *Let* $(E, \oplus, \otimes)$ *be a dioid and* $U \in M_k(E)(k \geq 2)$ *a* $k \times k$ *matrix assumed to be partitioned into blocks in the form:*

$$U = \begin{bmatrix} U_{11} & U_{12} \\ U_{21} & U_{22} \end{bmatrix}$$

with $U_{11} \in M_{k-1}(E)$ *and where:*

U_{12} *has dimensions* $(k-1) \times 1$, U_{21} *has dimensions* $1 \times (k-1)$ *and* $U_{22} \in E$. *It is assumed that the sub-matrix* U_{11} *is quasi-invertible with quasi-inverse* U_{11}^*, *that the entry* U_{22} *is quasi-invertible in E, and that the element* $\mu = U_{21} \otimes U_{11}^* \otimes U_{12} \oplus U_{22}$ *is quasi-invertible in E.*

Then U is quasi-invertible in $M_k(E)$ *and* $U^* = \begin{bmatrix} X_{11} & X_{12} \\ X_{21} & X_{22} \end{bmatrix}$ *is given by the formulae:*

$$X_{11} = U_{11}^* \oplus U_{11}^* \otimes U_{12} \otimes \mu^* \otimes U_{21} \otimes U_{11}^* \tag{46}$$

$$X_{12} = U_{11}^* \otimes U_{12} \otimes \mu^* \tag{47}$$

$$X_{21} = \mu^* \otimes U_{21} \otimes U_{11}^* \tag{48}$$

$$X_{22} = \mu^* = (U_{21} \otimes U_{11}^* \otimes U_{12} \oplus U_{22})^* \tag{49}$$

Proof. Let us show the existence of a minimal solution to the matrix equations:

$$X = U \otimes X \oplus I_k \tag{50}$$

$$X = X \otimes U \oplus I_k \tag{51}$$

where $X \in M_k(E)$ can be partitioned into blocks:

$$X = \begin{bmatrix} X_{11} & X_{12} \\ X_{21} & X_{22} \end{bmatrix}$$

(50) implies, in particular:

$$X_{12} = U_{11} \otimes X_{12} \oplus U_{12} \otimes X_{22} \tag{52}$$

$$X_{22} = U_{21} \otimes X_{12} \oplus U_{22} \otimes X_{22} \oplus e \tag{53}$$

Since U_{11} is quasi-invertible, (52) has a minimal solution in X_{12} for any value of X_{22}, which reads:

$$X_{12} = U_{11}^* \otimes U_{12} \otimes X_{22}$$

By using this relation, (53) is written:

$$X_{22} = (U_{21} \otimes U_{11}^* \otimes U_{12} \oplus U_{22}) \otimes X_{22} \oplus e$$

or, equivalently:

$$X_{22} = \mu \otimes X_{22} \oplus e$$

μ being assumed to be quasi-invertible, the equation above has a minimal solution which reads:

$$X_{22} = \mu^* \tag{49}$$

from which we deduce:

$$X_{12} = U_{11}^* \otimes U_{12} \otimes \mu^* \tag{47}$$

Similarly (51) leads to the four relations:

$$X_{11} = X_{11} \otimes U_{11} \oplus X_{12} \otimes U_{21} \oplus I_{k-1} \tag{54}$$
$$X_{21} = X_{21} \otimes U_{11} \oplus X_{22} \otimes U_{21} \tag{55}$$
$$X_{12} = X_{11} \otimes U_{12} \oplus X_{12} \otimes U_{22} \tag{56}$$
$$X_{22} = X_{21} \otimes U_{12} \oplus X_{22} \otimes U_{22} \oplus e \tag{57}$$

By using (47), (54) is rewritten:

$$X_{11} = X_{11} \otimes U_{11} \oplus U_{11}^* \otimes U_{12} \otimes \mu^* \otimes U_{21} \oplus I_{k-1}$$

and, since U_{11} is quasi-invertible, the latter equation has a minimal solution in X_{11}, which is written:

$$\begin{aligned} X_{11} &= (U_{11}^* \otimes U_{12} \otimes \mu^* \otimes U_{21} \oplus I_{k-1}) \otimes U_{11}^* \\ &= U_{11}^* \oplus U_{11}^* \otimes U_{12} \otimes \mu^* \otimes U_{21} \otimes U_{11}^* \end{aligned} \tag{46}$$

Finally, by using (49) and the quasi-invertibility of U_{11}, through (55) one obtains the minimal solution in X_{21}:

$$X_{21} = \mu^* \otimes U_{21} \otimes U_{11}^* \tag{48}$$

We have thus obtained the expressions (46)–(49) by exploiting two of the relations resulting from (50) and two of the relations resulting from (51). It remains to verify that X thus obtained satisfies for example the two other relations (56) and (57) deduced from (51). The expression $X_{11} \otimes U_{12} \oplus X_{12} \otimes U_{22}$ is written:

$$\begin{aligned} &(U_{11}^* \oplus U_{11}^* \otimes U_{12} \otimes \mu^* \otimes U_{21} \otimes U_{11}^*) \otimes U_{12} \oplus U_{11}^* \otimes U_{12} \otimes \mu^* \otimes U_{22} \\ &= U_{11}^* \otimes U_{12} \oplus (U_{11}^* \otimes U_{12} \otimes \mu^*) \otimes (U_{21} \otimes U_{11}^* \otimes U_{12} \oplus U_{22}) \\ &= U_{11}^* \otimes U_{12} \otimes (e \oplus \mu^* \otimes \mu) \\ &= U_{11}^* \otimes U_{12} \otimes \mu^* \\ &= X_{12} \text{ according to (47)} \end{aligned}$$

The expression $X_{21} \otimes U_{12} \oplus X_{22} \otimes U_{22} \oplus e$ is written

$$
\begin{aligned}
&\mu^* \otimes U_{21} \otimes U_{11}^* \otimes U_{12} \oplus \mu^* \otimes U_{22} \oplus e \\
&= \mu^* \otimes (U_{21} \otimes U_{11}^* \otimes U_{12} \oplus U_{22}) \oplus e \\
&= \mu^* \otimes \mu \oplus e = \mu^* = X_{22} \quad \text{according to (49).}
\end{aligned}
$$

X given by (46)–(49) therefore clearly satisfies (51).

Furthermore, by construction, it is clearly a minimal solution. We deduce that U is quasi-invertible with $U^* = X$. □

To determine the quasi-inverse of a matrix A (assumed to be quasi-invertible) one can then deduce directly from Proposition 5.3.1 the following general algorithm.

Algorithm 5 *(generalized "escalator" method)*

Determination of A^* *for* $A \in M_n(E)$ *assumed to be quasi-invertible.*

Set $A^*_{\{1\}} = a^*_{11}$

For $k = 2, \ldots, n$ *do:*

$$
A_{\{k\}} = \left[\begin{array}{c|c} A_{\{k-1\}} & v_{k-1} \\ \hline w_{k-1} & a_{kk} \end{array}\right]
$$

where:

- v_{k-1} *is the column-vector formed by the* $k-1$ *first entries of the* k^{th} *column of A.*
- w_{k-1} *is the row-vector formed by the* $k-1$ *first entries of the* k^{th} *row of A.*

Compute:

$$\mu^* = (w_{k-1} \otimes A^*_{\{k-1\}} \otimes v_{k-1} \oplus a_{kk})^*$$

$$\bar{v}_{k-1} = A^*_{\{k-1\}} \otimes v_{k-1} \tag{58}$$

$$\bar{w}_{k-1} = w_{k-1} \otimes A^*_{\{k-1\}} \tag{59}$$

then deduce:

$$
A^*_{\{k\}} = \left[\begin{array}{c|c} A^*_{[k-1]} \oplus \bar{v}_{k-1} \otimes \mu^* \otimes \bar{w}_{k-1} & \bar{v}_{k-1} \otimes \mu^* \\ \hline \mu^* \otimes \bar{w}_{k-1} & \mu^* \end{array}\right]
$$

Endfor

At the end of the iterations, we obtain $A^* = A^*_{\{n\}}$.

Let us observe that in the expression of $A^*_{\{k\}}$ the vectors $\bar{v}_{k-1} \otimes \mu^*$ and $\mu^* \otimes \bar{w}_{k-1}$ deduced from (58) and (59) correspond to (47) and (48) respectively. Moreover, the expression $A^*_{\{k-1\}} \oplus \bar{v}_{k-1} \otimes \mu^* \otimes \bar{w}_{k-1}$ can be rewritten:

$$A^*_{\{k-1\}} \oplus A^*_{\{k-1\}} \otimes v_{k-1} \otimes \mu^* \otimes w_{k-1} \otimes A^*_{\{k-1\}}$$

which is none other than relation (46).

Proposition 5.3.2. *Algorithm 5 requires* $\mathcal{O}(n^3)$ *operations* $\oplus$ *and* $\otimes$*, as well as* n *computations of the quasi-inverse of an element of E.*

Proof. On the whole, the algorithm performs a total of n computations of the quasi-inverse of an element of E. Moreover, at each iteration k the algorithm performs on matrices of order $k-1$ and vectors of dimension $k-1$:

- Three matrix-vector multiplications requiring $\mathcal{O}$ (k^2) operations $\oplus$ and $\otimes$;
- A scalar product of two vectors requiring $\mathcal{O}$ (k) operations $\oplus$ and $\otimes$;
- A product of a column-vector- $(\bar{v}_{k-1})$ by a row-vector $(\mu^* \otimes \bar{w}_{k-1})$ followed by a sum of two matrices, which requires $\mathcal{O}(k^2)$ operations $\oplus$ and $\otimes$.

The result is deduced by summation on k from 2 to n. □

A interesting special case is the one where the graph G(A), associated with matrix A, has no 0-absorbing circuit.

In this case at each iteration of Algorithm 5, we have $\mu^* = e$, and Algorithm 5 can be reformulated in the following simplified form:

Algorithm 5′ *Computation of* A^* *starting from* A *by the generalized "escalator" method (case where* G(A) *has no 0-absorbing circuit).*

$a_{11} \leftarrow e$;
For $k = 2$ *to n do:*

(a) $a_{kk} \leftarrow e$
(b) *For* $i = 1, \ldots, k-1$:

$$a_{i,k} \leftarrow \sum_{j=1}^{k-1} a_{i,j} \otimes a_{j,k};$$

$$a_{k,i} \leftarrow \sum_{j=1}^{k-1} a_{k,j} \otimes a_{j,i};$$

Endfor
(c) *For* $i = 1, \ldots, k-1$
For $j = 1, \ldots, k-1$
$a_{ij} \leftarrow a_{ij} \oplus a_{ik} \otimes a_{kj}$;
Endfor
Endfor
Endfor

In the case of the shortest path problem ($\oplus$ = Min, $\otimes$ = +) algorithm 5′ is recognized as Dantzig's algorithm (1966).

6. Examples of Application: An Overview of Path-finding Problems in Graphs

In this section we discuss some interesting examples of applications of the computation of the quasi-inverse of a matrix related to path-finding problems in graphs.

Table 1 on the following page summarizes the main features of these examples.

Table 1 Main features of path-finding problems in graphs

Problem types	Problems solved	E	⊕	⊗	ε	e
Existence	Problems of connectivity	$\{0,1\}$	Max	Min	0	1
Enumeration	Enumeration of elementary paths	$\mathcal{P}(X^*)$	Union	Latin multiplication	∅	X
	Multicriteria problems	$\mathcal{P}(\mathbb{R}^P)$	Set of efficient paths of the union	Set of efficient paths of the sum	$(+\infty)^P$	$(0)^P$
	Generation of regular languages (Kleene)	Set of words	Union	Concatenation	∅	The empty word
Optimization	Maximum capacity path	$\mathbb{R}_+ \cup \{+\infty\}$	Max	Min	0	$+\infty$
	Minimum spanning tree, Hierarchical classification	$\overline{\mathbb{R}},\ \mathbb{R}_+$	Min	Max	$+\infty$	$-\infty, 0$
	Minimum cardinality path	$\mathbb{N} \cup \{+\infty\}$	Min	$+$	$+\infty$	0
	Shortest path	$\mathbb{R} \cup \{+\infty\}$	Min	$+$	$+\infty$	0
	Longest path	$\mathbb{R} \cup (+\infty\}$	Max	$+$	$-\infty$	0
	Maximum reliability path	$\{a \mid 0 \le a \le 1\}$	Max	$\times$	0	1
	Reliability of a network	Idempotent polynomials	symmetrical difference	$\times$	0	1
Counting	Path counting	$\mathbb{N}$	$+$	$\times$	0	1
	Markov chains	$\{a \mid 0 \le a \le 1\}$	$+$	$\times$	0	1
Optimization and	kth-shortest path problem	Cone of $\overline{\mathbb{R}}^k$	k smallest terms of the 2 vectors	k smallest terms of the sums of pairs	$(+\infty)^k$	$(0, +\infty, \ldots, +\infty)$
Post-optimization	η-optimal paths	Ordered sequence of terms of $\mathbb{R}$ of amplitude η	Sequence formed by the η-smallest terms of the two sequences	Sequence formed by the η-smallest sums of pairs of elements of the two sequences	$(+\infty)$	(0)

6.1. Problems of Existence and Connectivity

To solve existence problems, we will use classical *Boolean algebra*, i.e. the structure:

$$E = \{0, 1\}, \quad \oplus = \max, \quad \otimes = \min, \quad \varepsilon = 0 \quad \text{and} \quad e = 1.$$

For a given graph $G = [X, U]$, the matrix A is defined as:

$$a_{ij} = 1 \quad \text{if} \quad (i, j) \in U, \quad a_{ij} = 0 \quad \text{otherwise}.$$

By interpreting the results of Sect. 3, we then obtain the following properties:

- There exists a path containing k arcs between i and j, if and only if $(A^k)_{ij} = 1$.
- There exists a path taking at most k arcs between i and j, if and only if $(A^{(k)})_{ij} = 1$.
- For any $a \in E$, $e \oplus a = \max(1, a) = 1 = e$, therefore any circuit is 0-stable; from this we deduce (theorem 1) the existence of A^*, which can be interpreted as the incidence matrix of the *transitive closure of the graph* G.

6.2. Path Enumeration Problems

To solve enumeration problems, we typically have to take for E the power set of some associated set, with set *union* as addition $\oplus$.

To be more specific, suppose we want to enumerate all the *elementary paths* of a graph (Kaufmann and Malgrange, 1963).

Given a graph $G = [X, U]$, let X^* be *the set of ordered sequences of* elements of $X = \{1, 2 \ldots, n\}$ satisfying a given property P. Each element of X^* will be referred to as a path. Here we require that the paths be elementary (property P).

The empty set $\emptyset$ will be considered as an element of X^*. We will take as ground set E the power set of X^*, i.e. $E = \mathcal{P}(X^*)$.

The $\oplus$ law is taken as the set *union*, hence $\varepsilon = \emptyset$. The $\otimes$ law will be the so-called *Latin multiplication* defined as:

- $u \otimes \emptyset = \emptyset \otimes u \qquad \forall u \in S$
- if $u_\alpha = (u_{\alpha_i})$ with $u_{\alpha_i} \in X^*$
  $u_\beta = (u_{\beta j})$ with $u_{\beta j} \in X^*$

  then:

$$u_\alpha \otimes u_\beta = \{(u_{\alpha i} \otimes u_{\beta j})\}$$

with:

$$\text{if} \quad u_{\alpha_i} = (i_1 i_2, \ldots i_k)$$
$$\text{and} \quad u_{\beta_j} = (j_1 j_2, \ldots j_l)$$

$$u_{\alpha_i} \otimes u_{\beta_j} = \begin{cases} \cdot\ (i_1, i_2, \ldots, i_k, j_2, j_3, ; \ldots, j_k) \quad \text{if} \quad i_k = j_1 \\ \text{and if this sequence satisfies property P} \\ \cdot\ \emptyset \quad \text{otherwise} \end{cases}$$

The neutral element for $\otimes$ is the set formed of all the individual elements of X, i.e. $e = X$.

With each arc (i, j) we associate $a_{ij} = \{i, j\} \in E$.

By interpreting the results of Sect. 3, we deduce the following properties:

- $(A^k)_{ij}$ represents the set of elementary paths between i and j containing exactly k arcs.
- The weight of any circuit is equal to $\emptyset$, therefore there is no zero-absorbing circuit. We deduce (Theorem 1) the existence of $A^* = A^{(n-1)}$
- $(A^{n-1})_{ij}$ represents the set of all *Hamiltonian paths* between i and j.

6.3. The Maximum Capacity Path Problem and the Minimum Spanning Tree Problem

Here we consider the following lattice (doubly idempotent dioid):

$$E = \mathbb{R}_+ \cup \{+\infty\}, \quad \oplus = \max, \quad \otimes = \min, \quad \varepsilon = 0 \quad \text{and} \quad e = +\infty.$$

With each arc (i, j), we associate its capacity $a_{ij} \geq 0$. The capacity represents the maximum flow which can be sent on arc (i, j). The problem is to find a path μ from i to j: $(i, i_1, i_2, \ldots, i_k, j)$, say such that $\underline{\sigma}(\mu) = \text{Min}\{a_{i,i_1}, a_{i_1,i_2}, \ldots, a_{i_k,j}\}$ is maximized. The quantity $\underline{\sigma}(\mu)$ is also referred to as the *inf-section* of the path μ. By interpreting the results of Sect. 3, we deduce the following properties:

- The maximum capacity of a path containing k arcs between i and j is $(A^k)_{ij}$.
- For any $a \in E$, $e \oplus a = \max(+\infty, a) = +\infty = e$, therefore there is no zero-absorbing circuit. We deduce (Theorem 1) the existence of $A^* = A^{(n-1)}$.
- The maximum capacity of a path between i and j is $(A^*)_{ij}$.

A problem closely related to the above concerns the search of a path μ minimizing the sup-section $\bar{\sigma}(\mu) = \text{Max}\{a_{i,i_1}, a_{i_1,i_2}, a_{i_k j}\}$. In the search for a path minimizing $\bar{\sigma}(\mu)$, the dioid $(\overline{\mathbb{R}}, \text{Min}, \text{Max})$ will be considered.

From a classical result (Hu, 1961) in the symmetric case (i.e.: $a_{ij} = a_{ji}$ for all i, j), the minimum spanning tree of a graph provides the set of optimal paths minimizing the sup-section for all pairs of nodes.

6.4. Minimum Cardinality Paths

Here we consider the following structure:

$$E = \mathbb{N} \cup \{+\infty\}, \quad \oplus = \min, \quad \otimes = +, \quad \varepsilon = +\infty \quad \text{and} \quad e = 0.$$

With each arc (i, j), we associate $a_{ij} = 1$.

By interpreting the results of Sect. 3, we obtain the following properties:

- $(A^k)_{ij} = k$ if there exists a path taking k arcs from i to j, $(A^k)_{ij} = +\infty$ otherwise.
- For any $a \in E$, $e \oplus a = \min(0, a) = 0 = e$, therefore there is no zero-absorbing circuit. We deduce (Theorem 1) the existence of $A^* = A^{(n-1)}$.
- $(A^*)_{ij}$ represents the number of arcs in the minimum cardinality path between i and j.

We observe that the resulting matrix A^* contains the relevant information required to determine the centre, radius, diameter, etc. of a graph.

6.5. The Shortest Path Problem

For this example, which has already been extensively discussed (see Sect. 2 above), we take:

$$E = \mathbb{R} \cup \{+\infty\}, \quad \oplus = \min, \quad \otimes = +, \quad \varepsilon = +\infty \quad \text{and} \quad e = 0.$$

With each arc (i, j), we associate its length a_{ij}.

By interpreting the results of Sect. 3, we obtain the following properties:

- $(A^k)_{ij}$ represents the length of the shortest path between i and j taking k arcs.
- If a is nonnegative, $e \oplus a = \min(0, a) = 0 = e$ and a is 0-regular. Therefore if there does not exist a negative length circuit, A^* exists (Theorem 1) and $(A^*)_{ij}$ correspond to the length of the shortest path between i and j.

6.6. Maximum Reliability Path

$$E = \{a/0 \le a \le 1\}, \quad \oplus = \max, \quad \otimes = \times, \quad \varepsilon = 0 \quad \text{and} \quad e = 1.$$

With each arc (i, j) we associate the probability $0 \le p_{ij} \le 1$ of being able to pass from i to j. The problem is, given two arbitrary vertices i_o and j_o, to find the probability of the path from i_o to j_o which is the most likely to exist. (We assume independence of the random events attached to the various arcs).

By interpreting the results of Sect. 3, we obtain the following properties:

- The maximum reliability of a path taking k arcs between i and j is $(A^k)_{ij}$.
- For any $a \in E$, $e \oplus a = \max(1, a) = 1 = e$, therefore there is no zero-absorbing circuit. We deduce (Theorem 1) the existence of $A^* = A^{(n-1)}$.
- The maximum reliability of a path between i and j is $(A^*)_{ij}$.

6.7. Multicriteria Path Problems

For the various optimization problems addressed in the previous sub-sections (6.3–6.6) one can define a *multicriteria* version of the problem which we formulate here in the case of the shortest path problem (one can define an equivalent formulation for the other optimization problems of Sects. 6.3–6.6).

Given a directed graph $G = [X, U]$, with each arc (i, j), we associate p lengths $v_{ij}^1, v_{ij}^2, \ldots, v_{ij}^k, \ldots v_{ij}^p$, with, $\forall k = 1, \ldots, p$:

$$v_{ij}^k \in \hat{\mathbb{R}} = \mathbb{R} \cup \{+\infty\}.$$

A vector $v \in \hat{\mathbb{R}}^p$ is said to be *efficient* with respect to a subset F of $\hat{\mathbb{R}}^p$ if there does not exist in F a vector $v' \neq v$ which has all its components smaller than or equal to the corresponding components of v. The problem is then to find all the efficient paths between two arbitrarily fixed vertices i and j, of G. We will therefore take for E: $\mathcal{P}(\hat{\mathbb{R}}^p)$, the power set of $\hat{\mathbb{R}}^p$. The operations $\oplus$ and $\otimes$ will be defined as follows:

$$\text{if} \quad u_\alpha = \{u_{\alpha_i}\} \quad \text{with} \quad u_{\alpha i} \in \hat{\mathbb{R}}^p \quad \text{and} \quad u_\beta = \{u_{\beta j}\} \quad \text{with} \quad u_{\beta j} \in \hat{\mathbb{R}}^p$$

then:

$$u_\alpha \oplus u_\beta = \text{set of efficient vectors of } u_\alpha \cup u_\beta$$
$$u_\alpha \otimes u_\beta = \text{set of efficient vectors in the set of vectors of the form } u_{\alpha_i} + u_{\beta j}.$$

It is easily verified that these two laws endow E with a dioid structure having neutral elements:

$$\varepsilon = (+\infty)^p \quad \text{and} \quad e = (0)^p.$$

By interpreting the results of Sect. 3, we then obtain the following properties:

- $(A^k)_{ij}$ represents the set of values of the efficient paths from i to j taking exactly k arcs.
- For any $v \in \hat{\mathbb{R}}^p$ where all the components are positive, we have $e \oplus v = e$ and v is 0-regular. Therefore, if there does not exist a circuit of negative length with respect to each of the p valuations of the graph, we deduce from Theorem 1 the existence of $A^* = A^{(n-1)}$.
- In this case, $(A^*)_{ij}$ represents the set of values of the efficient paths between i and j in G.

One can generalize the multicriteria problems thus defined in many ways. For example, one can consider multicriteria problems mixing various criteria such as capacity, length, reliability of a path, etc.

More generally, considering a partial order relation on a set T and an operation endowing T with a commutative monoid structure compatible with the order relation, one will be able to determine in $E = \mathcal{P}(T)$ the efficient paths which correspond to the minimal paths in the sense of the order relation.

6.8. The Kth Shortest Path Problem

For the various optimization problems discussed in Sects. 6.3–6.6., one can also seek to determine the k best paths between two given vertices i and j.

We consider here the case of the search for the k shortest paths (similar models for the other optimization problems of Sects. 6.3–6.6 could easily be defined).

Let us denote $\hat{\mathbb{R}} = \mathbb{R} \cup \{+\infty\}$ and let E be the cone of $\hat{\mathbb{R}}^k$ defined as follows:

$$u = (u_1, u_2, \ldots, u_i, \ldots, u_k) \in E$$

if and only if, for any i, $u_i \in \overline{\mathbb{R}}$ and: $u_1 \leq u_2 \ldots \leq u_k$.

With each arc (i, j), we associate the k-tuple $v_{ij} = (l_{ij}, +\infty, +\infty, \ldots, +\infty) \in E$ where l_{ij} represents the length of the arc (i, j).

In the case where a multigraph is being considered, there could exist several arcs from i to j. One then associates with arc (i, j) the vector v_{ij} where the components correspond to the lengths of the k smallest lengths of the arcs from i to j ordered according to nondecreasing values (and completed if necessary by components equal to $+\infty$).

The operation $\oplus$ is the operation $\mathrm{Min}_{(k)}$ (see Chap. 8, Sect. 1.3.1) defined as follows

$$\begin{aligned} \text{if} \quad u_\alpha &= (u_{\alpha_1}, u_{\alpha_2}, \ldots, u_{\alpha_k}) \quad \text{with} \quad u_{\alpha_i} \in \hat{\mathbb{R}} \\ u_\beta &= (u_{\beta 1}, u_{\beta_2}, \ldots, u_{\beta_k}) \quad \text{with} \quad u_{\beta_i} \in \hat{\mathbb{R}} \end{aligned}$$

then:

$$u_\alpha \oplus u_\beta = u_\gamma = (u_{\gamma_1}, u_{\gamma_2}, \ldots, u_{\gamma_k})$$

where the components of u_γ are the k smallest terms of $(u_{\alpha_1}, u_{\alpha_2}, \ldots, u_{\alpha_k}, u_{\beta_1}, u_{\beta_2}, \ldots, u_{\beta_k}))$ ordered according to non-decreasing values.

The operation $\otimes$ is the operation $\overset{(k)}{+}$ (see Chap. 8, Sect. 1.1.5) defined as:

$$u_\alpha \otimes u_\beta = u_\gamma = (u_{\gamma_1}, u_{\gamma_2}, \ldots, u_{\gamma_k})$$

where the components of u_γ are the k smallest terms of the form $u_{\alpha_i} + u_{\beta_j}$ with $i = 1, \ldots, k$ and $j = 1, \ldots, k$

These two laws endow E with a dioid structure with neutral elements: $\varepsilon = (+\infty)^k$ and $e = (0, +\infty, +\infty, \ldots, +\infty,)$ (see Chap. 8, Sect. 4.3.1).

Note that the operation $\oplus$ is not *idempotent*. For example if $u = (2, 3, 4, 4)$ then $u \oplus u = (2, 2, 3, 3) \neq u$.

Concerning computational complexity issues, it is easy to show that the operation $\oplus$ requires k comparisons and that the operation $\otimes$ can be implemented with $k \log_2 k$ comparisons and ordinary additions.

By interpreting the results of Sect. 3, we then deduce the following properties:

- $(A^p)_{ij}$ represents the values of the k shortest paths from i to j traversing exactly p arcs.
- $(A^{(p)})_{i,j}$ represents the values of the k shortest paths between i and j traversing at most p arcs.

  (Observe that the k shortest paths thus obtained are not all necessarily elementary).
- If $u \in E$ has all its components nonnegative, then u is $(k - 1)$ stable (see Chap. 3 Sect. 7). Therefore if the graph does not contain a circuit of negative length, the weight of each circuit is $(k - 1)$ stable and since the multiplication $\otimes$ is

commutative, we deduce from Theorem 2 of the present chapter the existence of $A^* = A^{(n_{k-1})}$. Here n_{k-1} denotes the maximum number of arcs in a path traversing each elementary circuit of G(A) no more than k-1 times (see Sect. 3.3, Theorem 2).

- $(A^*)_{ij}$ represents the values of the k best paths from i to j if there is no negative length circuit in the graph.

6.9. The Network Reliability Problem

Let G = [X, U] be a directed graph. With each arc (i, j) $\in$ U, we associate the Boolean variable y_{ij}, and the probability p_{ij} of the existence of arc (i, j). Assuming that the random variables associated with the various arcs are independent, we wish to determine the probability that two arbitrary given vertices i_o and j_o are linked by a path.

We will take for E the set of polynomials with entries in $\mathbb{Z}$, idempotent for ordinary multiplication (therefore these are polynomials in Boolean variables); addition is taken as the symmetrical difference ($a \oplus b = a + b - ab$), and multiplication is just ordinary multiplication. We therefore have $\varepsilon = 0$ and $e = 1$.

Let us quickly verify that these two laws endow E with a dioid structure. Two properties are not straight forward: the closure of $\oplus$ and the distributivity of $\otimes$ with respect to $\oplus$. Let us first check the closure of $\oplus$: if $P \in E$ and $Q \in E$, we have $P^2 = P$ and $Q^2 = Q$; then:

$$
\begin{aligned}
(P \oplus Q)^2 &= (P + Q - P \cdot Q)^2 = P^2 + Q^2 + P^2Q^2 + 2PQ - 2P^2Q - 2PQ^2 \\
&= P + Q + PQ + 2PQ - 2PQ - 2PQ = P + Q - PQ = P \oplus Q.
\end{aligned}
$$

Let us then check the distributivity of $\otimes$:

$$
\begin{aligned}
P \otimes (Q \oplus R) &= P(Q + R - QR) = PQ + PR - PQR = PQ + PR - PQ \cdot PR \\
&= PQ \oplus PR.
\end{aligned}
$$

Each polynomial P in E will be represented by its *reduced form*, i.e. such that each monomial of P be of degree at most 1 with respect to each Boolean variable y_{ij}.

By interpreting the results of Sect. 3, we then obtain:

- $(A^k)_{ij}$ is a polynomial in y, denoted $(A^k)_{ij}(y)$, such that $(A^k)_{ij}(p)$ represents the probability that j is linked to i by a path of length k;
- $(A^{(k)})_{ij}(p)$ is a polynomial representing the probability that j is linked to i by a path of length at most k;
- For any $P \in E$, $e \oplus P = 1 + P - P = e$, therefore there is no zero-absorbing circuit. We deduce (Theorem 1) the existence of $A^* = A^{(n-1)}$;
- $(A^*)_{ij}$ is a polynomial in p such that $(A^*)_{ij}(p)$ represents the probability that j is linked to i.

6.10. The η-Optimal Path Problem

For all optimization problems in Sects. 6.3–6.6 one can search for the set of η-optimal paths (i.e. optimal to within η) between two given vertices i and j.

We consider here the case of the search for the η-shortest path, leaving it to the reader to extend this model to the other optimization problems addressed in Sects. 6.3–6.6.

Let us denote $\hat{\mathbb{R}} = \mathbb{R} \cup \{+\infty\}$ and let E be the set of nondecreasing sequences of elements of $\hat{\mathbb{R}}$ defined as follows:

$$u = (u_1, \dots, u_i \dots, u_q) \in E \quad \text{if and only if:}$$

$q \geq 1$, $u_i \in \hat{\mathbb{R}}$ and $u_1 \leq u_2 \leq \dots \leq u_i \leq \dots \leq u_q$ and $u_q \leq u_1 + \eta$ where η is a given positive constant.

$\eta = 0$ will correspond to the problem of determining the set of optimal paths.

With each arc (i, j), we associate the sequence (consisting of a single term) $v_{ij} = (l_{ij}) \in E$, where l_{ij} represents the length of arc (i, j).

In the case where the graph under consideration is a multigraph, several arcs from i to j could exist. We then associate with the pair (i, j) the sequence v_{ij} of the lengths of the arcs from i to j with length deviating by at most η from the length of the shortest arc from i to j.

The operation ⊕ is the operation $\text{Min}_{(\leq\eta)}$ (see Chap. 8, Sect. 1.3.2) defined as follows:

$$\text{if} \quad u_\alpha = (u_{\alpha_i}) \in E$$
$$u_\beta = (u_{\beta_j}) \in E$$

then: $u_\alpha \oplus u_\beta = u_\gamma$ where u_γ, represents the ordered sequence of the terms of the sequences (u_{α_i}) and (u_{β_j}) smaller than or equal to $\min(u_{\alpha_1}, u_{\beta_1}) + \eta$.

The operation ⊗ is the operation $\overset{(\leq\eta)}{+}$ (see Chap. 8, Sect. 1.1.6) defined as: $u_\alpha \otimes u_\beta = u_\gamma$, where u_γ represents the ordered sequence of the terms of the form $u_{\alpha_i} + u_{\beta_j}$ smaller than or equal to $u_{\alpha_1} + u_{\beta_1} + \eta$.

We verify (see Chap. 8, Sect. 4.3.2) that these two laws endow E with a dioid structure with $\varepsilon = (+\infty)$ (the sequence formed of a single term equal to $+\infty$) and $e = (0)$ (the sequence formed of a single term equal to 0).

By interpreting the results of Sect. 3, we obtain the following properties:

- $(A^k)_{ij}$ represents the lengths of the paths from i to j taking k arcs and having a length deviating by at most η from that of the shortest path from i to j taking k arcs. (Let us observe, as previously, that the paths thus obtained are not necessarily elementary).

  If $u \in E$ is such that $u_1 > 0$, then u is p-stable with $p = \lceil \eta / u_1 \rceil$, see Chap. 3 Sect. 7.

Therefore if the graph does not contain a circuit of length negative or zero, the value of each circuit is p-stable (with p = integer rounding of [η/the length of the shortest circuit]) and since the multiplication ⊗ is commutative, we deduce from Theorem 2 the existence of $A^* = A^{(n_p)}$.

- $(A^*)_{ij}$ represents the lengths of the η-shortest paths from i to j if there does not exist a circuit of negative or zero length in the graph.

Let us finally discuss a few additional examples related to specific applications.

6.11. The Multiplier Effect in Economy

Let us consider Leontief's model of global balance by sectors (Leontief 1963). Each of the industrial sectors $j = 1, 2, \ldots, n$ of an economy is associated with a *type of product*.

Let $A = (a_{ij})$ be the matrix of the technical coefficients (or "input-output" coefficients): a_{ij} units of product i are required to manufacture a unit quantity of product in sector j.

If $d = (d_i)$ is the vector representing the final demand, the production x must satisfy the equation:

$$x = Ax + d$$

therefore

$$x = (I - A)^{-1}d = (I + A + A^2 + \cdots)d = A^*d$$

A^* is the quasi-inverse of the matrix A (transitive closure). A similar regularization is observed in the multiplier effect of investment according to Keynes.

The underlying algebraic structure in this example here is $(\mathbb{R}, +, \times)$, the field of real numbers endowed with the standard operations of addition and multiplication.

6.12. Markov Chains and the Theory of Potential

Let us consider a Markov chain having transient states T and a recurrent class C_a. Then if x_j is the probability of being absorbed in the class C_a, when starting from state j, the vector $x = (x_j)_{j\in T}$ must verify the equation:

$$x = Qx + r$$

where Q is the restriction of the transition matrix P of the Markov chain to transient states T and where, $\forall j \in T$:

$$r_j = \sum_{k\in C_a} p_{jk}$$

Then $x = Q^*r$ where Q^* is the so-called matrix of potentials.

The underlying algebraic structure here is $([0, 1], +, \times)$.

A more complex case is that of semi-Markovian processes. If $x_{ij}(t)$ is the probability of being in state i at time t knowing that we were in state j at time 0, the matrix

$x(t) = (x_{ij}(t))$ satisfies the equations.

$$x_{ij}(t) = \sum_{k=1}^{n} a_{kj}(t) * x_{ik}(t) + \delta_{ij} b_i(t)$$

where $*$ corresponds to the convolution product of two functions, a_{ij} (t) corresponds to the joint probability of leaving state j in the interval [0, t] and of passing to state i in a single transition, l-$b_i(t)$ corresponds to the unconditional distribution function of the time in state i.

Therefore:

$$x(t) = b(t) * (I + A(t) + A^2(t) + \cdots) = b(t) * A^*(t).$$

The underlying algebraic structure here is $((L^1(\mathbb{R}))^+, +, *)$ and the matrix $A^*(t)$ is the so-called *potential function*.

6.13. Fuzzy Graphs and Relations

It is said that we have the graph of a *fuzzy relation* whenever we are given a graph G valued on a set E, set on which we have defined two laws $\oplus$ and $\otimes$ corresponding respectively to set union and to set intersection. We associate with each arc (i, j) of G a value represented by a given element $a_{ij} \in E$. In many cases (see Table 2) $(E, \oplus, \otimes)$ is a dioid, and the *closure* of the fuzzy relation is the matrix A^*, the quasi-inverse of the matrix $A = (a_{ij})$.

For an extensive presentation of fuzzy systems and possibility theory, see for example Dubois and Prade (1980, 1987). For the extension to fuzzy integrals, see Sugeno (1977) who extends Choquet's capacity theory (1953). A survey of connections between dioids and fuzzy set theory can be found in Gondran and Minoux (2007).

Table 2 Examples of dioids associated with fuzzy relations

E	$\oplus$	$\otimes$	ε	e
[0, 1]	Max	Min	0	1
[0, 1]	Max	$\times$	0	1
Distributive lattice	Sup	Inf	Smallest element	Largest element
$\mathcal{P}([0, 1])$	$\cup$	$\cap$	$\emptyset$	[0, 1]

6.14. The Algebraic Structure of Hierarchical Clustering

Let us consider n objects. It is said that we have a *dissimilarity* index on these objects if we have assigned to each pair (i, j) a number $d_{ij} \in \mathbb{R}_+$ which will be all the larger that the objects i and j are dissimilar ($d_{ij} = 0$ if and only if i is identical to j).

This index is referred to as an *ultrametric distance* if it satisfies *the triangular ultrametric inequality*, i.e. if and only if, ∀i, j, k:

$$d_{ij} \le \max\{d_{ik}, d_{kj}\}$$

Considering the dioid $(E, \oplus, \otimes) = (\overline{\mathbb{R}}_+, \min, \max)$, we can state:

Proposition 6.14.1. *A matrix* $A \ge 0$ *with zero diagonal corresponds to an ultrametric distance distance if and only if we have, in the dioid* $(\overline{\mathbb{R}}_+, \min, \max)$*:*

$$A = A^2$$

Proof. Indeed, A will represent an ultrametric distance if and only if ∀i, j, k:

$$a_{ij} \le a_{ik} \otimes a_{kj}$$

therefore, since $a_{ii} = 0$, if and only if ∀i, j:

$$a_{ij} = \min_k (a_{ik} \otimes a_{kj}) = \sum_k^{\oplus} a_{jk} \otimes a_{kj}. \quad \square$$

A classification tree corresponds to a nested set of partitions of the set of objects (for an example, refer to Chap. 6, Sect. 6.1).

One can then show (see for example Benzecri (1974)) that there in a one-to-one correspondence between ultrametrics and their associated indexed classification trees (or indexed hierarchical clustering).

To construct a classification tree of n objects from a given dissimilarity matrix D, we therefore have to approximate the dissimilarity index by an ultrametric distance. There are many ways of constructing such an approximation. Let us consider *lower ultrametric distances*, i.e. such that:

$$A \le D$$

where A is a matrix corresponding to an ultrametric. We will say that A is the matrix associated with a *lower ultrametric* distance.

Proposition 6.14.2. *(Gondran, 1976a,b)*

The set of lower ultrametric distances has a largest element D^**, called sub-dominant ultrametric distance, which satisfies:*

$$D^* = D^{n-1} = D^n = D^{n+1} = \cdots$$

Proof. Any lower ultrametric distance A satisfies, by definition, $A \leq D$, hence we deduce $A^k \leq D^k$ for any k.

Now according to Theorem 1, in the dioid ($\overline{\mathbb{R}}_+$, Min, Max), the sequence D^k converges as soon as $k = n - 1$ towards:

$$D^* = D^{n-1} = D^n = D^{n+1} = \cdots$$

A being an ultrametric distance, this yields, according to Proposition 6.14.1: $A^k = A$. We deduce for any A, $A \leq D^*$.

Since:

$$(D^*)^2 = D^{2n-2} = D^{n-1} = D^*$$

D^* corresponds to an ultrametric which is therefore larger than any other ultrametric. □

Thus, through algebraic means, we find again a property well known in classification theory (see for example Benzecri, 1974).

Further properties of this algebraic structure in clustering, will be investigated in Chap. 6 Sect. 6.1, where the levels of a hierarchical clustering are interpreted in terms of eigenvalues and eigenvectors of the dissimilarity matrix on the dioid ($\overline{\mathbb{R}}_+$, Min, Max).

Exercises

Exercise 1. p-nilpotency for the shortest path problem with time constraints

Consider again the definitions in Example 3 of Sect. 4.4.3.

Let us define on S the following total preorder relation:

$$E \leq F \Leftrightarrow \min\{t/t \in E) \leq \min\{t/t \in F).$$

We will denote H' the set of given endomorphisms, i.e.:

$$H' = \left\{\varphi_{ij}/(i, j) \in U\right\}$$

and we set: $d_{min} = \min\{d_{ij}/(i, j) \in U\}$.

(1) Show that for any $\varphi_{ij} \in H'$ and for any $E \in S$, the following holds:

$$\varphi_{ij}(E) \geq d_{min} \top E.$$

(2) Let t_{max} be the largest of the coefficients involved in V_{ij} and W_i.
Show that one can always assume $t_{max} < +\infty$.

Let $p = \lceil t_{max}/d_{min} \rceil$

Show then that $\forall h_1, h_2, \ldots, h_p \in H'$:

$$\bar{h}(E) = (h_1 \circ h_2 \circ \cdots \circ h_p)(E) \geq t_{max} \top E.$$

Deduce the p-nilpotency of the endomorphisms of H'.

[*Answers*: see Minoux (1976)].

Exercise 2. Convergence of Algorithm 3′ of Sect. 4.4.4

(1) Show that E is minimal as soon as r is transferred into X_1.
(2) Show that in the case where the total preorder relation is a total order relation, we always have $X_1 \cap X_2 = \emptyset$.

[*Answers*: see Minoux (1976)].

Exercise 3. Constrained shortest path problem

Let $G = [X, U]$ be a directed graph where each arc $u \in U$ is endowed with two numbers l_u and α_u (e.g. l_u will be a distance, and α_u a transit time). Let Q be the set of all the elementary paths between two particular vertices s and t. For β, a given real number, let $Q' \subset Q$ be the subset of paths π satisfying the additional constraint:

$$\alpha(\pi) = \sum_{u \in \pi} \alpha_u \le \beta.$$

We wish to solve the shortest path problem with constraint between s and t:

$$(P):\ \underset{\pi \in Q'}{\text{Min}} \left\{ \ell(\pi) = \sum_{u \in \pi} \ell_u \right\}.$$

It is assumed that for any circuit μ of G we have:

$$l(\mu) > 0 \quad \text{and} \quad \alpha(\mu) > 0.$$

(1) Provide an algorithm to test whether $Q' \neq \emptyset$. Assuming this condition fulfilled, show that an optimal solution to (P) is necessarily an *elementary* path.
(2) A possible approach to problem (P) consists in associating with each vertex $j \in X$ a list L_j of $\bar{\nu}_j$ pairs of real numbers

$$\rho_j^\nu = \ell(\pi_\nu), \sigma_j^\nu = \alpha(\pi_\nu), \quad \text{for} \quad \nu = 1, \ldots, \bar{\nu}_j$$

Clearly, all the σ_j^ν satisfy: $\sigma_j^\nu \le \beta$
If ν_1 dominates ν_2, i.e. if

$$\rho_j^{\nu_1} \le \rho_j^{\nu_2} \quad \text{and} \quad \sigma_j^{\nu_1} \le \sigma_j^{\nu_2}$$

then ν_2 can be eliminated from the list: we obtain a reduced list. A list that can no longer be reduced is said to be *irreducible*.

Let $\mathcal{L}$ be the set of all the finite irreducible lists endowed with the law of internal composition $\oplus$ defined as:

$L_1 \in \mathcal{L}, L_2 \in \mathcal{L} \Rightarrow L_1 \oplus L_2 =$ reduced union of the lists L_1 and L_2.

This law is idempotent, and the zero element is the empty list: $\varepsilon = \emptyset$.

With each arc $u = (i, j)$ of G we associate the endomorphism φ_{ij} of $(\mathcal{L}, \oplus)$ where $L_j = \varphi_{ij}(L_i)$ is defined as:

$$L_j = \left(\rho_j^\nu, \sigma_j^\nu\right) \quad \nu = 1, \ldots, \bar{\nu}_j$$

L_i is the reduced list formed by all the pairs $\left(\rho_i^\nu + \ell_u, \sigma_i^\nu + \alpha_u\right)$ such that

$$\sigma_i^\nu + \alpha_u \leq \beta.$$

If $\mathcal{F}$ is the set of endomorphisms of $\mathcal{L}$, show that $(\mathcal{L}, \mathcal{F}, \oplus, \circ)$ is an algebra of endomorphisms on $\mathcal{L}$. Show next that one can use a Generalized Dijkstra's Algorithm on this structure to solve the problem (P).

State precisely this algorithm, and show that one can deduce (by limiting the size of the lists) a family of approximate methods.

[*Answers*: see Minoux (1975, 1976)].

Exercise 4. Right dioid and shortest path with gains or losses

Let $G = [X, U]$ be a directed graph on which a given type of product circulates. With each arc $u = (i, j) \in U$ two numbers are associated:

- c_{ij} representing the unit transportation cost of the product between i and j.
- $m_{ij} > 0$ representing the loss coefficient (if $m_{ij} < 1$) or gain (if $m_{ij} > 1$) of the product during transport from i to j; in other words, if q_i is the quantity of product available in i, the quantity available in j after traversing arc (i, j) is $m_{ij}\, q_i$.

Let us consider two arbitrary vertices $i \in X$ and $j \in X$ and $\mu \in P_{ij}$ a path joining these two vertices:

$$\mu = \{(i_0, i_1)(i_1, i_2), \ldots (i_{p-1}, i_p)\} \quad \text{(with} \quad i = i_0 \quad \text{and} \quad j = i_p).$$

Let us denote c (μ) the transport cost of a product unit between i and j via the path μ, and m (μ) the overall gain (or loss) coefficient along the path μ.

These quantities are defined by induction as follows:

If $\mu = \emptyset$: $c(\mu) = 0$; $m(\mu) = 1$.
If μ is a path between i and j and $\mu' = \mu \cup \{(j, k)\}$ a path between i and k, then

$$c(\mu') = c(\mu) + m(\mu)c_{jk}$$
$$m(\mu') = m(\mu) \cdot m_{jk}$$

The shortest path problem with gains (or losses) is to determine the path of minimum unit cost between two given vertices i and j, in other words to minimize the ratio $\frac{c(\mu)}{m(\mu)}$ on the set P_{ij} of paths μ from i to j.

(1) We consider the set $E = \mathbb{R} \times (\mathbb{R}_+ \backslash \{0\})$ endowed with the operations $\oplus$ and $\otimes$ defined as follows:

$$\begin{pmatrix} c \\ m \end{pmatrix} \oplus \begin{pmatrix} c' \\ m' \end{pmatrix} = \begin{cases} \begin{pmatrix} c \\ m \end{pmatrix} \text{if } \frac{c}{m} < \frac{c'}{m'}, \text{ or if } \frac{c}{m} = \frac{c'}{m'} \text{ and } m = \text{Max}\{m, m'\} \\ \begin{pmatrix} c' \\ m \end{pmatrix} \text{if } \frac{c}{m} > \frac{c'}{m'}, \text{ or if } \frac{c}{m} = \frac{c'}{m'} \text{ and } m' = \text{Max}\{m, m'\} \end{cases}$$

$$\begin{pmatrix} c \\ m \end{pmatrix} \otimes \begin{pmatrix} c' \\ m' \end{pmatrix} = \begin{pmatrix} c + mc' \\ mm' \end{pmatrix}$$

It will be observed that $\otimes$ is not commutative and only right distributive with respect to $\oplus$.

Verify that $(E, \oplus, \otimes)$ is a *right dioid*. We will denote ε and e the neutral elements of $\oplus$ and $\otimes$, respectively.

(2) The incidence matrix A of G is the $N \times N$ matrix ($N = |X|$) defined as:

$$\begin{cases} a_{ij} = \begin{pmatrix} c_{ij} \\ m_{ij} \end{pmatrix} & \text{if} \quad (i, j) \in U \\ a_{ij} = \varepsilon & \text{if} \quad (i, j) \notin U \\ a_{ii} = e & \qquad \forall i \in X \end{cases}$$

The weight w (μ) of an arbitrary path $\mu \in P_{ij}, \mu = \{(i_0, i_1), (i_1, i_2) \ldots (i_{p-1}, i_p)\}$ is defined as the product (in this order):

$$w(\mu) = a_{i_0,i_1} \otimes a_{i_1,i_2} \otimes \cdots \otimes a_{i_{p-1},i_p}$$

By using only right distributivity of $\otimes$ with respect to $\oplus$, show that we indeed have:

$$\left(A^k\right)_{i,j} = \sum_{\mu \in p_{ij}^k} w(\mu)$$

and:

$$\left(A^{(k)}\right)_{i,j} = \sum_{\mu \in p_{ij}^{(k)}} w(\mu)$$

(3) G is said to have no right 0-absorbing circuit if, for every elementary circuit γ of G, we have, $\forall a \in E$: $a \oplus a \otimes w(\gamma) = a$.

G being assumed to have no right 0-absorbing circuit, show that the following holds:

$$\left(A^k\right)_{i,j} = \sum_{\mu \in p_{ij}^k(0)} w(\mu)$$

and

$$\left(A^{(n-1)}\right)_{i,j} = \sum_{\mu \in p_{ij}(0)} w(\mu)$$

(where P_{ij}^k (0) is the set of elementary paths with exactly k arcs and P_{ij} (0) the set of elementary paths between i and j).

Deduce from the above:

$$A^{(n-1)} = \lim_{k \to \infty} A^{(k)} = A^*$$

where A^* satisfies:

$$A^* = A^* \otimes A \oplus I$$

(4) Show that, when some c_{ij} can be < 0, G has no right 0-absorbing circuit if and only if any circuit γ satisfies $c(\gamma) \geq 0$ and $m(\gamma) = 1$.

When all the c_{ij} are ≥ 0, show that G has no right 0-absorbing circuit, if and only any circuit γ satisfies $m(\gamma) \leq 1$.

(5) Verify then that the generalized algorithms of Jacobi, Gauss–Seidel and Gauss Jordan (see Sect. 4.1, 4.2 and 5 of Chap. 4) can be used to solve the shortest path problem with gains (or losses) in a graph without right 0-absorbing circuit.

(6) Taking into account the fact that $\oplus$ is selective and, under the additional assumption that $e = \binom{0}{1}$ is the largest element of E ($\forall a \in E$: $a \oplus e = e$), show that the generalized Dijkstra Algorithm (see Sect. 4.3) also applies.

Verify that the above assumptions are satisfied when, for every arc (i, j) of G:

$$c_{ij} \geq 0$$
$$0 < m_{ij} \leq 1.$$

[*Answers*: see Charnes and Raike (1966), Bako (1974), Gondran (1976b)].

Chapter 5
Linear Dependence and Independence in Semi-Modules and Moduloids

1. Introduction

The present chapter is devoted to problems of linear dependence and independence in semi-modules (and moduloids). The semi-module structure (resp. the moduloid structure) is the one which arises naturally in the properties of sets of vectors with entries in a semiring (resp. in a dioid). Thus, they turn out to be analogues for algebraic structures on semirings and dioids to the concept of a module for rings.

Section 2 introduces the main basic notions such as morphisms of semi-modules, definitions of linear dependence and independence, generating families and bases in semi-modules. As opposed to the classical case, it will be shown that, in many cases, when a semi-module has a basis, it is unique.

Section 3 is then devoted to studying the links between the *bideterminant of a matrix* and the concepts of linear dependence and independence previously introduced. Several classical results of linear algebra over vector fields are generalized here to semi-modules and moduloids, in particular those related to *selective-invertible dioids* and *MAX-MIN dioids*.

2. Semi-Modules and Moduloids

The concept of *semi-module* generalizes that of module when the reference set is a semiring instead of a ring.

2.1. Definitions

Definition 2.1.1. *(semi-module)*

Let (E, $\oplus$, $\otimes$) *be a commutative semiring where* ε *and* e *denote the neutral elements of* $\oplus$ *and* $\otimes$*, respectively. We refer to as a* semi-module *on* E *a set* M *endowed with an internal law* $\square$ *and an external law* $\perp$ *satisfying the following conditions:*

(a) $(\mathrm{M}, \square)$ *is a commutative monoid where the neutral element is denoted* 0*;*
(b) $\perp$ *is an external law on* M *which, with any* $\lambda \in \mathrm{E}, \mathrm{x} \in \mathrm{M}$ *associates* $\lambda \perp \mathrm{x} \in \mathrm{M}$ *satisfying:*

(b1) $\forall \lambda \in \mathrm{E}, \forall (\mathrm{x}, \mathrm{y}) \in \mathrm{M}^2$:

$$\lambda \perp (\mathrm{x} \square \mathrm{y}) = (\lambda \perp \mathrm{x}) \square (\lambda \perp \mathrm{y})$$

(b2) $\forall (\lambda, \mu) \in \mathrm{E}^2, \forall, \mathrm{x} \in \mathrm{M}$:

$$(\lambda \oplus \mu) \perp \mathrm{x} = (\lambda \perp \mathrm{x}) \square (\mu \perp \mathrm{x})$$

(b3) $\forall (\lambda, \mu) \in \mathrm{E}^2, \forall, \mathrm{x} \in \mathrm{M}$

$$\lambda \perp (\mu \perp \mathrm{x}) = (\lambda \otimes \mu) \perp \mathrm{x}$$

(b4) $\forall \mathrm{x} \in \mathrm{M}$,

$$\mathrm{e} \perp \mathrm{x} = \mathrm{x}$$
$$\varepsilon \perp \mathrm{x} = 0$$

(b5) $\forall \lambda \in \mathrm{E} \quad \lambda \perp 0 = 0$

When the reference set $(\mathrm{E}, \oplus, \otimes)$ is a *non-commutative* semiring, it is necessary to distinguish between the concept of *left semi-module* (where the operation $\perp$ represents the multiplication on the left of a vector by a scalar) and the concept of *right semi-module* (where the operation $\perp$ represents the multiplication on the right of a vector by a scalar). When $\otimes$ is commutative, the concepts of left semi-module and right semi-module coincide.

The following definition corresponds to the generalization of the concept of module when the reference set is a dioid instead of a ring.

Definition 2.1.2. *(moduloid)*
A semi-module on E *is referred to as a* moduloid *when* $(\mathrm{E}, \oplus, \otimes)$ *is a dioid and* $(\mathrm{M}, \square)$ *is a canonically ordered commutative monoid.*

Example 2.1.3. $(\mathrm{E}, \oplus, \otimes)$ being a semiring, let us consider E^n, the set of n-vectors with components on E endowed with the operations $\square$ and $\perp$ defined as:

$$\mathrm{x} = (\mathrm{x}_i)_{i=1,\ldots,n} \quad \mathrm{y} = (\mathrm{y}_i)_{i=1,\ldots,n}$$
$$\mathrm{x} \square \mathrm{y} = \mathrm{z} = (\mathrm{z}_i)_{i=1,\ldots,n} \quad \text{where } \forall i\text{: } \mathrm{z}_i = \mathrm{x}_i \oplus \mathrm{y}_i$$
$$\lambda \in \mathrm{E} \qquad \lambda \perp \mathrm{x} = \mathrm{u} = (\mathrm{u}_i)_{i=1,\ldots,n}$$
$$\text{where, } \forall i\text{:} \quad \mathrm{u}_i = \lambda \otimes \mathrm{x}_i.$$

It is easily verified that the set $(\mathrm{E}^n, \square, \perp)$ thus defined is a semi-module. According to common practice regarding modules on $\mathbb{Z}$ and vector fields on $\mathbb{R}$, the law □ (addition of vectors) introduced above, will be denoted ⊕ and the external law ⊥ (left multiplication of a vector by a scalar) will be denoted ⊗.

Let us also observe that, in reference to this classical example, the elements of a semi-module M are referred to as *vectors*. ||

Example 2.1.4. Let us consider again Example 2.1.3. above, but now assuming that (E, ⊕) is a dioid. Then $(E^n, \square)$ is canonically ordered and $(E^n, \square, \perp)$ is a *moduloid*. ||

Example 2.1.5. Let (M, ⊕) be a commutative monoid, with neutral element ε, and let us consider the dioid $(\mathbb{N}, +, \times)$ (see Chap. 8, Sect. 4.4.1).

Let us define the external law ⊥ operating on the elements of ℕ by:

$\forall n \in \mathbb{N}, \forall x \in M, n \perp x = x \oplus x \oplus \ldots \oplus x$ (n times) with the convention that $0 \perp x = \varepsilon$.

It is easily verified that (M, ⊕, ⊥) is a semi-module on $(\mathbb{N}, +, \times)$. If (M, ⊕) is canonically ordered, it is recognized as a *moduloid*. ||

Example 2.1.6. Let (E, ⊕, ⊗) be a commutative semiring, and $A \in M_n(E)$ a square $n \times n$ matrix with entries in E. Let $\lambda \in E$ and $V \in E^n$ such that:

$$A \otimes V = \lambda \otimes V$$

(V is said to be an eigenvector of A for the eigenvalue λ).

The set $\mathcal{V}_\lambda$ of all the eigenvectors of A for the eigenvalue $\lambda \in E$ is a *semi-module*. Indeed, $\forall V \in \mathcal{V}_\lambda, W \in \mathcal{V}_\lambda, (\alpha, \beta) \in E^2$:

$$\begin{aligned} A \otimes (\alpha \otimes V \oplus \beta \otimes W) &= (\alpha \otimes \lambda) \otimes V \oplus (\beta \otimes \lambda) \otimes W \\ &= \lambda \otimes (\alpha \otimes V \oplus \beta \otimes W) \end{aligned}$$

Hence we deduce that $\alpha \otimes V \oplus \beta \otimes W \in \mathcal{V}_\lambda$.

Chapter 6 will be devoted to the detailed study of eigen-semi-modules or eigen-moduloids associated with matrices. ||

2.2. Morphisms of Semi-Modules or Moduloids. Endomorphisms

Definition 2.2.1. *Let* U *and* V *be two semi-modules on the same semiring* (E, ⊕, ⊗). *The internal laws are denoted* □ *and* □′ *respectively and the external laws* ⊥ *and* ⊥′ *respectively. We call* morphism of semi-modules *from* U *to* V *any mapping* φ: U → V *satisfying the following conditions:*

(i) $\forall (x, y) \in U^2 \quad \varphi(x \square y) = \varphi(x) \square' \varphi y)$
(ii) $\forall (\lambda, x) \in E \times U$: $\varphi(\lambda \perp x) = \lambda \perp' \varphi(x)$

A morphism of semi-modules from U *to itself is referred to as an* endomorphism *(of a semi-module).*

When the reference set is a dioid we refer to morphisms or endomorphisms of moduloids.

According to common practice, morphisms of semi-modules can also be referred to as *linear mappings*.

2.3. Sub-Semi-Module. Quotient Semi-Module

Definition 2.3.1. *(sub-semi-module)*

Let $(M, \Box, \perp)$ *be a semi-module on a semiring* $(E, \oplus, \otimes)$. *We refer to as a sub-semi-module of* M *any subset* $M' \subset M$ *containing 0 and stable for the laws induced by* $\Box$ *and* $\perp$.

Definition 2.3.2. *(quotient semi-module)*

Let $(M, \Box, \perp)$ *be a semi-module on* E, $(M', \Box, \perp)$ *a sub-semi-module of* M *and* M/M' *the quotient set of* M *with respect to the equivalence relation* $\mathfrak{R}$ *defined as:*

$$x \,\mathfrak{R}\, y \Leftrightarrow \exists(u, v) \in M'^2$$
$$\text{such that:} \quad x \Box u = y \Box v$$

M/M' *is referred to as the quotient semi-module of* M *by* M′.

It is easily verified that the equivalence relation $\mathfrak{R}$ is compatible with the laws $\Box$ and $\perp$ of M. Indeed:

$$x_1 \,\mathfrak{R}\, y_1 \Leftrightarrow x_1 \Box u_1 = y_1 \Box v_1 \quad \text{with} \quad u_1 \in M', v_1 \in M'$$
$$x_2 \,\mathfrak{R}\, y_2 \Leftrightarrow x_2 \Box u_2 = y_2 \Box v_2 \quad \text{with} \quad u_2 \in M', v_2 \in M'$$

hence we deduce $(x_1 \Box x_2) \,\mathfrak{R}\, (y_1 \Box y_2)$ since $u_1 \Box u_2 \in M'$ and $v_1 \Box v_2 \in M'$.

Moreover, for $\lambda \in E$,

$$x \,\mathfrak{R}\, y \Leftrightarrow x \Box u = y \Box v \quad \text{with} \quad u \in M', v \in M'$$

hence we deduce:

$$(\lambda \perp x) \Box (\lambda \perp u) = (\lambda \perp y) \Box (\lambda \perp v)$$

which shows that $(\lambda \perp x) \,\mathfrak{R}\, (\lambda \perp y)$ since $\lambda \perp u \in M'$ and $\lambda \perp v \in M'$

It follows from the above that the *canonical surjection* φ (which, to any element x of M, lets correspond its equivalence class in M/M') *is a morphism of semi-modules* (or: linear mapping).

2.4. Generated Sub-Semi-Module. Generating Family of a (Sub-) Semi-Module

Definition 2.4.1. *(sub-semi-module generated by a family of elements)*

Let $(M, \Box, \perp)$ *be a semi-module on* E *and* $X = (x_i)_{i \in I}$ *an arbitrary non-empty family (whether finite or infinite) of elements of* M. *We call a* sub-semi-module generated by X, *denoted* Sp(X), *the smallest sub-semi-module of* M *containing* X. *If* Sp(X) = M, X *is said to be a* generating family *(or: generator) of* M.

We easily prove:

Proposition 2.4.2. *Let* $(M, \Box, \bot)$ *be a semi-module on* E *and* $X = (x_i)_{i \in I}$ *an arbitrary non-empty family (whether finite or infinite) of elements of* M.

Then $Sp(X)$ *is the set* Y *of all the elements* $y \in M$ *of the form:*

$$y = \sum_{j \in J} \lambda_j \bot x_j \tag{1}$$

(summation in the sense of $\Box$*) where* $J \subset I$ *is a finite subset of indices and,* $\forall, j \in J$, $\lambda_j \in E$.

Proof. Clearly Y, the set of y's obtained by (1), is a semi-module, the axioms (a) and (b1) — (b5) being satisfied. This set contains X (to obtain x_i it suffices to take $J = \{i\}$ and $\lambda_i = e$) and 0 the neutral element of M (for that it suffices to take arbitrary $i \in I$ and $J = \{i\}$ $\lambda_i = \varepsilon$). Y is therefore a sub-semi-module of M containing X and consequently $Sp(X) \subseteq Y$.

Moreover, it can be observed that any sub-semi-module of M containing $X = (x_i)_{i \in I}$ contains all the linear combinations of the x_i's. Therefore

$$Y \subseteq Sp(X)$$

We deduce that $Y = Sp(X)$ and Y is the smallest sub-semi-module of M containing X. □

2.5. Concept of Linear Dependence and Independence in Semi-Modules

In this section we propose a definition of the concepts of linear dependence and independence in semi-modules, which constitutes an extension of the corresponding concepts in standard linear algebra. Links with alternative definitions suggested by other authors, will also be mentioned.

Let us consider an arbitrary non-empty family (whether finite or infinite) $X = (x_i)_{i \in I}$ of elements in a semi-module $(M, \Box, \bot)$.

For any subset of indices $J \subset I$ we will denote X_J the sub-family of X restricted to the elements $x_j, j \in J$, and $Sp(X_J)$ the sub-semi-module generated by X_J.

Definition 2.5.1. *The family* $X = (x_i)_{i \in I}$ *is said to be* dependent *if and only if, there exist two finite disjoint subsets of indices* $I_1 \subset I$ *and* $I_2 \subset I$ *together with values* $\lambda_i \in E \backslash \{\varepsilon\}$ $(i \in I_1 \cup I_2)$, *such that:*

$$\sum_{i \in I_1} \lambda_i \bot x_i = \sum_{i \in I_2} \lambda_i \bot x_i \tag{2}$$

A family which is not dependent will be said to be independent, *this property being expressed by the condition:*

$$\begin{aligned}&\forall I_1 \subset I,\ \forall, I_2 \subset I, I_1 \cap I_2 = \emptyset: \\ &Sp(X_{I_1}) \cap Sp(X_{I_2}) = \{0\}\end{aligned} \tag{3}$$

It follows directly from the previous definition that any sub-family of an independent family is independent.

The concept of dependence in semi-modules as defined by (2) was introduced and studied by Gondran and Minoux (1977, 1978, 1984).

Remark: concepts of redundant and quasi-redundant families

Alternative concepts related to dependence and independence in semi-modules have been studied by other authors (Cuninghame-Green 1979; Cohen et al. 1985; Moller 1987; Wagneur 1991) who proposed them as possible definitions of dependence and independence. Nonetheless, these concepts correspond to much stronger notions of dependence than (2), which limits the range of applications (e.g. they would not make it possible to obtain the equivalent of our Theorem 2 Sect. 3.4 below). This is why, in what follows, we have chosen to give these concepts different names.

We will say that the family $X = (x_i)_{i \in I}$ is *redundant* if and only if there exists $i \in I$ and $I_1 \subset I$ $(i \notin I_1)$ such that:

$$x_i \in Sp(X_{I_1}) \tag{4}$$

In the opposite case, the family X will be said to be *non-redundant.*

We will say that the family $X = (x_i)_{i \in I}$ is *quasi-redundant* if and only if there exists $i \in I$, $\lambda \in E \backslash \{\varepsilon\}$ and $I_1 \subset I$ $(i \notin I_1)$ such that:

$$\lambda \perp x_i \in Sp(X_{I_1}) \tag{5}$$

In the opposite case, X will be said to be *non-quasi-redundant.*

The concept of quasi-redundancy (corresponding to (5)) was proposed as a definition of dependence by Wagneur (1991) and the concepts of redundancy and non-redundancy corresponding to (4) were introduced and studied first by Cuninghame-Green (1979), then by Cohen et al. (1985), Moller (1987) and Wagneur (1991).

It is easy to see that for a family $X = (x_i)_{i \in I}$ of elements in a semi-module, independence in the sense of Definition 2.5.1. implies non-quasi-redundancy, which implies non-redundancy (see Exercise 1). Moreover, when $(E, \otimes)$ has a group structure, then the concepts of redundancy and quasi-redundancy coincide. ||

Definition 2.5.2. *We refer to as a* basis *of a semi-module* $(M, \Box, \perp)$ *an independent generating family.*

We are going to show that, under specific conditions, if a semi-module $(M, \Box, \perp)$ has a basis, the latter is unique.

In order to do so let us first introduce the concept of *reducibility.*

Definition 2.5.3. *Let* $(M, \Box, \perp)$ *be a semi-module on* $(E, \oplus, \otimes)$.

Given a set of vectors $V = (V_k), k \in K$, *with* $V_k \in M (\forall k)$, *it is said that a vector* $x \in M$ *is* reducible *on* $Sp(V)$ *if and only if there exists* $y \neq x$ *and* $z \neq x$, $y \in Sp(V)$, $z \in Sp(V)$ *such that:* $x = y \Box z$.

In the opposite case, x will be said to be irreducible.

Remark: *x reducible on* $Sp(V) \Rightarrow x \in Sp(V)$ ||

As an immediate consequence of this definition, we have the following property.

Property 2.5.4. If x is irreducible on Sp(V) then one (and only one) of the two following conditions is satisfied:

(i) $x \notin Sp(V)$
(ii) $x = y \Box z$ with $y \in Sp(V)$ and $z \in Sp(V)$
$\Rightarrow y = x$ or $z = x$

We can now establish:

Proposition 2.5.5. *Let* $(E, \oplus, \otimes)$ *be a dioid.* ε *and e denoting the neutral elements for* $\oplus$ *and* $\otimes$ *respectively, it is assumed that* $a \oplus b = e \Rightarrow a = e$ *or* $b = e$ *(observe that this assumption holds in selective dioids).*

Let $(M, \Box, \perp)$ *be a moduloid on* E, *canonically ordered by* $\Box$. *We denote* $\propto$ *the canonical order relation on* M.

It is assumed moreover that for $u \in M$, $v \in M$, $\lambda \in E$ *with* $v \neq u$ *and* $v \neq 0$

$$v = \lambda \perp v \Box u \Rightarrow \lambda = e \quad \textit{(see remark 2.5.6 below)}$$

Under the above assumptions, if $X = (x_i)_{i \in I}$ *is an independent family of elements of* M *(with* $x_i \neq 0, \forall, i \in I$*), then* $\forall j \in I$, x_j *is irreducible on* $Sp(X)$.

Proof. Clearly, for all $j \in I$, $x_j \in Sp(X)$. We thus have to prove 2.5.4 (ii).

Let us assume that $x_j = y \Box z$ with $y \in Sp(X)$ and $z \in Sp(X)$. Observe that this implies:

$y \propto x_j$ and $z \propto x_j$. Now:
$y \in Sp(X) \Rightarrow$
$\exists \lambda_i \in E\backslash\{\varepsilon\}, \exists I_1 \subset I$: $y = \sum\limits_{i \in I_1} \lambda_i \perp x_i$ (summation in the sense of $\Box$).

Similarly

$$z \in Sp(X) \Rightarrow$$
$$\exists \mu_i \in E\backslash\{\varepsilon\}, \exists I_2 \subset I: z = \sum_{i \in I_2} \mu_i \perp x_i$$

By agreeing to set $\lambda_i = \varepsilon$ for $i \in I_2\backslash I_1$ and $\mu_i = \varepsilon$ for $i \in I_1\backslash I_2$ we therefore have:

$$x_j = \sum_{i \in I_1 \cup I_2} (\lambda_i \oplus \mu_i) \perp x_i$$

Let us observe that, necessarily, $j \in I_1 \cup I_2$ (otherwise the hypothesis of independence would be contradicted). Consequently:

$$x_j = (\lambda_j \oplus \mu_j) \perp x_j \,\Box \sum_{\substack{i \in I_1 \cup I_2 \\ i \neq j}} (\lambda_i \oplus \mu_i) \perp x_i$$

with $\lambda_j \oplus \mu_j \neq \varepsilon$. By setting $\lambda = \lambda_j \oplus \mu_j$ and $u = \sum_{\substack{i \in I_1 \cup I_2 \\ i \neq j}} (\lambda_i \oplus \mu_i) \perp x_i \in$ $Sp(X \backslash \{x_j\})$, we obtain:

$$x_j = \lambda \perp x_j \,\Box\, u \tag{6}$$

with $u \in Sp(X \backslash \{x_j\})$.

We have $x_j \neq 0$, and, because of the independence, we must have $x_j \neq u$, hence $\lambda \neq \varepsilon$ follows. The hypotheses of Proposition 2.5.5 then imply $\lambda = e$.

Since $\lambda = \lambda_j \oplus \mu_j$, we must have either $\lambda_j = e$, or $\mu_j = e$.

Let us assume for example $\lambda_j = e$.

Then y is rewritten:

$$y = x_j \,\Box \sum_{i \in I_1 \backslash \{j\}} \lambda_i \perp x_i$$

hence one can deduce: $x_j \propto y$ and, given that $\propto$ is an order relation, this implies $y = x_j$.

In the case where $\mu_j = e$, we would similarly deduce that $z = x_j$.

We deduce the irreducibility of x_j $\Box$.

Remark 2.5.6. The assumption:

$$\left.\begin{array}{l} v = \lambda \perp v \,\Box\, u \\ v \neq u, \ v \neq 0 \end{array}\right\} \Rightarrow \lambda = e$$

is satisfied in many moduloids, in particular those associated with selective-invertible or selective-cancellative dioids.

For instance, let us consider a moduloid in which the elements are n-component vectors on $(E, \oplus, \otimes)$ with the usual laws.

$$(x_i)_{i=1\ldots n} \,\Box\, (y_i)_{i=1\ldots n} = (x_i \oplus y_i)_{i=1\ldots n}$$

$$\lambda \perp (x_i)_{i=1\ldots n} = (\lambda \otimes x_i)_{i=1\ldots n}$$

$$0 = \begin{pmatrix} \varepsilon \\ \varepsilon \\ \vdots \\ \varepsilon \end{pmatrix}$$

The hypothesis $v \neq u$ implies that there exists i such that $v_i \neq u_i$, and the relation $v = \lambda \perp v \,\Box\, u$ implies: $v_i = \lambda \otimes v_i \oplus u_i$.

Necessarily in this case $v_i \neq \varepsilon$. ($v_i = \varepsilon$ would imply $u_i = \varepsilon$ and contradict $v_i \neq u_i$).

Similarly, $\lambda \neq \varepsilon$. If $\oplus$ is selective, then we have $v_i = \lambda \otimes v_i$ with $v_i \neq \varepsilon$.

Consequently, if $(E, \otimes)$ is a group or a cancellative monoid, one can simplify with $v_i \neq \varepsilon$, which implies $\lambda = e$. ||

Proposition 2.5.7. *Assume that the assumptions of Proposition 2.5.5 hold and, moreover, that:* $a \in E, b \in E\ a \otimes b = e \Rightarrow a = e$ *and* $b = e$ *(see Remark 2.5.8 below) Then, if* $(M, \square, \perp)$ *has a basis, it is unique.*

Proof. We use the property of irreducibility of the elements of a basis (Proposition 2.5.5).

Consider $X = (x_i)_{i \in I}$ and $Y = (y_j)_{j \in J}$ two bases of M.

$\forall x_i \in X$ this yields:

$$x_i = \sum_{j \in J} \mu_j^i \perp y_j$$

But $Sp(X) = Sp(Y) = M$.

The independence of X implies that, $\forall_i \in I$, x_i is irreducible on Sp(X), therefore irreducible on Sp(Y). Consequently: there exists $j \in J$ such that:

$$x_i = \mu_j^i \perp y_j \quad \text{for} \quad \mu_j^i \in E \quad \text{and} \quad y_j \in Y.$$

In the same way, we prove that $y_j = \theta_k^j \perp x_k$ for $\theta_k^j \in E$ and $x_k \in X$.

Hence $x_i = \mu_j^i \otimes \theta_k^j \perp x_k$ and since X is an independent family, necessarily $i = k$ and (in view of the assumptions of Proposition 2.5.5) $\mu_j^i \otimes \theta_k^j = e$.

We deduce $\mu_j^i = e$ and $\theta_k^j = e$, which shows that: $x_i = y_j$.

Thus, for any $x_i \in X$, one can find $y_j \in Y$ such that $x_i = y_j$. We deduce $X = Y$, which proves uniqueness. □

Remark 2.5.8. The assumption $a \otimes b = e \Rightarrow a = e$ and $b = e$ is satisfied in particular: (1) in dioids for which e is the greatest element and where: $a \otimes b \leq a$ and $a \otimes b \leq b$; (2) when $\otimes$ is selective ($a \otimes b = e \Rightarrow a = e$ or $b = e \Rightarrow a = e$ and $b = e$). ||

3. Bideterminant and Linear Independence

In this section we discuss links between linear dependence and independence, and the concept of *bideterminant* for square matrices with elements in a semiring or a dioid. Thus, in a much more general framework, extensions to various known results of classical linear algebra will be obtained.

It is interesting to observe that, given that $(E, \oplus)$ is not a group, the proofs are very different from those known in classical linear algebra and generally require the use of combinatorial arguments and graph theoretical properties. In particular, the graph G(A) associated with a matrix $A \in M_n(E)$ will play an essential role.

3.1. Permanent, Bideterminant and Alternating Linear Mappings

Given a matrix $A \in M_n(E)$, $A = (a_{ij})_{\substack{i=1\dots n \\ j=1\dots n}}$, and σ a permutation of $\{1, \dots, n\}$, we refer to as the *weight of* σ, denoted $w(\sigma)$, the element of E defined as:

$$w(\sigma) = a_{1,\sigma(1)} \otimes a_{2,\sigma(2)} \otimes \dots \otimes a_{n,\sigma(n)}$$

We recall (see Chap. 2, Sect. 4.2) that the *bideterminant* of A is the pair

$$\Delta(A) = \begin{pmatrix} \det^+(A) \\ \det^-(A) \end{pmatrix}$$

where:

$$\det^+(A) = \sum_{\sigma \in Per^+(n)}^{\oplus} w(\sigma)$$

$$\det^-(A) = \sum_{\sigma \in Per^-(n)}^{\oplus} w(\sigma)$$

$Per^+(n)$ (resp. $Per^-(n)$) denoting the set of permutations of $\{1, \dots, n\}$ with signature $+1$ (resp. with signature -1). Since $\operatorname{sign}(\sigma^{-1}) = \operatorname{sign}(\sigma)$, it can be observed that A and A^T (the transposed matrix of A) have the same bideterminant: $\Delta(A^T) = \Delta(A)$

The *permanent* of A is defined as:

$$\operatorname{Perm}(A) = \det^+(A) \oplus \det^-(A) = \sum_{\sigma \in Per(n)}^{\oplus} w(\sigma)$$

Observe that if one multiplies a column (a row) j by $\lambda_j \in E$ ($\lambda_j \neq \varepsilon$), the permanent is then multiplied by λ_j.

As opposed to the case of standard linear algebra, the permanent of a matrix in a semiring or a dioid can often be efficiently computed and have interesting combinatorial interpretations, as the following examples show:

Example 3.1.1. (The permanent and the assignment problem)

$$E = \mathbb{R} \cup \{+\infty\}, \oplus = \min, \otimes = +.$$

The permanent of a matrix $A \in M_n(E)$ can be obtained in this case by solving the classical *assignment problem*.

Indeed, the weight of an arbitrary permutation σ is the sum (in the sense of ordinary addition on the reals) of the terms of the matrix A corresponding to the permutation σ and, since $\oplus$ = Min, the value of the permanent corresponds to the weight of the permutation of minimum weight, i.e. to the optimal solution of the assignment problem: how to select one and only one term of the matrix in each row and in each column while minimizing the sum of the selected terms. The assignment problem is a classical problem of graph theory which is solved efficiently (in polynomial time) by the so-called "Hungarian algorithm" or network flow algorithms (see for example Gondran and Minoux 1995, chap. 5; Ahuja, Magnanti and Orlin, 1993). ||

Example 3.1.2. (Permanent and bottleneck assignment)

$$E = \mathbb{R} \cup \{+\infty\}, \oplus = \min, \otimes = \max.$$

The permanent of a matrix $A \in M_n(E)$ can be obtained in this case by solving a "bottleneck" assignment problem.

Indeed, the weight of an arbitrary permutation σ is then the largest value of the terms of the matrix A corresponding to the permutation. The value of the permanent therefore corresponds to the permutation for which the largest of the terms covered by the permutation is the smallest possible. Like the classical assignment problem, the "bottleneck" assignment problem is solved efficiently (in polynomial time) by network flow algorithms (see for example Gondran and Minoux 1995, chap. 5; Ahuja, Magnanti and Orlin 1993). ||

Definition 3.1.3. *(Alternating linear mapping)*

Given $x^1, x^2, \dots x^n$, n *vectors of* E^n, *an application* f: $E^{n^2} \to E^2$ *is said to be an* alternating linear mapping *if and only if the mapping:*

$$f\left(x^1, x^2, \dots, x^n\right) = \begin{pmatrix} f_1\left(x^1, x^2, \dots, x^n\right) \\ \\ f_2\left(x^1, x^2, \dots, x^n\right) \end{pmatrix}$$

is such that:

- f_1 *and* f_2 *are linear mappings:* $E^{n2} \to E$
- $f\left(x^1, \dots, x^i, \dots, x^j, \dots, x^n\right) = \begin{pmatrix} f_1 \\ f_2 \end{pmatrix}$

implies: $f\ \left(x^1, \dots, x^j, \dots, x^i, \dots, x^n\right) = \begin{pmatrix} f_2 \\ f_1 \end{pmatrix}$.

Proposition 3.1.4. *The bideterminant* $\Delta(A) = \begin{pmatrix} \det^+(A) \\ \det^-(A) \end{pmatrix}$ *is an alternating linear mapping.*

Proof. It readily follows from the fact that one transposition changes the sign of a permutation. □

As for the determinant in standard linear algebra, one can establish a formula of *expansion of the bideterminant with respect to a row or a column* of the matrix.

For a matrix $A \in M_n(E)$ let us denote $\bar{A}_i^j$ the $n \times n$ matrix obtained (from A) by replacing all the terms of the ith row and the jth column by ε (neutral element of $\oplus$) except the term a_{ij} which is replaced by e (neutral element of $\otimes$)

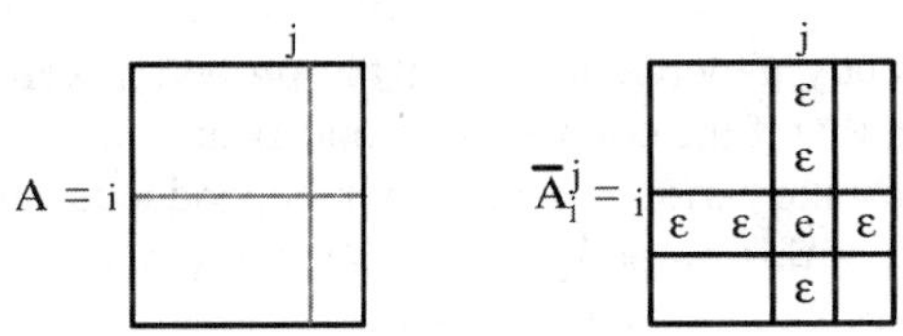

Given the linearity of the mappings $\det^+(A)$ and $\det^-(A)$, one can then write (expansion with respect to the ith row):

$$\det^+(A) = \sum_{j=1}^{n} a_{ij} \otimes \det^+(\bar{A}_i^j)$$

$$\det^-(A) = \sum_{j=1}^{n} a_{ij} \otimes \det^-(\bar{A}_i^j)$$

or equivalently, in vector notation:

$$\Delta(A) = \sum_{j=1}^{n} a_{ij} \otimes \Delta(\bar{A}_i^j)$$

An analogous formula would clearly be obtained by expanding with respect to a given column.

From the above, we easily deduce:

Property 3.1.5. If the matrix $A \in M_n(E)$ has a column (or a row) with all entries equal to ε (neutral element of $\oplus$) then $\det^+(A) = \det^-(A) = \varepsilon$.

Proof. Perform the expansion of the bideterminant with respect to the column (with respect to the row) and use the fact that ε is absorbing for $\otimes$. □

The converse of Property 3.1.5, is false as can be clearly seen from the following example:

$A \in M_4(S) \quad (n=4)$

	1	2	3	4
1	×	ε	ε	ε
2	×	ε	ε	ε
3	ε	×	ε	ε
4	ε	ε	×	×

Only the terms marked by a cross are different from ε. We clearly have: $\Delta_1(A) = \Delta_2(A) = \varepsilon$, but there exists an entry different from ε in each row and each column.

3.2. *Bideterminant of Matrices with Linearly Dependent Rows or Columns: General Results*

In this section, we study generalizations of the property, well known in standard linear algebra, stating that if the columns of a matrix are linearly dependent, then its determinant is zero. For the various concepts of independence introduced in Sect. 2.5, a generalized version of this property will be shown to hold, expressed here by the equality of the two terms of the bideterminant.

Let us state first the following elementary property:

Property 3.2.1. Let $(E, \oplus, \otimes)$ be a semiring, and consider a matrix $A \in M_n(E)$. If the matrix A has two identical columns (or rows) then:

$$\det^+(A) = \det^-(A)$$

Proof. For $j = 1 \ldots n$, let us denote A^j the jth column of A and let us assume for example that $A^i = A^j$ with $i \neq j$. Given that the bideterminant $\Delta(A)$ is a alternating linear mapping this yields:

$$\Delta(A^1, A^2, \ldots A^i, \ldots A^j, \ldots A^n) = \begin{pmatrix} \det^+(A) \\ \det^-(A) \end{pmatrix}$$

and:

$$\Delta(A^1, A^2, \ldots A^j, \ldots A^i, \ldots A^n) = \begin{pmatrix} \det^-(A) \\ \det^+(A) \end{pmatrix}$$

Since $A^i = A^j$, this implies clearly:

$$\det^+(A) = \det^-(A). \quad \square$$

Proposition 3.2.2. *Let* $(E, \oplus, \otimes)$ *be a semiring,* $A \in M_n(E)$ *and let us assume that the columns of* A *form a redundant family of* E^n *(see Sect. 2.5).*

$$\text{Then } \det^+(A) = \det^-(A).$$

Proof. Since the columns $A^1, A^2, \ldots A^n$ form a redundant family, there exists a column, A^1 say, which is a linear combination of the others. We therefore have:

$$A^1 = \sum_{j=2}^{n} \lambda_j \otimes A^j$$

with $\lambda_j \in E$ $(j = 2 \ldots n)$

By using the linearity of the bideterminant:

$$\Delta(A) = \sum_{j=2}^{n} \lambda_j \otimes \Delta(A^j, A^2, A^3, \ldots . A^j, \ldots . A^n)$$

Using Property 3.2.1, for any $j = 2, \ldots n$ we obtain: $\det^+(A^j, A^2, \ldots A^n) = \det^-(A^j, A^2, \ldots A^n)$ from which we can deduce: $\det^+(A) = \det^-(A)$. $\square$

Proposition 3.2.3. *Let* $(E, \oplus, \otimes)$ *be a semiring,* $A \in M_n(E)$ *and let us assume that the columns (the rows) of* A *form a quasi-redundant family of elements of* E^n.

Then:

(i) There exists $\alpha \in E$, $\alpha \neq \varepsilon$, *such that:*

$$\alpha \otimes \det^+(A) = \alpha \otimes \det^-(A).$$

(ii) If $(E, \oplus, \otimes)$ *is such that all the elements of* $E\backslash\{\varepsilon\}$ *are cancellative for* $\otimes$*, then* $\det^+(A) = \det^-(A)$

Proof. The quasi-redundancy of the family of columns $A^1, \dots A^n$ implies the existence of a column of A, e.g. A^1, and of a subset of indices $J \subset \{2, \dots n\}$ such that:

$$Sp(A^1) \cap Sp(A^J)$$

contains a vector different from $0 = \begin{pmatrix} \varepsilon \\ \cdot \\ \cdot \\ \varepsilon \end{pmatrix}$

This vector is necessarily of the form $\alpha \otimes A^1$ with $\alpha \in E\backslash\{\varepsilon\}$ and there exists $\lambda_j \in E (j \in J)$ such that:

$$\alpha \otimes A^1 = \sum_{j \in J} \lambda_j \otimes A^j$$

The columns of matrix $A' = (\alpha \otimes A^1, A^2, \dots A^n)$ therefore form a redundant family, thus, according to Proposition 3.2.2, we have: $\det^+(A') = \det^-(A')$.

Hence: $\alpha \otimes \det^+(A) = \alpha \otimes \det^-(A)$ which proves (i). (ii) is then immediately deduced. □

Let us now study the case where the columns of the matrix A are linearly dependent in the sense of Definition 2.5.1.

We first establish an initial result modulo quite restrictive assumptions (regularity of the elements of E for $\oplus$ and $\otimes$) to obtain the equality of the two terms of the bideterminant. Thereafter, we will investigate other types of less restrictive assumptions.

Proposition 3.2.4. *Let us assume that* $(E, \oplus, \otimes)$ *is a semiring such that all the elements of* E *are cancellative for* $\oplus$ *and all the elements of* $E\backslash\{\varepsilon\}$ *are cancellative for* $\otimes$

If the columns of $A \in M_n(E)$ *are linearly dependent, then:*

$$\det^+(A) = \det^-(A).$$

Proof. Since the columns of A are linearly dependent, there exist $I_1 \subset \{1, \dots n\}$ and $I_2 \subset \{1, \dots n\}$, $I_1 \neq \emptyset$, $I_2 \neq \emptyset$, $I_1 \cap I_2 = \emptyset$ such that:

$$\sum_{j \in I_1} \lambda_j \otimes A^j = \sum_{j \in I_2} \lambda_j \otimes A^j$$

with $\lambda_j \in E\backslash\{\varepsilon\}$ for $j \in I_1 \cup I_2$

It is not restrictive to assume that $1 \in I_1$. Let A' be the matrix deduced from A by replacing the column A^1 by $\sum\limits_{j \in I_1} \lambda_j \otimes A^j$.

The matrix A' is such that its first column is a linear combination of other columns, A^j for $j \in I_2$. According to Proposition 3.2.2, we therefore have:

$$\det^+(A') = \det^-(A')$$

Moreover, by using the linearity of the bideterminant, we can write:

$$\det^+(A') = \lambda_1 \otimes \det^+(A) \oplus \sum_{j \in I_1 \setminus \{1\}} \lambda_j \otimes \det^+(A^j, A^2, \ldots A^j, \ldots A^n)$$

and:

$$\det^-(A') = \lambda_1 \otimes \det^-(A) \oplus \sum_{j \in I_1 \setminus \{1\}} \lambda_j \otimes \det^-(A^j, A^2, \ldots A^j, \ldots A^n)$$

Since, $\forall j \in J_1\setminus\{1\}, \det^+(A^j, A^2, \ldots A^j, \ldots A^n) = \det^-(A^j, A^2, \ldots A^j, \ldots A^n)$ (see Property 3.2.1) and that any element is cancellative for $\oplus$, we deduce that:

$$\lambda_1 \otimes \det^+(A) = \lambda_1 \otimes \det^-(A)$$

Since $\lambda_1 \neq \varepsilon$ is cancellative for $\otimes$, this implies: $\det^+(A) = \det^-(A)$. □

There exist however many examples of semirings and dioids for which the assumptions of Proposition 3.2.4 do not hold. We study in the following section the links between linear dependence and bideterminant for the sub-class of *selective dioids*.

3.3. Bideterminant of Matrices with Linearly Dependent Rows or Columns: The Case of Selective Dioids

In this section, it will be assumed that $(E, \oplus, \otimes)$ is a selective dioid, i.e. that the operation $\oplus$ is such that, $\forall a \in E, \forall b \in E$:

$$a \oplus b = a \text{ or } b$$

We recall that, in this case, the canonical preorder relation $\leq$ is a *total order relation*. (see Chap. 1, Sect. 3.4).

Let $A \in M_n(E)$, $I = \{1, 2, \ldots n\}$ be the set of indices of the rows, $J = \{1, 2, \ldots n\}$ the set of indices of the columns, and let us consider a dependence relation among the columns, of the form:

$$\sum_{j \in J_1} A^j = \sum_{j \in J_2} A^j$$

with $J_1 \neq \emptyset \quad J_2 \neq \emptyset \quad J_1 \cap J_2 = \emptyset$

We observe that one can always assume that A does not have a column (or a row) with all entries equal to ε. (Indeed, if that was the case, one would immediately deduce $\det^+(A) = \varepsilon = \det^-(A)$, see Sect. 3.1, Property 3.1.5).

Since $a \oplus b = a$ or b ($\forall a \in E, \forall b \in E$), for any row i of A, we will have:

$$a_{i,j_1(i)} = \sum_{j \in J_1} a_{ij} \quad \text{for some index } j_1(i) \in J_1$$

Similarly:

$$a_{i,j_2(i)} = \sum_{j \in J_2} a_{ij} \quad \text{for some index } j_2(i) \in J_2$$

Let us consider now the bipartite graph $\overline{\overline{G}}$ (equality graph) where the set of vertices is: $X \cup Y$ with:

$$X = \{x_1, \ldots, x_n\} \qquad \text{(corresponding to the set of rows)}$$
$$Y = \{y_1, y_2, \ldots, y_n\} \quad \text{(corresponding to the set of columns)}$$

and where there exists an edge (x_i, y_j) if and only if

$$j = j_1(i) \quad \text{or} \quad j = j_2(i).$$

Example 3.3.1. Let the following 5×5 matrix (where we have marked with a cross the terms $j_1(i)$ and $j_2(i)$ for every row i).

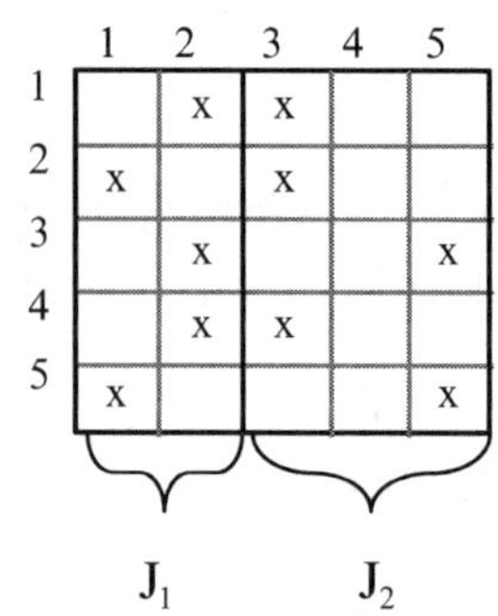

The corresponding equality graph $\overline{\overline{G}}$ is shown in Fig. 1. ||

We observe that the graph $\overline{\overline{G}}$ is not necessarily connected (as in Example 3.3.1 above where the vertex y_4 is isolated).

We will denote: $Y_1 = \{y_j / j \in J_1\}$ and $Y_2 = \{y_j / j \in J_2\}$.

Let us consider the *complete bipartite* graph $\mathcal{G}(A)$ constructed on the sets of vertices X and Y. Each arc (x_i, y_j) of $\mathcal{G}(A)$ corresponds to a term a_{ij} of the matrix A, and conversely.

With any permutation σ of $\{1, 2, \ldots, n\}$ one can associate one and only one *perfect matching* (see Berge, 1970) of $\mathcal{G}(A)$ and conversely (one-to-one correspondence).

Given two permutations σ_1, σ_2 of $\{1, \ldots, n\}$, K_1 and K_2 the corresponding (perfect) matchings of $\mathcal{G}(A)$, the set of edges $(K_1 \backslash K_2) \cup (K_2 \backslash K_1)$ forms a partial graph

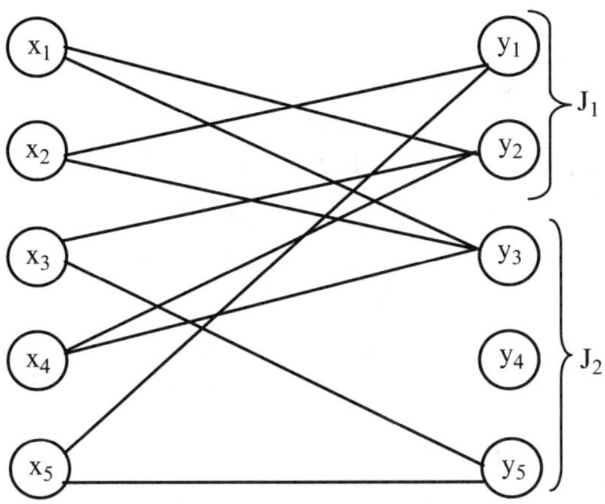

Fig. 1 Example of an equality graph associated with a dependence relation

of $\mathcal{G}(A)$ where each connected component is an *even* elementary cycle alternating in K_1 and K_2 (see Berge 1970; Lemma p. 118).

Denote $\gamma_1, \gamma_2, \ldots, \gamma_r$ these connected components having cardinalities: $|\gamma_1| = 2q_1, |\gamma_2| = 2q_2, \ldots, |\gamma_r| = 2q_r$.

Now, consider the permutation $\sigma = \sigma_2^{-1} \circ \sigma_1$.

To each cycle $\gamma_i (i = 1 \ldots r)$ there corresponds a cycle μ_i of $G(\sigma)$ (the graph associated with the permutation σ, see Chap. 2 Sect. 4) of cardinality:

$$|\mu_i| = \frac{|\gamma_i|}{2} \geq 1$$

(the cycles of $G(\sigma)$ which do not correspond to a cycle γ_i are loops, of cardinality 1).

The parity of σ is therefore the parity of:

$$\sum_{i=1}^{r} (|\mu_i| - 1) = \sum_{i=1}^{r} \left(\frac{|\gamma_i|}{2} - 1 \right)$$

σ_1 and σ_2 are of opposite parity if and only if the permutation σ is odd, i.e. if and only if the associated graph $G(\sigma)$ contains an odd number of circuits of even cardinality. We deduce:

Lemma 3.3.2. *A necessary and sufficient condition for* $\sigma = \sigma_2^{-1}$ *o* σ_1 *to be odd (i.e. for* σ_1 *and* σ_2 *to have opposite parities) is that, in* $\mathcal{G}(A)$*, the partial graph generated by* $(K_1 \backslash K_2) \cup (K_2 \backslash K_1)$ *contains an odd number of cycles of cardinality multiple of* 4.

One can then state:

Theorem 1. *(Gondran and Minoux, 1977)*

Let $(E, \oplus, \otimes)$ *be a selective dioid (i.e. in which the* $\oplus$ *law satisfies* $a \oplus b = a$ *or* b, $\forall a \in E, \forall b \in E$) *and let* $A \in M_n(E)$.

If the columns of A *satisfy a dependence relation of the form:*

$$\sum_{j \in J_1} A^j = \sum_{j \in J_2} A^j$$

with: $J_1 \neq \emptyset$ *and* $J_2 \neq \emptyset$. $J_1 \cap J_2 = \emptyset$,
then: $\det^+(A) = \det^-(A)$.

Proof. Let σ be a permutation of $\{1, \ldots, n\}$ such that:

$$w(\sigma) = \prod_i a_{i,\sigma(i)} = \text{perm}(A) = \det^+(A) \oplus \det^-(A)$$

(product in the sense of $\otimes$)

We observe that we have:

$$w(\sigma) = \underset{\pi \in Per(n)}{\text{Max}} \{w(\pi)\}$$

where the maximum is taken in the sense of the (total) order relation of the dioid $(E, \oplus, \otimes)$. (In other words, σ is an optimal solution to a problem of the "assignment" type. See Examples 3.1.1 and 3.1.2 above).

Let us consider the equality graph $\overline{\overline{G}}$ and add to $\overline{\overline{G}}$ the edges of the matching K of $\mathcal{G}(A)$ associated with the permutation σ. We obtain the graph G'.

Let us consider in G' the following path construction.

We start from a vertex y_{k_1} in $Y_1 (k_1 \in J_1)$; y_{k_1} is the endpoint of an edge in K and let x_{i_1} be the other endpoint.

There exists an edge of $\overline{\overline{G}}$ incident to x_{i_1} and having as its other endpoint $y_{k_2} \in Y_2 (k_2 = j_2(i_1))$. Observe that, necessarily, the edge (x_{i_1}, y_{k_2}) is not in K. (since $k_1 \in J_1$ and $k_2 \in J_2$).

When one is in y_{k_2}, there exists an edge of K incident to y_{k_2}. Let x_{i_2} be the other endpoint. Necessarily $x_{i_2} \neq x_{i_1}$. And so on...

At some stage, we have constructed a sequence of vertices of G':

$$y_{k_1}, x_{i_1}, y_{k_2}, x_{i_2}, \ldots x_{i_{p-1}}, y_{k_p}$$

We thus reach y_{k_p} right after visiting $x_{i_{p-1}}$ using an edge $\notin$ K. Moreover, $k_p \in J_1$ if p is odd, $k_p \in J_2$ if p is even.

Then there exists an edge of K incident to y_{k_p} and let x_{i_p} be the other endpoint. Necessarily $x_{i_p} \neq x_{i_{p-1}}$ since $(x_{i_{p-1}}, y_{k_p}) \notin K$.

From x_{i_p} there exists an edge of $\overline{\overline{G}}$ incident to x_{i_p} and such that the other endpoint is $y_{k_{p+1}}$ with $y_{k_{p+1}} \in Y_2$ if $y_{k_p} \in Y_1$ and $y_{k_{p+1}} \in Y_1$ if $y_{k_p} \in Y_2$. Necessarily, $(x_{i_p}, y_{k_{p+1}}) \notin K$.

We can therefore see that this path construction can be pursued indefinitely. Therefore, in a finite number of steps, one necessarily finds oneself back at a vertex already visited, and this vertex can only be one of the y_{k_i}. One has then detected a cycle γ of G' and γ contains as many vertices of Y_1 as of Y_2. Let q be this number. It is seen that:

$$|\gamma| = 4q.$$

Moreover, when one runs through the cycle γ, one alternatively traverses edges which are in K and edges which are not in K.

Consider then the matching K' obtained from K by replacing the edges of $K \cap \gamma$ by the edges of $\gamma\backslash K$: $K' = (K\backslash\gamma) \cup (\gamma\backslash K)$.

Let σ' be the permutation associated with K'. According to Lemma 3.3.2., σ and σ' have opposite parity.

Moreover, each term $a_{i,\sigma(i)}$ is replaced in the permutation σ' by a term of greater or equal weight: indeed: $a_{i,\sigma'(i)} = a_{i,\sigma(i)}$ for any i such that $\sigma(i) \notin J_1 \cup J_2$ and $a_{i,\sigma'}(i) \geq a_{i,\sigma(i)}$ for $\sigma(i) \in J_1 \cup J_2$ because: $a_{i,\sigma'(i)} = \underset{j\in J_1 \cup J_2}{\text{Max}}\{a_{i,j}\}$

We therefore have:

$$w(\sigma') \geq w(\sigma).$$

But since σ was chosen as a permutation of maximum weight and since $\geq$ is a total order relation, this necessarily yields:

$w(\sigma') = w(\sigma)$ and consequently, if one had for example: $w(\sigma) = \det^+(A)$ then $w(\sigma') = \det^-(A)$. From all this we can deduces $\det^+(A) = \det^-(A)$.

(Observe that the cycle γ exhibited in the above proof is not necessarily unique) □

Corollary 3.3.3. *(i) In a selective dioid* $(E, \oplus, \otimes)$, *if* $A \in M_n(E)$ *has linearly dependent columns, then there exists* $\alpha \in E\backslash\{\varepsilon\}$ *such that* $\alpha \otimes \det^+(A) = \alpha \otimes \det^-(A)$

(ii) If, moreover, any element of $E\backslash\{\varepsilon\}$ *is* regular *for* $\otimes$ *(case of selective-regular dioids) then:* $\det^+(A) = \det^-(A)$.

Proof. By hypothesis, there exists J_1 and J_2 ($J_1 \cap J_2 = \emptyset$) and $\lambda_j \neq \varepsilon$ ($j \in J_1 \cup J_2$) such that:

$$\sum_{j\in J_1} \lambda_j \otimes A^j = \sum_{j\in J_2} \lambda_j \otimes A^j$$

Let us consider the matrix $\bar{A} = (\bar{A}^1, \ldots, \bar{A}^n)$ such that:

$$\bar{A}^j = \lambda_j \otimes A^j, \quad j \in J_1 \cup J_2,$$

$$\bar{A}^j = A^j, \quad \forall j \notin J_1 \cup J_2.$$

Let us set: $\alpha = \prod_{j\in J_1 \cup J_2} \lambda_j$

We obtain: $\det^+(\bar{A}) = \alpha \otimes \det^+(A)$

$$\det^-(\bar{A}) = \alpha \otimes \det^-(A)$$

Matrix $\bar{A}$ therefore satisfies the assumptions of Theorem 1, hence: $\det^+(\bar{A}) = \det^-(\bar{A})$ and the first part of the corollary is deduced.

If $\alpha \neq \varepsilon$ is regular, we deduce the second part of the corollary: $\det^+(A) = \det^-(A)$. □

3.4. Bideterminant and Linear Independence in Selective-Invertible Dioids

Selective-invertible dioids form an interesting sub-class of dioids which Corollary 3.3.3 of Sect. 3.3 applies to: for a matrix A with linearly dependent columns, the two terms of the bideterminant are equal. In this paragraph, we prove a *converse* of this result: if A is a matrix with entries in a commutative selective-invertible dioid satisfying $\det^+(A) = \det^-(A)$, then one can construct a linear dependence relation involving the columns of A (see Theorem 2 below). Throughout this section, we therefore consider that $\oplus$ is selective and that $(E, \otimes)$ is a commutative group. Let us observe that, in this case, the canonical preorder on E is a total order (this property is used in the sequel). As in Sect. 3.3, for a given matrix $A \in M_n(E)$, the weight $w(\sigma)$ of a permutation of $\{1, 2, \ldots, n\}$ is defined as the product $w(\sigma) = a_{1,\sigma(1)} \otimes a_{2,\sigma(2)} \otimes \cdots \otimes a_{n,\sigma(n)}$. We begin by establishing two useful preliminary results.

Lemma 3.4.1. *Let* $\sigma_1 \in \mathrm{Per}^-(n)$ *and* $\sigma_2 \in \mathrm{Per}^+(n)$ *be two permutations of* $\{1, 2, \ldots, n\}$ *of opposite parities such that:*

$$w(\sigma_1) = w(\sigma_2) = \sum_{\sigma \in \mathrm{Per}(n)} w(\sigma) = \underset{\sigma \in \mathrm{Per}(n)}{\mathrm{Max}} \{w(\sigma)\}$$

(maximum taken in the sense of the total order relation of the dioid).

There exists then an even permutation $\bar{\sigma}_2 \in \mathrm{Per}^+(n)$ *such that:*

(i) $w(\bar{\sigma}_2) = w(\sigma_2)$

(ii) The permutations σ_1 *and* $\bar{\sigma}_2$ *only differ by a cycle in* $\mathcal{G}(A)$*, in other words the graph associated with* $\bar{\sigma}_2^{-1} \circ \sigma_1$ *only contains one even circuit of length* ≥ 2 *and loops.*

Proof. Let us consider the complete bipartite graph $\mathcal{G}(A)$ and K_1 and K_2 the (perfect) matchings of $\mathcal{G}(A)$ associated with the permutations σ_1 and σ_2. We recall that X denotes the set of vertices of $\mathcal{G}(A)$ corresponding to the rows of A, and Y, the set of vertices corresponding to the columns of A.

Since σ_1 and σ_2 are of opposite parities, the partial graph of $\mathcal{G}(A)$ induced by $(K_1 \backslash K_2) \cup (K_2 \backslash K_1)$ contains at least one connected component which is a cycle μ of cardinality multiple of 4: $|\mu| = 4q$.

Let $X' \subset X$ be the set of vertices x_i (rows) belonging to μ, and $Y' \subset Y$ the set of vertices y_j (columns) belonging to μ. Let us set:

$$X'' = X \backslash X' \quad Y'' = Y \backslash Y'.$$

Observe that:

$$|X'| = 2q \quad |Y'| = 2q \quad \text{and} \quad |\mu| = |X'| + |Y'|$$

Let us consider the partial graph $\overline{G}$ of $\mathcal{G}(A)$ induced by $K_1 \cup K_2$ (corresponding to the set of elements of the matrix appearing in at least one of the two permutations σ_1 or σ_2). μ is a connected component of $\overline{G}$.

The permutation σ_1 may then be decomposed into two one-to-one correspondences $\sigma'_1: X' \to Y'$ and $\sigma''_1: X'' \to Y''$.

We denote $\sigma_1 = (\sigma'_1, \sigma''_1)$ and similarly $\sigma_2 = (\sigma'_2, \sigma''_2)$

According to Lemma 3.3.2., σ'_1 and σ'_2 have opposite parities.

Let us show that:

$$w(\sigma'_1) = w(\sigma'_2)$$

We can write: $w(\sigma_1) = w(\sigma'_1) \otimes w(\sigma''_1)$

$$w(\sigma_2) = w(\sigma'_2) \otimes w(\sigma''_2)$$

If one had, for example:

$$w(\sigma'_1) < w(\sigma'_2)$$

then the permutation:

$\bar{\sigma} = (\sigma'_2, \sigma''_1)$ would have a weight $w(\bar{\sigma}) > w(\sigma_1)$, in contradiction with the fact that σ_1 is of maximum weight.

Similarly we cannot have: $w(\sigma'_1) > w(\sigma'_2)$.

Consequently, $w(\sigma'_1) = w(\sigma'_2)$ (because $\geq$ is a total order relation).

The desired permutation $\bar{\sigma}_2$ is thus

$$\bar{\sigma}_2 = (\sigma'_2, \sigma''_1). \quad \square$$

Lemma 3.4.2. *Let* $\sigma_1 \in \mathrm{Per}^-(n)$ *and* $\sigma_2 \in \mathrm{Per}^+(n)$ *be two permutations of maximum weight* $w(\sigma_1) = w(\sigma_2)$ *and let us denote* K_1 *and* K_2 *the associated matchings of* $\mathcal{G}(A)$. *We assume:*

- *That* $a_{i,\sigma_1(i)} = a_{i,\sigma_2(i)}$ *for* $i = 1, 2, \ldots, n$
- *That the partial graph* $\overline{\overline{G}}$ *induced by* $(K_1 \backslash K_2) \cup (K_2 \backslash K_1)$ *is connected and formed by a cycle* μ *of cardinality* $4q$ *(q integer) possibly with pendent arborescences of even cardinality.*

Then for any $x_i \in \mu$ *and for any* $y_k \in \mu$*:*

$$a_{i,k} \leq a_{i,\sigma_1(i)} = a_{i,\sigma_2(i)}$$

Proof. The proof is given by contradiction, assuming that there exists $x_i \in \mu$ and $y_k \in \mu$ such that:

$$a_{i,k} > a_{i,\sigma_1(i)} = a_{i,\sigma_2(i)}$$

and then showing that we can exhibit a permutation σ^* such that

$$w(\sigma^*) > w(\sigma_1) = w(\sigma_2).$$

It is easy to see (refer to Fig. 2) that there exists a perfect matching for the vertices of μ using the edge (x_i, y_k) and such that the other edges all belong to μ (indeed, the cardinalities of the two paths between a row i and a column k on the cycle μ are odd).

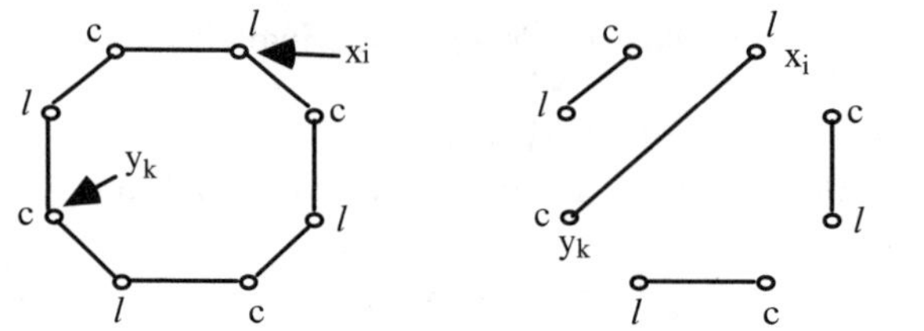

(5a) Cycle μ *(5b) Perfect matching*

Fig. 2 Illustration of the proof of Lemma 3.4.2

There therefore exists a perfect matching for the vertices of $\overline{\overline{G}}$ using the edge (x_i, y_k) and such that the other edges all belong to $\overline{\overline{G}}$ (the even pendent arborescences do not create any problem).

Let σ^* be the permutation corresponding to this perfect matching.

We then have:

$$a_{ik} = a_{i,\sigma^*(i)} > a_{i,\sigma_1(i)} = a_{i,\sigma_2(i)}$$

and for any $\ell = \{1 \ldots n\} \backslash \{i\}$

$$a_{\ell,\sigma^*(\ell)} = a_{\ell,\sigma_1(\ell)} = a_{\ell,\sigma_2(\ell)}$$

We deduce:

$$w(\sigma^*) > w(\sigma_1) = w(\sigma_2)$$

which contradicts the optimality of σ_1 and of σ_2. □

(Remark: above we have used the fact that, in a selective-invertible dioid, for any $c \neq \varepsilon$ one can write: $a < b \Rightarrow a \otimes c < b \otimes c$.

Indeed, $\leq$ is compatible with the law $\otimes$ therefore: $a < b \Rightarrow a \otimes c \leq b \otimes c$.

But we cannot have equality because $c \neq \varepsilon$ being regular, $a \otimes c = b \otimes c \Rightarrow a = b$ which is contrary to the hypothesis. Therefore: $a \otimes c < b \otimes c$).

We can then state:

Theorem 2. *(Gondran and Minoux, 1977)*
Let $(E, \oplus, \otimes)$ *be a commutative selective-invertible dioid and* $A \in M_n(E)$.

$$\text{If} \quad \det^+(A) = \det^-(A)$$

then the columns of A *(and similarly the rows of* A*) are* linearly dependent. *In other words, there exist* J_1 *and* $J_2 (J_1 \neq \emptyset, J_2 \neq \emptyset, J_1 \cap J_2 = \emptyset)$ *and coefficients* $\lambda_j \in E \backslash \{\varepsilon\}$ *(for* $j \in J_1 \cup J_2$*) such that:*

$$\sum_{j \in J_1} \lambda_j \otimes A^j = \sum_{j \in J_2} \lambda_j \otimes A^j \tag{7}$$

Proof. (1) Since $\oplus$ is selective, there exist two permutations σ_1 and σ_2 with opposite parities, such that:

$$w(\sigma_1) = w(\sigma_2) = \sum_{\sigma \in Per(n)} w(\sigma) = \underset{\sigma \in Per(n)}{Max} \{w(\sigma)\}$$

We observe that the permutations σ_1 and σ_2 remain optimal when one or several columns A^j of A are multiplied by $\lambda_j (\lambda_j \neq \varepsilon)$.
From Lemma 3.4.1 one can assume that σ_2 differs from σ_1 only by one even cycle.

(2) If one uses a cross ($\times$) to represent the elements a_{ij} of the matrix A such that $j = \sigma_1(i)\, or\, j = \sigma_2(i)$ and after a possible rearrangement of the rows and columns, we obtain a configuration such as the one shown below:

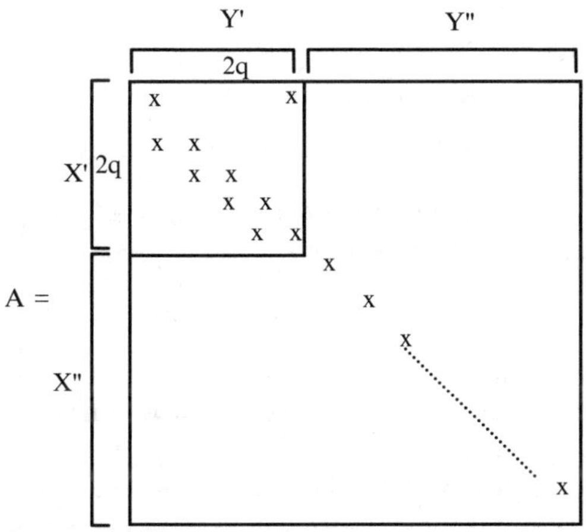

Let us then consider the sub-matrix A' of A formed by the 2q first lines (X'), the 2q first columns (Y') and by the elements marked by a cross (the other elements of A' being ε). In the example below, the cycle μ of Lemma 3.4.1 successively encounters $x_1, y_1, x_2, y_2, x_3, y_3, x_4, y_4, x_1$ (its cardinality is $|\mu| = 8$).

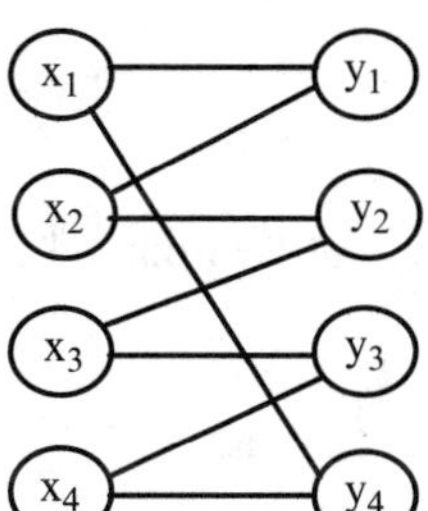

A'=

	y_1	y_2	y_3	y_4
x_1	a_{11}	ε	ε	a_{14}
x_2	a_{21}	a_{22}	ε	ε
x_3	ε	a_{32}	a_{33}	ε
x_4	ε	ε	a_{43}	a_{44}

Let us try and determine $\lambda_1, \lambda_2, \lambda_3, \lambda_4 \in E\backslash\{\varepsilon\}$ and the sets J_1 and $J_2 \subset \{1, 2, \ldots, 2q\}$ so that we have, for the matrix A':

$$\sum_{j \in J_1} \lambda_j \otimes A'j = \sum_{j \in J_2} \lambda_j \otimes A'j \tag{8}$$

We can choose one of the λ_j values arbitrarily (since $\otimes$ is invertible), e.g. $\lambda_1 = e$. We construct the sets J_1 and J_2, by running through the cycle μ while placing the columns (vertices y) successively encountered alternatively in J_1 and in J_2.

Thus, in the example, starting from x_1 we arrive at y_1 which we assign to J_1; then from x_2 we arrive at y_2 which we assign to J_2; then from x_3 we arrive at y_3 which we assign to J_1; finally from x_4 we arrive at y_4 which we assign to J_2. We therefore obtain: $J_1 = \{1, 3\}$ $\quad J_2 = \{2, 4\}$. The λ_j values are then determined so as to satisfy the relation (8). If $\lambda_1 = e$, we must necessarily have:

$$\begin{aligned} &\lambda_1 \otimes a_{11} = \lambda_4 \otimes a_{14} \quad \text{hence:} \quad \lambda_4 = a_{11} \otimes (a_{14})^{-1} \\ \text{then} \quad &\lambda_4 \otimes a_{44} = \lambda_3 \otimes a_{43} \quad \text{hence:} \quad \lambda_3 = \lambda_4 \otimes a_{44} \otimes (a_{43})^{-1} \\ \text{then} \quad &\lambda_3 \otimes a_{33} = \lambda_2 \otimes a_{32} \quad \text{hence:} \quad \lambda_2 = \lambda_3 \otimes a_{33} \otimes (a_{32})^{-1} \end{aligned}$$

Since $w(\sigma'_1) = w(\sigma'_2)$ we obtain $a_{11} \otimes a_{22} \otimes a_{33} \otimes a_{44} = a_{14} \otimes a_{21} \otimes a_{32} \otimes a_{43}$ and we check that the second relation:

$$\lambda_2 \otimes a_{22} = \lambda_1 \otimes a_{21}$$

is automatically satisfied.

The above readily generalizes to the case where μ has a cardinality $|\mu| = 4q$ with arbitrary $q(q > 1)$.

(3) Denote $A_{X'}^{Y'}$ the sub-matrix of A induced by the subset of rows X' and the subset of columns Y'. Now, let us multiply each column j of $A_{X'}^{Y'}$ by the value λ_j thus determined. We obtain a matrix $\bar{A} = (\bar{a}_{ij})$ of dimensions $(2q \times 2q)$ such that:

$$\forall i\colon \bar{a}_{i,\sigma_1(i)} = \bar{a}_{i,\sigma_2(i)} \quad \text{with} \quad \sigma_1(i) \in J_1 \quad \text{and} \quad \sigma_2(i) \in J_2.$$

By this transformation an optimal permutation remains optimal, σ_1 and σ_2 therefore remain permutations of maximum weight.

Let us now prove that we clearly have:

$$\sum_{j \in J_1} \bar{A}^j = \sum_{j \in J_2} \bar{A}^j$$

In order to do so, it suffices to observe that in view of Lemma 3.4.2., any term $\bar{a}_{ij}$ of $\bar{A}$ is less than or equal to $\bar{a}_{i,\sigma_{1(i)}} = \bar{a}_{i,\sigma_{2(i)}}$.

(4) Let us now return to the initial matrix A.

For every row $x_i \in X''$, therefore for $i = 2q + 1, \ldots, n$, let:

$$b_i = \max_{y_j \in Y'} \{a_{ij} \otimes \lambda_j\} = \sum_{y_j \in Y'} a_{ij} \otimes \lambda_j$$

Let us then show that there exists $n - 2q$ coefficients λ_j associated with the columns of Y'' (and the rows of X'') such that, $\forall x_i \in X''$:

$$a_{ii} \otimes \lambda_i = \sum_{\substack{y_j \in Y'' \\ j \neq i}} a_{ij} \otimes \lambda_j \oplus b_i \tag{9}$$

System (9) is equivalent to the system:

$$\lambda_i = \sum_{\substack{y_j \in Y'' \\ j \neq i}} a_{ij} \otimes a_{ii}^{-1} \otimes \lambda_j \oplus a_{ii}^{-1} \otimes b_i \quad \forall x_i \in X'' \tag{10}$$

Since the identity permutation is an optimal permutation for the matrix $A_{X''}^{Y''}$ it follows that the graph associated with the matrix $(a_{ij} \otimes a_{ii}^{-1})_{\substack{i \in X'' \\ j \in Y''}}$ has all its circuits of weight smaller than or equal to e. This property remains true if one replaces all the diagonal terms of this matrix by ε, i.e. for the matrix of system (10).

Then, according to Theorem 1 of Chap. 4, the matrix of system (10) is *quasi-invertible* and system (10) has a smallest solution λ.

With the values λ thus determined, we have, $\forall x_i \in X''$:

$$a_{ii} \otimes \lambda_i = \sum_{\substack{y_j \in Y \\ j \neq i}} a_{ij} \otimes \lambda_j \tag{11}$$

For every $i = 2q + 1, \ldots n$, there therefore exists an index $\varphi(i) \neq i$ such that:

$$a_{ii} \otimes \lambda_i = a_{i,\varphi(i)} \otimes \lambda_{\varphi(i)} = \sum_{\substack{y_j \in Y \\ j \neq i}} a_{ij} \otimes \lambda_j \tag{12}$$

(5) Let us denote $\widetilde{G}$ the partial graph of $\mathcal{G}(A)$ defined as follows:

- $\forall x_i \in X'$, $\widetilde{G}$ contains the two edges of the cycle μ incident to x_i.
- $\forall x_i \in X''$, $\widetilde{G}$ contains the two edges (x_i, y_i) and $(x_i, y_{\varphi(i)})$.

Since $\widetilde{G}$ has as many edges as vertices, each of its connected components contains exactly one elementary cycle and possibly pendent arborescences. Let us show that it is not restrictive to assume that $\widetilde{G}$ is connected.

If $\widetilde{G}$ is not connected, it contains a connected component $\mathcal{T}$ having all its vertices in $X'' \cup Y''$.

For any $x_i \in \mathcal{T}$ we cannot have the strict inequality.

$$a_{ii} \otimes \lambda_i = a_{i,\varphi(i)} \otimes \lambda_{\varphi(i)} > \sum_{\substack{y_j \in Y \\ y_j \notin \mathcal{T}}} a_{ij} \otimes \lambda_j \tag{13}$$

Indeed, if (13) is satisfied $\forall x_i \in \mathcal{T}$, then it would be possible to reduce all the λ_j values, for $y_j \in \mathcal{T}$, while satisfying (9), which would contradict the fact that λ is the minimal solution to (9).

Consequently, there exists $x_i \in \mathcal{T}$ and $y_k \notin \mathcal{T}$ such that:

$$a_{ii} \otimes \lambda_i = a_{i,\varphi(i)} \otimes \lambda_{\varphi(i)} = a_{ik} \otimes \lambda_k$$

One can then replace in $\widetilde{G}$ the edge $(x_i, y_{\varphi(i)})$ by (x_i, y_k), which reduces by one unit the connectivity number of $\widetilde{G}$. If necessary, the previous reasoning will be repeated as long as the connectivity condition is not satisfied, to obtain in a finite number of steps a connected graph $\widetilde{G}$.

Figure 4b shows the (connected) graph $\widetilde{G}$ corresponding to the matrix of Fig. 4a where, in each row, the crosses correspond to the terms a_{ii} and $a_{i,\varphi(i)}$ of relation (12).

(a)

		y_1	y_2	y_3	y_4	y_5	y_6	y_7	y_8	y_9	y_{10}
X'	x_1	**x**	**x**								
	x_2		**x**	**x**							
	x_3			**x**	**x**						
	x_4	**x**			**x**						
X''	x_5	x				x					
	x_6						x		x		
	x_7				x			x			
	x_8			x					x		
	x_9				x					x	
	x_{10}								x		x
		Y'				Y''					

(b)

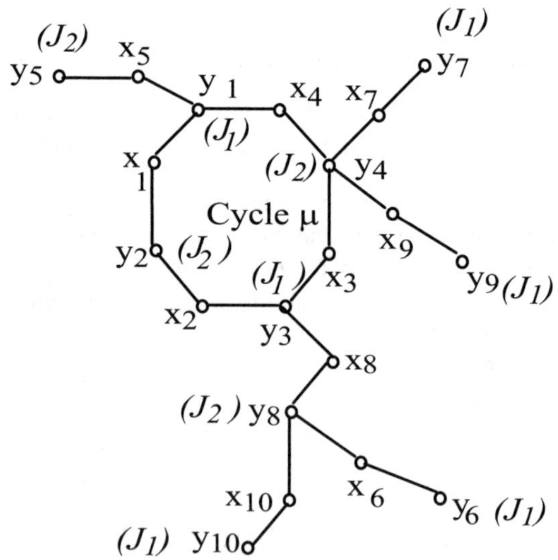

Fig. 4 An example illustrating equation (12) in the proof of Theorem 2

(6) Since each vertex of $\widetilde{G}$ associated with a row has degree exactly 2, it can be observed that:

- The pendent arborescences of $\widetilde{G}$ are necessarily attached to μ in vertices $y_j \in Y'$ (columns);
- The pendent vertices (i.e. those having degree 1) of these pendent arborescence are necessarily vertices $y_j \in Y''$ (columns).

The columns of Y'' can then easily be assigned either to J_1 or to J_2 by proceeding as follows.

Each pendent arborescence is run through starting from its vertex of attachment to μ, which already belongs either to J_1, or to J_2. Between this vertex and each pendent vertex, the vertices y_j are alternatively assigned to J_1 and J_2. Thus, for the example of Fig. 4b (where we already know that $J_1 = \{1, 3\}$ and $J_2 = \{2, 4\}$) running through the branch attached to y_3, column 8 is added to J_2 and columns 6 and 10 to J_1.
We finally obtain:

$$J_1 = \{1, 3, 6, 7, 9, 10\} \quad J_2 = \{2, 4, 5, 8\}$$

(7) With the values λ_j and the sets J_1 and J_2 resulting from the above, we satisfy the relations:
$\forall i = 1, 2, \ldots 2q$:

$$\sum_{j \in J_1 \cap Y'} a_{ij} \otimes \lambda_j = \sum_{j \in J_2 \cap Y'} a_{ij} \otimes \lambda_j \tag{14}$$

and $\forall i = 2q + 1 \ldots n$:

$$\sum_{j \in J_1} a_{ij} \otimes \lambda_j = \sum_{j \in J_2} a_{ij} \otimes \lambda_j \tag{15}$$

It remains to check that the relation:
$\forall i = 1, \ldots 2q$:

$$\sum_{j \in J_1} a_{ij} \otimes \lambda_j = \sum_{j \in J_2} a_{ij} \otimes \lambda_j \tag{16}$$

is satisfied and to do so, we are going to show that:
$\forall x_i \in X'$ and $y_k \in Y''$:

$$\underset{j \in J_1 \cap Y'}{\text{Max}} \left\{ \overline{a_{ij}} \right\} = \underset{j \in J_2 \cap Y'}{\text{Max}} \left\{ \overline{a_{ij}} \right\} \geq \overline{a_{ik}}$$

This will be obtained through an argument similar to the one used in the proof of Lemma 3.4.2. Let us assume that for $x_i \in X'$ and $y_k \in Y''$, we have that $\bar{a}_{ik} > \bar{a}_{i\sigma_1(i)} = \bar{a}_{i\sigma_2(i)}$. The edge (x_i, y_k) joins a vertex $x_i \in \mu$ to a vertex $y_k \notin \mu$.

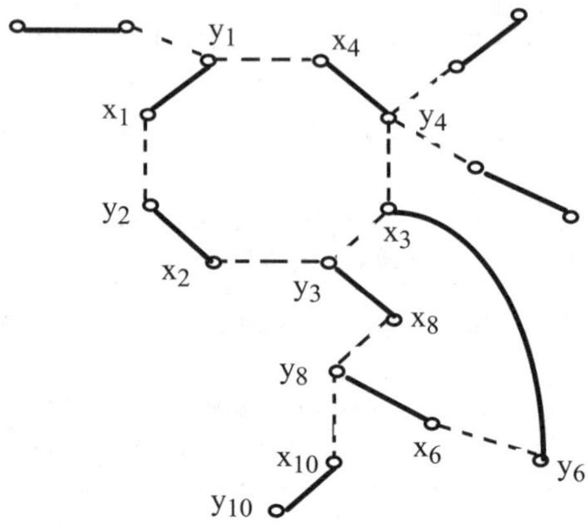

Fig. 5 Illustration of part 7 of the proof of Theorem 2

Let $y_j \in \mu$ be the point of attachment of the arborescent branch of $\widetilde{G}$ containing y_k. We construct a perfect matching of $\widetilde{G} \cup \{(x_i, y_k)\}$ by selecting:

- All the edges of the maximum matching of μ leaving x_i and y_j unsaturated;
- The maximum matching of the chain L joining y_j and y_k leaving the vertex y_k unsaturated;
- The edge (x_i, y_k);
- All the edges of the form (x_i, y_i) for all the vertices i not belonging to μ, nor to L.

Let σ^* be the permutation corresponding to this perfect matching (Fig. 5 illustrates the matching obtained on the example of Fig. 4, taking $x_i = x_3$ and $y_k = y_6$).

We then have $\bar{a}_{ik} = \bar{a}_{i,\sigma^*(i)} > \bar{a}_{i,\sigma_1(i)} = \bar{a}_{i,\sigma_2\,(i)}$ and, $\forall l \in \{1, \ldots, n\} \backslash \{i\}$:

$$\bar{a}_{l,\sigma^*(l)} = \bar{a}_{l,\sigma_1(l)} = \bar{a}_{l,\sigma_2(l)}$$

hence we deduce $w(\sigma^*) > w(\sigma_1) = w(\sigma_2)$, which contradicts the optimality of σ_1 and of σ_2.

Thus relation (16) is deduced and the theorem is proven. □

A direct consequence of this result and of Corollary 3.3.3 is the characterization of singular matrices on commutative selective-invertible dioids:

Corollary 3.4.3. *Let* $(E, \oplus, \otimes)$ *be a commutative selective-invertible dioid.* $A \in M_n(E)$ *is singular if and only if* $\det^+(A) = \det^-(A)$.

3.5. Bideterminant and Linear Independence in Max-Min or Min-Max Dioids

In this section we investigate properties of the bideterminant for another important sub-class of dioids, namely MAX-MIN dioids (or MIN-MAX dioids). This sub-class belongs to the intersection of *doubly selective* dioids and *distributive lattices*.

A MAX-MIN dioid is therefore a selective dioid $(E, \oplus, \otimes)$ with the additional property:

$$\forall a \in E, \forall b \in E\colon a \otimes b = \mathrm{Min}\{a, b\}$$

(the minimum above is to be understood in the sense of the total order relation deriving from the law $\oplus$ on E).

It is therefore seen that a MAX-MIN dioid can be defined more directly from a totally ordered set E endowed with the laws $\oplus$ and $\otimes$ as follows:

$$\begin{aligned}&\forall a \in E, b \in E\colon a \oplus b = \mathrm{Max}\{a, b\}\\ &a \otimes b = \mathrm{Min}\{a, b\}\end{aligned}$$

It is also observed that it is a special type of doubly selective dioid. The definition of a MIN-MAX dioid is analogous, with the roles of MIN and MAX interchanged. In the literature, MAX-MIN or MIN-MAX dioids have also been investigated under the names of "Minimax algebras" (Cuninghame-Geene, 1979) or "Bottleneck algebras" (see e.g. Cechlárová, 1992, Cechlárová & Plávka, 1996).

The following example shows that, for a matrix A with elements in a MAX-MIN dioid $(E, \oplus, \otimes)$, the existence of a relation of linear dependence of the form:

$$\sum_{j \in J_1} \lambda_j \otimes A^j = \sum_{j \in J_2} \lambda_j \otimes A^j$$

(with $J_1 \neq \emptyset$, $J_2 \neq \emptyset$, $J_1 \cap J_2 = \emptyset$ and $\lambda_j \neq \varepsilon$ for $j \in J_1 \cup J_2$) does not necessarily imply $\det^+(A) = \det^-(A)$.

Example 3.5.1. On the dioid $(\mathbb{R}_+, \mathrm{Max}, \mathrm{Min})$ let us consider the matrix:

$$A = \begin{bmatrix} 4 & 1 & 1 \\ 1 & 3 & 2 \\ 1 & 3 & 4 \end{bmatrix}$$

For $\lambda_2 = 2$ and $\lambda_3 = 2$ we clearly have the dependence relation:

$$\lambda_2 \otimes A^2 = \lambda_3 \otimes A^3 = \begin{pmatrix} 1 \\ 2 \\ 2 \end{pmatrix}$$

However it can be observed that: $\det^+(A) = 3$ (the weight of the even permutation σ of maximum weight defined as: $\sigma(1) = 1$; $\sigma(2) = 2$; $\sigma(3) = 3$), and: $\det^-(A) = 2$ (the weight of the odd permutation σ' of maximum weight defined as: $\sigma'(1) = 1$; $\sigma'(2) = 3$; $\sigma'(3) = 2$).

We therefore have in this example, $\det^+(A) \neq \det^-(A)$, despite the linear dependence of the columns. ||

It will be observed that the situation highlighted in the previous example can be explained by the fact that the coefficients λ_2 and λ_3 were chosen too small.

More generally, if A^j and A^k are two arbitrary distinct columns where all the terms are > 0, by choosing λ_j and λ_k satisfying.

$$0 < \lambda_j = \lambda_k \leq \underset{i=1\ldots n}{\text{Min}} \left\{\text{Min}\left\{a_{i,j};\ a_{i,k}\right\}\right\}$$

then this yields: $\lambda_j \otimes A^j = \lambda_k \otimes A^k$, even for an arbitrary matrix such that $\det^+(A) \neq \det^-(A)$.

The following result shows that choosing all the $\lambda_j \geq \text{perm}(A)$ in the dependence relation is sufficient to guarantee the equality of the two terms of the bideterminant.

Property 3.5.2. Let $(E, \oplus, \otimes)$ be a MAX-MIN dioid and $A \in M_n(E)$.

If the columns of A satisfy a linear dependence relation of the form:

$$\sum_{j \in J_1} \lambda_j \otimes A^j = \sum_{j \in J_2} \lambda_j \otimes A^j$$

with, $\forall j$: $\lambda_j \geq \text{perm}(A) = \det^+(A) \oplus \det^-(A)$ then $\det^+(A) = \det^-(A)$.

Proof. Let $\bar{A} = (\bar{A}^1, \bar{A}^2, \ldots \bar{A}^n)$ be the matrix such that $\bar{A}^j = A^j$ if $j \notin J_1 \cup J_2$, and $\bar{A}^j = \lambda_j \otimes A^j$ if $j \in J_1 \cup J_2$.

By using the Corollary 3.3.3. of Sect. 3.3, we obtain:

$$\det^+(\bar{A}) = \alpha \otimes \det^+(A)$$
$$\det^-(\bar{A}) = \alpha \otimes \det^-(A)$$

with $\det^+(\bar{A}) = \det^-(\bar{A})$ and $\alpha = \prod_{j \in J_1 \cup J_2} \lambda_j$ (product in the sense of $\otimes$).

The condition:

$$\alpha = \underset{j \in J_1 \cup J_2}{\text{Min}} \left\{\lambda_j\right\} \geq \text{perm}(A) = \text{Max}\left\{\det{}^+(A),\ \det{}^-(A)\right\}$$

then implies:

$$\alpha \otimes \det^+(A) = \det^+(A)$$
$$\alpha \otimes \det^-(A) = \det^-(A)$$

which yields the desired result. □

Let us now study the converse of the previous property.

The example below shows that, as opposed to the case of selective-invertible dioids (see Sect. 3.4) $\det^+(A) = \det^-(A)$ does not necessarily imply the existence of a relation of linear dependence of the form

$$\sum_{j \in J_1} \lambda_j \otimes A^j = \sum_{j \in J_2} \lambda_j \otimes A^j \tag{17}$$

$$\text{with,} \quad \forall j,\ \lambda_j \geq \text{perm}(A) \tag{18}$$

Example 3.5.3. In the dioid ($\mathbb{R}_+$, Max, Min) let us consider the matrix

$$A = \begin{bmatrix} 2 & 1 & 0 \\ 1 & 2 & 2 \\ 1 & 2 & 2 \end{bmatrix}$$

which satisfies $\det^+(A) = \det^-(A) = 2$ and $\mathrm{perm}(A) = 2$.

It can be observed that, for any $j = 1, \ldots, 3$, if $\lambda_j \geq 2$, then $\lambda_j \otimes A^j = A^j$

Consequently, if there exists a dependence relation satisfying (17) and (18), it is necessarily of the form

$$\sum_{j \in J_1} A^j = \sum_{j \in J_2} A^j$$

with $J_1 \neq \emptyset, J_2 \neq \emptyset, J_1 \cap J_2 = \emptyset$.

We easily observe for our example that this is impossible if $|J_1| = |J_2| = 1$ (because the three columns of A are distinct) and also impossible if $|J_1| = 1$ and $|J_2| = 2$ (because, in the first row, none of the coefficients is equal to the maximum of the two others). ||

From the previous example, it is seen that in a MAX-MIN dioid the singularity of a matrix $A \in M_n(E)$ cannot be characterized by the equality $\det^+(A) = \det^-(A)$.

We now proceed to show that for matrices $A \in M_n(E)$ satisfying an additional condition, referred to as *non-degeneracy*, it is nonetheless possible to characterize singularity of matrices in MAX-MIN dioids in terms of *coloration of hypergraphs*.

Definition 3.5.4. $A \in M_n(E)$ *is said to be* non-degenerate *if there does not exist a row i containing two terms* a_{ij} *and* a_{ik} *such that* $a_{ij} = a_{ik} < \mathrm{perm}(A)$

Remark. We observe that the non-degeneracy of a matrix can be checked in polynomial time since the computation of perm(A) can be reduced to a "Max-Min" or "bottleneck" assignment problem. (refer to Example 3.1.2) ||

For non-degenerate matrices, we can state the following property.

Property 3.5.5. Let $A \in M_n(E)$ be a non-degenerate matrix on a MAX-MIN dioid $(E, \oplus, \otimes)$.

If there exists a dependence relation implying a subset of columns of A, then there exists a *complete dependence relation*, i.e. one in which *all the columns* of A are involved.

Proof. Let us assume that there exists a dependence relation of the form

$$\sum_{j \in J_1} \lambda_j \otimes A^j = \sum_{j \in J_2} \lambda_j \otimes A^j$$

with $J_1 \neq \emptyset, J_2 \neq \emptyset \quad J_1 \cap J_2 = \emptyset, J_1 \cup J_2 \neq J = \{1, \ldots, n\}$, and $\lambda_j \geq \mathrm{perm}(A), \forall j \in J_1 \cup J_2$.

By multiplying both sides of the above relation by $\bar{\lambda} = \text{perm}(A)$ and by observing that for $\lambda_j \geq \text{perm}(A)$ $\bar{\lambda} \otimes \lambda_j = \bar{\lambda}$ we obtain:

$$\forall i \in I = \{1, 2, \ldots, n\}$$
$$\sum_{j \in J_1} \bar{\lambda} \otimes a_{ij} = \sum_{j \in J_2} \bar{\lambda} \otimes a_{ij} = v_i$$

We necessarily have $v_i \leq \bar{\lambda}$. Indeed, if for $i \in I$, we have that: $v_i < \bar{\lambda}$, this implies that there exists $j_1 \in J_1$ and $j_2 \in J_2$ such that:

$$a_{i,j_1} = a_{i,j_2} = v_i < \bar{\lambda}$$

which implies a contradiction with the non degeneracy assumption.

We therefore have, $\forall i \in I$: $v_i = \bar{\lambda}$ and consequently by setting $\lambda'_j = \bar{\lambda}$ (for any $j \in J$)

$$J'_1 = J_1 \quad J'_2 = J \backslash J_1 \quad \text{we have the relation:}$$
$$\forall i \in I\colon \sum_{j \in J'_1} \lambda'_j \otimes a_{ij} = \sum_{j \in J'_2} \lambda'_j \otimes a_{ij}$$

which is a *complete dependence relation* on the set of columns of A. □

Let us now introduce the concept of the *skeleton-hypergraph* of a given matrix A:

Definition 3.5.6. *Let* $A \in M_n(E)$ *and* $\bar{\lambda} = \text{perm}(A)$.

We refer to as the skeleton-hypergraph *of* A *the hypergraph* $H(A) = [J, \mathcal{S}(A)]$ *having* $J = \{1, \ldots, n\}$ *as set of vertices and where the set of edges is:*

$$S(A) = \{S_1(A), S_2(A), \ldots S_n(A)\}$$
$$\textit{where,} \quad \forall i \in I, S_i(A) = \{j/\bar{\lambda} \otimes a_{ij} = \bar{\lambda}\} = \{j/a_{ij} \geq \text{perm}(A)\}.$$

In $H(A)$, each edge corresponds to a row of A, and contains at least one vertex. Indeed, since $\bar{\lambda} = \text{perm}(A)$, there exists at least one permutation σ of $\{1, \ldots, n\}$ such that:

$$w(\sigma) = a_{1,\sigma(1)} \otimes a_{2,\sigma(2)} \otimes \ldots \otimes a_{n,\sigma(n)} = \bar{\lambda}.$$

Which implies that, $\forall i \in I$: $a_{i,\sigma(i)} \geq \bar{\lambda}$ hence $\bar{\lambda} \otimes a_{i,\sigma(i)} = \bar{\lambda}$ which shows that $\sigma(i) \in S_i(A)$.

The *chromatic number* of a hypergraph H (see Berge, 1970) is the minimum number of colors required to color the vertices of H so that for every edge having at least two vertices, all the vertices do not have the same color. A hypergraph with chromatic number 2 is said to be *bicolorable*. One can then state the following characterization of non-degenerate singular matrices in a MAX-MIN dioid:

Theorem 3. *(Minoux, 1982)*

Let $(E, \oplus, \otimes)$ *be a MAX-MIN dioid. A necessary and sufficient condition for a non-degenerate matrix* $A \in M_n(E)$ *to be singular is that the skeleton-hypergraph* $H(A)$ *satisfies the two conditions:*

(i) Each edge of H(A) *has at least two vertices*
(ii) H(A) *is bicolorable.*

Proof. (a) For $A \in M_n(E)$ with $\bar{\lambda} = \text{perm}(A)$, let us denote $\bar{A} = (\bar{a}_{ij})$ the matrix where the coefficients are:

$$\bar{a}_{ij} = \bar{\lambda} \otimes a_{ij}$$

Clearly, A is non-degenerate if and only if $\bar{A}$ is non-degenerate.
Consequently, using Property 3.5.5 a non-degenerate matrix $A \in M_n(E)$ is singular (has a subset of dependent columns) if and only if there exists J_1 and J_2, such that:

$$\left.\begin{array}{l} \sum\limits_{j\in J_1} \overline{A}^j = \sum\limits_{j\in J_2} \overline{A}^j \\ \text{with} \\ J_1 \neq \emptyset;\ J_2 \neq \emptyset:\ J_1 \cap J_2 = \emptyset \\ J_1 \cup J_2 = J \end{array}\right\} \quad (19)$$

(b) Let us now assume that A is singular. (19) then shows that each edge of H(A) contains at least two vertices, one in J_1 and one in J_2. Moreover, by attributing a color to the vertices of H(A) corresponding to the columns $j \in J_1$ and another color to the vertices of H(A) corresponding to the columns $j \in J_2$ we obtain a bicoloring of H(A).

Conversely, let us assume that H(A) is bicolorable and that each edge contains at least two vertices. This means that $J = \{1, \dots, n\}$ can be partitioned into $J_1 \neq \emptyset$ (the set of vertices having color 1) and $J_2 \neq \emptyset$ (the set of vertices having color 2), so that each row i of $\bar{A}$ contains at least two terms $\bar{a}_{ij_1}$ and $\bar{a}_{ij_2}$ equal to $\bar{\lambda}$ with $j_1 \in J_1$ and $j_2 \in J_2$.

We therefore have a complete dependence relation of the form (19) for the columns of $\bar{A}$, and this shows that A is singular. □

The problem of 2-colorability of a hypergraph being NP-complete (see Garey and Johnson, 1979, Appendix A3, p. 221), Theorem 3 therefore shows that testing the singularity of a (non-degenerate) matrix in a MAX-MIN dioid is a difficult problem. If, moreover, we recall that the computation of the permanent in such a dioid is easy (it reduces to solving a "bottleneck" assignment problem, see Sect. 3.1, Example 3.1.2), one can observe a notable difference from the point of view of *computation*, with standard algebra (where checking singularity of a matrix can be done in polynomial time, whereas computing the permanent is difficult).

Exercises

Exercise 1. Let $X = (x_i)_{i\in I}$ be a family of elements in a semi-module $(M, \Box, \perp)$ on $(E, \oplus, \otimes)$.

(a) Show that if X is independent, then it is non-quasi-redundant.
(b) Show that if X is non-quasi-redundant, then it is non-redundant.
(c) Show that if $(E, \otimes)$ has a group structure, the concepts of non-redundancy and non-quasi-redundancy are equivalent.

Exercise 2. Let $(E, \oplus, \otimes)$ be the dioid $(\mathbb{N}, \text{Max}, \times)$ with $\varepsilon = 0$, $e = 1$, and $M = E^2$, the set of vectors with two components on E.

For $\lambda > 1$ integer, let us define:

$$x_1 = \begin{pmatrix} \lambda \\ 0 \end{pmatrix} \quad x_2 = \begin{pmatrix} 0 \\ \lambda \end{pmatrix} \quad \text{and} \quad x_3 = \begin{pmatrix} 1 \\ 0 \end{pmatrix} \oplus \begin{pmatrix} 0 \\ 1 \end{pmatrix} = \begin{pmatrix} 1 \\ 1 \end{pmatrix}$$

Show that the family $X = [x_1, x_2, x_3]$ is non-redundant but that it is quasi-redundant.

Exercise 3. Let $M = E^4$, the set of vectors with four components on $(\mathbb{R}_+, \text{Max}, \times)$ and let us consider $X = [x_1, x_2, x_3, x_4]$ where:

$$x_1 = \begin{pmatrix} 1 \\ 1 \\ 0 \\ 0 \end{pmatrix} \quad x_2 = \begin{pmatrix} 1 \\ 0 \\ 1 \\ 0 \end{pmatrix} \quad x_3 = \begin{pmatrix} 0 \\ 1 \\ 0 \\ 1 \end{pmatrix} \quad x_4 = \begin{pmatrix} 0 \\ 0 \\ 1 \\ 1 \end{pmatrix}$$

Show that this family is non-quasi-redundant but not independent (in the sense of Definition 2.5.1).

Exercise 4. Show that the result of Proposition 2.5.5 remains valid, under the same assumptions, for non-redundant families $X = (x_i)_{i \in I}$ or similarly for non-quasi-redundant families.

Deduce that if $(M, \Box, \perp)$ has a non-redundant generating family (resp. non-quasi-redundant), then the latter is unique.

[*Indication*: verify that the proofs of Propositions 2.5.5 and 2.5.7 apply].

Chapter 6
Eigenvalues and Eigenvectors of Endomorphisms

1. Introduction

The celebrated Perron–Frobenius theorem, which applies to real nonnegative matrices, may be viewed as the first result stating the existence of an eigenvalue and associated eigenvector on matrices with coefficients in the dioid $(\mathbb{R}_+, +, \times)$. Indeed, it asserts that such a matrix has an eigenvalue in this dioid, with an associated eigenvector having all components in the dioid; moreover, it establishes a special property for this eigenvalue, as compared with the other eigenvalues on the field of complex numbers: it is actually the one having the largest modulus.

The importance of this largest eigenvalue is well-known as it is often related to stability issues for dynamical systems (Lyapounov coefficient), or to asymptotic behavior of systems (see, e.g. Exercise 2 at the end of this chapter).

The present chapter is devoted to the characterization of eigenvalues and eigenvectors for endomorphisms of semi-modules and of moduloids in finite dimensions. Extension to functional operators in infinite dimensions will be studied in Exercise 3 of this chapter (for Max+ dioids) and in Chap. 7, Sect. 4 (for Min–Max dioids).

Conditions guaranteeing the existence of eigenvalues and eigenvectors are studied in Sect. 2. These conditions involve the quasi-inverse A^* of the matrix A associated with the endomorphism under consideration.

In Sect. 3, we present for various classes of idempotent dioids, results characterizing the eigen-semi-module associated with a given eigenvalue.

Section 4 focuses on the important special case of dioids with multiplicative group structure. An analogue to the classical Perron–Frobenius theorem is obtained for a subclass of selective-invertible dioids. Section 5 investigates the links between eigenvalues and the characteristic bipolynomial of a matrix.

A number of noteworthy applications of eigenvalues and eigenvectors in dioids are presented in detail in Sects. 6 and 7: hierarchical clustering, preference analysis, theory of linear dynamical systems in the dioid $(\mathbb{R} \cup \{-\infty\}, \text{Max}, +)$. An interesting application to the Ising model in statistical Physics is also presented in Exercise 2.

2. Existence of Eigenvalues and Eigenvectors: General Results

Let $(E, \oplus, \otimes)$ be a semiring (where ε is the zero element and e the unit element) and denote $M = E^n$ the semi-module formed by the set of vectors with n components in E. The internal operation of M is the operation $\oplus$ defined as:

$$\begin{pmatrix} x_1 \\ x_2 \\ \cdot \\ \cdot \\ x_n \end{pmatrix} \oplus \begin{pmatrix} y_1 \\ y_2 \\ \cdot \\ \cdot \\ y_n \end{pmatrix} = \begin{pmatrix} x_1 \oplus y_1 \\ \cdot \\ \cdot \\ \cdot \\ x_n \oplus y_n \end{pmatrix}$$

and the external operation, denoted $\otimes$, is defined as:

$$\forall \lambda \in E: \lambda \otimes \begin{pmatrix} x_1 \\ x_2 \\ \vdots \\ x_n \end{pmatrix} = \begin{pmatrix} \lambda \otimes x_1 \\ \lambda \otimes x_2 \\ \vdots \\ \lambda \otimes x_n \end{pmatrix}$$

Let $h: M \to M$ be an endomorphism of M. Since any vector $x \in M$ can be written as: $x = x_1 \otimes e_1 \oplus x_2 \otimes e_2 \oplus \ldots \oplus x_n \otimes e_n$ (where, $\forall j = 1, \ldots n$, e_j denotes the vector whose jth component is e and all the other components ε) it is seen than h is perfectly defined by specifying the n vectors $h(e_1), h(e_2) \ldots h(e_n)$, or, equivalently, by the matrix $A = (a_{ij})_{\substack{i=1\ldots n \\ j=1\ldots n}}$ with columns $h(e_1), \ldots h(e_n)$.

Thus we can write, $\forall x = \begin{pmatrix} x_1 \\ \cdot \\ \cdot \\ x_n \end{pmatrix} \in M: h(x) = A \otimes x$

where the product of the matrix $A \in M_n(E)$ by the vector $x \in M$ is defined by:

$$\forall i: (A \otimes x)_i = \sum_{j=1}^{n} a_{ij} \otimes x_j$$

(where the sum above is in terms of the operation $\oplus$ of the semiring). Therefore, for finite dimensional semi-modules, there is a one-to-one correspondence between the endomorphisms and $n \times n$ square matrices with coefficients in E.

Now, given a matrix $A \in M_n(E)$ (i.e. an endomorphism of E^n), we say that $\lambda \in E$ is an eigenvalue of A if there exists $V \in E^n$, $V \neq \begin{pmatrix} \varepsilon \\ \varepsilon \\ \cdot \\ \cdot \\ \varepsilon \end{pmatrix}$ such that:

$$A \otimes V = \lambda \otimes V$$

V is referred to as the *eigenvector* of A for the eigenvalue λ.

If the operation $\otimes$ is commutative, it is easily checked that the set of eigenvectors of A for a given eigenvalue λ, denoted $\mathcal{V}(\lambda)$, is a sub-semi-module of E^n called *eigen-semi-module*. If the operation $\otimes$ is not commutative, then $\mathcal{V}(\lambda)$ is a *right* sub-semi-module, in other words, $\forall \alpha \in E, \beta \in E, V \in \mathcal{V}(\lambda)\ W \in \mathcal{V}(\lambda)$:

$$V \otimes \alpha \oplus W \otimes \beta \in \mathcal{V}(\lambda).$$

Property 2.1. Assuming that the operation $\otimes$ is idempotent and commutative, if V is an eigenvector of A for the eigenvalue e, then $\lambda \otimes V$ is an eigenvector of A for the eigenvalue λ.

Proof. We have $A \otimes V = V$

$$\begin{aligned} \text{thus: } A \otimes (\lambda \otimes V) &= \lambda \otimes (A \otimes V) \\ &= \lambda \otimes V \\ &= \lambda^2 \otimes V \\ &= \lambda \otimes (\lambda \otimes V) \quad \square \end{aligned}$$

We recall (see Chap. 4, Sect. 3.2) that the graph G(A) associated with a matrix $A \in M_n(E)$ is the directed graph having $X = \{1, 2, \ldots n\}$ as its vertex set, and where (i, j) is an arc if and only if $a_{ij} \neq \varepsilon$. For every arc (i, j) in G(A), the coefficient a_{ij} is called the *weight of* arc (i, j).

For $k \in \mathbb{N}$, we will denote A^k the kth power of A and:

$$A^{[k]} = A \oplus A^2 \oplus \ldots \oplus A^k$$

When the matrices $A^{[k]}$ have a limit as $k \to \infty$, this limit will be denoted A^+ and we define $A^* = I \oplus A^+$ (I denotes the identity matrix in $M_n(E)$). It is easily seen that the matrices A^* and A^+ satisfy:

$$A \otimes A^* = A^* \otimes A = A^+ \tag{1}$$

We also recall that the term (i, i) of A^+ can be expressed as:

$$[A^+]_{i,i} = \sum_{\gamma \in P_{ii}} w(\gamma) \tag{2}$$

where P_{ii} denotes the set of circuits of cardinality ≥ 1 containing node i in G(A) and $w(\gamma)$ the weight of circuit $\gamma \in P_{ii}$ (see Chap. 4, Property 3.2.1).

$$(\text{if } \gamma = \{(i_1, i_2), (i_2, i_3), \ldots (i_k, i_1)\} \text{ then: } w(\gamma) = a_{i_1,i_2} \otimes a_{i_2,i_3} \otimes \ldots \otimes a_{i_k,i_1})$$

We recall that, if $\otimes$ is not commutative, the weight of a circuit depends on which node is chosen as the starting node to traverse the circuit and we need to refer to the notion of *pointed circuit* (see Chap. 4, Sect. 3.2). In this case, P_{ii} in (2) will denote the set of pointed circuits having i as starting node.

Below, we denote $[A^+]^i$ and $[A^*]^i$ the ith column of matrices A^+ and A^* respectively. We can then state:

Theorem 1. *(Gondran and Minoux 1977)*

Assuming that A^* *exists (note that in this case the existence of* $A^+ = A \otimes A^*$ *follows) then the two following conditions are equivalent:*

(i)
$$\left(\sum_{\gamma \in P_{ii}} w(\gamma)\right) \otimes \lambda \oplus \lambda = \left(\sum_{\gamma \in P_{ii}} w(\gamma)\right) \otimes \lambda \tag{3}$$

for some $i \in [1, n]$ *and* $\lambda \in E$.

(ii) $[A^+]^i \otimes \lambda$ *(together with* $[A^*]^i \otimes \lambda$*) is an eigenvector of* A *for the eigenvalue* e.

Proof. Let us show that (i) ⇒ (ii)

From (1) we get:

$$\begin{aligned}[A^+]^i &= A \otimes [A^*]^i \\ &= A \otimes ([I]^i \oplus [A^+]^i)\end{aligned}$$

where $[I]^i$ denotes the ith column of the identity matrix I.

From condition (3) we can write:

$$\begin{aligned}[A^*]^i \otimes \lambda &= ([I]^i \oplus [A^+]^i) \otimes \lambda \\ &= [A^+]^i \otimes \lambda\end{aligned}$$

from which (1) implies that $[A^+]^i \otimes \lambda = [A^*]^i \otimes \lambda$ is an eigenvector of A for the eigenvalue e.

Now we show that (ii) ⇒ (i).

Assume that $[A^*]^i \otimes \lambda$ is an eigenvector of A for the eigenvalue e.

Thus:

$$\begin{aligned}A \otimes [A^*]^i \otimes \lambda &= [A^*]^i \otimes \lambda \\ &= ([I]^i \oplus [A^+]^i) \otimes \lambda\end{aligned} \tag{4}$$

Since $A \otimes A^* = A^+$ we deduce from (4):

$$[A^*]^i \otimes \lambda = [A^+]^i \otimes \lambda$$

which then implies that:

$$[A^+]^i \otimes \lambda = ([I]^i \oplus [A^+]^i) \otimes \lambda$$

For the ith component, the above equality implies:

$$[A^+]_{i,i} \otimes \lambda = \lambda \oplus [A^+]_{i,i} \otimes \lambda$$

which is none other than (3). □

Several interesting results will now be deduced from this theorem.

Corollary 2.2. *If* A^* *exists and if there exists* $\mu \in E$ *and a pointed circuit* γ *originating at* i *such that:*

$$w(\gamma) \otimes \mu \oplus \mu = w(\gamma) \otimes \mu$$

then $[A^*]^i \otimes \mu$ *is an eigenvector of* A *for the eigenvalue* e.

Proof. We need simply observe that in view of the assumptions of Corollary 2.2, relation (3) is satisfied. □

Theorem 1 and Corollary 2.2 will often be used in the special situation where $\lambda = e$ or $\mu = e$.

Corollary 2.3. *If* A^* *exists and if* $I \oplus A = A$, *then all columns of* A^* *are eigenvectors of* A *for the eigenvalue* e.

Proof. It follows directly from Corollary 2.2 taking $\mu = e$ and $w(\gamma) = a_{ii}$. □

Corollary 2.4 below is a consequence of Theorem 1 in the case where $(E, \otimes)$ is a group (the inverse of an element $\lambda \in E$ for $\otimes$ being denoted λ^{-1}).

Corollary 2.4. *(i) If* $(E, \otimes)$ *is a group, if* $(\lambda^{-1} \otimes A)^*$ *exists and if:*

$$\left(\lambda^{-1} \otimes A\right)^*_{i,i} \oplus e = \left(\lambda^{-1} \otimes A\right)^*_{i,i} \tag{5}$$

then $[(\lambda^{-1} \otimes A)^*]^i$ *is an eigenvector of* A *for the eigenvalue* λ.

(ii) If, furthermore, $\otimes$ *is commutative, then condition* (5) *can be replaced by:*

$$\sum_{\gamma \in P_{ii}} w(\gamma) \otimes (\lambda^{-1})^{|\gamma|} \oplus e = \sum_{\gamma \in P_{ii}} w(\gamma) \otimes (\lambda^{-1})^{|\gamma|} \tag{6}$$

where $|\gamma|$ *denotes the number of arcs of circuit* γ *and where* P_{ii} *denotes the set of circuits of* G(A) *containing* i.

Proof. (i) Relation (5) shows that Theorem 1 applies to the matrix $\lambda^{-1} \otimes A$ and that, as a result, $[(\lambda^{-1} \otimes A)^*]^i$ is an eigenvector of $\lambda^{-1} \otimes A$ for the eigenvalue e. It follows that this vector is also an eigenvector of A for the eigenvalue λ.

(ii) In the case where $\otimes$ is commutative, the weight of a circuit γ of $G(\lambda^{-1} \otimes A)$ is none other than $w(\gamma) \otimes (\lambda^{-1})^{|\gamma|}$ where $w(\gamma)$ is the weight of this circuit in G(A). □

The following result concerns another important special case, in which both operations $\oplus$ and $\otimes$ are idempotent.

Corollary 2.5. *Assume that* $\oplus$ *and* $\otimes$ *are both idempotent, that* A^+ *exists, and define* $\mu \in E$ *as:*

$$\mu = [A^+]_{i,i} = \sum_{\gamma \in P_{ii}} w(\gamma) \tag{7}$$

Then $[A^*]^i \otimes \mu$ *is an eigenvector of* A *for the eigenvalue* e.

Proof. Since $\oplus$ and $\otimes$ are idempotent, we have $\mu^2 \oplus \mu = \mu^2$ and, μ being defined by (7), we observe that relation (3) is satisfied with $\lambda = \mu$; the result then follows from Theorem 1. □

In the case where, in addition to the idempotency of $\oplus$ and $\otimes$, we assume the commutativity of $\otimes$, the existence of A^+ is guaranteed and we obtain the following corollary.

Corollary 2.6. *If* $\oplus$ *is idempotent,* $\otimes$ *idempotent and commutative, then:*

(i) A^+ *and* A^* *exist:*
(ii) by setting:

$$\mu = [A^+]_{i,i} = \sum_{\gamma \in P_{ii}} w(\gamma)$$

for any $\lambda \in E$, $\lambda \otimes \mu \otimes [A^*]^i$ *is an eigenvector of* A *for the eigenvalue* λ.

Proof. (i) In the case where $\oplus$ and $\otimes$ are idempotent, any element $a \in E$ is 1-stable and has a quasi-inverse $a^* = e \oplus a$ (see Sect. 7 in Chap. 3). If, furthermore, the law $\otimes$ is commutative, according to Theorem 2 of Chap. 4, any square matrix A has a quasi-inverse A^*: the existence of A^+ and A^* is therefore guaranteed.
(ii) By applying Corollary 2.5 we immediately deduce that $[A^*]^i \otimes \mu = \mu \otimes [A^*]^i$ is an eigenvector of A for the eigenvalue e. It then follows from Property 2.1, that $\forall \lambda \in E$, $\lambda \otimes \mu \otimes [A^*]^i$ is an eigenvector of A for the eigenvalue λ. □

3. Eigenvalues and Eigenvectors in Idempotent Dioids: Characterization of Eigenmoduloids

In this section we will present a number of results characterizing eigen-semi-modules $\mathcal{V}(\lambda)$ associated with eigenvalues λ.

All these results will be obtained by assuming that $\oplus$ is either *idempotent* or *selective*, in other words, correspond to cases where $(E, \oplus, \otimes)$ are *dioids* (either idempotent or selective). The eigen-semi-modules under consideration in this section are therefore *moduloids*.

We start by studying $\mathcal{V}(e)$. Thereafter we will show how results for $\mathcal{V}(\lambda)$, for arbitrary λ, can be deduced.

Lemma 3.1. *Let* $(E, \oplus, \otimes)$ *be an idempotent dioid and* $A \in M_n(E)$ *having the eigenvalue* e, *and such that* A^* *exists with* $A^* = A^{(p)} = I \oplus A \oplus \ldots \oplus A^p$ *for some integer* $p \in \mathbb{N}$.

Then: $V \in \mathcal{V}(e) \Rightarrow V = A^* \otimes V$

Proof. We can write:

$$\begin{aligned} V &= V \\ A \otimes V &= V \\ A^2 \otimes V &= A \otimes V = V \\ &\vdots \\ &\vdots \\ A^k \otimes V &= A \otimes V = V \end{aligned}$$

for any $k \in \mathbb{N}$. It follows that: $(I \oplus A \oplus \ldots \oplus A^k) \otimes V = A^{(k)} \otimes V = V$ and consequently if $A^* = A^{(p)}$ for $p \in \mathbb{N}$, we deduce:

$$A^* \otimes V = V. \quad \square$$

The previous result thus shows that, when the reference set is an *idempotent dioid* (and even more so in the case of a *selective dioid*), any vector of $\mathcal{V}(e)$ is a linear combination of the columns of A^*. Nevertheless, the columns of A^* do not necessarily belong to $\mathcal{V}(e)$ themselves (see Theorem 1 of Sect. 2 and its corollaries).

With the stronger assumption of selectivity for $\oplus$, we now show that the set of vectors of the form $[A^*]^i \otimes \mu_i$ which belong to $\mathcal{V}(e)$ form a generator for $\mathcal{V}(e)$.

Theorem 2. *(Gondran and Minoux 1977)*

Let $(E, \oplus, \otimes)$ *be a selective dioid and* $A \in M_n(E)$ *having the eigenvalue* e *and such that* A^* *exists with* $A^* = A^{(p)}$, *for some integer* $p \in \mathbb{N}$.

Then there exists a subset $\{i_1, i_2, \ldots, i_K\}$ *of* $\{1, 2, \ldots, n\}$ *and coefficients* $\mu_{i_k} \in E (k = 1 \ldots K)$ *such that*

$$V \in \mathcal{V}(e) \Rightarrow V = \sum_{k=1}^{K} [A^*]^{i_k} \otimes \mu_{i_k}$$

$$\text{with,} \quad \forall k\colon\ \left[A^*\right]^{i_k} \otimes \mu_{i_k} \in \mathcal{V}(e)$$

Proof. $\oplus$ is idempotent, thus $V = A^* \otimes V$ from Lemma 3.1 and consequently:

$$V = \sum_{i=1}^{n} \left[A^*\right]^i \otimes V_i \tag{8}$$

Moreover, since $V \in \mathcal{V}(e)$:

$$\forall i = 1, \ldots n\colon \quad \sum_{j=1}^{n} a_{ij} \otimes V_j = V_i$$

Since $\oplus$ is selective, with every index $i \in \{1, \ldots n\}$ we can associate an index $j = \varphi(i)$ such that: $a_{i,\varphi(i)} \otimes V_{\varphi(i)} = V_i$ (if several indices j exist such that $a_{ij} \otimes V_j = V_i$, we arbitrarily choose one of these indices for $\varphi(i)$).

The partial graph H of G(A), formed by the subset of arcs of the form $(i, \varphi(i))$ for $i = 1, \ldots n$, contains n vertices and n arcs. Its cyclomatic number is therefore equal to its connectivity number. Moreover, as any vertex of H has an out-degree exactly equal to 1, each connected component of H contains a unique circuit. Let us denote $H^1, H^2, \ldots H^K$ the connected components of H, and, for $k \in \{1, \ldots K\}$, let us denote γ^k the circuit contained in H^k. Any vertex $i \in H^k$ either belongs to γ^k or is connected to γ^k by a unique path originating at i and terminating at $j \in \gamma^k$.

Let us assume that γ^k has vertex set $\{i_1, i_2, \ldots i_q\}$.

We have:

$$
\begin{aligned}
a_{i_1 i_2} \otimes V_{i_2} &= V_{i_1} \\
a_{i_2 i_3} \otimes V_{i_3} &= V_{i_2} \\
&\vdots \\
&\vdots \\
a_{i_q i_1} \otimes V_{i_1} &= V_{i_q}
\end{aligned}
$$

thus we deduce, for any vertex i of γ^k (for example $i = i_1$):

$$w(\gamma^k) \otimes V_i = V_i$$

We can thus write:

$$w(\gamma^k) \otimes V_i \oplus V_i = V_i = w(\gamma^k) \otimes V_i$$

We then observe that, as a result, the assumptions of Theorem 1 are satisfied with $\lambda = V_i$, which implies:

$$[A^*]^i \otimes V_i \in \mathcal{V}(e)$$

Let us now consider an arbitrary vertex j of H^k, $j \notin \gamma^k$, and show that, in expression (8), the term $[A^*]^j \otimes V_j$ is absorbed by the term $[A^*]^i \otimes V_i$ $(i \in \gamma^k)$.

In H^k there exists a unique path π_{ji} joining j to i. On each arc (s, t) of this path we have the relation:

$$a_{st} \otimes V_t = V_s$$

thus we deduce:

$$w(\pi_{ji}) \otimes V_i = V_j$$

We can thus write:

$$[A^*]^j \otimes V_j \oplus [A^*]^i \otimes V_i = [[A^*]^j \otimes w(\pi_{ji}) \oplus [A^*]^i] \otimes V_i$$

$w(\pi_{ji})$ comes into play in the term (j, i) of the matrix A^r, where r is the number of arcs of the path π_{ji}. From the elementary property: $A^* \oplus A^* \otimes A^r = A^*$ (which follows from the idempotency of $\oplus$) we deduce that:

$$[A^*]^j \otimes w(\pi_{ji}) \oplus [A^*]^i = [A^*]^i$$

which shows that:

$$[A^*]^i \otimes V_i \oplus [A^*]^j \otimes V_j = [A^*]^i \otimes V_i$$

As a result, it is enough in expression (8) to retain a unique term of the form $[A^*]^{i_k} \otimes V_{i_k}$, with $i_k \in \gamma^k$, for each connected component H^k of H, and we have:

$$V = \sum_{k=1}^{K} [A^*]^{i_k} \otimes V_{i_k}$$

and $[A^*]^{i_k} \otimes V_{i_k} \in \mathcal{V}(e)$ for $k = 1 \ldots K$, which proves the theorem. □

We can deduce from Theorem 2 the following consequences.

Corollary 3.2. *Let* $(E, \oplus, \otimes)$ *be a selective-invertible dioid and* $A \in M_n(E)$ *such that* A^* *exists with* $A^* = A^{(p)} (p \in \mathbb{N})$

Then, if e *is an eigenvalue of* A, $\mathcal{V}(e)$ *is the (right) moduloid generated by those columns of* A^* *which are eigenvectors of* A *for* e.

Proof. According to Theorem 2, if $V \in \mathcal{V}(e)$, then V has the form:

$$V = \sum [A^*]^k \otimes \mu_k$$

with, $\forall k$, $[A^*]^k \otimes \mu_k \in \mathcal{V}(e)$ (sum on a subset of indices from $\{1, 2, \ldots, n\}$).

Since $(E, \otimes)$ is a group, $\mu_k \neq \varepsilon$ has an inverse for $\otimes$ and:

$A \otimes [A^*]^k \otimes \mu_k = [A^*]^k \otimes \mu_k$ shows (via right-multiplication by $(\mu_k)^{-1}$) that we also have $[A^*]^k \in \mathcal{V}(e)$

Hence, Corollary 3.2 follows. □

In Corollary 3.3 and Theorem 3 below, we will assume that $\oplus$ is selective, that $\otimes$ is idempotent and that e is the greatest element of E. Then, we know that the set E is totally ordered by the canonical order relation and it is easy to see that:

$$\forall a \in E, b \in E, a \otimes b = \text{Min}\{a, b\} \tag{9}$$

Indeed, we have:

$$a \otimes b \oplus a = a \otimes (b \oplus e) = a$$
$$a \otimes b \oplus b = (a \oplus e) \otimes b = b$$

which implies $a \otimes b \leq a$ and $a \otimes b \leq b$

If now we assume $a \geq b$, we have:

$a \oplus b = a$ and we can write:

$$a \otimes b = (a \oplus b) \otimes b = a \otimes b \oplus b$$

thus we deduce $a \otimes b \geq b$ and (by the antisymmetry of $\geq$) $a \otimes b = b$. Similarly: $a \leq b \Rightarrow a \otimes b = a$.

Corollary 3.3. *Let* $(E, \oplus, \otimes)$ *be a selective dioid for which:*

- *the* $\otimes$ *law is idempotent;*
- e *is the greatest element (i.e.* $\forall a \in E, e \oplus a = e$*)*

Let $A \in M_n(E)$ *having* e *as an eigenvalue. Then* A^* *exists and* $\mathcal{V}(e)$ *is the (right) moduloid generated by the set of vectors of the form* $[A^+]^i \otimes \mu_i$ *(for* $i = 1 \ldots n$*) where,* $\forall i$: $\mu_i = \sum_{\gamma \in P_{ii}} w(\gamma) = [A^+]_{i,i}$

Proof. In view of the assumptions, for any pointed circuit γ of G(A) we have:

$$w(\gamma) \oplus e = e$$

which shows that G(A) is without 0-absorbing circuit.

According to Theorem 1 of Chap. 4, we deduce the existence of A^* (note that here we do not assume the commutativity of $\otimes$). Furthermore, $A^* = A^{(n-1)}$.

According to Corollary 2.5, the vectors $[A]^i \otimes \mu_i$ (with $\mu_i = [A^+]_{i,i}$) are elements of $\mathcal{V}(e)$.

Let us now apply *Theorem 2*: any vector $V \in \mathcal{V}(e)$ can be written as:

$$V = \sum [A^*]^k \otimes \alpha_k \tag{10}$$

with $[A^*]^k \otimes \alpha_k \in \mathcal{V}(e)$.

(Sum on a subset of indices from $\{1, 2, \ldots, n\}$).

According to Theorem 1, α_k satisfies (3) in other words:

$$\mu_k \otimes \alpha_k \oplus \alpha_k = \mu_k \otimes \alpha_k$$

Moreover:

$$\begin{aligned} \mu_k \otimes \alpha_k \oplus \alpha_k &= (\mu_k \oplus e) \otimes \alpha_k \\ &= \alpha_k \quad \text{(because } \mu_k \oplus e = e\text{)} \end{aligned}$$

Thus, we have: $\mu_k \otimes \alpha_k = \alpha_k$ and (10) can be rewritten:

$$V = \sum ([A^*]^k \otimes \mu_k) \otimes \alpha_k$$

This shows that the set of vectors of the form $[A^*]^i \otimes \mu_i (i = 1, \ldots n)$ is a generator of $\mathcal{V}(e)$. □

According to Corollary 3.3, the set of distinct vectors of the form $[A^*]^i \otimes [A^+]_{i,i}$ is a generator of $\mathcal{V}(e)$. The following result is fundamental, because, under the same assumptions, it shows that it is a *minimal* generator, and that it is *unique*.

Theorem 3. *Let* $(E, \oplus, \otimes)$ *be a selective dioid for which the* $\otimes$ *law is idempotent and where* e *is the greatest element. Let* $A \in M_n(E)$ *having* e *as eigenvalue.*

Then A^* *exists and* $G = \underset{i=1\ldots n}{\cup} \left\{\overline{V}^i\right\}$, *the set of distinct vectors of the form* $\overline{V}^i = [A^*]^i \otimes [A^+]_{i,i}$ *(for* $i = 1, \ldots n$*), is the only minimal generator of* $\mathcal{V}(e)$.

Proof. The existence of A^* and the fact that $\overline{V}^i (i = 1, \ldots, n)$ form a generator of $\mathcal{V}(e)$ follow from Corollary 3.3.

Let us show, therefore, that G is a minimal generator and that it is unique. Let us proceed by contradiction and assume that there exists another minimal generator $G' = \underset{k \in K}{\cup} \left\{W^k\right\}$ of $\mathcal{V}(e)$ composed of vectors W^k, k running through a finite set of indices K. Some vectors $\overline{V}^i$ can coincide with vectors of G′, however there necessarily exists at least one index i such as: $\overline{V}^i \neq W^k$ for all $k \in K$. Since G′ is a generator of $\mathcal{V}(e)$, there exist coefficients $\gamma_k \in E$ such that:

$$\overline{V}^i = \sum_{k \in K} W^k \otimes \gamma_k \tag{11}$$

The component i of $\overline{V}^i$ is $[A^+]_{i,i}$

(Indeed, we have:

$$[A^*]_{i,i} \otimes [A^+]_{i,i} = (e \oplus [A^+]_{i,i}) \otimes [A^+]_{i,i} = [A^+]_{i,i}$$

since e is the greatest element of E).

As $\oplus$ is selective, there exists an index $k' \in K$ such that:

$$\overline{V}^i_i = [A^+]_{i,i} = W^{k'}_i \otimes \gamma_{k'} \tag{12}$$

As $W^{k'} \in \mathcal{V}(e)$, from Lemma 3.1, it can be expressed as:

$$W^{k'} = \sum_{j=1}^{n} [A^*]^j \otimes W^{k'}_j$$

thus we deduce:

$$W^{k'} \geq [A^*]^i \otimes W^{k'}_i$$

Since, according to (12), $W^{k'}_i \geq [A^+]_{i,i}$,
we deduce:

$$W^{k'} \geq [A^*]^i \otimes [A^+]_{i,i} = \overline{V}^i \tag{13}$$

Furthermore, from relation (11) we can write:

$$\overline{V}^i \geq W^{k'} \otimes \gamma_{k'}$$

and furthermore, by noting that (from (12)) $\gamma_{k'} \geq [A^+]_{i,i}$:

$$\overline{V}^i \geq W^{k'} \otimes [A^+]_{i,i} \tag{14}$$

Now multiplying (13) by $[A^+]_{i,i}$ and noting that, as a result of the idempotency of $\otimes$:

$$\overline{V}^i \otimes [A^+]_{i,i} = \overline{V}^i$$

we obtain:

$$W^{k'} \otimes [A^+]_{i,i} \geq \overline{V}^i \tag{15}$$

The inequalities (14) and (15) then imply:

$$\overline{V}^i = W^{k'} \otimes [A^+]_{i,i} \tag{16}$$

The above reasoning shows that, for any vector $\overline{V}^i$ not coinciding with one of the vectors W^k, there exists an index $k' \in K$ such that the vectors $\overline{V}^i$ and $W^{k'}$ satisfy (16) (in other words are "colinear").

Furthermore, as the set of $W^k (k \in K)$ was assumed to be minimal, it cannot contain any vector $W^t (t \in K)$ which does not correspond to one of the $\overline{V}^i$ (indeed, in this case, a generator strictly smaller in the sense of inclusion would be obtained by eliminating the vector W^t).

From the above we deduce that the sets G and G′ are in one-to-one correspondence. For any $k \in K$, let us denote $\alpha(k) = i$, the index of the vector $\overline{V}$ corresponding to W^k. Thus we have, $\forall k \in K$:

$$W^k = \overline{V}^{\alpha(k)} \quad \text{or} \quad W^k \otimes \left[A^+\right]_{\alpha(k),\alpha(k)} = \overline{V}^{\alpha(k)}.$$

In all cases, we thus have:

$$W^k \geq \overline{V}^{\alpha(k)} \quad \text{for any } k.$$

This shows that, among all minimal generators (in the sense of inclusion) in one-to-one correspondence with $G = \bigcup_{i=1\ldots n} \left\{\overline{V}^i\right\}$, G is the one which contains the least vectors (in the sense of the order relation on the vectors of E^n). G is therefore the unique minimal generator with this property. □

Let us note that Theorem 3 above generalizes a result obtained by Gondran (1976a,b) for the special case of the dioid ($\mathbb{R}$, Min, Max).

In Chap. 7, Sect. 6.4, an extension of Theorem 3 to infinite dimensions (functional semi-modules) will be found for the case where the basic dioid is ($\overline{\mathbb{R}}$, Min, Max) (see also Gondran and Minoux 1997, 1998).

The following results characterize $\mathcal{V}(\lambda)$ for $\lambda \neq e$ in the case of selective dioids with idempotent multiplication.

Lemma 3.4. *Let* $(E, \oplus, \otimes)$ *be a dioid with* $\otimes$ *commutative and assuming* e *to be the greatest element of* $E(\forall a \in E: e \oplus a = e)$. *Let* $A \in M_n(E)$.

Then A^* *exists, and if* $\lambda \in E$ *is an eigenvalue of* A:

$$V \in \mathcal{V}(\lambda) \Rightarrow V = A^* \otimes V$$

Proof. By using the commutativity of $\otimes$, $A \otimes V = \lambda \otimes V$ implies, $\forall k \in \mathbb{N}$:

$$A^k \otimes V = \lambda^k \otimes V$$

Since e is the greatest element of E, we have: $e \oplus \lambda \oplus \lambda^2 \oplus \ldots \oplus \lambda^k = e$ thus, $\forall k$:

$$(I \oplus A \oplus A^2 \oplus \ldots \oplus A^k) \otimes V = V$$

As $A^{(k)} = A^*$ as soon as $k \geq n - 1$ (see Chap. 4, Theorem 1) we deduce the result. □

We can then state:

Corollary 3.5. *(Gondran and Minoux 1977)*
Let $(E, \oplus, \otimes)$ *be a selective dioid for which:*

- *the* $\otimes$ *law is idempotent and commutative;*
- e *is the greatest element (i.e.* $\forall a \in E$: $e \oplus a = e$*)*

Let $A \in M_n(E)$. *Then:*

(i) A^* *exists and any* $\lambda \in E$ *is an eigenvalue of* A;

(ii) $V \in \mathcal{V}(\lambda) \Rightarrow V = \sum_{k=1}^{K} [A^*]^{i_k} \otimes \lambda \otimes \mu_{i_k}$

with $K \leq n$ *and* $\forall k$: $[A^*]^{i_k} \otimes \lambda \otimes \mu_{i_k} \in \mathcal{V}(\lambda)$
with $\mu_{i_k} = [A^+]_{i_k, i_k}$

Proof. (i) follows directly from Corollary 2.6. Let us therefore prove (ii).

The assumptions of Lemma 3.1 being satisfied, we have: $V = A^* \otimes V = \sum_{i=1}^{n} [A^*]^i \otimes V_i$

On the other hand, we have:

$$\forall i = 1, \ldots n: \quad \sum_{j=1}^{n} a_{ij} \otimes V_j = \lambda \otimes V_i$$

As in the proof of Theorem 2, we can construct the partial graph H of G(A) whose arcs have the form $(i, \varphi(i))$ where, $\forall i = 1, \ldots n$:

$$a_{i,\varphi(i)} \otimes V_{\varphi(i)} = \lambda \otimes V_i \tag{13}$$

Each connected component H^k contains a circuit γ^k. By writing the relations (13) along the circuit γ^k, and by taking into account the fact that $\otimes$ is idempotent, we obtain for $i \in \gamma^k$:

$$w(\gamma^k) \otimes V_i = \lambda \otimes V_i$$

We can then write:

$$w(\gamma^k) \otimes \lambda \otimes V_i \oplus \lambda \otimes V_i = w(\gamma^k) \otimes \lambda \otimes V_i$$

thus we deduce, from Corollary 1, that: $[A^*]^i \otimes \lambda \otimes V_i \in \mathcal{V}(e)$

We then note that we also have $[A^*]^i \otimes \lambda \otimes V_i \in \mathcal{V}(\lambda)$ (indeed, $\otimes$ being idempotent and commutative, $A \otimes U = U$ implies: $A \otimes (\lambda \otimes U) = \lambda \otimes A \otimes U = \lambda \otimes U = \lambda^2 \otimes U$, which shows that $U \in \mathcal{V}(e) \Rightarrow \lambda \otimes U \in \mathcal{V}(\lambda)$).

As in the proof of *Theorem* 2 we also show that any term of the form $[A^*]^j \otimes \lambda \otimes V_j$ $(j \neq i, j \in H^k)$ is absorbed by $[A^*]^i \otimes \lambda \otimes V_i$.

Then by choosing a vertex $i_k \in \gamma^k$ in each connected component H^k of H, we can write, denoting by K the connectivity number of H:

$$V = \sum_{k=1}^{K} [A^*]^{i_k} \otimes \lambda \otimes V_{i_k}$$

with, $\forall k$, $[A^*]^{i_k} \otimes \lambda \otimes V_{i_k} \in \mathcal{V}(\lambda)$, which proves (ii). □

4. Eigenvalues and Eigenvectors in Dioids with Multiplicative Group Structure

In this section we will investigate the special case, which is important for applications, of dioids $(E, \oplus, \otimes)$ for which $(E, \otimes)$ is a group.

For matrices with entries in such dioids, we will see that well-known properties of irreducible matrices with positive entries in ordinary algebra will thus be found again. We will then use these properties to establish an analogue to the classical Perron–Frobenius theorem for matrices with entries in some *selective-invertible dioids*.

We recall that a matrix $A \in M_n(\mathbb{R}_+)$ is said to be *irreducible* if and only if the associated graph G(A) is *strongly connected*. In classical linear algebra, the Perron–Frobenius theorem is stated as:

Theorem. *(Perron, Frobenius) Let* $A \in M_n(\mathbb{R}_+)$ *be an irreducible matrix and* $\rho(A)$ *its spectral radius (the modulus of the eigenvalue having the largest modulus). Then* $\rho(A)$ *is an eigenvalue of* A, *and the associated eigenspace is generated by an eigenvector whose components are all positive.*

For a proof and further discussion of this theorem, see Exercise 1.

In the case of dioids featuring multiplicative group structure we start Sect. 4.1 by establishing a few preliminary results.

4.1. Eigenvalues and Eigenvectors: General Properties

Lemma 4.1.1. *Let* $(E, \oplus, \otimes)$ *be a dioid. Then:*

$$a \oplus b = \varepsilon \Rightarrow a = b = \varepsilon \tag{14}$$

If furthermore, $(E, \otimes)$ *is a group, then:*

$$a \otimes b = \varepsilon \Rightarrow a = \varepsilon \text{ or } b = \varepsilon \tag{15}$$

Proof. Equation (14) follows from Proposition 3.4.8 of Chap. 1.

Let us now assume that $(E, \otimes)$ is a group and that $a \otimes b = \varepsilon$.

Let us show that $a \neq \varepsilon$ and $b \neq \varepsilon$ is impossible. If $a \neq \varepsilon$, then a^{-1}, the inverse of a for $\otimes$ exists, from which we deduce $b = a^{-1} \otimes \varepsilon = \varepsilon$ (because of the property of absorption), which exhibits a contradiction. From the above (15) is deduced. □

Lemma 4.1.2. *(Gondran and Minoux 1977)*

Let $(E, \oplus, \otimes)$ *be a dioid such that* $(E, \otimes)$ *is a group, and* $A \in M_n(E)$ *be an irreducible matrix* (G(A) *strongly connected). Then, if* $V = (V_i)_{i=1\ldots n}$ *is an eigenvector of* A *for the eigenvalue* λ *we have:*

$$\lambda > \varepsilon \tag{16}$$

and:

$$\forall i = 1, \ldots n: \; V_i > \varepsilon \tag{17}$$

Proof. Let us first prove (16). We necessarily have $\varepsilon \leq \lambda$ (ε is the least element in the sense of the order relation). If $\lambda = \varepsilon$ then we have:

$$A \otimes V = \begin{pmatrix} \varepsilon \\ \varepsilon \\ \vdots \\ \varepsilon \end{pmatrix}$$

therefore, $\forall i = 1 \ldots n$:

$$\sum_{j=1}^{n} a_{ij} \otimes V_j = \varepsilon$$

According to Lemma 4.1.1, this implies:

$$\forall i, \forall j: a_{ij} \otimes V_j = \varepsilon$$

and therefore:

$$a_{ij} = \varepsilon \quad \text{or} \quad V_j = \varepsilon.$$

Since V is an eigenvector, $V \neq \begin{pmatrix} \varepsilon \\ \varepsilon \\ \vdots \\ \varepsilon \end{pmatrix}$ there hence exists j_o such that $V_{j_o} \neq \varepsilon$.

Then the relation $a_{ij_o} \otimes V_{j_o} = \varepsilon$ implies $a_{ij_o} = \varepsilon$, and this is so for all i. This leads to a contradiction with the strong connectivity of G(A). Consequently, we cannot have $\lambda = \varepsilon$, and (16) is proven.

To prove (17) let us again proceed by contradiction. Let us assume that V has some components $V_j = \varepsilon$. The set $X = \{1, 2, \ldots n\}$ can then be partitioned into:

$X_1 = \{j/V_j = \varepsilon\}$ and $X_2 = X \backslash X_1$ and we have that $X_1 \neq \emptyset$ $X_2 \neq \emptyset$.

By reordering the rows and columns of A and the components of V if necessary, we can put V into the form $V = \begin{pmatrix} V_1 \\ V_2 \end{pmatrix}$ (where V_1 corresponds to the components of V equal to ε and V_2 to the components of V different from ε) and A in the form:

$$A = \begin{pmatrix} A_{11} & A_{12} \\ A_{21} & A_{22} \end{pmatrix}$$

where $\forall l \in \{1, 2\}$ and $\forall k \in \{1, 2\}$ A_{lk} is the submatrix induced by the subset of rows X_l and the subset of columns X_k. The relation $A \otimes V = \lambda \otimes V$ then implies $A_{12} \otimes V_2 = \begin{pmatrix} \varepsilon \\ \varepsilon \\ \vdots \\ \varepsilon \end{pmatrix}$. V_2 having all its components $\neq \varepsilon$, we can deduce, in a way similar to the above, that all the terms of the submatrix A_{12} are equal to ε. This contradicts the irreducibility of the matrix A, and proves (17). □

This result will now be used to study the properties of the eigenvalues of matrices with entries in dioids featuring multiplicative group structure.

Lemma 4.1.3. *(Gondran and Minoux 1977)*

Let $(E, \oplus, \otimes)$ *be a dioid where* $(E, \otimes)$ *is a commutative group,* $A \in M_n(E)$ *an irreducible matrix and* $\lambda \in E$ *an eigenvalue of* A. *Then:*

(i) For any circuit γ *of* G(A) *we have:*

$$w(\gamma) \leq \lambda^{|\gamma|} \tag{18}$$

(where $|\gamma|$ *denotes the cardinality of circuit* γ*).*

(ii) If $\oplus$ *is selective, there exists an elementary circuit* γ *of* G(A) *such that:*

$$w(\gamma) = \lambda^{|\gamma|} \tag{19}$$

(iii) If $\oplus$ *is idempotent, the matrix* $(\lambda^{-1} \otimes A)^*$ *exists.*

Proof. (i) Let us consider an arbitrary circuit $\gamma = \{i_1, i_2, \ldots i_k, i_1\}$ of G(A) where $|\gamma| = k$.

If $V = (V_j)_{j=1\ldots n}$ is an eigenvector associated with the eigenvalue λ, we have:

$$\sum_j a_{i_1 j} \otimes V_j = \lambda \otimes V_{i_1}$$

$$\sum_j a_{i_2 j} \otimes V_j = \lambda \otimes V_{i_2}$$

$$\vdots$$

$$\vdots$$

$$\sum_j a_{i_k j} \otimes V_j = \lambda \otimes V_{i_k}$$

This implies:

$$\left.\begin{array}{l} a_{i_1 i_2} \otimes V_{i_2} \leq \lambda \otimes V_{i_1} \\ a_{i_2 i_3} \otimes V_{i_3} \leq \lambda \otimes V_{i_2} \\ \dots \\ a_{i_k i_1} \otimes V_{i_1} \leq \lambda \otimes V_{i_k} \end{array}\right\} \quad (20)$$

By multiplying the first inequality by λ (and by using the compatibility of $\leq$ with $\otimes$) we obtain:

$$a_{i_1 i_2} \otimes a_{i_2 i_3} \otimes V_{i_3} \leq a_{i_1 i_2} \otimes \lambda \otimes V_{i_2} \leq \lambda^2 \otimes V_{i_1}$$

Similarly, by multiplying the latter inequality by λ and by using the relation: $a_{i_3 i_4} \otimes V_{i_4} \leq \lambda \otimes V_{i_3}$ we obtain:

$$a_{i_1 i_2} \otimes a_{i_2 i_3} \otimes a_{i_3 i_4} \otimes V_{i_4} \leq \lambda^3 \otimes V_{i_1}$$

and by iterating this process k times:

$$a_{i_1 i_2} \otimes \dots \otimes a_{i_k i_1} \otimes V_{i_1} \leq \lambda^{|\gamma|} \otimes V_{i_1} \quad (21)$$

Since A is irreducible we have $\lambda \neq \varepsilon$ and $V_j \neq \varepsilon$ $(j = 1 \dots n)$ (see Lemma 4.1.2) therefore $V_{i_1} \neq \varepsilon$. Then by multiplying the two sides of the above inequality by $\left(V_{i_1}\right)^{-1}$ we obtain:

$$a_{i_1 i_2} \otimes a_{i_2 i_3} \otimes \dots \otimes a_{i_k i_1} \leq \lambda^{|\gamma|}$$

which proves (18).

(ii) If $V = (V_j)_{j=1\dots n}$ is an eigenvector of A for the eigenvalue λ, we have,

$$\forall i = 1 \dots n: \sum_j a_{ij} \otimes V_j = \lambda \otimes V_i$$

Since $\oplus$ is assumed to be selective, with each index $i \in \{1, \dots n\}$ we can associate an index $\varphi(i)$ such as:

$$a_{i,\varphi(i)} \otimes V_{\varphi(i)} = \lambda \otimes V_i$$

(if there exist several indices j such that $a_{i,j} \otimes V_j = \lambda \otimes V_i$, we arbitrarily choose one of these indices for φ (i)).

The partial graph H of G(A) formed by the subset of arcs in the form $(i,\varphi(i))$ contains n vertices and n arcs. In H, each vertex has an out-degree equal to 1, hence there exists an elementary circuit γ. Along this circuit, relations (20) are satisfied *with equality*, and relation (21) reads:

$$w(\gamma) \otimes V_{i_1} = \lambda^{|\gamma|} \otimes V_{i_1}$$

Since V_{i_1} is invertible ($V_{i_1} \neq \varepsilon$ in view of Lemma 4.1.2) we can deduce from this relation (19).

(iii) The existence of $(\lambda^{-1} \otimes A)^*$ follows directly from (18). Indeed, let γ be an arbitrary circuit of G(A). Its weight with respect to $\lambda^{-1} \otimes A$ is:

$$\theta(\gamma) = w(\gamma) \otimes (\lambda^{-1})^{|\gamma|}$$

(18) then implies that $\theta(\gamma) \leq e$ and, consequently, for any circuit γ of G(A) we have:

$$\theta(\gamma) \oplus e \leq e \oplus e = e. \quad \text{As} \quad \theta(\gamma) \oplus e \geq e \quad \text{we deduce} \quad \theta(\gamma) \oplus e = e$$

which shows that $\theta(\gamma)$ is 0-*stable*. The existence of $(\lambda^{-1} \otimes A)^*$ can then be directly deduced from Theorem 1 of Chap. 4. □

From Lemma 4.1.3 above we can then deduce:

Corollary 4.1.4. *Let* (E, ⊕, ⊗) *be a selective-invertible dioid with* ⊗ *commutative, and* $A \in M_n(E)$ *an irreducible matrix. Then if* A *has an eigenvalue* λ, *this eigenvalue is unique.*

Proof. Let λ_1 and λ_2 be two distinct eigenvalues of A.

From Lemma 4.1.3 there exist two elementary circuits γ_1 and γ_2 in G(A) such that:

$$w(\gamma_1) = \lambda_1{}^{|\gamma_1|}$$
$$w(\gamma_2) = \lambda_2{}^{|\gamma_2|}$$

In addition, again from Lemma 4.1.3, we can write:

$$w(\gamma_2) \leq \lambda_1{}^{|\gamma_2|}$$

and

$$w(\gamma_1) \leq \lambda_2{}^{|\gamma_1|}$$

which leads to:

$$\lambda_1{}^{|\gamma_1|} \leq \lambda_2{}^{|\gamma_1|} \quad (22)$$

and:

$$\lambda_2{}^{|\gamma_2|} \leq \lambda_1{}^{|\gamma_2|} \quad (23)$$

Since ⊕ is selective, ≤ is a total order relation, therefore if $\lambda_1 \neq \lambda_2$ we have, either $\lambda_1 < \lambda_2$ or $\lambda_2 < \lambda_1$.

If, for example, we have $\lambda_1 < \lambda_2$ then we can deduce:

$$\lambda_1{}^{|\gamma_2|} < \lambda_2{}^{|\gamma_2|}$$

which is incompatible with (23).

Similarly, if $\lambda_2 < \lambda_1$ it follows:

$$\lambda_2{}^{|\gamma_1|} < \lambda_1{}^{|\gamma_1|}$$

which is incompatible with (22).

$\lambda_2 \neq \lambda_1$ therefore leads to a contradiction, which proves the property. □

For a matrix A with entries in a selective-invertible dioid having a unique eigenvalue λ, the following result characterizes the minimal generators of the eigenmoduloid $\mathcal{V}(\lambda)$.

Theorem 4. *Let* $(E, \oplus, \otimes)$ *be a selective-invertible-dioid with* $\otimes$ *commutative, and* $A \in M_n(E)$ *be an irreducible matrix having* λ *as unique eigenvalue. Let us denote* $G_c(A)$ *(critical graph) the partial subgraph of* $G(A)$ *induced by the set of vertices and arcs belonging to at least one circuit* γ *of weight* $w(\gamma) = \lambda^{|\gamma|}$ *(critical circuit). Let* $H_1, H_2, \dots H_p$ *be the strongly connected components of* $G_c(A)$ *and in each component let us choose a particular vertex* $j_1 \in H_1, j_2 \in H_2, \dots j_p \in H_p$. *Then denoting* $\widetilde{A} = (\lambda^{-1}) \otimes A$, *the family of vectors* $F = \{[\widetilde{A}^*]^{j_1}, [\widetilde{A}^*]^{j_2}, \dots [\widetilde{A}^*]^{j_p}\}$ *is a minimal generator of the eigenmoduloid* $\mathcal{V}(\lambda)$.

Proof. We will provide the proof for the case $\lambda = e$. The general case of a matrix A with a unique eigenvalue $\lambda \in E$, $\lambda \neq e$, is easily deduced by considering the matrix $\widetilde{A} = (\lambda^{-1}) \otimes A$. Let us also observe that, according to Lemma 4.1.3, there exists at least one critical circuit, therefore G_c contains at least one non empty strongly connected component.

(a) First let us show that F is a generator of $\mathcal{V}(e)$.

Let us denote γ_1 a critical circuit of H_1 containing vertex j_1, γ_2 a critical circuit of H_2 containing vertex j_2, etc.

Let $V = (V_j)_{j=1\dots n}$, be an eigenvector corresponding to the eigenvalue e.

Let us note that, A being irreducible, according to Lemma 4.1.2: $\forall j = 1, \dots n$ $V_j > \varepsilon$.

By using the proof of Lemma 4.1.3 we have, along each of the circuits $\gamma_1, \dots \gamma_p$:

$$\begin{cases} a_{i_1 i_2} \otimes V_{i_2} \leq V_{i_1} \\ a_{i_2 i_3} \otimes V_{i_3} \leq V_{i_2} \\ \vdots \\ a_{i_k i_1} \otimes V_{i_1} \leq V_{i_k} \end{cases} \tag{24}$$

($i_1, i_2, \dots i_k, i_1$ denoting the succession of vertices visited along the circuit). If we had strict inequality in at least one of the relations above, this would imply:

$$a_{i_1 i_2} \otimes a_{i_2 i_3} \otimes \dots \otimes a_{i_k i_1} \neq e$$

thus a contradiction would result with the fact that the circuits $\gamma_1, \gamma_2 \dots \gamma_p$ are critical ($w(\gamma_1) = w(\gamma_2) = \dots w(\gamma_p) = e$).

For each of the circuits $\gamma_1, \gamma_2, \dots \gamma_p$, the relations (24) are therefore all equalities.

Then, by using the proof of Theorem 2 we first deduce:

$$[A^*]^{j_1} \otimes V_{j1} \in \mathcal{V}(e)$$
$$[A^*]^{j_2} \otimes V_{j2} \in \mathcal{V}(e)$$
$$\vdots$$
$$\vdots$$
$$[A^*]^{j_p} \otimes V_{j_p} \in \mathcal{V}(e)$$

and, since $(E, \otimes)$ is a group and each component of V is distinct from ε, this implies:

$$[A^*]^{j_1} \in \mathcal{V}(e), [A^*]^{j_2} \in \mathcal{V}(e) \ldots [A^*]^{j_p} \in \mathcal{V}(e).$$

Furthermore, again according to the proof of Theorem 2, we know that relation (8) is satisfied and that, in the expression $\sum_{i=1}^{n} [A^*]^i \otimes V_i$, it is enough to retain a single term of the form $[A^*]^{j_k} \otimes V_{j_k}$ for each circuit $\gamma_k (k = 1, \ldots, p)$.
From the above we deduce that $V \in \mathcal{V}(e)$ can be written as the expression:

$$V = \sum_{k=1}^{p} [A^*]^{j_k} \otimes V_{j_k}$$

which proves that F is a generator of $\mathcal{V}(e)$.

(b) Let us now check that F is a minimal generator of $\mathcal{V}(e)$.
To do so, we will show that for any $\alpha \in [1, \ldots p]$, none of the vectors $[A^*]^i$, $i \in H_\alpha$, can be expressed as a linear combination of the vectors $[A^*]^j$ with $j \in H_q$, $q \neq \alpha$.

Let us proceed by contradiction and let us assume that $[A^*]^i$, $i \in H_\alpha$, is a linear combination of other columns of A^* taken in strongly connected components distinct from one another and distinct from H_α, and let us denote $K \subset \{1, 2, \ldots, n\}$ the set of indices of these columns.

Let us then consider the submatrix B deduced from A^* by eliminating all the rows and columns whose indices do not belong to $K \cup \{i\}$.

It is clear that, by construction, the columns of submatrix B are *linearly dependent*, so in view of Corollary 3.3.3 of Chap. 5 (Sect. 3.3):

$$\det^+(B) = \det^-(B) \tag{25}$$

Moreover, each diagonal term of B corresponds to a diagonal term $[A^*]_{j,j}$ where j is a vertex of the critical graph $G_c(A)$. Consequently $[A^*]_{j,j}$ (the weight of the maximum weight circuit through j) is equal to e and $[A^*]_{j,j} = e \oplus [A^*]_{j,j} = e$. It follows that all the diagonal terms of B are equal to e and $\det^+(B) = e$.

From relation (25) we then deduce the existence of an odd permutation of the indices of $K \cup \{i\}$ with weight equal to e. The decomposition into circuits of this odd permutation then features at least one elementary circuit (which is not a loop) and having weight e in G(B). This circuit would correspond to a critical circuit (of weight e) in G(A) joining vertices belonging to distinct strongly connected components of the critical graph. We are thus lead to a contradiction, which proves the theorem. □

We are now going to use the above properties to derive an analogue to the Perron–Frobenius theorem in some selective-invertible dioids.

4.2. The Perron–Frobenius Theorem for Some Selective-Invertible Dioids

In this section we consider a special class of selective-invertible dioids: those in which the calculation of the pth root of an element (for a natural number p) is always possible.

We therefore assume, throughout this section, that $(E, \oplus, \otimes)$ is a selective-invertible-dioid with $\otimes$ commutative having the following additional property (π):

$$(\pi)\begin{cases}\forall p \in \mathbb{N}, \forall a \in E, \quad \text{the equation:}\\ x^p = a\\ \text{has a unique solution in E, denoted } a^{1/p}\end{cases}$$

Example 4.2.1. A typical example of a selective invertible-dioid enjoying the above property is the dioid $(\mathbb{R}, \text{Max}, +)$. The operation $\otimes$ being the addition of real numbers, for any $a \in \mathbb{R}$ the equation $x^p = a$ $(p \in \mathbb{N}, p \neq 0)$ has a unique solution which is the real number a/p (usual quotient of the real number a by the integer p). ||

We can now define the spectral radius $\rho(A)$ of a matrix $A \in M_n(E)$.

Definition 4.2.2. *(spectral radius)*

Let $(E, \oplus, \otimes)$ *be a selective invertible-dioid with the property* (π). *The spectral radius of* $A \in M_n(E)$ *is the quantity:*

$$\rho(A) = \sum_{k=1}^{n} \left(\text{tr}(A^k)\right)^{\frac{1}{k}} \tag{26}$$

(sum in the sense of $\oplus$*) where* $\text{tr}(A^k)$ *denotes the trace of the matrix* A^k, *in other words the sum (in the sense of* $\oplus$*) of its diagonal elements.*

The following property shows that the spectral radius thus defined can be re-expressed simply in terms of the weights of the elementary circuits of the graph G(A).

Property 4.2.3. Let $(E, \oplus, \otimes)$ be a selective-invertible-dioid with the property (π).

Let $A \in M_n(E)$ and $\rho(A)$ be its spectral radius. Then:

$$\rho(A) = \sum_{\gamma\in\Gamma} (w(\gamma))^{\frac{1}{|\gamma|}} \tag{27}$$

where Γ denotes the set of elementary circuits of G(A).

Proof. The ith diagonal term of the matrix A^k is the sum of the weights of the circuits of length k (whether elementary or not) through i in G(A). As $\oplus$ is selective, $(A^k)_{ii}$ is therefore the weight of the maximum weight circuit of length k through i (maximum in the sense of the total order relation of the dioid).

Consequently $tr(A^k)$ is the weight of the maximum weight circuit of length k in G(A).

We can therefore rewrite $\rho(A)$ in the form:

$$\rho(A) = \sum (w(\gamma))^{\frac{1}{|\gamma|}} \tag{28}$$

where the sum extends to all circuits γ of G(A) with cardinality between 1 and n.

Let us show that, in this sum, only the elementary circuits have to be taken into account.

Let us assume that γ is a non-elementary circuit which can be decomposed into two elementary circuits γ_1 and γ_2.

Since $\leq$ is a total order relation, we can always assume that

$$w(\gamma_2)^{\frac{1}{|\gamma_2|}} \leq w(\gamma_1)^{\frac{1}{|\gamma_1|}}$$

Let us then show that:

$$w(\gamma)^{\frac{1}{|\gamma|}} \leq w(\gamma_1)^{\frac{1}{|\gamma_1|}}$$

Let us denote $a = w(\gamma)^{\frac{1}{|\gamma|}}$ $a_1 = w(\gamma_1)^{\frac{1}{|\gamma_1|}}$ $a_2 = w(\gamma_2)^{\frac{1}{|\gamma_2|}}$

Since:

$$w(\gamma) = w(\gamma_1) \otimes w(\gamma_2)$$

we have:

$$a^{|\gamma|} = a_1^{|\gamma_1|} \otimes a_2^{|\gamma_2|}$$

and as $a_2 \leq a_1$, this implies:

$$a^{|\gamma|} \leq a_1^{|\gamma_1|} \otimes a_1^{|\gamma_2|}$$

thus: $a \leq a_1$.

Consequently, in the expression (28), any term of the form $w(\gamma)^{\frac{1}{|\gamma|}}$ where γ is a non-elementary circuit is dominated by a term of the form $w(\gamma_1)^{\frac{1}{|\gamma_1|}}$ where γ_1 is an elementary circuit. We thus deduce the desired property. □

Remark 4.2.4. The definition given above for the spectral radius of a matrix of $M_n(E)$ is consistent with the usual definition for real matrices. Indeed, in standard algebra, for a matrix A whose eigenvalues are positive reals, the spectral radius of A is equal to $\lim_{k\to\infty} \left(\text{tr } A^k\right)^{\frac{1}{k}}$ (as we can clearly see by putting A in diagonal form).

Furthermore, in expression (26) the sum can be extended from $k = 1$ to $k = +\infty$ as demonstrated by the proof of Property 4.2.3 (the sum of the weights of the non-elementary circuits thus added is absorbed by the sum of the weights of the elementary circuits).

Hence, we observe that expression (26) is clearly the analogue to the usual spectral radius. ||

We can then state the following result which is an analogue to the Perron–Frobenius theorem.

Theorem 5. *Let* (E, ⊕, ⊗) *be a selective invertible-dioid with* ⊗ *commutative and satisfying the property* (π). *If* $A \in M_n(E)$ *is an irreducible matrix with spectral radius*

$$\rho(A) = \sum_{k=1}^{n} \left(tr(A^k)\right)^{\frac{1}{k}}$$

(i) ρ(A) *is an eigenvalue of* A

(ii) ρ(A) *is the unique eigenvalue of* A

(iii) if $V = (V_i)_{i=1\ldots n}$ *is an eigenvector of* A *for the eigenvalue* ρ(A), *then* $\rho(A) > \varepsilon$ *and* $\forall i = 1\ldots n$: $V_i > \varepsilon$.

Proof. (i) According to Property 4.2.3 there exists an elementary circuit γ_o of G(A) such that:

$$\rho(A) = \left(w\left(\gamma_0\right)\right)^{\frac{1}{|\gamma_0|}}$$

and, for any circuit γ of G(A)

$$\left(w(\gamma)\right)^{\frac{1}{|\gamma|}} \leq \left(w(\gamma_0)\right)^{\frac{1}{|\gamma_0|}} \tag{29}$$

We are going to see that the Corollary 2.4 applies for λ = ρ(A).

Let us show first that $(\lambda^{-1} \otimes A)^*$ exists, and, in order to do so, let us show that in $G(\lambda^{-1} \otimes A)$ the weight of any circuit is 0-stable (see Chap. 4, Sect. 3.3). Let γ be an arbitrary circuit of G(A). Its weight with respect to $\lambda^{-1} \otimes A$ is:

$$\theta(\gamma) = w(\gamma) \otimes (\lambda^{-1})^{|\gamma|} = w(\gamma) \otimes \left[w(\gamma_0)^{-1}\right]^{\frac{|\gamma|}{|\gamma_0|}}$$

According to inequality (29) we have:

$$w(\gamma) \leq (w(\gamma_0))^{\frac{|\gamma|}{|\gamma_0|}}$$

and consequently $\theta(\gamma) \leq \left(w(\gamma_0)\right)^{\frac{|\gamma|}{|\gamma_0|}} \otimes \left[w(\gamma_0)^{-1}\right]^{\frac{|\gamma|}{|\gamma_0|}} = e$

Therefore, for any circuit γ of $G(\lambda^{-1} \otimes A)$:

$$\theta(\gamma) \oplus e = e$$

and consequently θ(γ) is 0-stable. From this we deduce that $[\lambda^{-1} \otimes A]^*$ exists (see Theorem 1 in Chap. 4).

Now let us show that if i is a vertex of γ_0, then relation (6) of Corollary 2.4 (Sect. 2) is satisfied.

According to the above, the left hand side of (6) is equal to e and we have:

$$\sum_{\gamma \in P_{ii}} w(\gamma) \otimes (\lambda^{-1})^{|\gamma|} \leq e$$

Moreover, for circuit γ_0 through i we have: $w(\gamma_0) \otimes (\lambda^{-1})^{|\gamma_0|} = \lambda^{|\gamma_0|} \otimes (\lambda^{-1})^{|\gamma_0|} = e$ which shows that the right hand side of (6) is also equal to e.

This shows that $\lambda = \rho$ (A) is an eigenvalue of A and that, for any $i \in \gamma_0$, $[(\lambda^{-1} \otimes A)^*]^i$ is an associated eigenvector.

(ii) The uniqueness of the eigenvalue ρ(A) follows from Corollary 4.1.4.

(iii) This follows directly from Lemma 4.1.2. □

Among the applications of Theorem 5, we can mention those which concern the dioid $(\overset{\vee}{\mathbb{R}}, \text{Max}, +)$, which is at the basis of new models in automatic control for discrete event systems (these will be discussed in more detail in Sect. 7 at the end of the present chapter).

Example 4.2.1. **(continued) Calculation of the eigenvalue in the dioid** $(\overset{\vee}{\mathbb{R}}, \text{Max}, +)$

In the dioid $(\overset{\vee}{\mathbb{R}}, \text{Max}, +)$, an irreducible matrix $A = (a_{ij})$ has thus a unique eigenvalue $\lambda = \rho(A)$ whose value, according to Property 4.2.3 can be written as:

$$\lambda = \rho(A) = \underset{\gamma \in \Gamma}{\text{Max}} \left\{ \frac{w(\gamma)}{|\gamma|} \right\}$$

where Γ is the set of elementary circuits of G(A), and where, for each elementary circuit γ, $\dfrac{w(\gamma)}{|\gamma|}$ denotes the *average weight* of circuit γ (weight divided by the number of arcs in the circuit, with division in the sense of ordinary algebra).

The calculation of the eigenvalue λ is therefore reduced to the determination in G(A) of the (elementary) circuit of maximum average weight.

The problem of determining a circuit of average minimal length has been studied by Dantzig et al. (1967) and by Karp (1978) who described a polynomial algorithm of complexity $\mathcal{O}$ (mn). This algorithm can be directly adapted to the maximum average weight circuit problem thus leading, for the calculation of the eigenvalue in the dioid $(\overset{\vee}{\mathbb{R}}, \text{Max}, +)$, to the following algorithm.

Algorithm 1 *Calculation of the eigenvalue of a matrix on the dioid* $(\overset{\vee}{\mathbb{R}}, \text{Max}, +)$:

(a) Let $A = (a_{ij})$ *be an irreducible matrix on* $(\overset{\vee}{\mathbb{R}}, \text{Max}, +)$ *and* G(A) *its (strongly connected) associated graph.*

Initialize the labels of the vertices $i = 1 \ldots n$ *of* G(A) *by:*

$$\pi^\circ(1) = 0$$

$$\pi^\circ(i) = \begin{cases} a_{1,i} & \text{if} \quad i \in \Gamma_1 \\ -\infty & \text{otherwise} \end{cases}$$

(b) For (k = 1, 2, ... n) *do:*
Compute ∀j = 1, 2, ... n:

$$\pi^k(j) = \underset{i \in \Gamma_j^{-1}}{\mathrm{Max}} \left\{ \pi^{k-1}(i) + a_{ij} \right\}$$

End for
(c) Determine the eigenvalue λ by:

$$\lambda = \underset{j=1\ldots n}{\mathrm{Max}} \; \underset{0 \le k \le n-1}{\mathrm{Min}} \left\{ \frac{\pi^n(j) - \pi^k(j)}{n-k} \right\}$$

The complexity of the above algorithm is $\mathcal{O}(m\,n)$, where m is the number of arcs of the graph G(A) associated with matrix A.

The justification of this algorithm is based on the following result, due to Karp (1978):

For any fixed i $(1 \le i \le n)$:

$$\lambda = \underset{j=1\ldots n}{\mathrm{Max}} \; \underset{0 \le k \le n-1}{\mathrm{Min}} \left\{ \frac{(A^n)_{i,j} - (A^k)_{i,j}}{n-k} \right\}$$

where A^k denotes the kth power of A in $(\overset{\vee}{\mathbb{R}}, \mathrm{Max}, +)$ and where the division by $n-k$ is the ordinary division of real numbers.

5. Eigenvalues, Bideterminant and Characteristic Bipolynomial

In this section we investigate the links between the notions of eigenvalues/eigenvectors and:

- the notion of dependence (in the sense of definition 2.5.1, Chap. 5);
- the concepts of bideterminant and of characteristic bipolynomial.

Let us begin by stating a general result valid in dioids:

Theorem 6. *Let* (E, ⊕, ⊗) *be a dioid with* ⊗ *commutative and* $A \in M_n(E)$. I *denoting the* n × n *identity matrix, for any* λ ∈ E, *let* $\bar{A}(\lambda)$ *be the* 2 n × 2 n *matrix:*

$$\bar{A}(\lambda) = \left[\begin{array}{c|c} A & \lambda \otimes I \\ \hline I & I \end{array} \right]$$

Then λ *is an eigenvalue of* A *if and only if the columns of* $\bar{A}(\lambda)$ *are dependent.*

Proof. (i) If $V = (V_1, V_2, \ldots, V_n)^T \in E^n$ is an eigenvector of A for the eigenvalue λ, then by choosing $J_1 = [1, 2, \ldots n\}$, $J_2 = \{n+1, \ldots, 2n\}$ and the coefficients:

$$\mu_j = V_j \quad (j = 1, \ldots, n)$$
$$\mu_j = V_{j-n} \quad (j = n+1, \ldots, 2n)$$

the relation $A \otimes V = \lambda \otimes V$ implies, on the columns of $\bar{A}(\lambda)$, the dependence relation:

$$\sum_{j\in J_1} \mu_j \otimes \left[\bar{A}(\lambda)\right]^j = \sum_{j\in J_2} \mu_j \otimes \left[\bar{A}(\lambda)\right]^j \tag{30}$$

(ii) Conversely, let us assume the columns of $\bar{A}(\lambda)$ to be linearly dependent (in the sense of definition 2.5.1 in Chap. 5).

By denoting the weights associated with the columns of $\bar{A}(\lambda)$ as $(\mu_1, \mu_2, \ldots \mu_n, \mu_{n+1}, \ldots \mu_{2n})$, we have a relation of type (30) with $J_1 \neq \emptyset$ $J_2 \neq \emptyset$ $J_1 \cap J_2 = \emptyset$ and $\mu_j \neq \varepsilon$ for $j \in J_1 \cup J_2$. (we agree to set, $\mu_j = \varepsilon$ for $j \notin J_1 \cup J_2$).

By using (30) on the components $n+1$ to $2n$, we observe that, for any $j \in [1, n]$, the indices j and $n+j$ cannot both belong to J_1, nor both to J_2 (indeed, assuming the contrary, we would have $\mu_j \oplus \mu_{n+j} = \varepsilon$; $(E, \oplus)$ being canonically ordered, from Proposition 3.4.3 of Chap. 1 this would imply $\mu_j = \mu_{n+j} = \varepsilon$).

Consequently, if $j \in J_1$ then necessarily $n+j \in J_2$ and the dependence relation (30) implies: $\mu_{n+j} = \mu_j$.

As a result, in the dependence relation (30), the indices j and $n+j$ are interchangeable and we can therefore always assume $J_1 \subset \{1, 2, \ldots n\}$.

Then, by setting:

$$V_j = \mu_j \quad \text{for} \quad j \in J_1,\ 1 \leq j \leq n$$
$$V_j = \varepsilon \quad \text{for} \quad j \in [1, n] \backslash J_1$$

relation (30) on the first n components of the columns of $\bar{A}(\lambda)$ reads:

$$\sum_{j=1}^{n} \mu_j \otimes A^j = \sum_{j=1}^{n} A^j \otimes V_j = \begin{bmatrix} \lambda \otimes V_1 \\ \lambda \otimes V_2 \\ \vdots \\ \lambda \otimes V_n \end{bmatrix}$$

which shows that λ is an eigenvalue of A and $V = [V_1 \ldots V_n]^T$ an associated eigenvector. □

Remark 5.1. In the case of usual linear algebra where $(E, \oplus)$ is a group:

$$\det \left[\begin{array}{c|c} A & \lambda I \\ \hline I & I \end{array}\right] = \det \left[\begin{array}{c|c} A - \lambda I & \lambda I \\ \hline 0 & I \end{array}\right] = \det(A - \lambda I)$$

the condition of Theorem 6 clearly yields the classical result: λ is an eigenvalue of A if and only if $\det(A - \lambda I) = 0$ ||

Now, by considering λ as a variable, each term of the bideterminant of the matrix $\bar{A}(\lambda)$ can be considered as a polynomial in λ. The characteristic bipolynomial of $A \in M_n(E)$ introduced in Sect. 4.3 in Chap. 2, is then defined as the pair: $(P^+(\lambda), P^-(\lambda))$ where:

$$P^+(\lambda) = \det{}^+(\bar{A}(\lambda))$$
$$P^-(\lambda) = \det{}^-(\bar{A}(\lambda))$$

By using the characteristic bipolynomial, we can then state the following result, which extends, to some classes of selective dioids, the classical characterization of eigenvalues as roots of the characteristic polynomial.

Proposition 5.2. *Let* $(E, \oplus, \otimes)$ *be a selective dioid, with* $\otimes$ *commutative,* $A \in M_n(E)$ *and* $(P^+(\lambda), P^-(\lambda))$ *its characteristic bipolynomial.*

(i) If every element of $E\backslash\{\varepsilon\}$ *is cancellative for* $\otimes$*, then any eigenvalue* λ *of* A *satisfies the characteristic equation* $P^+(\lambda) = P^-(\lambda)$

(ii) If $(E, \otimes)$ *is a commutative group, then any* λ *satisfying the characteristic equation:*
$P^+(\lambda) = P^-(\lambda)$ *is an eigenvalue.*

Proof. (i) If λ is an eigenvalue of A, in view of Theorem 6, the columns of $\bar{A}(\lambda)$ are linearly dependent, and, according to Corollary 3.3.3 of Chap. 5 (Sect. 3.3) we have:

$$\det{}^+(\bar{A}(\lambda)) = \det{}^-(\bar{A}(\lambda))$$

(ii) If $(E, \oplus, \otimes)$ is a commutative selective-invertible dioid, then by using Theorem 2 of Chap. 5, $\det^+(\bar{A}(\lambda)) = \det^-(\bar{A}(\lambda))$ implies the existence of a linear dependence relation on the columns of $\bar{A}(\lambda)$; according to Theorem 6 we can then deduce that λ is an eigenvalue of A. □

6. Applications in Data Analysis

This section is devoted to the presentation of some important applications in Data Analysis of the calculation of eigen-elements in dioids.

In *hierarchical clustering* the starting point is to assume that we are given a dissimilarity matrix between objects. Then, considering the dioid $(\mathbb{R}_+ \cup \{+\infty\}$, Min, Max), we show in Sect. 6.1 that, with each level λ of the clustering, we can associate a set of eigenvectors of the dissimilarity matrix associated with the eigenvalue λ; furthermore, this set constitutes the (unique) minimal generator of the eigen-semi-module $\mathcal{V}(\lambda)$. This exhibits an interesting link with another classical approach to Data Analysis, namely Factor Analysis.

In *Preference Analysis* the aim is to order a set of objects given a matrix of *preferences* (deduced from pairwise comparisons). We show in Sect. 6.2. that several approaches to this problem can then be interpreted, in a unifying framework, in terms of the search for the eigenvalues and eigenvectors of the preference matrix in dioids such as $(\mathbb{R}_+, +, \times)$ $(\mathbb{R}_+ \cup \{+\infty\}, \text{Max}, \text{Min})$, $(\mathbb{R}_+ \cup \{+\infty\}, \text{Max}, \times)$.

6.1. Applications in Hierarchical Clustering

Given a set of n objects $X = \{1, 2, \dots n\}$, let us assume that, for each pair of objects (i, j), we can define an index or a measure of *dissimilarity* $d_{ij} \in \mathbb{R}_+$ (let us note that this index does not necessarily satisfy the axioms of a distance). The dissimilarity d_{ij} will take on small values if the two objects i and j are very similar (in particular, $d_{ij} = 0$ if i and j are identical); conversely, d_{ij} will take on large values if the objects i and j are very dissimilar.

Thus, with any set of objects $X = \{1, 2, \dots n\}$, we can associate a dissimilarity matrix

$$D = \left(d_{ij}\right)_{\substack{i=1,\dots,n\\ j=1,\dots,n}}$$

Observe that this matrix is symmetric with zero diagonal ($\forall i \in X$: $d_{ii} = 0$).

A *clustering* of the n objects in X consists in determining a partition of X into subsets (classes) so that within the same class, objects are as similar as possible, and that on the contrary, objects belonging to distinct classes are strongly dissimilar.

As we can give many different meanings to the notion of the proximity or homogeneity of a subset of objects, there exists a very wide variety of clustering methods. In *hierarchical clustering* one proceeds as follows. The matrix $D = (d_{ij})$ can be considered as the adjacency matrix of an undirected complete graph $G = [X, U]$ whose vertices correspond to objects, and whose edges are assigned the d_{ij} values.

For any real number $\lambda \geq 0$ we consider the partial graph of G *at threshold* λ, denoted: $G_\lambda = [X, U_\lambda]$ where $U_\lambda = \{(i, j) \in U / d_{ij} \leq \lambda\}$.

The connected components of G_λ form a partition of the set of objects X. The elements of this partition form the classes of the clustering of X at threshold λ.

In view of the above, two vertices i and j are in a same class at threshold λ if and only if there exists in G a chain joining i and j with all edges having valuations $\leq\lambda$. Equivalently, i and j are in a same class at threshold λ if and only if there exists in G a path of *sup-section* $\leq\lambda$ from i to j, the sup-section of a path $\pi = \{i_o = i, i_1, i_2, \dots i_p = j\}$ being defined as:

$$\bar{\sigma}(\pi) = \underset{k=0,\dots,p-1}{\text{Max}} \left\{d_{i_k i_{k+1}}\right\}$$

(see Gondran and Minoux 1995 Chap. 4, Sect. 2.9).

Thus, for i and j to be in a same class at threshold λ, it is necessary and sufficient that $d^*_{ij} = \underset{\pi \in P_{ij}}{\text{Min}} \{\bar{\sigma}(\pi)\} \leq \lambda$

(where P_{ij} denotes the set of paths between i and j in G).

Since d^*_{ij} can be written: $d^*_{ij} = \sum\limits_{\pi \in P_{ij}}^{\oplus} \left(\prod\limits_{k}^{\otimes} d_{i_k i_{k+1}} \right)$ (with $\oplus$ = Min and $\otimes$ = Max) we observe that the matrix $D^* = (d^*_{ij})$ is none other than the quasi-inverse of D in the dioid $(\mathbb{R}_+ \cup \{+\infty\}, \text{Min}, \text{Max})$.

According to a result due to Hu (1961), a chain of minimal sup-section between two vertices in G corresponds to the chain of the *minimum weight spanning tree* of G joining these two vertices. The matrix D^* can therefore be efficiently determined using a minimum spanning tree algorithm (see for example Collomb and Gondran 1977).

The clustering tree is then directly deduced by considering all possible values of the threshold λ (at most $n - 1$ distinct values are to be considered, those which correspond to the valuations of the $n - 1$ edges of the minimum spanning tree).

If we have p distinct values: $\lambda_1 > \lambda_2 > \ldots > \lambda_p$ $(p \le n - 1)$, the classes of the partition of level λ_{i+1} are included in the classes of the partition of level λ_i $(i = 1, \ldots, p - 1)$ (hence the name of *hierarchical clustering*).

Example 1. On the set of 7 objects $X = \{1, 2, 3, 4, 5, 6, 7\}$, let us consider the following dissimilarity matrix:

$$D = \begin{bmatrix} 0 & 7 & 5 & 8 & 10 & 8 & 10 \\ 7 & 0 & 2 & 10 & 9 & 9 & 10 \\ 5 & 2 & 0 & 7 & 11 & 10 & 9 \\ 8 & 10 & 7 & 0 & 8 & 4 & 11 \\ 10 & 9 & 11 & 8 & 0 & 9 & 5 \\ 8 & 9 & 10 & 4 & 9 & 0 & 10 \\ 10 & 10 & 9 & 11 & 5 & 10 & 0 \end{bmatrix}$$

The corresponding minimum spanning tree is given in Fig. 1. The matrix D^* is easily deduced from this spanning tree.

For example, d^*_{27} is the sup-section of the unique chain of the tree joining the vertices 2 and 7, therefore:

$$d^*_{27} = \text{Max}\{d_{23}, d_{34}, d_{45}, d_{57}\} = 8$$

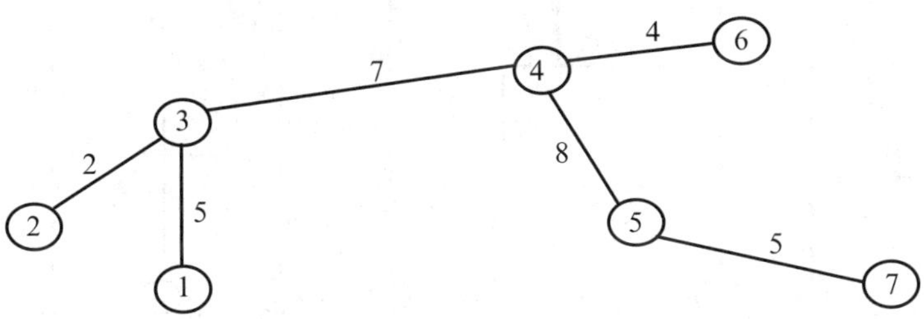

Fig. 1 Minimum weight spanning tree corresponding to the dissimilarity matrix D

We obtain:

$$D^* = \begin{bmatrix} 0 & 5 & 5 & 7 & 8 & 7 & 8 \\ 5 & 0 & 2 & 7 & 8 & 7 & 8 \\ 5 & 2 & 0 & 7 & 8 & 7 & 8 \\ 7 & 7 & 7 & 0 & 8 & 4 & 8 \\ 8 & 8 & 8 & 8 & 0 & 8 & 5 \\ 7 & 7 & 7 & 4 & 8 & 0 & 8 \\ 8 & 8 & 8 & 8 & 5 & 8 & 0 \end{bmatrix} \tag{31}$$

At threshold $\lambda = 5$, we have, for example:

$$d^*_{23} \leq \lambda \quad d^*_{13} \leq \lambda \quad d^*_{46} \leq \lambda$$

which shows that the vertices 1, 2, 3 belong to the same class; in the same way, vertices 4 and 6 are in a same class.

By contrast, the vertices 3 and 5 are not in a same class because

$$d^*_{35} = 8 > 5$$

The hierarchical clustering tree for the above example is shown in Fig. 2.

To obtain the partition into classes corresponding to a given level λ, it is enough to "cut" the clustering tree of Fig. 2 by a horizontal line with ordinate λ. Thus, for example, the clustering on the level $\lambda = 5$ is the partition $\{1, 2, 3\}$ $\{4, 6\}$ $\{5, 7\}$; on the level $\lambda = 4$: $\{1\}$ $(2, 3\}$ $\{4, 6\}$ $\{5\}$ $\{7\}$. ||

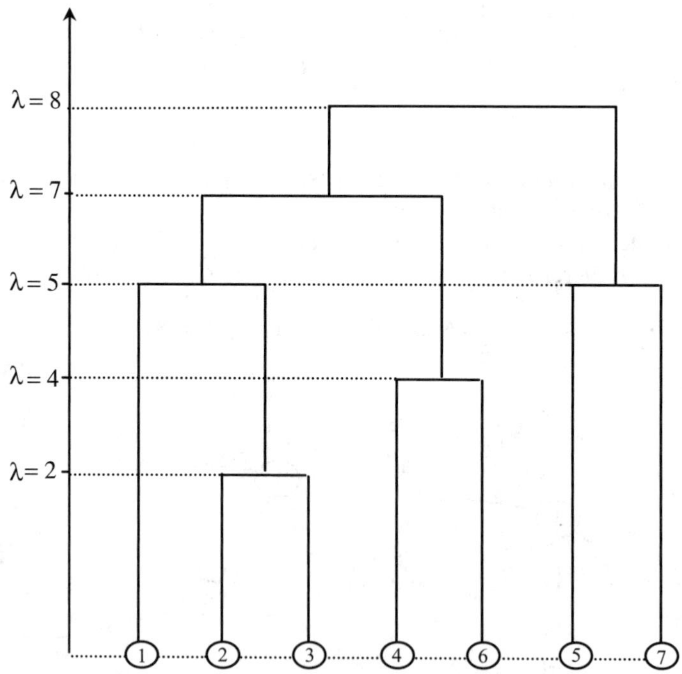

Fig. 2 Clustering tree corresponding to Example 1

As all the diagonal terms of D are equal to e = 0 (the zero element of ⊗) this yields $D = I \oplus D$ and consequently:

$$D^* = D^{n-1}$$

(see Chap. 4 Sect. 3 Proposition 1).

In the dioid ($\mathbb{R}_+ \cup \{+\infty\}$, Min, Max) this therefore yields: $(D^*)^2 = D^*$ which shows that D^* is an *ultrametric distance*, i.e. that it satisfies, $\forall i, j, k$:

$$d^*_{ij} \le \mathrm{Max}\left\{d^*_{i_k}, d^*_{k_j}\right\}$$

We recall that D^* is the subdominant ultrametric distance, i.e. the greatest element of the set of ultrametrics less than or equal to D (see Chap. 4, Sect. 6.14).

We are now going to see that there exist interesting links between hierarchical clustering and the structure of the eigenvectors of the dissimilarity matrix D. The dioid ($\mathbb{R}_+ \cup \{+\infty\}$, Min, Max) is *selective* with *commutative* and *idempotent* multiplication. Moreover, e = 0 is the greatest element (in the sense of the canonical order relation) for, $\forall a \in \mathbb{R}_+$: $e \oplus a = \mathrm{Min}\{0, a\} = e$.

The main results from Sect. 3 (Theorem 3 and Corollary 8) can thus be applied. In particular (taking into account that, $\forall i$: $d^*_{ii} = e = 0$):

- The set of distinct vectors of the form $\bar{V}^i = [D^*]^i$ constitutes the (unique) minimal generator of $\mathcal{V}(e)$;
- Any $\lambda \in \mathbb{R}_+$ is an eigenvalue of D;
- For any $\lambda \in \mathbb{R}_+$, the set of distinct vectors of the form $\lambda \otimes [D^*]^i$ constitutes the (unique) minimal generator of $\mathcal{V}(\lambda)$.

The following result shows that, for each level λ of the clustering, the set of eigenvectors for the eigenvalue λ contains all the information required to define the classes at level λ.

Theorem 7. *(Gondran 1976)*

At each level λ of a hierarchical clustering w.r.t. a dissimilarity matrix $D = (d_{ij})$, *two objects i and j belong to a same class if and only if the two eigenvectors* $\lambda \otimes [D^*]^i$ *and* $\lambda \otimes [D^*]^j$ *are equal. The distinct vectors of the form* $\lambda \otimes [D^*]^i$ *for* $i = 1, \dots, n$ *form the unique minimal generator of* $\mathcal{V}(\lambda)$.

Example 1. (continued)

Let us illustrate the above results on the matrix D^* given by (31).

On the level λ = 5 for example, a first class is formed by the objects {1, 2, 3}.

We then check that we have:

$$\lambda \otimes \left[D^*\right]^1 = \lambda \otimes \left[D^*\right]^2 = \lambda \otimes \left[D^*\right]^3 = \begin{bmatrix} 5 \\ 5 \\ 5 \\ 7 \\ 8 \\ 7 \\ 8 \end{bmatrix} \tag{32}$$

A second class is formed by the objects {4, 6}, and we check that:

$$\lambda \otimes [D^*]^4 = \lambda \otimes [D^*]^6 = \begin{bmatrix} 7 \\ 7 \\ 7 \\ 5 \\ 8 \\ 5 \\ 8 \end{bmatrix} \tag{33}$$

The third class is formed by {5, 7}, and this yields:

$$\lambda \otimes [D^*]^5 = \lambda \otimes [D^*]^7 = \begin{bmatrix} 8 \\ 8 \\ 8 \\ 8 \\ 5 \\ 8 \\ 5 \end{bmatrix} \tag{34}$$

The three vectors given by (32)–(34) constitute (the unique) minimal generator of the eigenmoduloid $\mathcal{V}(5)$. ||

The interpretation of the classes on an arbitrary level λ in terms of generators of the eigenmoduloid associated with the eigenvalue λ, establishes an interesting analogy to another classical approach to Data Analysis, namely: Factor Analysis (see Benzecri 1974, for example).

In both cases, the relevant information serving as the basis of the analysis is provided by the eigen-elements of the dissimilarity matrix. Only the underlying algebraic structures are different.

6.2. Applications in Preference Analysis: A Few Answers to the Condorcet Paradox

Given n objects, we seek to establish a preference order on these objects through pairwise comparisons of objects produced, for example, by referees or judges (the case of a tournament) or by consumers (the case of a market analysis).

We therefore assume as given the matrix of preferences on the pairs of objects, denoted $A = (a_{ij})_{\substack{i=1,\dots,n \\ j=1,\dots,n}}$

Thus, for any (ordered) pair of objects (i, j), a_{ij} is equal to the number of judges (consumers) having preferred i to j. We agree that the weight of a judge is counted as $\frac{1}{2}$ in both a_{ij} and a_{ji} in the case of indifference between the objects i and j. In the case of a tournament, a_{ij} will be equal to the number of games won on j during meetings between i and j.

Various approaches to ordering objects from a matrix of preferences have been proposed, each constituting a possible answer to the Condorcet paradox. We are going to see that, for each of them, the relevant information used to compare the objects is provided by the eigenvalues and eigenvectors of the preference matrix *in a well chosen dioid.*

The "Mean Order" Method and the Dioid $(\mathbb{R}_+, +, \times)$

This method, proposed in Berge (1958), consists in ordering the objects (the players in a tournament, in the example studied by Berge) according to the values of the components of the eigenvector associated with the (real) eigenvalue having greatest modulus of the matrix A (this eigenvalue being simple, the associated eigen-subspace has dimension 1 (see Berge 1958, Chap. 14). Since, according to the Perron–Frobenius theorem, this eigenvector has components that are all positive, the proposed order – also referred to as the *mean order* – simply corresponds to that of nonincreasing values of the components.

Example 2. Let us consider, for a set of 4 objects, and 6 judges, the following matrix of preferences:

$$A = \begin{bmatrix} 0 & 3 & 4 & 3.5 \\ 3 & 0 & 4 & 1 \\ 2 & 2 & 0 & 5 \\ 2.5 & 5 & 1 & 0 \end{bmatrix}$$

A being a real matrix whose coefficients are all nonnegative, its spectral radius $\rho(A)$ is equal to the largest real positive eigenvalue, here $\lambda = 8.92$.

We cheek that the associated eigenvector is:

$$V = \begin{bmatrix} 0.56 \\ 0.46 \\ 0.50 \\ 0.47 \end{bmatrix}$$

The resulting mean order on the objects is: 1, 3, 4, 2 (object 1 is the first, object 3 is the second one, etc.). ||

The "mean order" method thus exploits the information provided by a particular eigenvector (the one corresponding to the largest real nonnegative eigenvalue) in the dioid $(\mathbb{R}_+, +, \times)$.

The Method of Partial Orders and the Dioid $(\mathbb{R}_+ \cup \{+\infty\}, \text{Max}, \text{Min})$

A second method, proposed by Defays (1978) proceeds by determining a hierarchy of (partial) order relations on the objects, whose equivalence classes are nested. More precisely, for any $\lambda \in \mathbb{R}$, the classes of level λ are defined as the strongly connected components of the graph:

$G_\lambda = [X, U_\lambda]$, where X is the set of objects and where the set of arcs is:

$$U_\lambda = \{(i, j)/a_{ij} \geq \lambda\}$$

If we denote $C_\lambda(i)$ the class of level λ containing i, then for any $\lambda' \leq \lambda$,

$$C_{\lambda'}(i) \supset C_\lambda(i).$$

By denoting $\mathcal{R}_\lambda$ the equivalence relation: i $\mathcal{R}\lambda$ j $\Leftrightarrow$ i and j belong to the same class of level λ then, for each level λ, the quotient graph $G_\lambda/\mathcal{R}_\lambda$ is a circuitless graph, therefore the graph associated with a (partial) order relation. When λ takes all possible real values we then obtain a set of partial orders on nested classes.

As in hierarchical clustering (see Sect. 6.1), the classes obtained on each level λ (therefore the clustering tree) are deduced from the structure of the eigenvectors associated with the eigenvalue λ in the dioid $(\mathbb{R}_+ \cup \{+\infty\}, \text{Max}, \text{Min})$.

The following example illustrates the method, and shows that the matrix A^+ actually contains more useful information than required to construct the clustering tree alone: it can be used to obtain in addition the quotient graphs, in other words the partial orders on each level. This information may be quite useful for interpreting the results of the analysis.

Example 3. Take the same matrix as in Example 2, this time considered in the dioid $(\mathbb{R}_+ \cup \{+\infty\}, \text{Max}, \text{Min})$. The matrix $A^+ = A \oplus A^2 \oplus A^3$ is:

$$A^+ = \begin{bmatrix} 0 & 4 & 4 & 4 \\ 3 & 0 & 4 & 4 \\ 3 & 5 & 0 & 5 \\ 3 & 5 & 4 & 0 \end{bmatrix}$$

The clustering thus contains three levels $\lambda_1 = 5$, $\lambda_2 = 4$ and $\lambda_3 = 3$.

Figure 3 displays the graphs G_{λ_1} G_{λ_2} and G_{λ_3} and the quotient graphs obtained on the various levels of this clustering.

On the level $\lambda_1 = 5$ we have four classes {1} {2} {3} {4}. On the level $\lambda_2 = 4$, we have the two classes {1} {2 3 4} and on the level $\lambda_3 = 3$, we have a unique class {1, 2, 3, 4,}.

Observe that on level $\lambda_3 = 3$, the four objects appear as being equivalent. Nevertheless, examining level $\lambda_2 = 4$ refines the analysis, with the quotient graph $G_{\lambda_2}/\mathcal{R}_{\lambda_2}$ showing that on this level, object 1 is preferred to the other three. On level λ_2, the three objects 2, 3, 4 appear as being equivalent, but on the lowest level ($\lambda_1 = 5$) they can be further differentiated: 3 is preferred to 4, which itself is preferred to 2. ||

The "Circuit of Least Consensus" Method and the Dioid $(\mathbb{R}_+ \cup \{+\infty\}, \text{Max}, \times)$

This method uses the information provided by the eigenvalue and the associated eigenvectors of the matrix of preferences A in the dioid $(\mathbb{R}_+ \cup \{+\infty\}, \text{Max}, \times)$.

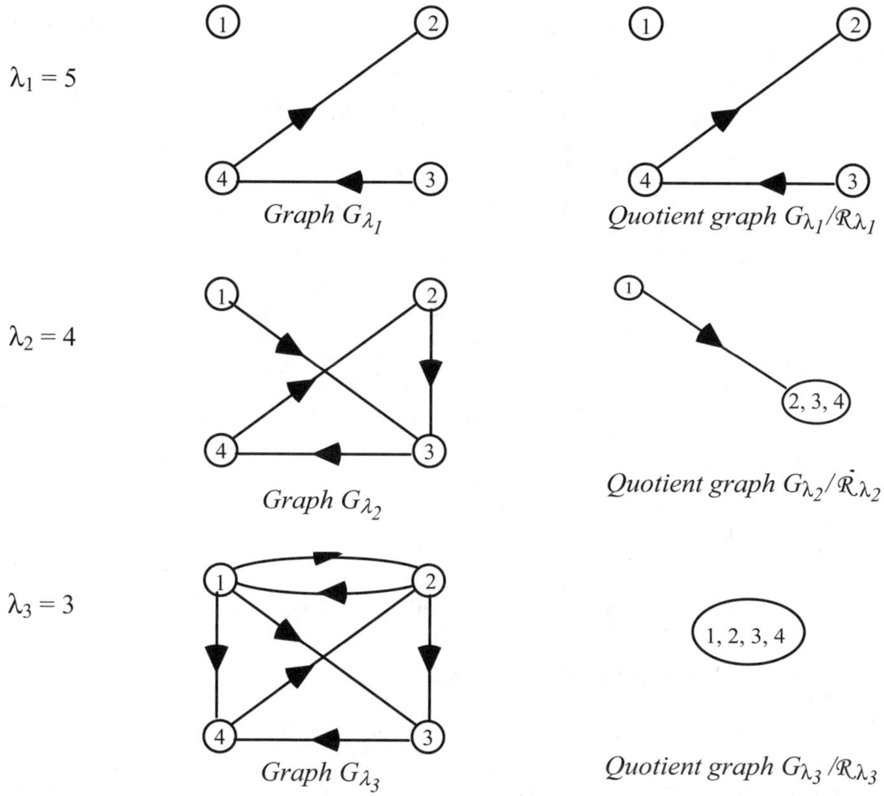

Fig. 3 The graphs G_λ and the quotient graphs $G_\lambda/\mathcal{R}_\lambda$ obtained on the different levels of clustering in Example 3

According to Lemma 4.1.3 and Corollary 4.1.4 of Sect. 4, A has a unique eigenvalue which corresponds to the circuit γ_o of G(A) such that:

$$\lambda = w(\gamma_o)^{\frac{1}{|\gamma_o|}} = \underset{\gamma}{\text{Max}}\left\{ w(\gamma)^{\frac{1}{|\gamma|}} \right\}$$

(maximum taken on the set of elementary circuits of the graph) where, for each circuit γ, $|\gamma|$ denotes the number of arcs of the circuit and $w(\gamma)$ the product of the valuations a_{ij} of the arcs of γ.

The circuit γ_o is therefore the circuit of the graph G(A) for which the geometric mean of the valuations is the largest. This circuit can be interpreted as the set of objects for which the consensus is the worst (see Gondran 1995).

Example 4. For the preference matrix in Example 1, now considered in the dioid $(\mathbb{R}_+ \cup \{+\infty\}, \text{Max}, \times)$, we have:

$$A^2 = \begin{pmatrix} 9 & 17.5 & 12 & 20 \\ 8 & 9 & 12 & 20 \\ 12.5 & 25 & 8 & 7 \\ 15 & 7.5 & 20 & 8.75 \end{pmatrix}$$

$$A^3 = \begin{pmatrix} 52.5 & 100 & 70 & 60 \\ 50 & 100 & 36 & 60 \\ 75 & 37.5 & 100 & 43.75 \\ 40 & 43.75 & 60 & 100 \end{pmatrix}$$

$$A^4 = \begin{pmatrix} 300 & 300 & 400 & 350 \\ 300 & 300 & 400 & 180 \\ 200 & 225 & 300 & 500 \\ 250 & 500 & 175 & 300 \end{pmatrix}$$

Consequently, the eigenvalue of A is:

$$\lambda = \text{Max}\left\{\sqrt{9};\ \sqrt[3]{100};\ \sqrt[4]{300}\right\} = \sqrt[3]{100} = 4.64$$

and this eigenvalue corresponds to the circuit

$$\gamma = \{(2,3)(3,4)(4,2)\} \quad \text{of weight} \quad w(\gamma) = 4 \times 5 \times 5 = 100.$$

From the point of view of preference analysis, this circuit of least consensus represents a subset of objects or individuals which appear to be difficult to distinguish. We note on the example that this result is consistent with the one previously deduced from Max-Min analysis (see Fig. 3, for the case $\lambda_2 = 4$). ||

7. Applications to Automatic Systems: Dynamic Linear System Theory in the Dioid $(\mathbb{R} \cup \{-\infty\}, \text{Max}, +)$

Dynamic linear system theory is a classical and very active branch of the automation field. Many problems of observability, controllability, stability and optimal control are well-solved in this class of systems. This contrasts with the case of nonlinear systems where theory is often lacking.

Nevertheless, one of the remarkable results obtained in recent years has been to exhibit particular subclasses of non-linear problems (in the sense of classical theory) that can be tackled by linear algebraic techniques, provided their state equations are written in appropriate algebraic structures, other than standard algebra on $\mathbb{R}$.

A characteristic example is that of the dioid $(\mathbb{R} \cup \{-\infty\}, \text{Max}, +)$ which has turned out to be the basic algebraic structure for modeling some types of *discrete event* systems. This has thus made possible to transpose many classical results concerning automatic systems into this new context and to set up a theoretical framework of system theory in the dioid "(Max, +)."

After a brief reminder of classical models in linear system theory (Sect. 7.1) and Petri nets (Sect. 7.3), we show in Sect. 7.4 how the dynamic behavior of a particular class of Petri nets (more specifically: timed event graphs) can be modeled by linear state equations in the dioids ($\mathbb{R}_+$, Max, +) or ($\mathbb{R}_+$, Min, +). In Sect. 7.5, we look at the maximum possible performances that can be obtained from such systems when operating in autonomous mode. We show that the maximum production rate is related to the unique eigenvalue in the dioid ($\mathbb{R} \cup \{-\infty\}$, Max, +) of the matrix involved in the explicit expression of the solution for the state equation.

7.1. Classical Linear Dynamic Systems in Automation

Dynamic linear system theory is concerned with systems whose behavior over time is described (in the case of a discrete time model) by an evolution equation of the form:

$$\begin{cases} x(t) = A \cdot x(t-1) + B \cdot u(t) & (35) \\ y(t) = C \cdot x(t) & (36) \end{cases}$$

where:

$x(t) = \begin{bmatrix} x_1(t) \\ x_2(t) \\ \cdot \\ \cdot \\ \cdot \\ x_n(t) \end{bmatrix}$ denotes the *state* vector of the system at the instant t;

$y(t) = \begin{bmatrix} y_1(t) \\ y_2(t) \\ \cdot \\ \cdot \\ \cdot \\ y_m(t) \end{bmatrix}$ denotes the *observation* vector on the system at the instant t;

$u(t) = \begin{bmatrix} u_1(t) \\ u_2(t) \\ \cdot \\ \cdot \\ \cdot \\ u_p(t) \end{bmatrix}$ denotes the *control* vector applied to the system at the instant t;

A, B, C are real matrices of dimensions $n \times n$, $n \times p$, $m \times n$ respectively (when the coefficients of these matrices are varying over time, we have the case of a non-stationary system. When they do not depend on time, we then say that the system is stationary).

The evolution equations (35) and (36), which use the operations of standard algebra in $\mathbb{R}^n$, constitute one of the basic models for the theory of dynamic systems, and they have numerous applications: control of industrial processes, trajectory control problems, signal processing, etc. (see for example Kwakernaak and Sivan 1972; Kailath 1980).

There exist however many other applications which cannot be encompassed by these classical linear models. This is the case, in particular, for "dynamic scheduling" problems.

7.2. Dynamic Scheduling Problems

Dynamic scheduling problems appear in many applications, such as industrial automation (optimization of flexible manufacturing processes, for example) or the architecture of parallel or distributed computer systems (compilation of parallel applications, task placement, etc.).

With such problems, the aim is typically to process a flow of input tasks, by assigning these tasks to processing units (machines, manufacturing processes) under various constraints (mutual exclusion, synchronization, maximum processing time, etc.). Among all the possible solutions, we often consider:

- Particular solutions (cyclic scheduling, for example);
- Solutions optimizing a criterion such as *production rate* (number of units processed per unit of time).

Problems of this kind cannot be reduced to linear equations of the form (35), (36) in standard algebra. On the contrary, many studies such as those of Cuninghame-Green (1962, 1979, 1991), Cuninghame-Green and Meijer (1988), Cohen and co-authors (1983, 1985, 1989), Baccelli and co-authors (1992), Gaubert (1992, 1995a, b) have shown that some dynamic scheduling problems could be represented by appropriate *linear models in dioids* such that $(\mathbb{R} \cup \{-\infty\}, \text{Max}, +)$ or $(\mathbb{R} \cup \{+\infty\}, \text{Min}, +)$. As will be seen in Sect. 7.4, these are essentially systems whose evolution over time correspond to the behavior of Petri nets of a special kind, namely: *timed event graphs*. We first recall below some basic notions concerning Petri nets.

7.3. Modeling Discrete Event Systems Using Petri Nets

We recall (see for instance Peterson 1981, Brams 1983) that a Petri net is an oriented and valued *bipartite graph* $R = [P \cup T, U]$, where the set of vertices is formed by a set of *places* P and a set of *transitions* T, and where the set of arcs U includes arcs joining places to transitions and arcs joining transitions to places (the graph being *bipartite*, there is no arc joining a place to another place, nor any arc joining a transition to another transition). A *valuation* $\alpha\,(i, j) \in \mathbb{N}$ is associated with each arc $u = (i, j)$ of R. Figure 4 gives an example of a Petri net where, according to common practice, the places are indicated by circles and the transitions by elongated rectangles.

A Petri net R is said to be *labeled* when a natural number M(p), called *label of p*, is associated with each place $p \in P$.

In applications, the label of a place typically corresponds to the amount of a resource available at a certain point in the system. The label of a place is often represented by *tokens*, we then say that the place p contains M(p) tokens.

The evolution over time of place labels in a Petri net occurs according to the process of the *activation* (or firing) of the transitions.

A transition $t \in T$ is said to be activable (or fireable) w.r.t. a given labeling M if and only if:

$$\forall p \in \Gamma^{-}(t) = \{i/i \in P, (i,t) \in U\}, \quad \text{we have:} \quad M(p) \geq \alpha(p,t)$$

When this condition is satisfied, the activation (or firing) of transition t leads to a new labeling M′ defined, $\forall p \in P$, by:

$$\begin{cases} M'(p) = M(p) & \text{if} \quad p \notin \Gamma^{-}(t) \cup \Gamma^{+}(t) \\ M'(p) = M(p) - \alpha(p,t) & \text{if} \quad p \in \Gamma^{-}(t) \\ M'(p) = M(p) + \alpha(t,p) & \text{if} \quad p \in \Gamma^{+}(t) \end{cases}$$

Thus, for example, in the case of the net in Fig. 4, starting from the labeling $M = \begin{pmatrix} 3 \\ 3 \\ 1 \end{pmatrix}$ (where $M(p_1) = 3, M(p_2) = 3, M(p_3) = 1$), we reach, through the firing of transition t_1 the labeling $M' = \begin{pmatrix} 2 \\ 4 \\ 0 \end{pmatrix}$; then from M′, through the firing of transition t_2, we obtain the labeling $M'' = \begin{pmatrix} 1 \\ 2 \\ 2 \end{pmatrix}$ and so on.

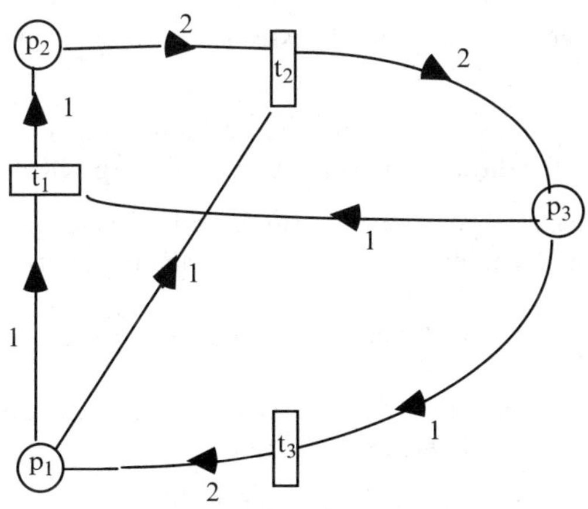

Fig. 4 An example of a Petri net where $P = \{p_1, p_2, p_3\}$ is the set of places and $T = \{t_1, t_2, t_3\}$ is the set of transitions. The valuations are indicated next to each arc

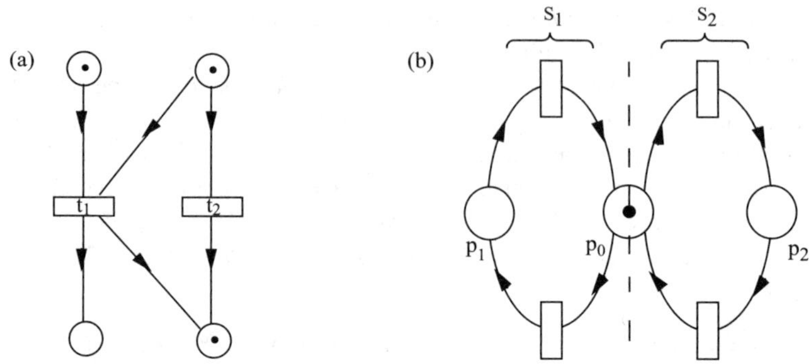

Fig. 5 Two characteristic examples of behavior that can be modeled using a Petri net. Fig. 5(**a**) shows a situation of conflict between two transitions t_1 and t_2: only one of these two transitions can be activated. Fig. 5(**b**) shows an example of a situation of mutual exclusion where a common resource has to be shared between two subsystems S_1 and S_2

The formalism of general Petri nets is very powerful, in the sense that it can accurately and appropriately represent a huge variety of types of behavior in real systems. Without being exhaustive, we mention:

- *Conflict* between two (or several) series of actions. Figure 5a provides an example: from the indicated labeling, each of the two transitions t_1 and t_2 is activable, but one and only one of the two transitions t_1 and t_2 can be activated.
- *Mutual exclusion* (sharing of resources, for example). Figure 5b provides an example where the two subsystems S_1 and S_2 cannot evolve independently from each other, given that they share the resource represented by the central place p_o. One of the two subsystems starts operating and consumes the token initially present in p_o; the other subsystem must therefore wait until a token has returned to p_o to start operating.
- *Synchronization* between several processes (this will be explained into detail in Sect. 7.4 below).

Obviously, the expressive power enjoyed by the formalism of general Petri nets has one counterpart: the difficulty in solving many basic problems such as:

- *Accessibility*: does a sequence of transitions exist where the transitions can be activated sequentially in order to reach some labeling M' starting from a given labeling M?
- *Liveness*: from any accessible labeling, is it possible, for any transition t, to reach a state where t is activable?
- *Boundedness*: does a natural integer B exist such that, for any place $p \in P$ and any accessible labeling M, we have $M(p) \leq B$?

For all these problems (and a few more), no practical, efficient algorithms exist, except for particular subclasses of Petri nets such as: state machines, event graphs, free choice and extended free choice networks (see for instance Peterson, 1981). We are going to see that, among these particular subclasses, the dynamic behavior of *timed event graphs* can be represented by *linear equations in dioids*.

7.4. Timed Event Graphs and Their Linear Representation in $(\mathbb{R} \cup \{-\infty\}, Max, +)$ and $(\mathbb{R} \cup \{+\infty\}, \min, +)$

We refer to an *event graph* as a Petri net having the special property that each place has a single input arc and a single output arc (we are limiting ourselves to the case where all the arcs have valuation 1).

Figure 6 shows an example of an event graph (drawn from Baccelli et al. 1992) which will subsequently serve as an illustration.

Observe that event graphs cannot be used to model situations of conflict or mutual exclusion such as those illustrated by Fig. 5a,b. Nevertheless, this subclass of Petri nets is interesting for many applications where the aim is essentially to model *synchronization* constraints between several processes.

We say that an event graph is *timed* if we associate with each place $p \in P$ a time $\theta(p)$ interpreted as the minimal time a token has to stay in the place. (We can also associate with each transition $t \in T$ a time $\theta(t)$ – the minimal activation duration of the transition t – but it is easy to show that it is always possible to reduce this to the case where only the places are timed). Figure 6 provides an example of a timed event graph.

The dynamic behavior of a system such as the one in Fig. 6 can be represented algebraically in different ways.

A natural way consists in associating with each transition t an increasing function $x_t\colon \mathbb{N} \to \mathbb{R}_+$ where, for any $n \in \mathbb{N}$, $x_t(n)$ denotes the date on which the nth firing of transition t occurs (date counted from the instant 0, considered as the instant the system starts operating). We are going to see that the variables $x_t(n)$ must satisfy a set of inequalities know as "equations of timers."

Let us illustrate first the basic ideas for writing the state equations on the small examples of Fig. 7, where we have three transitions and two places p_1 and p_2 timed with $\theta(p_1) = 2$ and $\theta(p_2) = 3$.

In Fig. 7a, no token is present initially in the places. The earliest termination date for the nth activation of transition t_3 therefore depends on the date of the nth activation of t_1 and the date of the nth activation of t_2. If we take into account the minimal residence time of a token in p_1 we must have:

$$x_3(n) \geq 2 + x_1(n)$$

and in the same way, taking into account the time delay of p_2:

$$x_3(n) \geq 3 + x_2(n)$$

We obtain therefore for $x_3(n)$ the inequality $x_3(n) \geq \mathrm{Max}\{2 + x_1(n);\ 3 + x_2(n)\}$.

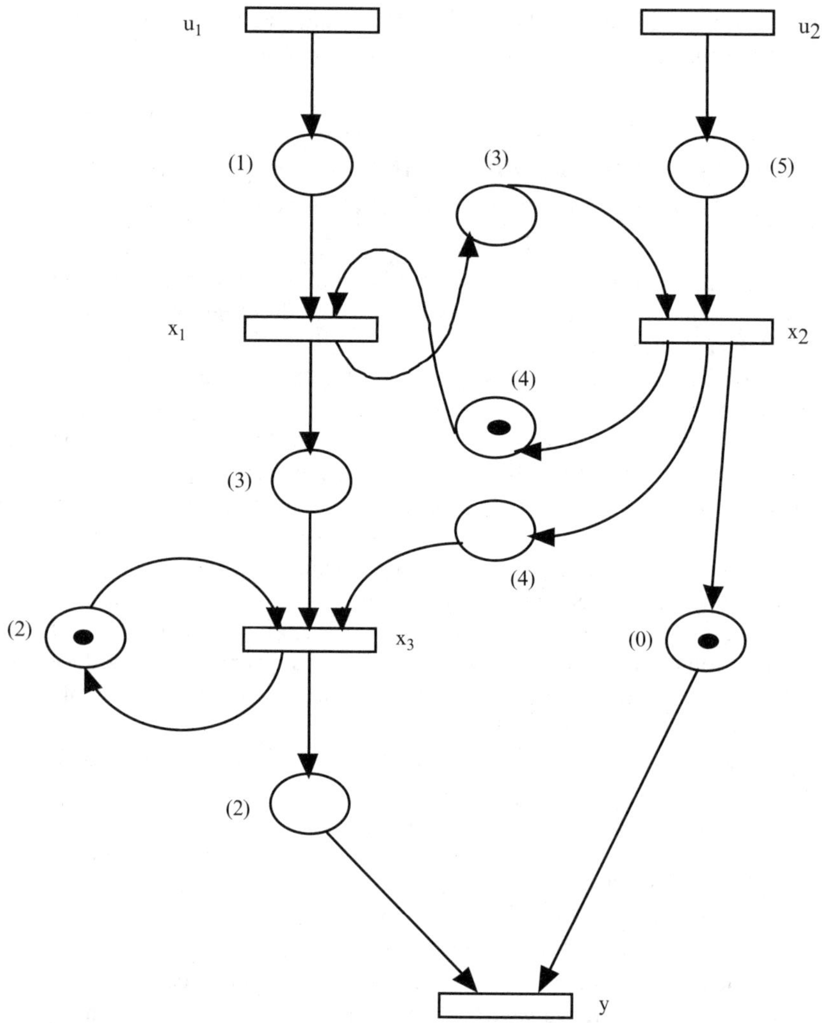

Fig. 6 Example of a (timed) event graph. The time delays indicated between brackets next to each place, represent the minimal time a token has to stay in the place. We have also indicated the tokens in the initial labeling considered

Figure 7b represents a more general situation where tokens can be present in some places at the initial instant. In this case, the earliest termination date of the nth activation of t_3 depends on the earliest termination date of the $(n-1)^{th}$ activation of t_1 (because a token is present initially in p_1) and on the earliest termination date of the $(n-2)^{th}$ activation of t_2 (because two tokens are initially present in p_2). The inequality on $x_3(n)$ can then be written:

$$x_3(n) \geq \text{Max}\{2 + x_1(n-1);\ 3 + x_2(n-2)\}$$

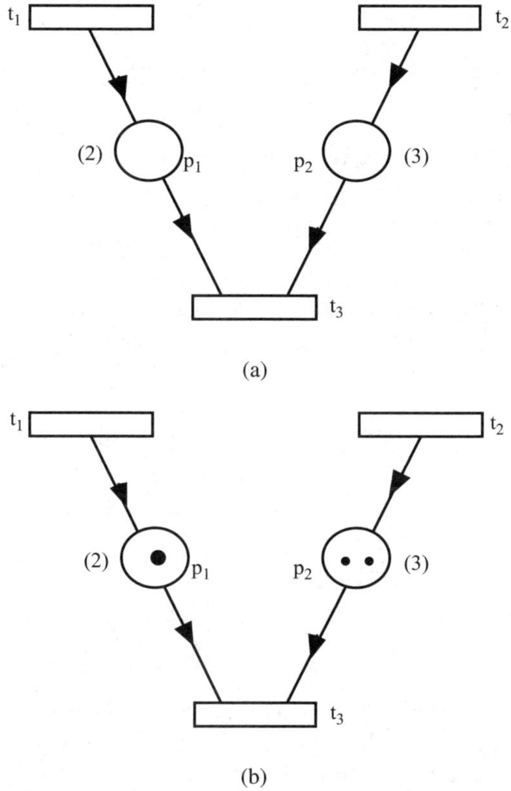

Fig. 7 Writing the state equation of a timed event graph. (**a**) The state equation for this example is $x_3(n) \geq \text{Max}\{2 + x_1(n);\ 3 + x_2(n)\}$. (**b**) The state equation for this example is: $x_3(n) \geq \text{Max}\{2 + x_1(n-1); 3 + x_2(n-2)\}$

Now, by applying this principle to the whole network of Fig. 6, we obtain the system:

$$\begin{cases} x_1(n) \geq \text{Max}\{4 + x_2(n-1);\ 1 + u_1(n)\} \\ x_2(n) \geq \text{Max}\{3 + x_1(n);\ 5 + u_2(n)\} \\ x_3(n) \geq \text{Max}\{3 + x_1(n);\ 4 + x_2(n);\ 2 + x_3(n-1)\} \\ \ y(n) \geq \text{Max}\{x_2(n-1);\ 2 + x_3(n)\} \end{cases}$$

(where the inequalities should be understood in the sense of the standard order relation on $\mathbb{R}$). Observe that, in the above set of relations, we obtain earliest termination dates by ensuring that each inequality is satisfied with equality.

In the dioid $(\mathbb{R} \cup \{-\infty\}, \text{Max}, +)$ where $\oplus = \text{Max}$ and $\otimes = +$, $\varepsilon = -\infty$, $e = 0$, the above system takes the form:

$$\begin{cases} x_1(n) \geq 4 \otimes x_2(n-1) \oplus 1 \otimes u_1(n) \\ x_2(n) \geq 3 \otimes x_1(n) \oplus 5 \otimes u_2(n) \\ x_3(n) \geq 3 \otimes x_1(n) \oplus 4 \otimes x_2(n) \oplus 2 \otimes x_3(n-1) \\ y(n) \geq 2 \otimes x_3(n) \oplus e \otimes x_2(n-1) \end{cases}$$

or equivalently, expressed in matrix form:

$$\begin{cases} x(n) \geq A_o \otimes x(n) \oplus A_1 \otimes x(n-1) \oplus B \otimes u(n) & (37) \\ y(n) \geq C_o \otimes x(n) \oplus C_1 \otimes x(n-1) & (38) \end{cases}$$

with:

$$A_o = \begin{bmatrix} \varepsilon & \varepsilon & \varepsilon \\ 3 & \varepsilon & \varepsilon \\ 3 & 4 & \varepsilon \end{bmatrix} \quad A_1 = \begin{bmatrix} \varepsilon & 4 & \varepsilon \\ \varepsilon & \varepsilon & \varepsilon \\ \varepsilon & \varepsilon & 2 \end{bmatrix}$$

$$B = \begin{bmatrix} 1 & \varepsilon \\ \varepsilon & 5 \\ \varepsilon & \varepsilon \end{bmatrix}$$

$$C_o = \begin{bmatrix} \varepsilon & \varepsilon & 2 \end{bmatrix} \quad C_1 = \begin{bmatrix} \varepsilon & e & \varepsilon \end{bmatrix}$$

By analogy to the classical case ((35) and (36) of Sect. 7.1) matrix equation (37) is the state equation, (38) is the so-called equation of observation, x is the state vector (here having three components), u the control vector (here having two components) and y the vector of observations or output of the system (here having one component).

To solve the state equation (37), we can use the general results of Chap. 4 concerning the resolution of linear systems of the form $X = A \otimes X \oplus B$, by applying it to:

$$x(n) = A_0 \otimes x(n) \oplus A_1 \otimes x(n-1) \oplus B \otimes u(n) \quad (37)'$$

(in other words (37) in which all inequalities are replaced by equalities).

Modulo the existence of A_0^*, quasi-inverse of A_0 (in other words, by assuming the absence of a positive circuit in $G(A_o)$) the minimal solution of $(37)'$ can be written:

$$x(n) = A_0^* \otimes A_1 \otimes x(n-1) \oplus A_0^* \otimes B \otimes u(n) \quad (39)$$

Since x(n) given by (39) also satisfies (37), we observe that it is also the minimal solution to (37). In this solution, all the inequalities are satisfied at equality: (39) therefore defines the set of earliest termination dates when running the system. Therefore, given the sequence of control vectors u(1) u(2) ... and the "initial state" x(o), (39) successively determines all the values x(1), x(2), ... of the state vector.

Each component i of the state vector corresponds to a transition t_i of the system and, in the solution expressed by (39), $x_i(1)$ represents the earliest termination date of the first activation of transition t_i, $x_i(2)$ the earliest termination date of the second activation of t_i, etc.

Timed event graphs can also be represented as dynamic linear systems in the dioid $(\mathbb{R} \cup \{+\infty\}, \mathrm{Min}, +)$ by considering another form of state equation called "*equation of counters*" instead of "the equation of timers" (37)–(38). Thus, for the example of Fig. 6, by associating with each transition x_i (resp. u_i, y) a variable $x_i(\tau)$ (resp. $u_i(\tau)$, $y(\tau)$) representing the number of firings of the transition x_i (resp. u_i, y) up to the instant τ, we obtain the following system of inequalities:

$$\begin{cases} x_1(\tau) \le \mathrm{Min}\{1 + x_2(\tau-4); u_1(\tau-1)\} \\ x_2(\tau) \le \mathrm{Min}\{x_1(\tau-3); u_2(\tau-5)\} \\ x_3(\tau) \le \mathrm{Min}\{1 + x_3(\tau-2); x_1(\tau-3); x_2(\tau-4)\} \\ \ y(\tau) \le \mathrm{Min}\{1 + x_2(\tau); x_3(\tau-2)\} \end{cases}$$

It is a linear system similar to (37)–(38), but in the dioid $(\mathbb{R} \cup \{+\infty\}, \mathrm{Min}, +)$.

7.5. Eigenvalues and Maximum Throughput of an Autonomous System

Let us consider a timed event graph whose state equation is of the form (37) and, consequently, whose minimal solution (earliest termination dates) is expressed by (39):

$$x(n) = \bar{A}_1 \otimes x(n-1) \oplus \overline{B} \otimes u(n)$$

with $\bar{A}_1 = A_o^* \otimes A_1$ and $\overline{B} = A_o^* \otimes B$.

Let us study the evolutions of this system operating in *autonomous* mode, in other words with controls u (n) not constraining the evolutions of the system (this corresponds to choosing for example $\forall n$: $u(n) = \begin{bmatrix} -\infty \\ -\infty \\ \vdots \\ -\infty \end{bmatrix}$).

Thus we can write, $\forall n$:

$$x(n) = \bar{A}_1 \otimes x(n-1) \tag{40}$$

thus:

$$x(n) = (\bar{A}_1)^n \otimes x(0) \tag{41}$$

By assuming, for the sake of simplicity, that the matrix $\bar{A}_1$ is *irreducible* ($G(\bar{A}_1)$ strongly connected) we know from Theorem 4 (Sect. 4.1 of this chapter) that it has a unique eigenvalue $\lambda = \rho(\bar{A}_1)$ (spectral radius). Then, for any eigenvector w associated with $\lambda = \rho(\bar{A}_1)$, by choosing $x(0) = w$ we obtain from (40):

$$x(n) = \lambda \otimes x(n-1)$$

We therefore have, for any transition i the network:

$$x_i(n) = \lambda + x_i(n-1) \tag{42}$$

which shows that an additional activation of each transition of the network takes place every λ units of time. If the system represents, for example, a production workshop, (42) can then be interpreted as the production of a new unit every λ units of time; λ is therefore the inverse of the *production rate*.

According to the results of Sect. 4, the eigenvalue $\lambda = \rho(\bar{A}_1)$ corresponds, in the graph $G(\bar{A}_1)$, to the average weight of the circuit of maximum average weight (the weight of a circuit being the sum of the weights of the arcs which compose it). By analogy to the notion of critical path (for standard scheduling problems of the PERT type), such a circuit is called a *critical circuit* of the system. Critical circuits are those which limit the performances of the system and the value $\frac{1}{\lambda} = \frac{1}{\rho(\bar{A}_1)}$ is the maximum production rate which can be obtained. In addition, the eigenvectors associated with the eigenvalue $\lambda = \rho(\bar{A}_1)$ provide the various possible initial conditions enabling to obtain these maximum performances from the system.

For the effective calculation of λ, we can use Karp's algorithm (1978) to find the circuit of maximum average weight in $G(\bar{A}_1)$ (see above, Algorithm 1, Sect. 4.2).

For further details concerning the applications of the dioid $(\mathbb{R} \cup \{-\infty\}, \text{Max}, +)$ to discrete dynamic systems that can be represented by event graphs, we refer to Baccelli and co-authors (1992) and to Gaubert (1992, 1994, 1995a,b).

Exercises

Exercise 1. Proof of the Perron–Frobenius Theorem

A real square matrix $A = (a_{ij})$ is said to be nonnegative (denoted $A \geq 0$) or positive (denoted $A > 0$) if all its entries are nonnegative ($a_{ij} \geq 0$) or positive ($a_{ij} > 0$).

(1) A $n \times n$ matrix $A = (a_{ij})$ is said to be *reducible* if there exists a permutation matrix P such that:

$$PAP^T = \begin{bmatrix} B & 0 \\ C & D \end{bmatrix}$$

where B and D are square matrices. Otherwise A is said to be *irreducible*.

(a) *Lemma 1*

Show that $A \geq 0$ is an irreducible $n \times n$ matrix if and only if we have:

$$(I + A)^{n-1} > 0.$$

(b) Deduce from the above that if we denote $A^p = (a_{ij}^{(p)})$ the pth power of A, $A \geq 0$ is irreducible if and only if, for any i and j, there exists an integer $q \leq n$ such as $a_{ij}^{(q)} > 0$.

(c) We denote G(A) the directed graph associated with the matrix A; this is a graph having n vertices $s_1, s_2, \ldots, s_n$ with an arc (s_i, s_j) if and only if $a_{ij} \neq 0$.

Show that A is irreducible if and only if G(A) is strongly connected.

(2) Perron Theorem (1907)

We define an order relation on the $n \times n$ matrices by $A \geq B$ if and only if $a_{ij} \geq b_{ij}$ for any i and j. We denote $A > B$ if and only if $a_{ij} > b_{ij}$ for any i and j.

To study the spectrum of an irreducible matrix $A \geq 0$, let us consider for any vector $x = (x_1, \ldots, x_n) \geq 0$ $(x \neq 0)$, the number

$$r_x = \min_{x_i \neq 0} \left\{ \frac{(Ax)_i}{x_i} \right\}$$

It is clear that $r_x \geq 0$ and that r_x is the greatest real number ρ for which:

$$\rho x \leq Ax.$$

(a) Show that the function: $x \to r_x$ has a maximum value r for at least one vector $z > 0$.

(b) Let $r = r_z = \mathrm{Max}_{x>0}\ \{r_x\}$. Show that

$$r > 0, \quad A\,z = r\,z \quad \text{and} \quad z > 0.$$

(c) Show that the modules of all the eigenvalues of A are not greater than r.

Deduce that there exists a unique eigenvector associated with r.

(3) Frobenius Theorem (1912)

For any complex matrix B, let |B| denote the matrix whose entries are $|b_{ij}|$.

(a) *Lemma 2*

If A is irreducible and $A \geq |B|$, show that for any eigenvalue γ of B, we have

$$r \geq |\gamma|.$$

Show that we have the equality $r = |\gamma|$ if and only if,

$$B = e^{i\varphi}\, D\, A\, D^{-1}$$

where $e^{i\varphi} = \gamma / r$ and D is a diagonal matrix whose diagonal elements have modules equal to 1 (i.e. $|D| = I$).

(b) If a matrix $A \geq 0$ is irreducible and has h eigenvalues of modulus $\rho(A) = r$, show that these values are equal to $\lambda_k = r\, e^{i\frac{2k\,\pi}{h}}$ (for k running from 0 to $h - 1$).

These are the roots of the equation $\lambda^h = r^h$.

(c) Finally, show that if $h > 1$, there exists a permutation matrix P such that

$$P\,A\,P^T = \begin{bmatrix} 0 & A_{12} & 0 & \ldots & 0 \\ 0 & 0 & A_{23} & \ldots & 0 \\ \vdots & & & & \vdots \\ 0 & \ldots & \ldots & 0 & A_{h-1,h} \\ A_{h1} & 0 & \ldots & \ldots & 0 \end{bmatrix}$$

[*Answers*: see Gantmacher (1966), Chap. 13.

(1) (a) A possible way is to show that, for any column vector $y \geq 0$ ($y \neq 0$), the inequality $(I + A)^{n-1}\, y > 0$ is satisfied and that it is equivalent to the fact that the vector $z = (I + A)\, y$ always has a number of zero components lower than the number of zero components of y.

(2) (a) Use the fact that while r_x is not necessarily continuous on the set of $x \geq 0$ and $\sum x_i^2 = 1$, on the contrary, r_y with $y = (I + A)^{n-1}\, x$ is continuous (see Lemma 1) on this set.

(b) Just consider r_u with $u = (1, 1, \ldots, 1)$.

(3) (a) Consider y such that $B\, y = \gamma\, y$. Then show successively that

$$|\gamma||y| \leq |C||y| \leq A|y|,\ |y| \leq r,\ \text{and when}\, |\gamma| = r,\ |C| = A,$$
$$y_j = |y_j| e^{i\varphi_j} \text{ yields } D = \{e^{i\varphi_1}, \ldots, e^{i\varphi_n}\}.$$

(b) Use Lemma 2 with $A\, z = rz$, $z > 0$ and $y^{(k)} = D_k z$.]

Exercise 2. Asymptotics of the Perron Eigenvalues and Eigenvectors

Let $\mathcal{A}_p$ be a $n \times n$ real matrix with nonnegative entries dependent on a large real parameter p.

We consider the spectral problem

$$(1) \quad \mathcal{A}_p U_p = \lambda_p U_p,\ U_p \in (\mathbb{R}_+)^n \backslash \{0\},\ \lambda_p \in \mathbb{R}_+.$$

In the case where $\mathcal{A}_p$ is irreducible, the Perron Frobenius theorem shows that λ_p is unique and that there exists a unique U_p satisfying $\sum (U_p)_i = 1$. The aim in this exercise is to determine the asymptotic values of λ_p and U_p from those of $\mathcal{A}_p$.

(1) We assume that, when p tends to $+\infty$, $(\mathcal{A}_p)_{ij}^{1/p}$ has for any $i, j = 1, \ldots, n$, a limit A_{ij} and that $A = (A_{ij})$ is irreducible.

(a) Show then that $(\lambda_p)^{1/p}$ has as a limit $\rho_{Max}(A)$, the eigenvalue of A in $(\mathbb{R}_+, \mathrm{Max}, \times)$, in other words the maximum value of the geometric mean of the weights of the circuits of the graph G(A) associated with A.

(b) We call critical circuit, a circuit for which this weight is reached. The critical graph CG(A) is the subgraph of G(A) restricted to the vertices and to the arcs which belong to a critical circuit. We set $\widetilde{A} = (\rho_{Max}(A))^{-1}\, A$. Show that if the critical graph CG(A) is strongly connected, then

$$\lim_{p \to +\infty} (U_p)_i^{1/p} = \frac{(\widetilde{A})_{ij}^*}{\mathrm{Max}_k (\widetilde{A})_{kj}^*}$$

where j is an arbitrary vertex of this critical class and where $(\widetilde{A})^*$ is the quasi-inverse of $\widetilde{A}$ in $(\mathbb{R}_+, \mathrm{Max}, \times)$.

(2) We consider now the case where the nonzero coefficients of $\mathcal{A}_p$ have an asymptotic expansion of the form

$$(2) \quad (A_p)_{ij} \sim a_{ij} A_{ij}^p$$

where A is irreducible.

(a) Show that we have

$$(3)\quad \lambda_p \sim \rho\left(a^{CG(A)}\right)(\rho_{Max}(A))^p$$

where $\rho_{Max}(A)$ is the eigenvalue of A in $(\mathbb{R}_+, \text{Max}, \times)$, $a^{CG(A)}$ the matrix obtained from the matrix $a = (a_{ij})$ by substituting 0 to the coefficients a_{ij} such that the arc (i, j) is not in the critical graph CG(A), and where $\rho(\cdot)$ denotes the Perron eigenvalue.

(b) Show on an example, that in general, (3) does not imply the convergence of $(U_p)^{1/p}$ when $p \to +\infty$.

(c) Show that if $a^{CG(A)}$ has only one basic class, then

$$(4)\quad (U_p)_i \sim u_i (U_i)^p$$

where $U = (U_i)$ is a column of $(\widetilde{A})^*$ corresponding to a vertex of the critical graph and where $u = (u_i)$ is the unique Perron eigenvector of $a^{CG(A)}$.

(3) Consider the transfer matrix of the simplest one-dimensional Ising model, (see Baxter 1982, Chap. 2):

$$\mathcal{A}_{1/T} = \begin{bmatrix} \exp((J+H)/T) & \exp(-J/T) \\ \exp(-J/T) & \exp((J-H)/T) \end{bmatrix}, \quad \text{with} \quad J > 0, H \in \mathbb{R}$$

where T is the temperature that we let decrease downwards to 0.
Show that:

$$\lambda_{1/T} \sim \text{Max}\left\{\exp\left(\frac{J+H}{T}\right), \exp\left(\frac{J-H}{T}\right)\right\}.$$

When $H > 0$, show that

$$U_{1/T} \sim \begin{pmatrix} 1 \\ \exp\left(\frac{-2J-H}{T}\right) \end{pmatrix}$$

and when $H < 0$

$$U_{1/T} \sim \begin{pmatrix} \exp\left(\frac{-2J-H}{T}\right) \\ 1 \end{pmatrix}.$$

[*Answers*: see Akian et al. (1998).

(1) From the spectral problem in $(\mathbb{R}_+, \text{Max}, \times)$.

(2) (a) From the spectral problem in the semi-field J_{Max}.

(b) $A_p = \begin{bmatrix} 1+\cos(p)e^{-p} & e^{-2p} \\ e^{-2p} & 1 \end{bmatrix}$,

$$\liminf \left(\frac{(U_p)_2}{(U_p)_1}\right)^{1/p} = e^{-1} < \limsup \left(\frac{(U_p)_2}{(U_p)_1}\right)^{1/p} = e.$$

(c) From the following theorem: an irreducible matrix $\mathcal{A} = (a, A) \in (J_{Max})^{n\times n}$ has a unique eigenvector (up to a given factor) if and only if it has a unique basic class.

The generalization to the case of several basic classes can be found in the above quoted reference.]

Exercise 3. Eigenvalues and Eigenvectors of Some Endomorphisms in Infinite Dimensions

The aim of this exercise is to extend to infinite dimensions the characterization of the eigenvalues and eigenvectors studied in this chapter for idempotent dioids featuring multiplicative group structure, such as the dioids $(\mathbb{R} \cup \{-\infty\}, \text{Max}, +)$ and $\{\mathbb{R} \cup \{+\infty\}, \text{Max}, \cdot\}$. We consider here the dioid $D = (\mathbb{R} \cup \{-\infty\}, \text{Max}, +)$.

Let X be a totally bounded metric space (in other words, for any $\varepsilon > 0$ there exists a finite covering of X with balls with radius ε).

We consider then the semi-module C(X, D) of continuous and bounded functions f: $X \to D$, where the dioid D is endowed with the metric $\rho(a, b) = |e^a - e^b|$, as well as an "integral operator of kernel a" by:

$$(Af)(x) = \sup_{y\in X} \{a(x, y) + f(y)\}$$

where a: $X \times X \to D$ is supposed to be given. In the case where the element $\sup\limits_{y\in X} a(y, y)$ exists, it will be called trace of A and denoted Tr(A).

(1) Show that if the kernel a: $X \times X \to D$ of the integral operator is a uniformly continuous bounded function of the first argument and equicontinuous w.r.t. the second argument, then A is a continuous endomorphism of the semi-module C (X, D).
(2) Show that if we assume that X is a totally bounded metric space and that the kernel a: $X \times X \to D$ of the endomorphism A is a uniformly continuous bounded function of the first argument and equicontinuous w.r.t. the second, then there exists a sub-semi-module J of the semi-module C (X, D) ($J \neq 0$) and an element $\lambda \in D$ such that:

$$(Af)(x) = \lambda + f(x)$$

for any $f \in J$, and that the maximal element of the set of such λ is defined as

$$\lambda = \mathop{\text{Sup}}_{i\in N} \lambda(i)$$

where $\lambda(i) = (\text{Tr}\, A^i)^{\frac{1}{i}}$.

[*Answer*: see Lesin and Samborskii (1992) (difficult).

For a similar study of eigenvalues and eigenvectors in infinite dimensions on the dioid $(\mathbb{R}, \text{Min}, \text{Max})$, see Gondran and Minoux (1997, 1998) and also Chap. 7, Sect. 6.4.]

Chapter 7
Dioids and Nonlinear Analysis

1. Introduction

Most of the problems dealt with in the preceding chapters have concerned finite dimensional or discrete problems. The aim of the present chapter is to show that the structures of dioids lend themselves to defining, in the *continuous domain*, new branches of *nonlinear analysis*.

The basic idea is to replace the classical field structure on the reals by a dioid structure. Thus, a new branch of nonlinear analysis will correspond to each type of dioid. This approach was pioneered by Maslov (1987b), Maslov and Samborskii (1992), under the name of *idempotent analysis*. (The underlying dioid structure considered being $(\mathbb{R} \cup \{+\infty\}, \text{Min}, +)$, the so-called MINPLUS dioid).

From an historical point of view, the concept of *capacity* due to Choquet (1953) may also be viewed as a starting point. We recall the definition of the Choquet capacity of a function f: $E \to \mathbb{R}$ on a subset $A \subset E$:

$$C_A(f) = \inf_{x \in A} \{f(x)\}.$$

Now, we observe that the above may be considered as an analogue to the Rieman integral, when the operation inf is taken in place of addition and the operation + is taken in place of multiplication. Indeed, the Rieman integral $\int_A f(x)\,dx$ on an interval $A = [\alpha, \beta]$ of $\mathbb{R}$, may be viewed as the limit of finite sums of the form $\sum_{i=1}^{n} f(x_i)\,\Delta(x_i)$ (with $x_0 = \alpha$, $x_n = \beta$, $\Delta(x_i) = x_i - x_{i-1}$) when n tends to infinity and $\Delta(x_i)$ tends to 0.

In the suggested substitution, the above expression becomes:

$$\lim_{\Delta x_i \to 0} \left\{ \inf_{i=1,\dots,n} \{f(x_i) + \Delta(x_i)\} \right\} = \inf_{x \in A} \{f(x)\} = C_A(f).$$

After Choquet this approach has been further developed in the context of fuzzy set theory with the introduction of the concept of fuzzy measures and the Sugeno integral (Sugeno 1977).

Special emphasis will be placed in the present chapter on the analyses related to the dioids MIN-PLUS ($\mathbb{R} \cup \{+\infty\}$, Min, +) and MIN-MAX ($\mathbb{R} \cup \{+\infty, -\infty\}$, Min, Max). It will be seen that, in a sense, the proposed approach bears a close similarity to the *theory of distributions* for the nonlinear case; here, the operator is "linear" and continuous with respect to the dioid structure, though *nonlinear* with respect to the classical structure ($\mathbb{R}$, +, ×).

The main interest of such extensions lies in the fact that the structures of a dioid and moduloid will turn out to play a role similar to the one played by fields and vector fields when proceeding from linear to nonlinear problems. This idea can be illustrated by considering three classical problems in Physics, namely:

– The heat transfer equation: find u(x, t) such that:

$$\begin{cases} \dfrac{\partial u}{\partial t} - \dfrac{\partial^2 u}{\partial x^2} = 0 & t > 0, x \in \mathbb{R} \\ u(x, 0) = u_0(x) & x \in \mathbb{R} \end{cases} \tag{1}$$

where $u_0(x)$ is supposed to be given for all $x \in \mathbb{R}$.

– The Halmilton–Jacobi equation in classical mechanics: find u(x, t) such that:

$$\begin{cases} \dfrac{\partial u}{\partial t} + \dfrac{1}{2}\left(\dfrac{\partial u}{\partial x}\right)^2 = 0 & t > 0, x \in \mathbb{R} \\ u(x, 0) = u_0(x) & x \in \mathbb{R} \end{cases} \tag{2}$$

where $u_0(x)$ is supposed to be given for all $x \in \mathbb{R}$.

– The following variant of the Bürgers equation in fluid mechanics: find u(x, t) such that:

$$\begin{cases} \dfrac{\partial u}{\partial t} + \dfrac{1}{2} u \left|\dfrac{\partial u}{\partial x}\right| = 0 & t > 0, x \in \mathbb{R} \\ u(x, 0) = u_0(x) & x \in \mathbb{R} \end{cases} \tag{3}$$

where $u_0(x)$ is given for all $x \in \mathbb{R}$.

For all the above equations, it is easily checked that, if $u_1(x, t)$ and $u_2(x, t)$ are any two solutions to (1) (resp. (2), (3)) and if λ and μ are two real constants, then:

$$\lambda u_1(x, t) + \mu u_2(x, t)$$

is a solution to the heat transfer equation (1);

$$\text{Min}\{\lambda + u_1(x, t); \mu + u_2(x, t)\}$$

is a solution to the Hamilton–Jacobi equation (2);

$$\text{Min}\{\text{Max}\{\lambda; u_1(x, t)\}; \text{Max}\{\mu; u_2(x, t)\}\}$$

is a solution to the Bürgers equation (3).

It is therefore realized that the solution sets for the Hamilton–Jacobi and Bürgers equations are no longer vector (sub)fields (as in the case of the heat transfer equation) but (sub) moduloids based on the MIN-PLUS and MIN-MAX dioids respectively. This is further confirmed by considering the various explicit solutions corresponding to (1)–(3), stated below:

$$u(x,t) = \int u_0(y) \frac{1}{2\sqrt{\pi t}} e^{-\frac{(x-y)^2}{4t}} dy \tag{4}$$

$$u(x,t) = \inf_y \left\{ u_0(y) + \frac{(x-y)^2}{2t} \right\} \tag{5}$$

$$u(x,t) = \inf_y \left\{ \mathrm{Max} \left\{ u_0(y); \left| \frac{(x-y)}{t} \right| \right\} \right\} \tag{6}$$

(4) is the classical solution to the heat transfer equation (see e.g. Dautray and Lions 1985); (5) is the so-called Hopf solution to Hamilton–Jacobi (see e.g. Lions 1982).

In each case, it is observed that the general solution is the *convolution product* of $u_0(x)$ with the so-called "elementary solution" to the corresponding equation $\left(\frac{1}{2\sqrt{\pi t}} e^{-x^2/4t} \text{ for}(1); \ \frac{x^2}{2t} \text{ for}(2); \ \frac{|x|}{t} \text{ for}(3) \right)$. Indeed, (5) is obtained by using the *MIN-PLUS convolution product*:

$$f \otimes g(x) = \inf_y \{g(y) + f(y-x)\}$$

and (6) is obtained by using the MIN-MAX convolution product:

$$f \otimes g(x) = \inf_y \{\mathrm{Max}\{g(y); f(y-x)\}\}$$

More generally, taking $\oplus = \mathrm{Min}$ and $\otimes$ as the ordinary addition for real numbers, and considering the functional space $\mathbb{R}^X$ (the set of functions: $X \to \mathbb{R}$) (5) can be seen as a special case of a mapping of the form:

$$(Ag)(x) = \inf_{y \in X} \{k(x,y) + g(y)\} \tag{7}$$

Such functional mappings are "linear" with respect to the MIN-PLUS dioid since, in this case:

$$A(\lambda \otimes f \oplus \mu \otimes g) = \lambda \otimes Af \oplus \mu \otimes Ag$$
$$(\forall \lambda, \mu \in \mathbb{R}, \forall f, g \in \mathbb{R}^X).$$

Indeed, relation (7) may be viewed as the application to g of an integral operator with kernel k, and could be formally denoted:

$$\int_X^{\oplus} k(x,y) \otimes g(y) dy$$

Going one step further, in the functional space $\mathbb{R}^{\times}$, we can define the MIN-PLUS scalar product:

$$\langle f, g \rangle = \inf_{x \in X} \{f(x) + g(x)\} \overset{\text{def}}{=} \int_X^{\oplus} f(x) \otimes g(x) dx$$

which can be taken as a starting point to construct an analogue to Hilbert analysis, derive analogues to the Riesz and Hahn-Banach theorems, to Fourier transforms, to distributions and measure theory, etc.

In particular, an analogue to the Fourier transform in MIN-PLUS analysis is recognized as the so-called Legendre–Fenchel transform:

$$\hat{f}(p) = \sup_x \{\langle p, x \rangle - f(x)\}.$$

This transform is known to have many applications in Physics: this is the one which sets the correspondence between the Lagrangian and the Hamiltonian of a physical system; which sets the correspondence between microscopic and macroscopic models; which is also at the basis of *multifractal analysis* relevant to modeling turbulence in fluid mechanics, etc.

Another useful transform, called the Cramer transform (for details, see Sect. 8 below) has been investigated, in particular by Quadrat (1990) and later Quadrat et al. (1994, 1995) who have exhibited the analogy between optimization and probability (this issue is the subject of Exercise 5 at the end of the present chapter). This analogy has its origin in the fact that the dioid $(\mathbb{R}_+, +, \times)$ is the one underlying measure and probability theory.

In Sect. 2 based on the MINPLUS scalar product, a special concept of equivalence among functions (the so-called inf-ϕ-equivalence) is introduced, which may be viewed as an analogue to almost everywhere equality (a.e. equality) of measurable functions. It is shown that a new derivation of lower-semi-continuous (l.s.c.) functions and convex analysis can be deduced.

In Sect. 3, we show how MINPLUS analysis turns out to provide an appropriate framework for the development of *nonlinear wavelet analysis*. This generalizes the classical linear wavelet analysis for L^2 functions to "linear" (in the sense of the MIN-PLUS dioid) wavelet analysis for l.s.c. and convex l.s.c. functions.

In Sect. 4, it is shown how the MINPLUS scalar product lends itself to defining weak convergence concepts, and to exhibiting their links to the so-called epiconvergence (or Γ-convergence) which turn out to be increasingly used in Physics and Applied Mathematics.

In Sect. 5, MINPLUS analysis is used to define weak solution concepts for first and second order partial differential equations. Some explicit (weak) solutions to first order nonlinear partial differential equations will then be derived in Sect. 6 in the context of MINPLUS analysis through the use of the so-called inf-convolution.

MINMAX analysis is considered in Sect. 7, where it is shown how, by introducing the MINMAX scalar product, the whole approach may be extended to account for *quasi-convex analysis*. The *Infmax-affine transform* is introduced which plays a role in MINMAX analysis which is similar to the one played by the Legendre–Fenchel transform in MINPLUS analysis.

Finally, relations between MINPLUS analysis, the Cramer transform and the theory of large deviations are studied in Sect. 8.

2. MINPLUS Analysis

Our primary aim will be to show that, by taking $(f, g) = \inf_x \{f(x) + g(x)\}$ as the "scalar product" of two functions, one can reconstruct and synthesize an entire branch of nonlinear analysis by means of a new formalism of the "variational" kind for generalized functions, and, in particular, for lsc (lower semi-continuous) and usc (upper semi-continuous) functions.

Let X be a Banach space having $\mathbb{R}$ as underlying set of scalars. We will denote $|.|$ the norm on X, X^* the topological dual of X and $< ., . >$ the product in the duality X^*, X. When X is a Hilbert space, we will identify X^* and X. We denote by Ω a locally compact open set of X.

We consider functions $f: \Omega \to \hat{\mathbb{R}} = \mathbb{R} \cup \{+\infty\}$. We will say that such a function is *proper* if it is bounded from below ($f(x) \geq m, \forall x \in \Omega$) and not identically equal to $+\infty$.

Definition 2.1. *(Maslov 1987a,b)*

For two functions f *and* g: $\Omega \to \hat{\mathbb{R}}$, *we define the* "MINPLUS scalar product" *denoted* (f, g) *as:*

$$(f, g) = \inf_{x \in \Omega} \{f(x) + g(x)\}.$$

Definition 2.2. *For any family* ϕ *of functions:* $\Omega \to \overline{\mathbb{R}}$ $(\phi \subset \overline{\mathbb{R}}^{\Omega})$, *we define the* inf-$\phi$- equivalence *of two functions* f *and* g *as:*

$$f \overset{\phi}{\approx} g \Leftrightarrow (f, \varphi) = (g, \varphi) \quad \forall \varphi \in \phi$$

and the bi-conjugate *of* f *with respect to* ϕ, *denoted* $P_\phi f$, *as:*

$$P_\phi f(x) = \sup_{\varphi \in \phi} [(f, \varphi) - \varphi(x)].$$

Our approach differs from that of Maslov in that we consider families ϕ of test-functions not limited to continuous or usc functions.

Also note that $P_\phi f$ can be considered as the *bi-conjugate* of f with respect to ϕ in the sense of Moreau (1970).

Given a proper function f, we consider the solutions u of the equation:

$$u \overset{\phi}{\approx} f$$

where $\overset{\phi}{\approx}$ denotes inf-ϕ-equivalence.

For any family of test functions ϕ, we can state:

Theorem 1. *(Gondran 1996)*
The set of functions inf-Φ- equivalent to a given proper function f *has a smallest element which will be referred to as the inf-solution, equal to* $P_\phi f$.

Proof. The proof will be given in three stages:

(a) The set of functions inf-ϕ- equivalent to f has a smallest element; (b) $P_\phi f$ is inf-ϕ- equivalent to f, (c) it is the smallest element.

(a) Let u_i, $i \in I$, be the set of functions inf-ϕ- equivalent to f, in other words such that $(u_i, \varphi) = (f, \varphi) \forall \varphi \in \phi$.
Let us show that the function u defined for any x as:

$$u(x) = \inf_{i \in I} u_i(x)$$

is also inf-ϕ- equivalent to f.
As $u(x) \le u_i(x) \forall x$, we have $(u, \varphi) \le (u_i, \varphi) = (f, \varphi)$, $\forall \varphi \in \phi$. If u is not inf-ϕ-equivalent to f, there exists $\varphi_0 \in \phi$ and $\varepsilon > 0$ such that:

$$(u, \varphi_0) + \varepsilon = (f, \varphi_0)$$

There then exists x_0, such that:

$$u(x_0) + \varphi_0(x_0) \quad \le (u, \varphi_0) + \frac{\varepsilon}{3}$$

and for this x_0, there exists $i_0 \in I$ such that:

$$u_{i_0}(x_0) \le u(x_0) + \frac{\varepsilon}{3}.$$

Thus, finally, we have:

$$(f, \varphi_0) = (u_{i_0}, \varphi_0) \le u_{i_0}(x_0) + \varphi_0(x_0) \le u(x_0) + \varphi_0(x_0) + \frac{\varepsilon}{3}$$
$$\le (u, \varphi_0) + \frac{2\varepsilon}{3} \le (f, \varphi_0) - \frac{\varepsilon}{3}$$

The inconsistency of the above shows that u is inf-ϕ- equivalent to f.

(b) For any $\varphi \in \phi$, we have:

$$(f, \varphi) - \varphi(x) = \inf_{y \in \Omega} \{f(y) + \varphi(y)\} - \varphi(x) \le f(x) + \varphi(x) - \varphi(x) = f(x),$$

therefore:

$$P_\phi f(x) = \sup_{\varphi \in \phi} \{(f, \varphi) - \varphi(x)\} \le f(x),$$

from which we deduce:

$$(P_\phi f, \varphi) \le (f, \varphi).$$

Conversely, we have:

$$\begin{aligned}(P_\phi f, \varphi) &= \inf_{x\in\Omega}\{P_\phi f(x) + \varphi(x)\} \\ &= \inf_{x\in\Omega}\left\{\sup_{\varphi_1\in\phi}\left[(f, \varphi_1) - \varphi_1(x)\right] + \varphi(x)\right\} \\ &\le \inf_{x\in\Omega}\{(f, \varphi) - \varphi(x) + \varphi(x)\} = \inf_{x\in\Omega}(f, \varphi) = (f, \varphi),\end{aligned}$$

which completes the proof that $P_\Phi f$ is inf-ϕ- equivalent to f.

(c) As $P_\Phi f$ is inf-ϕ- equivalent to f, to show that $u = P_\phi f$ it suffices to show that there cannot exist $x_0 \in \Omega$ such that

$$u(x_0) < P_\phi f(x_0)$$

Let us assume that there exists such an x_0 and let us set

$$\varepsilon = P_\phi f(x_0) - u(x_0) > 0$$

As u is inf-ϕ- equivalent to f, we have, as in (b), $P_\phi u(x) \le u(x)$ and as u is the smallest element of the set of functions inf-ϕ- equivalent to f, $P_\phi u = u$.

By definition of $P_\phi f(x_0)$ there then exists $\varphi_0 \in \phi$ such that:

$$P_\phi f(x_0) \le (f, \varphi_0) - \varphi_0(x_0) + \frac{\varepsilon}{2}$$

From the above we deduce:

$$\begin{aligned}(f, \varphi_0) - \varphi_0(x_0) + \frac{\varepsilon}{2} \ge P_\phi f(x_0) &= u(x_0) + \varepsilon = P_\phi u(x_0) + \varepsilon \\ &\ge (u, \varphi_0) - \varphi_0(x_0) + \varepsilon \\ &= (f, \varphi_0) - \varphi_0(x_0) + \varepsilon\end{aligned}$$

the inconsistency of the above implies that $u = P_\phi f$. □

This *smallest element* of the equivalence class of proper functions inf-ϕ- equivalent to f as defined by the equation $u \overset{\phi}{\approx} f$, will be referred to as the inf-ϕ- representative.

Let us now study some particular classes of test functions.

Example 2.3. Let Δ be the set of Dirac inf-functions δ_{x_0} defined as:

$$\delta_{x_0}(x) = \begin{Bmatrix} 0 & \text{if } x = x_0 \\ +\infty & \text{otherwise} \end{Bmatrix}$$

We then have:

$$(f, \delta_{x_0}) = \inf_x \left\{f(x) + \delta_{x_0}(x)\right\} = f(x_0)$$

and the *inf-Δ-equivalence* corresponds to the classical *pointwise equality* between functions.

The following examples concern lsc functions whose definition and some related properties are recalled [see Berge (1959) for further developments on these lsc functions].

Definition 2.4. *A function* f *is lower semi-continuous (lsc) at* x *if, for every sequence* x_n *converging towards* x, *we have:*

$$\liminf_{x_n \to x} f(x_n) \geq f(x)$$

We recall that the limit inferior (lim inf) of a sequence of reals corresponds to the smallest accumulation point of this sequence.

Proposition 2.5. *Let* f *be a function:* $X \to \overline{\mathbb{R}}$. *The following conditions are equivalent:*

(i) f is lsc for all $x \in X$,
(ii) the set $\{x | f(x) \leq \alpha\}$ *is closed for any* $\alpha \in \mathbb{R}$,
(iii) the epigraph of f, $\{(x, \alpha) | f(x) \leq \alpha\}$ *is a closed set of* $X \times \mathbb{R}$. □

Definition 2.6. *For any function* f: $X \to \overline{\mathbb{R}}$, *there exists a largest lsc function with* f *as upper bound.*

We will refer to it as the lsc closure of f *and we will denote it* f_**:*

$$f_*(x) = \sup\{g(x) \text{: g lsc and } g \leq f\}$$

In the same way, there exists a smallest upper semi-continuous (usc) function with f *as lower bound; this is the usc closure of* f, *which we will denote* f^* :

$$f^*(x) = \inf\{g(x) \text{: g usc and } g \geq f\}$$

Remark 2.7. We verify that f_* is the lsc closure of f if and only if, for any x, we have the two properties:

$$\forall x_n \to x, \liminf f(x_n) \geq f_*(x).$$
$$\exists \bar{x}_n \to x \text{ such that } \liminf f(\bar{x}_n) = f_*(x). \;||$$

For $\rho > 0$, let us denote:

$$_\rho f(x) = \inf\{f(y) \text{: } |y - x| < \rho\},$$
$$^\rho f(x) = \sup\{f(y) \text{: } |y - x| < \rho\}.$$

Proposition 2.8. $_\rho f(x)$ *(resp.* $^\rho f(x)$*) is nonincreasing (resp. nondecreasing) with respect to* ρ, *and for every* $\rho > 0$:

$$_\rho f(x) \leq f_*(x) \leq f(x) \leq f^*(x) \leq {}^\rho f(x)$$

and

$$f_*(x) = \lim_{\rho \to 0^+} {}_\rho f(x) = \sup_{\rho > 0} {}_\rho f(x), \quad f^*(x) = \lim_{\rho \to 0^+} {}^\rho f(x) = \inf_{\rho > 0} {}^\rho f(x). \quad \square$$

Remark 2.9. The continuity of f in x can be expressed as:

$$\forall\varepsilon > 0, \quad \exists\rho(x,\varepsilon) \quad \text{such that} \quad {}^{\rho}f(x) - \varepsilon \le f(x) \le_{\rho} f(x) + \varepsilon.$$

The lower semi-continuity of f in x can be expressed as:

$$\forall\varepsilon > 0, \quad \exists\rho(x,\varepsilon) \quad \text{such that} \quad f(x) \le_{\rho} f(x) + \varepsilon.$$

Example 2.10. Let $\tilde{\Delta}$ be the set of functions $\delta_{y,\rho}$ defined as:

$$\delta_{y,\rho}(x) = \begin{Bmatrix} 0 & \text{if } |x-y| < \rho \\ +\infty & \text{otherwise} \end{Bmatrix}$$

We then have

$$(f, \delta_{x_0,\rho}) = \inf\{f(x): \ |x - x_0| < \rho\} = {}_{\rho}f(x_0).$$

Proposition 2.11. *Two functions* f *and* g *are said to be inf-*$\tilde{\Delta}$*-equivalent if for every* x*:*

$${}_{\rho}f(x) = {}_{\rho}g(x) \forall\rho > 0, \forall x$$

this then yields:

$$P_{\tilde{\Delta}}f = f_* = g_* = P_{\tilde{\Delta}}g.$$

Proof. For every $\rho > 0$, we have:

$$\begin{aligned} P_{\tilde{\Delta}}f(x) &= \sup_{\rho,y}\left\{(f, \delta_{y,\rho}) - \delta_{y,\rho}(x)\right\} \\ &= \sup_{\rho,y}\left\{{}_{\rho}f(y) - \delta_{y,\rho}(x)\right\} = \sup_{y/|x-y|<\rho} {}_{\rho}f(y). \end{aligned}$$

Since, if $|x - y| < \rho$:

$${}_{2\rho}f(x) \le {}_{\rho}f(x)$$

and $\sup\limits_{\rho>0} {}_{2\rho}f(x) = f_*(x)$, we have:

$$f_*(x) \le P_{\tilde{\Delta}}f.$$

On the other hand,

$$P_{\tilde{\Delta}}f = \sup_{|x-y|<\rho}\{\rho f(y)\} \ge \sup_{\rho} {}_{\rho}f(x) = f_*(x)$$

from which we deduce:

$$P_{\tilde{\Delta}}f = f_*. \quad \square$$

Example 2.12. Let C be the set of continuous functions on Ω. ||

Proposition 2.13. *The smallest element inf-*C*-equivalent to a proper function* f *is* f_*, *its lsc closure. We therefore have* $P_C f = f_*$.

From the above we deduce that two functions f *and* g *are inf-*C*-equivalent if and only if they have the same lsc closure:* $f_* = g_*$.

Proof.

$$P_C f(x) = \sup_{\varphi \in C} \{(f, \varphi) - \varphi(x)\}$$

Now $(f, \varphi) - \varphi(x)$ is a continuous function bounded from above by f(x) and of the form $-\varphi(x) +$ constant. From this we deduce that:

$$P_C\, f(x) = \sup_{\psi} \{\psi(x): \psi \in C \quad \text{and} \quad \psi(x) \le f(x)\}$$

is the largest continuous function bounded from above by f, and therefore that $P_C f$ is the lsc closure of f. □

Remark 2.14. We obtain the same result by taking the lsc functions instead of the continuous functions as the set of test functions.

Indeed, in this case $(f, \varphi) - \varphi(x)$ is the largest lsc function of the form $-\varphi(x) +$ constant and bounded from above by f(x); see for example Moreau (1970).

Therefore, we have:

$$P_{lsc} f(x) = f_*.$$

Definition 2.15. *For any function* $f: \Omega \to \overline{\mathbb{R}}$, *we refer to as the inf and sup Moreau-Yosida transforms the functions* f_λ *and* f^λ *defined for* $\lambda > 0$ *as:*

$$f_\lambda(x) = \inf_{y \in \Omega} \left\{ f(y) + \frac{1}{2\lambda} |x - y|^2 \right\},$$

$$f^\lambda(x) = \sup_{y \in \Omega} \left\{ f(y) - \frac{1}{2\lambda} |x - y|^2 \right\}.$$

We have the following classical properties:

$$f_\lambda(x) \le f_*(x) \le f(x) \le f^*(x) \le f^\lambda(x),$$

$f_\lambda(x)$ (resp. $f^\lambda(x)$) is nonincreasing (resp. nondecreasing) with λ and

$$\lim_{\lambda \to 0+} f_\lambda(x) = \sup_{\lambda > 0} f_\lambda(x) = f_*(x); \qquad \lim_{\lambda \to 0+} f^\lambda(x) = \inf_{\lambda > 0} f^\lambda(x) = f^*(x).$$

Let us recall another property of the Moreau-Yosida transform which is extensively used in nondifferentiable optimization:

$$\forall \lambda > 0: \inf_{x \in \Omega} f(x) = \inf_{x \in \Omega} f_\lambda(x)$$

and, if f is lsc convex, f_λ is differentiable in any $x \in \Omega$ with

$$\nabla f_\lambda(x) = \frac{1}{\lambda}(x - \bar{y}(\lambda)) \quad \text{and} \quad \bar{y}(\lambda) = \arg\min_y \left\{ f(y) + \frac{1}{2\lambda} |x - y|^2 \right\}$$

(see Exercise 3 at the end of the present chapter).

Thus, at least in theory, one can replace the minimization of f which is not everywhere differentiable with the minimization of f_λ which is differentiable. This is the basis of a new class of efficient optimization algorithms for lsc convex functions, namely proximal algorithms, see Exercise 4.

Example 2.16. Let Q be the set of quadratic (strictly) convex functions on Ω.

Proposition 2.17. *The smallest element inf-*Q*-equivalent to a proper function* f *is* f_*, *its lsc closure.*

Proof. The functions $\varphi_{\lambda,y}(x) = \frac{1}{2\lambda}|x-y|^2$ form a special class $\widetilde{Q}$ of Q and:

$$(f, \varphi_{\lambda,y}) = \inf_{x\in\Omega}\left\{f(x) + \frac{1}{2\lambda}|x-y|^2\right\} = f_\lambda(y).$$

According to Definition 2.2, if $\phi_1 \subset \phi_2$ then $P_{\phi_1} f \leq P_{\phi_2} f$.

Since $\widetilde{Q} \subset Q \subset C$, this yields:

$$P_{\widetilde{Q}}\, f(x) \leq P_Q\, f(x) \leq P_C\, f(x) = f_*(x)$$

(according to Proposition 2.13). Now:

$$P_{\widetilde{Q}} f(x) = \sup_{\lambda>0, y\in\Omega}\left[f_\lambda(y) - \frac{1}{2\lambda}|x-y|^2\right] \geq \sup_{\lambda>0}[f_\lambda(x)] = f_*(x)$$

from which we deduce: $P_{\widetilde{Q}} f = P_Q\, f = f_*$. □

Example 2.18. Let $\widetilde{C}$ be the set of continuous and inf-compact functions, i.e. such that the level set $\{x|\varphi(x) \leq \lambda\}$ is compact for any finite λ.

This yields $\widetilde{Q} \subset \widetilde{C} \subset C$. We deduce $P_{\widetilde{Q}}\, f \leq P_{\widetilde{C}}\, f \leq P_C\, f$ and as

$$P_{\widetilde{Q}} f = P_C f = f_*, \quad \text{this implies} \quad P_{\widetilde{C}} f = f_*.$$

In the above examples, we considered the functions of an open set Ω of X in $\overline{\mathbb{R}}$. For the following example, we consider the case of functions: $X \to \overline{\mathbb{R}}$.

Definition 2.19. *For any proper function* $f\colon X \to \overline{\mathbb{R}}$, *we refer to as the Legendre–Fenchel transform the function* $\hat{f} = F(f)\colon X^* \to \overline{\mathbb{R}}$ *defined as:*

$$\hat{f}(y) = F(f)(y) = \sup_{x\in X}\{\langle x, y\rangle - f(x)\}.$$

This Legendre–Fenchel (or Fenchel) transform corresponds to the analogue of the Fourier transform when one goes from the space of the functions $L^2(\mathbb{R}^n)$ to the space of the lsc convex proper functions. For more details on this transform, see Rockafellar (1970), Attouch and Wets (1986) and Exercise 3.

The property to be highlighted here is that F(F(f)) is none other than the convex lsc closure of f which we will denote f_{**}. The Legendre–Fenchel transform is therefore an involution on the set of convex lsc proper functions.

Example 2.20. Let L be the set of continuous linear functions on X.

Proposition 2.21. *Two functions are inf-*L*-equivalent if they have the same Legendre–Fenchel transform and the smallest element inf-*L*-equivalent to a given proper function* f *is* f_{**}, *its lsc convex closure,* $P_L f = f_{**}$.

Proof. If $\varphi(x) \in L$ is written $\varphi(x) = -\langle p_\varphi, x\rangle$ with $p_\varphi \in X^*$, then:

$$(f, \varphi) = \inf_x \left(f(x) - \langle p_\varphi, x\rangle\right) = -\sup_x \left(\langle p_\varphi, x\rangle - f(x)\right) = -\hat{f}(p_\varphi)$$

where $\hat{f}$ is the Fenchel transform of f; this yields the first part of the proposition.

Furthermore, we have $P_L f(x) = \sup_{p_\varphi \in X^*} \left[-\hat{f}(p_\varphi) + \langle p_\varphi, x\rangle\right]$, an expression equal to f_{**}, the lsc convex closure of f, $f_{**}(x) = \sup\{\psi(x) : \psi(x) \leq f(x)$, and $\psi(x)$ convex and lsc$\}$, see for example Rockafellar (1970), Sect. 12. □

Remark 2.22. By observing that the Hamiltonian of a physical system is equal to the Legendre–Fenchel transform of its Lagrangian, one can say that *two Lagrangians* are inf-L-equivalent if they have the same Hamiltonian. One can therefore understand the importance of this inf-L-equivalence in Physics.

An interesting problem is to determine the set of *various distinct equivalence classes* which one can define from *MINPLUS* analysis. We have highlighted three of them: Example 2.3 with pointwise equality, the Examples 2.10, 2.12 and 2.16 corresponding to equality of lsc closures and Example 2.18 corresponding to the equality of lsc convex closures.

3. Wavelets in MINPLUS Analysis

Through wavelet transform, as introduced by Morlet (1983), we can analyze signals presenting several characteristic scales.

This wavelet transform of a function $f: \mathbb{R} \to \mathbb{R}$ is given for any $a \in \mathbb{R}_+$ and $b \in \mathbb{R}$ by:

$$T_f(a, b) = \frac{1}{\sqrt{a}} \int_{-\infty}^{+\infty} f(x)\Psi\left(\frac{x-b}{a}\right) dx$$

where ψ is the mother "wavelet" function also referred to as "analyzing function"; this is a function with zero mean and featuring some oscillations, see Grossmann and Morlet (1984).

These wavelet transforms thus enable a multiresolution analysis of L^2 functions, see for example Mallat (1989) and Meyer (1992). We discuss below a *multiresolution analysis of lsc functions* thanks to the introduction of new transforms analogous to the wavelets, but in a nonlinear framework.

Definition 3.1. *The* MINPLUS-wavelet transform *of a function* $f: \mathbb{R}^n \to \mathbb{R}$ *is given for any* $a \in \mathbb{R}_+$ *and* $b \in \mathbb{R}^n$ *by:*

$$T_f(a, b) = \inf_{x \in \mathbb{R}^n} \left(f(x) + h\left(\frac{b-x}{a}\right)\right)$$

where h *is an inf-compact usc function which, similarly to classical wavelet transforms, will be referred to as an "analyzing" function.*

Remark 3.2. Examples 2.10 and 2.16 correspond to the cases where the set ϕ of test functions is defined from an "analyzing" function h, equal respectively to

$$h(x) = \delta_{0,1}(x) \quad \text{and} \quad h(x) = \frac{1}{2}\,|x|^2 .$$

Theorem 1 and Propositions 2.11 and 2.21 provide the reconstruction formula for an lsc function f for Examples 2.10 and 2.16:

$$f(x) = \sup_{a\in\mathbb{R}_+, b\in\mathbb{R}^n} \left(T_f(a,b) - h\left(\frac{b-x}{a}\right)\right). \quad ||$$

Remark 3.3. We show that, in addition, we have for $h(x) = \delta_{0,1}(x)$, $h(x) = \frac{1}{2}\,|x|^2$ and $h(x) = |x|$,

$$f(x) = \sup_{a\in\mathbb{R}_+} T_f(a,x). \quad ||$$

In a way similar to the *MINPLUS* analysis, it is possible to introduce the *MAXPLUS* analysis based on the *MAXPLUS* scalar product defined as:

$$(f,g)_+ = \sup_{x\in\mathbb{R}^n} (f(x)+g(x)).$$

We then obtain analogous results, the lsc functions being replaced by usc functions and the convex functions by concave functions.

Finally, by simultaneously using *MINPLUS* analysis and *MAXPLUS* analysis, we now introduce new classes of generalized functions.

Definition 3.4. *For two functions* f *and* g: $\mathbb{R}^n \to \overline{\mathbb{R}}$, *we introduce the scalar biproduct by:*

$$((f,g)) = \{(f,g)_-, (f,g)_+\} = \left\{ \inf_{x\in\mathbb{R}^n}\{f(x)+g(x)\}\,,\ \sup_{x\in\mathbb{R}^n}\{f(x)-g(x)\}\right\}$$

Definition 3.5. *For any family* ϕ *of functions:* $\mathbb{R}^n \to \overline{\mathbb{R}}$, *we define the* ϕ*-equivalence of two functions bounded on* $\mathbb{R}^n$ *as:*

$$f \overset{\phi}{\approx} g \Leftrightarrow ((f,\varphi)) = ((g,\varphi)) \quad \forall\varphi\in\phi$$

We easily verify that the function f and g are $\tilde{C}$-equivalent if and only if $f_* = g_*$ and $f^* = g^*$, and are L-equivalent if and only if $f_{**} = g_{**}$ and $f^{**} = g^{**}$.

The classes of ϕ-equivalent functions can be considered as *distributions* in nonlinear analysis.

With every bounded function f and every usc inf-compact function h we can associate, for every $a \in \mathbb{R}_+$ and $b \in \mathbb{R}^n$ a simultaneous analysis of the lower envelopes of f represented as:

$$T_f^-(a,b) = \inf_{x\in\mathbb{R}^n}\left\{f(x) + h\left(\frac{b-x}{a}\right)\right\}$$

and of the upper envelopes of f represented as:

$$T_f^+(a, b) = \sup_{x \in \mathbb{R}^n} \left\{ f(x) - h\left(\frac{b - x}{a}\right) \right\}.$$

Thus for each of the analyzing functions

$$h(x) = \frac{1}{\alpha}|x|^\alpha \quad \text{with} \quad \alpha \geq 1, \quad \text{and} \quad h(x) = \delta_{0,1}(x)$$

we verify that:

$$T_f^-(a, x) \leq f_*(x) \leq f(x) \leq f^*(x) \leq T_f^+(a, x).$$

We can then define the Inf-Sup-Wavelets transform of a bounded function $f: \mathbb{R}^n \to \mathbb{R}$, for any $a \in \mathbb{R}_+$ and $b \in \mathbb{R}^n$, by the scalar biproduct:

$$T_f(a, b) = \left(\left(f(\cdot), h\left(\frac{\cdot - b}{a}\right) \right) \right) = \left\{ T_f^-(a, b), T_f^+(a, b) \right\}.$$

It is seen that the non invertibility of the MIN operator is compensated for by considering the pair $\{T_f^-, T_f^+\}$.

Now, by considering $W_a f(x) = T_f^+(a, x) - T_f^-(a, x)$, we can analyze the global and local regularity of the function f.

Proposition 3.6. *(Gondran, 1997)*

The function f *is Hölderian of exponent* H, $0 < H \leq 1$, *if and only if there exists a constant* C *such that for any a, we have one of the following conditions:*

(i) $W_a f(x) \leq C a^H$ *if* $h(x) = \delta_{0,1}(x)$

(ii) $W_a f(x) \leq C a^{\frac{H}{\alpha - H}}$ *if* $h(x) = \frac{1}{\alpha}|x|^\alpha$ *and* $\alpha > H$.

Proof. Case (i) is dealt with in Tricot et al. (1988). It corresponds to the limit of $h(x) = \frac{1}{\alpha}|x|^\alpha$ when $\alpha \to +\infty$, $W_a f(x) = {}^a f(x) - {}_a f(x)$ corresponding to the oscillation of f on $\{y: |y - x| < a\}$, named the *a-oscillation* and denoted osc $_a f(x)$. □

We will denote $W_a^\alpha f(x)$ the value of $W_a f(x)$ for $h(x) = \frac{1}{\alpha}|x|^\alpha$ and $\alpha > 1$.

The function f is said to be *fractal* if and only if we have:

$$\lim_{a \to 0^+} \frac{\text{osc}_a f(x)}{2a} = +\infty$$

uniformly with respect to x (see Tricot 1993) or if we have:

$$\lim_{a \to 0^+} \frac{W_a^\alpha f(x)}{2a^{\frac{\alpha}{\alpha - 1}}} = +\infty$$

uniformly with respect to x.

Proposition 3.7. *The function* f *is Hölderian in the point* x_0*, with exponent* H, $0 < H < 1$, *if and only if there exists a constant* C *such that for any* a, *we have one of the following conditions:*

(iii) $W_a f(x) \leq C\left(a^H + |x - x_0|^H\right)$ *if* $h(x) = \delta_{0,1}(x)$

(iv) $W_a f(x) \leq C\left(a^{\frac{H}{a-H}} + |x - x_0|^H\right)$ *if* $h(x) = \frac{1}{\alpha}|x|^\alpha$ and $\alpha > H$.

Here we obtain a converse ((iii) or (iv) implies f Hölderian in x_0) which does not exactly hold with wavelets, see Jaffard (1989).

4. Inf-Convergence in MINPLUS Analysis

We shall see that thanks to the MINPLUS (resp. MAXPLUS) scalar product we can define concepts of weak convergence, then show the equivalence of some of these convergences with the epiconvergence (or Γ-convergence) introduced by De Giorgi (1975) and increasingly used in continuum mechanics (small parameter problems, homogenization of composite environments, thin films, phase transitions and so on) stochastic optimization, theories of optimization and approximation, etc. See for example Attouch (1984), Attouch et al. (1989), Attouch and Thera (1993), and Dal Maso (1993).

Definition 4.1. *For any family* ϕ *of functions:* $\Omega \to \overline{\mathbb{R}}$, *we define the semi-inf-*ϕ*-convergence of a sequence of proper functions* f_n *uniformly bounded from below* $(f_n \geq m, \ \forall n)$ *towards proper* f *as:*

$$\liminf_{n\to+\infty} (f_n, \varphi) = (f, \varphi) \quad \forall \varphi \in \phi$$

Definition 4.2. *For any family* ϕ *of functions:* $\Omega \to \overline{\mathbb{R}}$, *we define the inf-*$\phi$*-convergence of a sequence of proper functions* f_n*, uniformly bounded from below, towards the proper function* f *as:*

$$\lim_{n\to+\infty} (f_n, \varphi) = (f, \varphi) \quad \forall \varphi \in \phi$$

In the case of Example 2.3, where $\phi = \Delta$, the inf-Δ-convergence corresponds exactly to the simple, classical convergence since we then have:

$$(f_n, \delta_y) = f_n(y) \to f(y) = (f, \delta_y) \quad \forall y \in \Omega.$$

Examples 2.10, 2.12, 2.16, 2.18 and 2.20 will lead us to retrieve two important classical convergence concepts in nonlinear analysis: epiconvergence (or Γ-convergence) and the Mosco-epiconvergence.

Definition 4.3. *Given a sequence of proper functions* f_n*, uniformly bounded from below, we call lsc envelope of* f_n*, the function:*

$$\underline{f}(x) = \liminf_{\substack{y\to x \\ n\to+\infty}} f_n(y),$$

the limit inf being taken when y *and* n *tend simultaneously towards* x *and* $+\infty$ *respectively. This lower limit will also be denoted liminf* $_*\mathrm{f_n}$.

Similarly, we call usc envelope of $\mathrm{f_n}$, *assumed to be uniformly upper bounded, the function:*

$$\bar{\mathrm{f}}(\mathrm{x}) = \limsup_{\substack{\mathrm{n}\to+\infty \\ \mathrm{y}\to\mathrm{x}}} \mathrm{f_n}(\mathrm{y})$$

We then prove, see for example Barles (1994), that the following holds:

$$\underline{\mathrm{f}}(\mathrm{x}) = \lim_{\mathrm{j}\to+\infty} \left\{ \inf \left\{ \mathrm{f_n}(\mathrm{y}) \quad \text{with} \quad \mathrm{n} \geq \mathrm{j} \quad \text{and} \quad |\mathrm{x}-\mathrm{y}| \leq \frac{1}{\mathrm{j}} \right\}\right\}.$$

Example 4.4. Consider the sequence of continuous functions $\mathrm{f_n}(\mathrm{x}) = \mathrm{e}^{-\mathrm{n}\,\mathrm{x}^2}$. We have:

$$\liminf_{\mathrm{n}\to\infty} \mathrm{f_n}(\mathrm{x}) = \begin{cases} 0 & \text{if} \quad \mathrm{x} \neq 0 \\ 1 & \text{if} \quad \mathrm{x} = 0 \end{cases}$$

However, the lsc envelope $\underline{\mathrm{f}}(\mathrm{x}) = \mathrm{liminf}_*\mathrm{f_n}$ is equal to 0 for any x. For this example we have a different result for the usc envelope of $\mathrm{f_n}$:

$$\bar{\mathrm{f}}(\mathrm{x}) = \lim\sup{}^*\mathrm{f}(\mathrm{x}) = \begin{Bmatrix} 0 & \text{if} \quad \mathrm{x} \neq 0 \\ 1 & \text{if} \quad \mathrm{x} = 0 \end{Bmatrix} = \limsup_{\mathrm{n}\to+\infty} \mathrm{f_n}(\mathrm{x}).$$

Example 4.5. On $\Omega = [0, 1]$ the sequence of continuous functions $\mathrm{f_n}(\mathrm{x}) = \mathrm{a}(\mathrm{x}) \cos \mathrm{n\,x}$, where a(x) is a continuous bounded function, has as lsc envelope:

$$\underline{\mathrm{f}}(\mathrm{x}) = -\,|\mathrm{a}(\mathrm{x})|$$

and as usc envelope:

$$\bar{\mathrm{f}}(\mathrm{x}) = |\mathrm{a}(\mathrm{x})|\,.$$

Theorem 2. *Every sequence of uniformly lower bounded proper functions* $\mathrm{f_n}$ *semi-inf-*$\widetilde{\mathrm{C}}$ *converges towards* $\underline{\mathrm{f}}$, *the lsc envelope of the* $\mathrm{f_n}$, *i.e. for any* $\varphi \in \widetilde{\mathrm{C}}$,

$$\lim\inf(\mathrm{f_n}, \varphi) = (\underline{\mathrm{f}}, \varphi)$$

Proof. Refer to Maslov (1987a) and Gondran (1996). □

Corollary 4.6. *If the sequence* $\{\mathrm{f_n}\}$ *inf-*$\widetilde{\mathrm{C}}$*-converges towards the lsc function* f, *then* f *is the lsc envelope of* $\{\mathrm{f_n}\}$, *i.e.* $\mathrm{f} \equiv \underline{\mathrm{f}}$.

Proof. Theorem 2 leads to:

$$(\underline{\mathrm{f}}, \varphi) = \lim\inf(\mathrm{f_n}, \varphi) = \lim(\mathrm{f_n}, \varphi) = (\mathrm{f}, \varphi) \quad \forall\varphi \in \widetilde{\mathrm{C}}$$

and Proposition 2.13 gives us the uniqueness $\mathrm{f} = \underline{\mathrm{f}}$. □

Corollary 4.7. *If the sequence* $\mathrm{f_n}$ *inf-*$\widetilde{\mathrm{C}}$*-converges towards* $\underline{\mathrm{f}}$, *then it inf-*$\widetilde{\mathrm{Q}}$*-converges and also inf-*$\widetilde{\Delta}$*-converges towards* $\underline{\mathrm{f}}$.

Proof. This is straightforward, as $\widetilde{Q} \subset \widetilde{C}$ and Theorem 2 is valid for the usc functions and: $\tilde{\Delta} \subset \widetilde{USC}$, where $\widetilde{USC}$ denotes the set of inf-compact usc functions on Ω.

The uniqueness is then deduced respectively from Propositions 2.11 and 2.17. □

Definition 4.8. *(De Giorgi and Attouch 1984; Dal Maso 1993)*

A sequence $\{f_n;\ n \in N\}$ *epiconverges towards* f, *lsc, if for any* x *we have the following two properties:*

(i) For any sequence $\{x_n;\ n \in N\}$ *converging towards* x, *lim inf* $f_n(x_n) \geq f(x)$,
(ii) There exists a sequence $\bar{x}_n$ *converging towards* x *such that* $f(x) \geq$ *lim sup* $f_n(\bar{x}_n)$.

This definition can be again written, denoting $\{S^\alpha\}$ the set of sequences $S^\alpha = \{x_n^\alpha\}$ converging towards x:

$$f(x) = \inf_{\{S^\alpha\}} \left(\liminf_{x_n^\alpha \to x} f_n(x_n) \right)$$

We then have the following theorem:

Theorem 3. *(Gondran 1996,b)*

In a Banach space, the epiconvergence implies the inf-$\widetilde{C}$*-convergence.*

Proof. Given $\varphi \in \widetilde{C}$, it must be shown that (f_n, φ) tends towards (f, φ). Since f is lsc and φ compact, there exists x_0 such that:

$$(f, \varphi) = f(x_0) + \varphi(x_0).$$

Since f_n converges towards f, there exists $\bar{x}_n \to x_0$ such that:

$$\limsup f_n(\bar{x}_n) \leq f(x_0).$$

Now:

$$(f_n, \varphi) \leq f_n(\bar{x}_n) + \varphi(\bar{x}_n)$$

and taking the *lim sup* of this inequality, we find:

$$\limsup\,(f_n, \varphi) \leq \limsup\ [f_n(\bar{x}_n) + \lim \varphi(\bar{x}_n)] \leq f(x_0) + \varphi(x_0)$$

and consequently:

$$\limsup\,(f_n, \varphi) \leq (f, \varphi).$$

The hypothesis $\forall x_n \to x$, $\liminf f_n(x_n) \geq f(x)$ implies that $\underline{f}(x) = \liminf_* f_n(x) \geq f(x)$.

Theorem 2 then implies that:

$$\liminf (f_n, \varphi) = (\underline{f}, \varphi) \geq (f, \varphi).$$

Finally, we clearly have:

$$\lim(f_n, \varphi) = (f, \varphi). \quad \square$$

Theorem 4. *(Gondran 1996,b)*

In $\mathbb{R}^N$*, the inf-*$\widetilde{C}$*-convergence, inf-*$\widetilde{Q}$*-convergence, inf-Q-convergence, inf-*$\tilde{\Delta}$*-convergence coincide with the epiconvergence.*

Proof. (a) First let us show that if f_n inf-$\widetilde{Q}$-converges towards f lsc then $f = \underline{f}$. Theorem 2 shows that $\lim\inf\,(f_n, \varphi) = (\underline{f}, \varphi) \forall\varphi \in \widetilde{C}$, hence $\forall\varphi \subset \widetilde{\widetilde{Q}} \subset \widetilde{C}$. Therefore if $\widetilde{f_n}$ inf-$\widetilde{Q}$-converges towards f lsc, we have $\lim\,(f_n, \varphi) = (\underline{f}, \varphi) = (f, \varphi)\ \forall\varphi \in \widetilde{Q}$ and according to Proposition 2.17 we have uniqueness, i.e. $f = \underline{f}$.

(b) Corollary 2.67 by Attouch (1984) shows that if f_n inf-$\widetilde{Q}$-converges towards $\underline{f}$, then f_n epiconverges towards $\underline{f}$. Theorem 3 then shows that f_n inf-$\widetilde{C}$-converges towards $\underline{f}$. We therefore clearly have the desired equivalence for $\widetilde{Q}$ and Q.

(c) If f_n inf-$\tilde{\Delta}$-converges towards f lsc, then $f = \underline{f}$ (same proof as in a)). It remains to be shown that if f_n inf-$\tilde{\Delta}$-converges towards $\underline{f}$, then f_n epiconverges towards $\underline{f}$.

Let us first show that for every sequence x_n converging towards x_0, lim inf $f_n(x_n) \geq \underline{f}(x_0)$.

We know that $\forall\varepsilon > 0$, there exists ρ_0 such that:

$$(\underline{f}, \delta_{x_0,\rho_0}) \geq {}_{\rho}\underline{f}(x_0) - \frac{\varepsilon}{2}.$$

Indeed, as

$$(\underline{f}, \delta_{x_0,\rho_0}) = {}_{\rho}\underline{f}(x_0) \quad \forall\rho > 0$$

and as ${}_{\rho}\underline{f}(x_0)$ converges towards $\underline{f}(x_0)$ when $\rho \to 0^+$ ($\underline{f}$ being lsc), $\exists\rho_0$ such that: $\rho > \rho_0$ implies:

$${}_{\rho}\underline{f}(x_0) \geq \underline{f}(x_0) - \frac{\varepsilon}{2}.$$

We will denote δ_{x_0,ρ_0} by φ_0:

After Theorem 2, since:

$$\lim\inf\,(f_n, \varphi_0) = (\underline{f}, \varphi_0)$$

as soon as $n \geq N(\varepsilon)$ we have:

$$(f_n, \varphi_0) \geq (\underline{f}, \varphi_0) - \frac{\varepsilon}{2}$$

hence:

$$f_n(x_n) + \varphi_0(x_n) \geq (f_n, \varphi_0) \geq (\underline{f}, \varphi_0) - \frac{\varepsilon}{2} \geq \underline{f}(x_0) - \varepsilon.$$

Taking the inf limit of the inequality we have:

$$\lim\inf f_n(x_n) + \varphi_0(x_0) \geq \underline{f}(x_0) - \varepsilon$$

in other words:

$$\lim\inf f_n(x_n) \geq \underline{f}(x_0) - \varepsilon \quad \forall\varepsilon > 0$$

hence the desired inequality is obtained.

– Let us now show that for any x_0, $\exists$ a sequence $\bar{x}_n$ converging towards x_0 and satisfying:

$$\limsup f_n(\bar{x}_n) \le f(x_0)$$

If $\underline{f}(x_0) = +\infty$, the result is true. If $\underline{f}(x_0) < +\infty$, then for sufficiently large n the $f_n(x)$ are bounded in a neighborhood of x_0.

For all $p \in N$, consider $\delta_{x_{0,\frac{1}{p}}}$. We have $\left(f_n, \delta_{x_0,\frac{1}{p}}\right)$ which converges towards $\left(\underline{f}, \delta_{x_{0,\frac{1}{p}}}\right)$.

For fixed p and $\varepsilon > 0$, $\exists$ therefore $N(p, \varepsilon)$ such that for $n \ge N(p, \varepsilon)$, we have:

$$\left(f_n, \delta_{x_0,\frac{1}{p}}\right) \le \left(\underline{f}, \delta_{x_0,\frac{1}{p}}\right) + \varepsilon.$$

Let $x_{n_p} = \arg\min\left\{f_n(x) + \delta_{x_0,\frac{1}{p}(x)}\right\}$. From this we deduce that:

$$\left|x_{n_p} - x_0\right| < \frac{1}{p}$$

so that $\left(f_n, \delta_{x_0,\frac{1}{p}}\right)$ is not infinite, and therefore we have:

$$f_n(x_{n_p}) = \left(f_n, \delta_{x_0,\frac{1}{p}}\right) \le \left(f_n, \delta_{x_0,\frac{1}{p}}\right) + \varepsilon \le \underline{f}(x_0) + \varepsilon.$$

By a diagonal method, we then extract from the double sequence x_{n_p} a sequence $\bar{x}_n$ converging towards x_0 and such that

$$f_n(\bar{x}_n) \le \underline{f}(x_0) + \varepsilon$$

which yields the desired conclusion. □

Closely related results can be found in Attouch (1984) and in Akian et al. (1995).

These theorems are very important because they form the link between the concepts of inf-convergence and epiconvergence and play a key role in the transfer of many properties studied in nonlinear analysis to MINPLUS analysis.

Conversely, let us demonstrate on an example how one can very simply derive properties on epiconvergence by proofs in MINPLUS analysis.

Proposition 4.9. *The inf-$\widetilde{C}$-convergence is preserved in the Moreau-Yosida transform, i.e.:*

If f_n *inf-$\widetilde{C}$-converges towards f, then* $(f_n)_\lambda$ *inf-$\widetilde{C}$-converges towards* f_λ. □

Proof. First we verify that for every pair of functions g and φ we have $\forall\lambda > 0$, $(f, \varphi_\lambda) = (f_\lambda, \varphi)$. From this we deduce that $\forall\varphi$, $((f_n)_\lambda, \varphi) = (f_n, \varphi_\lambda)$ and that according to the assumption (f_n, φ_λ) converges towards $(f, \varphi_\lambda) = (f_\lambda, \varphi)$; from this we deduce that $((f_n)_\lambda, \varphi)$ converges towards (f_λ, φ). □

Let us now study the case $\phi = L$, in other words the inf-L-convergence.

Proposition 4.10. *The sequence* $\{f_n\}$ *inf-L-converges towards the function* g *if and only if the sequence* $\left\{\hat{f}_n\right\}$ *of the Legendre transforms of* f_n *converges simply towards* $\hat{g}$, *the Legendre transform of* g.

Proof. Straightforward, since:

$$(f_n, -\langle p, x\rangle) = -\hat{f}_n(p). \quad \square$$

Lemma 4.11. *In a Hilbert space, for any proper function* f, *for any linear form* $\varphi(x) = -\langle p_\varphi, x\rangle$ *and for any* $\lambda > 0$, *we have:*

$$(f_\lambda, \varphi) = (f, \varphi) - \frac{\lambda\,|p_\varphi|^2}{2}.$$

Proof.

$$\begin{aligned}
(f_\lambda, \varphi) &= \inf_x\left[\inf_y\left(f(y) + \frac{1}{2\lambda}|x-y|^2\right) - \langle p_\varphi, x\rangle\right]\\
&= \inf_{x,y}\left[f(y) + \frac{1}{2\lambda}|x-y|^2 - \langle p_\varphi, x\rangle\right]\\
&= \inf_y\left[f(y) + \inf_x\left(\frac{1}{2\lambda}|x-y|^2 - \langle p_\phi, x\rangle\right)\right]\\
&= \inf_y\left[f(y) - \langle p_\varphi, y\rangle\right] - \lambda\frac{|p_\varphi|^2}{2} = (f, \varphi) - \lambda\frac{|p_\varphi|^2}{2}. \quad \square
\end{aligned}$$

Proposition 4.12. *In a Hilbert space, the inf-L-convergence of a sequence is equivalent to the inf-L-convergence of the Moreau-Yosida transform, i.e.:*

$$f_n \overset{\inf-L}{\longrightarrow} f \Leftrightarrow (f_n)_\lambda \overset{\inf-L}{\longrightarrow} f_\lambda$$

Proof. Straightforward in view of Lemma 4.11 because for any $\varphi \in L$, $(f_n, \varphi) \to (f, \varphi)$ if and only if $((f_n)_\lambda, \varphi) \to (f_\lambda, \varphi)$. $\square$

Definition 4.13. *(see Attouch 1984)*

In a reflexive Banach space, a sequence f_n *of convex lsc proper functions* Mosco-epiconverges towards f *if in all points* $x \in \Omega$, *the following two properties are satisfied:*

(i) for any sequence x_n *weakly converging (in the Banach topology) towards* x, $\liminf f_n(x_n) \geq f(x)$;
(ii) there exists a sequence $\bar{x}_n$, *strongly converging towards* x, *such that* $f(x) \geq \limsup f_n(\bar{x}_n)$.

Definition 4.14. *(Attouch 1984)*

In a Hilbert space, we will say that a sequence f_n *of convex lsc proper functions is equicoercive if it satisfies the following relation:*

$$f_n(x) \geq c(|x|) \quad \text{with} \quad \lim_{r \to \infty} \frac{c(r)}{r} = +\infty.$$

From the above we deduce the following theorem which corresponds to a reformulation of a theorem by Attouch (1984) in terms of inf-L-convergence.

Theorem 5. *In a Hilbert space, the Mosco-epiconvergence implies the inf-*L*-convergence. Conversely, if the functions* f_{n^*} *are equicovercive and inf-*L*-convergent towards* f*, then the* f_n *Mosco-epiconverge towards* f*.*

Proof. See Attouch (1984) and Gondran (1996b). □

Remark 4.15. One can generalize the MINPLUS analysis to point-to-set maps with values in $\overline{\mathbb{R}}$.

Thus, with the point-to-set map $x \to H(x)$ we can associate the function $\widetilde{H}(x) = \inf\{y : y \in H(x)\}$ and we define the MINPLUS scalar product as:

$$(H, g) = \inf_x \left(\widetilde{H}(x) + g(x)\right) = (\widetilde{H}, g).$$

Such a point-to-set map is lsc if and only if $\widetilde{H}$ is lsc, in other words if for any sequence x_n converging towards x, we have:

$$\liminf_{x_n \to x} H(x_n) \geq \widetilde{H}(x). \ ||$$

The epiconvergence (and the Mosco-epiconvergence) towards an lsc function (resp. convex lsc function) extends in the same way, as well as the *inf-ϕ-convergences* and the equivalence Theorems 3–5.

In addition to the epiconvergence (inf-$\widetilde{C}$-convergence) and the Mosco-epiconvergence (inf-L-convergence), we can define a new type of convergence, which we will call *semi-continuous convergence* (or *semi-convergence*) by simultaneously using the MINPLUS and MAXPLUS dioids, and therefore by using the "scalar biproduct" (Definition 3.4).

Definition 4.16. *For any family* ϕ *of test functions:* $\Omega \to \overline{\mathbb{R}}$*, we define the* ϕ*-convergence towards* f *of a sequence of functions* f_n *uniformly upper and lower bounded by:*

$$\lim_{n \to \infty} ((f_n, \varphi)) = ((f, \varphi)) \quad \forall \varphi \in \phi$$

where, by definition, we set

$$\lim_{n \to \infty} ((f_n, \varphi)) = \{\liminf (f_n, \varphi)_-, \limsup (f_n, \varphi)_+\}.$$

Definition 4.17. *A sequence of functions* f_n *uniformly upper and lower bound semi-converges towards* f *if for any* x *we have the following two properties:*

(i) $\forall x_n \to x, \ \lim\ \inf f_{n_*}(x_n) \geq f_*(x)$,
(ii) $\forall x_n \to x, \ \lim\ \sup f_n^*(x_n) \leq f^*(x)$.

We can then deduce the following result from Theorem 4*:*

Proposition 4.18. *In* $\mathbb{R}^N$*, the* $\tilde{\Delta}$*-convergence, the* $\widetilde{C}$*-convergence, the* $\widetilde{Q}$*-convergence are equivalent to the semi-convergence.*

In the case where the functions f_n *and* f *are continuous, then the semi-convergence is reduced to the following property:*

$$\forall x_n \to x, \qquad \lim_{n \to +\infty} f_n(_n) = f(x).$$

5. Weak Solutions in MINPLUS Analysis, Viscosity Solutions

Let us consider an equation with partial derivatives in u: $\bar{\Omega} \to \mathbb{R}$; for example the following Dirichet problem of second order:

$$\begin{aligned} H(x, u, Du, D^2u) &= 0 \quad \text{in } \Omega \\ u &= g \quad \text{on } \partial\Omega \end{aligned} \tag{5}$$

where Ω is an open set of R^N, $\bar{\Omega} = \Omega \cup \partial\Omega$, H (x, u, p, M) a continuous function on $\bar{\Omega} \times \mathbb{R} \times \mathbb{R}^N \times S^N$, where S^N is the set of symmetric N × N matrices, Du the vector $\left(\frac{\partial u}{\partial x_i}\right)$ and D^2u the matrix $\left(\frac{\partial^2 u}{\partial x_i \partial x_j}\right)$, and g a continuous function on $\partial\Omega$.

Moreover, we assume that H satisfies the following conditions of monotony:

$$H(x, u_1, p, M_1) \leq H(x, u_2, p, M_2)$$

when $u_1 \leq u_2$ and $M_1 \leq M_2$ where the order $\leq$ on S^N is the partial order corresponding to the condition: $M_2 - M_1$ positive definite. The first order equations correspond to the case where H does not depend on D^2u.

We will write the above system in the form:

$$G(x, u, Du, D^2u) = 0 \quad \text{on } \bar{\Omega} \tag{6}$$

where the function G is defined as:

$$G(x, u, p, M) = \begin{cases} H(x, u, p, M) & \text{in } \Omega \\ u(x) - g(x) & \text{on } \partial\Omega. \end{cases}$$

In the case of more general boundary condition of the form:

$$F(x, u, Du) = 0 \quad \text{on } \partial\Omega$$

where F is a continuous function on $\partial\Omega \times \mathbb{R} \times \mathbb{R}^N$ and nondecreasing in u, we set:

$$G(x, u, p, M) = \begin{cases} H(x, u, p, M) & \text{in } \Omega \\ F(x, u, p) & \text{on } \partial\Omega. \end{cases}$$

When H is nonlinear (w.r.t. u, p, M), the solution to such a problem is not differentiable and therefore it is necessary to consider a notion of generalized solution. The solution in the sense of the distributions is not sufficient and not well suited since H is nonlinear. Nor is the other classical answer considering solutions in $W^1_{loc}(\Omega)$ and satisfying (5) almost everywhere suitable here. We recall that $W^{1,p}(\Omega) = \{\upsilon \in L^p(\Omega), D\upsilon \in L^p(\Omega)\}$, and $W^{1,p}_{loc}(\Omega) = \{\upsilon \in W^{1,p}(\theta),\ \forall\theta$ open compact set of $\Omega\}$.

We are going to define several notions of *weak solutions* to (6) and for one of them we will show its connections with the viscosity solutions introduced by Crandall and Lions (1983); see Crandall et al. (1992) and Barles (1994) for a summary on viscosity solutions. For this link, we first introduce the sub and upper-differentials as well as new functional spaces.

Let us begin by recalling the definition of generalized subdifferentials and upper-differentials introduced by De Giorgi et al. (1980) and Lions (1985).

Definition 5.1. *We refer to as the subdifferential of a function* $f\colon \bar{\Omega} \to \overline{\mathbb{R}} = \mathbb{R}\cup\{+\infty\}$ *in the point* $x \in \bar{\Omega}$*, the set, denoted* $\partial^{1,-}_{\bar{\Omega}} f(x)$*, of all* $p \in \mathbb{R}^N$ *such that when* $n \to +\infty$:

$$\forall x_n \to x \text{ in } \bar{\Omega},\ \liminf \frac{f(x_n) - f(x) - (p, x_n - x)}{|x_n - x|} \geq 0;$$

$p \in \partial^{1,-}_{\bar{\Omega}} f(x)$ *is referred to as a subgradient.*

We refer to as the upper-differential of a function $f\colon \bar{\Omega} \to \mathbb{R} \cup \{-\infty\}$ *in the point* $x \in \bar{\Omega}$*, the set, denoted* $\partial^{1,+}_{\bar{\Omega}} f(x)$*, of all* $p \in \mathbb{R}^N$ *such that when* $n \to +\infty$:

$$\forall x_n \to x \text{ in } \bar{\Omega},\ \limsup \frac{f(x_n) - f(x) - (p, x_n - x)}{|x_n - x|} \leq 0;$$

$p \in \partial^{1,+}_{\bar{\Omega}} f(x)$ *is referred to as an upper-gradient.*

We observe that the sub and upper-differentials depend only on the function f for the interior points ($x \in \Omega$) but also depend on $\bar{\Omega}$ for the points of the border ($x \in \partial\Omega$).

We observe that if a function has a subgradient (resp. an upper-gradient) in x, it is lsc (resp. usc) in x.

In the case where f is convex, the above definition of the subdifferential coincides with the standard one defined in convex analysis, see for example Rockafellar (1970) and Aubin and Frankowska (1990), as:

$$\partial f(x) = \left\{p \in \mathbb{R}^N / \forall y\colon f(y) \geq f(x) + \langle p, y - x\rangle\right\}$$

Finally, let us note that a continuous function can have neither a subgradient nor an upper-gradient as in the case of the function $x \to \sqrt{|x|}\sin\left(\frac{1}{x^2}\right)$ extended in 0 by 0.

Definition 5.2. *We will say that an lsc function* $f\colon \theta \to \hat{\mathbb{R}}$ *is order* 1 *lower semi-differentiable, denoted* $LSD^1(\theta)$, *if the function* f *has a non empty subdifferential in any point* $x \in \theta$.

We will say that an usc function $f\colon \theta \to \mathbb{R} \cup \{-\infty\}$ *is order* 1 *upper semi-differentiable, denoted* $USD^1(\theta)$, *if the function* f *has a non empty super-differential in any point* $x \in \theta$.

We will say that a continuous function $f\colon \theta \to \mathbb{R}$ *is order* 1 *semi-differentiable, denoted* $SD^1(\theta)$, *if for any* $x \in \theta$ *the function* f *has an super-gradient or a subgradient, i.e. if* $\partial_\theta^{1,+} f(x) \cup \partial_\theta^{1,-} f(x) \neq \emptyset$.

If the function f *is not continuous, we will say that it is order* 1 *semi-differentiable denoted* $SD^1(\theta)$, *if for any* x *the l.s.c. function* f_* *has a subgradient or the usc closure* f^* *has a super-gradient, in other words if:*

$$\partial_\theta^{1,-} f_*(x) \cup \partial_\theta^{1,+} f^*(x) \neq \emptyset.$$

The class of LSD^1 *functions is important. It of course contains the convex lsc functions and functions such as* $f(x) = x - \lfloor x \rfloor$ ($\lfloor x \rfloor$ = *the integer part of* x). *and*

$$f(x) = \begin{cases} -x & \text{if} \quad x \le 0 \\ 1 & \text{if} \quad x > 0. \end{cases}$$

We easily verify that it also contains semi-convex functions, i.e. the functions f *for which there exists* $k > 0$ *such that* $f(x) + k\,|x|^2$ *is convex.*

More generally, LSD^1 *contains functions of the form* $\sup\limits_{b \in B} g(x, b)$ *where* $g(x, b)$ *is continuous w.r.t.* b *and LSD w.r.t.* x *(and, in particular, differentiable in* x*); other examples will be discussed in Exercise* 8.

Definition 5.3. *We refer to as the order* 2 *subdifferential of a function* $f\colon \bar{\Omega} \to \overline{\mathbb{R}}$ *in the point* $x \in \bar{\Omega}$, *the set, denoted* $\partial_{\bar{\Omega}}^{2,+} f(x)$, *of the pairs* (p, Y) *with* $p \in \mathbb{R}^N$ *and* $Y \in S^N$, *the set of symmetrical* $N \times N$ *matrices, such that:*

$$\forall x_n \to x \text{ in } \bar{\Omega},$$

$$\liminf \frac{f(x_n) - f(x) - \langle p, x_n - x\rangle - \frac{1}{2}\langle Y(s_n - x), x_n - x\rangle}{|x - x_n|^2} \ge 0;$$

(p, Y) *is called the order* 2 *subgradient.*

We refer to as the order 2 *upper-differential of a function* $f\colon \bar{\Omega} \to \mathbb{R} \cup \{-\infty\}$ *in the point* $x \in \bar{\Omega}$, *the set, denoted* $\partial_{\bar{\Omega}}^{2,+}(x)$, *of the pairs* (p, Y) *with* $p \in \mathbb{R}^N$ *and* $Y \in S^N$ *such that:*

$$\limsup \frac{f(x_n) - f(x) - \langle p, x_n - x\rangle - \frac{1}{2}\langle Y(s_n - x), x_n - x\rangle}{|x - x_n|^2} \le 0;$$

(p, Y) *is called the order* 2 *super-gradient.*

As previously, the order 2 sub and upper-differentials only depend on the function f for the interior points ($x \in \Omega$) but also depend on Ω for the border points ($x \in \partial\Omega$), see for example Crandall et al. (1992) who call them sub and upper "jets" of second order and denoted them $J_{\Omega}^{2,+}$ and $J_{\Omega}^{2,-}$. To make the notation easier to use, one can delete the index θ when θ is an open set and the upper index. Thus we will replace $\partial_{\theta}^{1,+}f(x)$ with $\partial^{+}f(x)$ if θ is an open set, and we clearly return, in this case, to the standard notation.

Definition 5.4. *We will say that a function* $f\colon \theta \to \overline{\mathbb{R}}$ *(resp.* $\theta \to \mathbb{R} \cup \{-\infty\}$*) is order 2 lower semi-differentiable, denoted* $LSD^2(\theta)$*, (order 2 upper semi-differentiable, denoted* $USD^2(\theta)$*) if the function f has a nonempty order 2 subdifferential (resp. upper-differential) in any point* $x \in \theta$.

We have of course the inclusions:

$$LSD^2(\theta) \subseteq LSD^1(\theta) \subseteq LSC(\theta) \quad \text{and} \quad USD^2(\theta) \subseteq USD^1(\theta) \subseteq USC(\theta).$$

We will say that a bounded function is order 2 semi-differentiable, denoted $SD^2(\theta)$*, if for any* x*, either the lsc closure* f_* *has an order 2 subgradient, or the usc closure* f^* *has an order 2 upper-gradient, i.e. if* $\partial_{\theta}^{2,-} f_*(x) \cup \partial_{\theta}^{2,+} f^*(x) \neq \emptyset$.

Proposition 5.5. *If* f_1 *and* $f_2 \in LSD^1(\theta)$ *(resp.* $\in LSD^2(\theta)$*) then* $f = \sup\{f_1, f_2\}$, $\in LSD^1(\theta)$ *(resp.* $\sup\{f_1, f_2\} \in LSD^2(\theta)$*).*

Proof. In the point $x \in \mathbb{R}^N$ we have either $f(x) = f_1(x)$ or $f(x) = f_2(x)$. Let us consider the first case. We then have for any $p_1 \in \partial^{1,-}f_1(x)$ and for any $y \neq x$,

$$\frac{f(y) - f(x) - \langle p_1, y - x\rangle}{|y - x|} \geq \frac{f_1(y) - f_1(x) - \langle p_1, y - x\rangle}{|y - x|}$$

and therefore

$$\liminf_{y \to x} \frac{f(y) - f(x) - \langle p_1, y - x\rangle}{|y - x|} \geq \liminf_{y \to x} \frac{f_1(y) - f_1(x) - \langle p_1, y - x\rangle}{|y - x|} \geq 0$$

which shows that $p_1 \in \partial^{1,-} f(x)$ and thus that $\partial^{1,-} f(x)$ is not empty. □

$LSD(\theta), LSD^1(\theta)$ and $LSD^2(\theta)$ (resp. $USD(\theta), USD^1(\theta), USD^2(\theta)$) will consequently play a similar role to the spaces $L^2(\theta), H^1(\theta)$ and $H^2(\theta)$ of Hilbertian analysis.

We will use the semi-inf-Φ-convergence defined in §4, Definition 4.1 and the set $\widetilde{C}$ of inf-compact continuous functions.

Definition 5.6. *We will say that a function* u *is a semi-inf-*$\widetilde{C}$*-solution (resp. semi-sup-*$\widetilde{C}$*-solution) of (6) if there exists a sequence of functions* $u_n \in LSD^2(\bar{\Omega})$ *(resp.* $u^n \in USD^2(\bar{\Omega})$*) such that* u_n *inf-*$\widetilde{C}$*-converges (resp.* u^n *sup-*$\widetilde{C}$*-converges) towards* u *and such that the point-to-set map* $H(x, u_n(x), p_n, |rmY_n)$ *(resp.* $H(x, u^n(x), p^n, Y^n)$ *with* $(p_n, Y_n) \in \partial_{\bar{\Omega}}^{2,+} u_n(x)$*) (resp.* $(p^n, Y^n) \in \partial_{\bar{\Omega}}^{2,-} u^n(x)$*) semi-inf-*$\widetilde{C}$*-converges (resp. semi-sup-*$\widetilde{C}$*-converges) towards* 0.

A $\widetilde{C}$*-solution is at the same time a semi-inf-*$\widetilde{C}$*-solution and a semi-sup-*$\widetilde{C}$*-solution.*

Theorem 6 will link the various classes of solutions to the viscosity solutions whose definition is recalled in the general case of discontinuous viscosity solutions.

We denote G_* and G^* the lsc and usc closures of G(x, u, p, M) with respect to the variables x, u, p, M and we will assume that G_* and G^* are continuous w.r.t. M.

For the case of (5) we have:

$$G_*(x, u, p, M) = \begin{cases} H(x, u, p, M) & \text{in } \Omega \\ \min(H(x, u, p, M), u - g) & \text{on } \partial\Omega \end{cases}$$

$$G^*(x, u, p, M) = \begin{cases} H(x, u, p, M) & \text{in } \Omega \\ \max(H(x, u, p, M), u - g) & \text{on } \partial\Omega \end{cases}$$

Definition 5.7. *(Crandall et al. 1992)*

A viscosity subsolution to

$$G(x, u, Du, D^2u) = 0 \quad \text{on } \bar{\Omega}$$

is a locally bounded function u *such that:*

$$G_*(u, u^*(x), p, M) \leq 0 \quad \forall x \in \bar{\Omega} \quad \text{and} \quad (p, M) \in \partial^{2,+}_{\bar{\Omega}} u^*(x)$$

where $\partial^{2,+}_{\bar{\Omega}} u^*(x)$ *is the order* 2 *upper-differential of* u^* *(the usc closure of* u*) in* x.

A viscosity upper-solution to (6) *is a locally bounded function* u *such that:*

$$G^*(u, u_*(x), p, M) \geq 0 \quad \forall x \in \bar{\Omega} \quad \text{and} \quad (p, M) \in \partial^{2,-}_{\bar{\Omega}} u_*(x).$$

Finally, a viscosity solution to (6) *is both a viscosity subsolution and upper-solution to* (6).

Theorem 6. *If u is a semi-inf-$\widetilde{C}$-solution (resp. semi-sup-$\widetilde{C}$-solution, $\widetilde{C}$-solution) of* (6), *then* u *is a viscosity upper-solution (resp. subsolution, viscosity solution) of* (6). *Conversely, if* u *is a viscosity upper-solution (resp. subsolution, viscosity solution) of* (6) *and if* G *is continuous w.r.t.* u, p, M *and uniformly continuous w.r.t.* x, *then* u *is a semi-inf-$\widetilde{C}$-solution (resp. semi-sup-$\widetilde{C}$-solution, $\widetilde{C}$-solution).*

Proof. (see Gondran 1998) □

We will now study a few special viscosity solutions, the *inf-ϕ-solutions* and the *episolutions*.

Definition 5.8. *For any family* ϕ *of function:* $\bar{\Omega} \to \overline{\mathbb{R}}$, *we will say that* u *is an inf-ϕ-solution (resp. sup-ϕ-solution) of* (6) *if, for the sequence of functions* $u_n(x) = \inf_y \left(u(y) + \frac{n}{2}|y - x|^2\right)$ *(resp.* $u^n(x) = \sup_y \left(u(y) - \frac{n}{2}|y - x|^2\right)$*, the point-to-set map* $G(x, u_n(x), p_n, Y_n)$ *(resp.* $G(x, u^n(x), p^n, Y^n)$ *with* $(p_n, Y_n) \in \partial^{2,-}_{\bar{\Omega}} u_n(x)$ *(resp.* $(p^n, Y^n) \in \partial^{2,+}_{\bar{\Omega}} u^n(x)$*) inf-Φ-converges (resp. sup-Φ-converges) towards* 0.

Definition 5.9. *A proper function* $u \in LSD^2(\bar{\Omega})$ *is an* episolution *to* (6) *if and only if, for every* $x \in \bar{\Omega}$*:*

$$\forall (p, Y) \in \partial^{2,-}_{\bar{\Omega}} u(x), \quad G(x, u, p, Y) \geq 0$$
$$\exists (p, Y) \in \partial^{2,-}_{\bar{\Omega}} u(x) \quad \text{such that} \quad G(x, u, p, Y) = 0.$$

An upper-bounded function $u \in USD^2(\bar{\Omega})$ *is a* hyposolution *to* (6) *if and only if for every* $x \in \bar{\Omega}$:

$$\forall (p, Y) \in \partial^{2,+}_{\bar{\Omega}} u(x), \quad G(x, u, p, Y) \leq 0$$
$$\exists (p, Y) \in \partial^{2,+}_{\bar{\Omega}} u(x) \quad \text{such that} \quad G(x, u, p, Y) = 0.$$

We verify that an episolution (resp. hyposolution) is a viscosity solution.

Proposition 5.10. $u \in LSD^2(\Omega)$ *is an inf-$\widetilde{C}$-solution to* (6) *if and only if* u *is an episolution to* (6).

Proof. See Gondran (1998). □

6. Explicit Solutions to Nonlinear PDEs in MINPLUS Analysis

We recall a few classical results for explicit solutions to the first order Hamilton-Jacobi equation. For a summary and the proofs of these results, one can refer to Lions (1982).

The objective is to show that such solutions can be naturally expressed with MINPLUS and MAXPLUS scalar products and the solutions belong to the LSD^1, USD^1, LSD^2 and USD^2 spaces which we have just defined in Sect. 5.

These spaces will play a similar role to the L^2, H^1 and H^2 spaces in Hilbertian analysis.

This will be illustrated successively on the Dirichlet problem and then on the Cauchy problem of the Hamilton–Jacobi equation.

6.1. The Dirichlet Problem for Hamilton–Jacobi

The aim is to find a function $u: \bar{\Omega} \to \bar{R}$ satisfying the equations:

$$\begin{cases} H(x, u, Du) = f & \text{in } \Omega \\ u = g & \text{on } \partial\Omega \end{cases} \tag{7}$$

where Ω is a regular open set of $\mathbb{R}^N$, H is a continuous numerical function on $\Omega \times \mathbb{R} \times \mathbb{R}^N$, generally referred to as the Hamiltonian, $Du = \left(\frac{\partial u}{\partial x_1}, \ldots, \frac{\partial u}{\partial x_N}\right)$ is the gradient u, f a function defined on $\bar{\Omega}$ (closure of Ω) and g a function defined on $\partial\Omega$ (boundary of Ω). In general $f \in (\bar{\Omega})$ and $g \in (\partial\Omega)$.

We will consider several special cases.

In each case, we will define a class of *elementary solutions*, then we will define the general solution from these special solutions.

Case 1. Solving the problem

$$\begin{cases} |\mathrm{Du}| = f(x) & \text{in } (\Omega) \\ u = g(x) & \text{on } (\partial\Omega) \end{cases} \tag{8}$$

where $f \in C(\bar{\Omega})$ and $f \geq 0$ in $\bar{\Omega}$.

Let us define $L(x, y)$ on $\bar{\Omega} \times \bar{\Omega}$ as $L(x, y) = \inf \left\{ \int_0^T f(\varphi(s))\, ds \right\}$, the inf being taken on all the φ and T such that φ is almost everywhere (a.e.) differentiable and such that

$$\varphi(0) = x,\ \varphi(T) = y,\ \left|\frac{d\varphi}{dt}\right| \leq 1 \quad \text{a.e.} \quad \text{in } [0, T],\ \varphi(t) \in \bar{\Omega}\ \forall t \in [0, T]$$

Proposition 6.1.1. *L defined above satisfies the following properties:*

(i) L *is a semi-distance on* $\bar{\Omega}$: $L(x, x) = 0$.

$$L(x, y) = L(y, x) \quad \text{and} \quad L(x, y) \leq L(x, r) + L(r, y) \quad \forall x, y, r \in \bar{\Omega}$$

(ii) $L(., z)$ *is a solution to the problem*

$$\begin{cases} |\mathrm{Du}| = f & \text{in } \Omega \backslash \{z\} \\ u(z) = 0 \end{cases}$$

$L(., z)$ *is a special solution to* (8). *We now turn to show the solutions to* (8) *can be expressed in terms of the special solutions* $L(., z)$.

Proposition 6.1.2. *If the condition:*

$$g(x) - g(y) \leq L(x, y) \quad \forall x, y \in \partial\Omega$$

is satisfied, then:

$$u(x) = \inf_{y \in \partial\Omega} \{g(y) + L(x, y)\}$$

$$v(x) = \sup_{y \in \partial\Omega} \{g(y) - L(y, x)\}$$

are respectively the USD[1] *and* LSD[1] *solutions to* (8). *Furthermore,* u *(resp.* v*) is the maximum element (resp. minimal element) of the set* U *(resp.* V*) of the subsolutions (resp. upper-solutions) to* (8)*:*

$$U = \{w \in W^{1,\infty}(\Omega),\ |Dw| \leq f, \quad \text{a.e. in } \Omega,\ w \leq g \text{ on } \partial\Omega\}$$

$$V = \{w \in W^{1,\infty}(\Omega),\ |Dw| \geq f, \quad \text{a.e. in } \Omega,\ w \geq g \text{ on } \partial\Omega\}$$

We recall that $W^{1,\infty}(\Omega)$ is the set of functions u, bounded and Lipschitzian.

For a proof of Propositions 6.1.1 and 6.1.2 see Lions (1982).

Remark 6.1.3. If Ω is convex and if $f \equiv 1$, we have $L(x, y) = |x - y|$. L is therefore the distance between x and y and this yields:

$$u(x) = \inf_{y \in \partial\Omega} \{g(y) + |x - y|\}$$
$$v(x) = \sup_{y \in \partial\Omega} \{g(y) - |x - y|\}.$$

In dimension 1, problem (8) becomes:

$$\begin{cases} |u'(x)| = 1 & \text{on }]-1, 1[\\ u(-1) = u(1) = 0 \end{cases}$$

and we have the USD[1] and LSD[1] solutions:

$$u(x) = 1 - |x| \quad \text{and} \quad v(x) = |x| - 1.$$

Case 2. Solving the problem

$$\begin{cases} H(Du) = f & \text{in } \Omega \\ u = g & \text{on } \partial\Omega \end{cases} \tag{9}$$

where H is convex, continuous and satisfies $H(p) \to +\infty$ when $p \to +\infty$ ($H(p) \geq \alpha|p| - C$ with α, C positive constants).

In addition, we assume that $f(x) \geq \inf_{p \in \mathbb{R}^N} H(p)$ for any $x \in \bar{\Omega}$.

Let us define L(x,y) on $\bar{\Omega} \times \bar{\Omega}$ as:

$$L(x, y) = \inf \left\{ \int_0^T \left\{ f(\varphi(s)) + \hat{H}\left(-\frac{d\varphi}{ds}\right) \right\} ds \right\}$$

the inf being taken on the set of all the a.e. differentiable φ and T such that:

$\varphi(0) = x$, $\varphi(T) = y$, $H\left(-\frac{d\varphi}{ds}\right) \leq +\infty$ a.e. in $[0, T]$, $\varphi(t) \in \bar{\Omega}$ $\forall t \in [0, T]$ and where $\hat{H}$ is the Legendre–Fenchel transform of H (see Definition 2.19).

Another equivalent definition of L can be:

$$L(x, y) = \inf \left\{ \int_0^1 \max_{H(p)=f(\varphi(t))} \left\langle -\frac{d\varphi}{dt}, p \right\rangle dt \right\}$$

the inf being taken on the set of all the a.e. differentiable φ such that:

$$\varphi(0) = x, \varphi(1) = y, \varphi(t) \in \bar{\Omega} \quad \forall t \in [0, 1], \frac{d\varphi}{dt} \in L^\infty(0, 1).$$

Proposition 6.1.4. L *defined above satisfies:*

$$L(x, x) = 0 \quad \text{and} \quad L(x, y) \leq L(y, r) + L(r, y) \quad \forall x, y, r \in \bar{\Omega}$$

and $L(\cdot, z)$ *(resp.* $L(z, \cdot)$*) is a solution to the problem*

$$\begin{cases} H(Du) = f & \text{in } \Omega \backslash \{z\} \\ u(z) = 0 \end{cases}$$

(resp. $H(-Du) = f$ *in* $\Omega \backslash \{z\}, u(z) = 0$*).*

If the condition

$$g(x) - g(y) \leq L(x, y) \quad \forall x, y \in \partial\Omega$$

is satisfied, then

$$u(x) = \inf_{y \in \partial\Omega} [g(y) + L(x, y)]$$

$$v(x) = \sup_{y \in \partial\Omega} [g(y) - L(y, x)]$$

are respectively USD^1 *and* LSD^1 *solutions to* (9).

(see Lions (1982) for a proof).

Case 3. Solving the problem

$$\begin{cases} H(x, Du) = 0 & \text{in } \Omega \\ u = g & \text{on } \partial\Omega \end{cases} \tag{10}$$

where $H(x, p) \in C(\bar{\Omega} \times \mathbb{R}^N)$ *is convex in* p, *satisfies* $H(x, p) \geq \alpha|p| - C$ *where* α *and* C *are positive constants and where* $\inf_{p \in \mathbb{R}^N} H(x, p) \leq 0$ in $\bar{\Omega}$.

Let us denote $\hat{H}(x, p)$ *the Lagrangian of* $H(x, p)$*, i.e. the Fenchel transform in* p *of* H:

$$\hat{H}(x, q) = \sup_{p \in \mathbb{R}^N} \{\langle p, q \rangle - H(x, p)\}.$$

We define a new function $L(x, y)$ *for* $x, y \in \bar{\Omega}$:

$$L(x, y) = \left\{ \inf \int_0^T \hat{H}\left(\xi, \frac{d\xi}{ds}\right) ds \right\}$$

the inf being taken on all the pairs (T, ξ) *such that* $\xi(0) = x, \xi(T) = y, \xi(t) \in \bar{\Omega}\ \forall t \in [0, T], \dfrac{d\xi}{dt} \in L^\infty(0, T)$ *or, equivalently:*

$$L(x, y) = \inf \left\{ \int_0^1 \max_{H(\xi(t), p) = 0} \left\{ -\left\langle \frac{d\xi}{dt}, p \right\rangle \right\} dt \right\}$$

the inf being taken on the ξ *such that* $\xi(0) = x, \xi(1) = y, \xi(t) \in \bar{\Omega}, \forall t \in [0, 1], \dfrac{d\xi}{dt} \in L^\infty(0, 1)$.

Proposition 6.4 is still valid and for (10) *we have the two following* USD^1 *and* LSD^1 *solutions:*

$$u(x) = \inf_{y \in \partial\Omega} \{g(y) + L(x, y)\}$$

$$v(x) = \sup_{y \in \partial\Omega} \{g(y) - L(y, x)\}.$$

Case 4. Solving the problem

$$\begin{cases} H(x,u,Du) = 0 & \text{in } \Omega \\ u = g & \text{on } \partial\Omega \end{cases} \tag{11}$$

where $H(x,u,p) \in C(\bar{\Omega} \times \mathbb{R} \times \mathbb{R}^N)$ *is convex in* u *and* p*, satisfies the following properties:*

$H(x,u,p) \to +\infty$ *when* $|p| \to +\infty$ *uniformly for* $x \in \bar{\Omega}$ *and bounded* u,

$H(x,u,p)$ *is nondecreasing in* u*, for any* $x \in \bar{\Omega}, p \in \mathbb{R}^N$.

We introduce the Lagrangian $\hat{H}(x,s,q)$*, Fenchel transform in* u *and* p *of* H:

$$\hat{H}(x,s,q) = \sup_{\substack{u\in\mathbb{R} \\ p\in\mathbb{R}^N}} \{su + \langle p,q\rangle - H(x,u,p)\}$$

and we define for $x \in \bar{\Omega}$, $y = \partial\Omega$ *a function:*

$$L(x,y) = \inf\left\{\int_0^T \hat{H}\left(\xi(t), v(t), -\frac{d\xi}{dt}(t)\right)\exp\left\{-\int_0^t v(s)ds\right\}dt \right.$$
$$\left. + g(y)\exp\left\{-\int_0^T v(s)ds\right\}\right\}$$

the inf being taken on all the triples (T, v, ξ) *such that* $\xi(0) = x, \xi(T) = y$,

$$\xi(t) \in \bar{\Omega} \quad \forall t \in [0,T], \frac{d\xi}{dt} \in L^\infty(0,T), v \in L^\infty(0,T).$$

Example 6.1.5. If $H(x,u,p) = H(x,p) = \lambda u$ $(\lambda > 0)$, then $\hat{H}(x,u,p) = \hat{H}(x,p)$ if $u = \lambda$, $\hat{H}(x,u,p) = +\infty$ if $u \neq \lambda$.

In this case we then obtain:

$$L(x,y) = \inf_{(T,\xi)}\left\{\int_0^T \hat{H}\left(\xi(t), -\frac{d\xi}{dt}(t)\right)e^{-\lambda t}dt + g(y)\,e^{-\lambda T}\right\}.$$

Proposition 6.1.6. *Under the conditions of Case 4 on* H, $L(\cdot, z)$ *is a solution to the problem:*

$$\begin{cases} H(x,u,Du) = 0 & \text{in } \Omega\backslash\{z\} \\ u(z) = 0 \end{cases}$$

If the condition

$$g(x) \le L(x,y) \quad \forall x,y \in \partial\Omega$$

is satisfied, then

$$u(x) = \inf_{y\in\partial\Omega} L(x,y)$$

$$v(x) = \sup_{y\in\partial\Omega} L(x,y)$$

are respectively the USD[1] *and* LSD[1] *solutions to* (11). *In addition,* u *(resp.* v*) is the maximum element (resp. minimum element) of the set of subsolutions (resp. upper-solutions) to* (8).

6.2. The Cauchy Problem for Hamilton–Jacobi: The Hopf–Lax Formula

The aim is to find a scalar solution $u(x, t)$ ($(x, t) \in \bar{\Omega}\times]0, T[$) satisfying the equations:

$$\begin{cases} \dfrac{\partial u}{\partial t} + H(x, t, u, Du) = 0 & \text{in } \Omega \times]0, T[\\ u(x, t) = g(x, t) & \text{on } \partial\Omega \times]0, T[\\ u(x, 0) = u_0(x) & \text{in } \Omega \end{cases} \tag{12}$$

If H does not depend exclusively on Du, we will see in this section how to obtain explicit "linear" solutions in the MINPLUS dioid. Other solutions will be given in Sect. 8.

In the case where H depends on u and on Du, we will see in Sect. 7.3 how to obtain explicit "linear" solutions in the MINMAX dioid.

Let us consider the following problem:

$$\begin{cases} \dfrac{\partial u}{\partial t} + H(Du) = f(x) & \text{in } \Omega \times [0, T] \\ u(x, t) = g(x, t) & \text{on } \partial\Omega \times [0, T] \\ u(x, 0) = u_0(x) & \text{in } \Omega \end{cases} \tag{13}$$

where Ω is a regular, bounded, convex open set, $f \in W^{1,\infty}(\Omega)$, $g \in C\,(\partial\Omega\; x[T, 0])$, $u_0 \in C\,(\bar{\Omega})$, $u_0(x) = g(x, 0)$ on $\partial\Omega$, $H \in C\,(\mathbb{R}^N)$, H convex on $\mathbb{R}^N$ and verifying $(H(p) \geq \alpha|p| - C(\alpha > 0)$.

For any s, t such that $0 \leq s < t \leq T$ and $x, y \in \bar{\Omega}$, let us define:

$$L(x, t; y, s) = \inf \left\{ \int_s^1 \left\{ f(\xi(\lambda)) + \hat{H}\left(\frac{d\xi}{d}\lambda(\lambda)\right)\right\} d\lambda \right\}$$

the inf being taken on the ξ such that:

$$\xi(s) = y,\ \xi(t) = x,\ \xi(\lambda) \in \bar{\Omega} \quad \forall\lambda \in [s, t],\ \frac{d\xi}{d\lambda} \in L^\infty(s, t)$$

and where $\hat{H}$, the Lagrangian, is the Fenchel transform of H:

$$\hat{H}(q) = \sup_{p\in\mathbb{R}^N} \{\langle p, q\rangle - H(p)\}.$$

Formally, we have for any solution to (16):

$$\begin{aligned} u(x, t) - u(y, s) &= \int_s^t \frac{d}{d\lambda}\{u(\xi(\lambda), \lambda)\}\, d\lambda \\ &= \int_s^t D_x u(\xi(\lambda), \lambda).\frac{d\xi}{d\lambda} + \frac{\partial u}{\partial t}(\xi(\lambda), \lambda) d\lambda \\ &\leq \int_s^t \left\{ \left(\frac{\partial u}{\partial t} + H(D_x u)\right)(\xi(\lambda), \lambda) + \hat{H}\left(\frac{d\xi}{d\lambda}\right)\right\} d\lambda \\ &\leq L(x, t; y, s). \end{aligned}$$

Theorem 7. *Under the above conditions, we have:*

(i) for any $x \in \bar{\Omega}$, $t > 0$: $L(x, t; y, s) \to 0$ *if* $s \uparrow t$, $y \to x$,

$$L(x, t; y, s) \leq L(x, t; z, \tau) + L(z, \tau; y, s) \ \forall x, y, z \in \bar{\Omega}, \forall 0 < s < \tau < t \leq T,$$

(ii) $L(\cdot, \cdot; y, s)$ *(resp.* $L(x, t; \cdot, \cdot)$*) is a solution to*

$$\frac{\partial u}{\partial t} + H(Du) = f \qquad \text{in } \Omega \times [s, t] \quad \text{and} \quad \lim_{t \downarrow s} u(y, t) = 0$$

$$(\text{resp.} \ -\frac{\partial u}{\partial t} + H(-Du) = f \quad \text{in } \Omega \times [0, t] \quad \text{and} \quad \lim_{s \uparrow t} u(x, s) = 0),$$

(iii) if the following conditions are satisfied:

$$\begin{aligned} g(x, t) - g(y, s) &\leq L(x, t; y, s) \quad \forall (x, t), (y, s) \in \partial\Omega \times [0, T], s < t \\ g(x, t) - u_0(y) &\leq L(x, t; y, 0) \quad \forall (x, t) \in \partial\Omega \times [0, T], y \in \bar{\Omega} \end{aligned}$$

then, for $(x, t) \in \bar{\Omega} \times [0, T]$*:*

$$u(x, t) = \inf \left\{ \inf_{y \in \bar{\Omega}} \{u_0(y) + L(x, t; y, 0)\}, \inf_{\substack{y \in \partial\Omega \\ 0 \leq s < t}} \{g(y, s) + L(x, t; y, s)\} \right\} \tag{14}$$

is the USD[1] *solution to the problem,*

(iv) if the following conditions are satisfied:

$$\begin{aligned} g(x, t) - g(y, s) &\leq L(x, t; y, s) \quad \forall (x, t), (y, s) \in \partial\Omega \times [0, T], s > t \\ u_0(x) - g(y, s) &\leq L(x, 0; y, s) \quad \forall (y, s) \in \partial\Omega \times [0, T], y \in \bar{\Omega} \end{aligned}$$

then, for $(x, t) \in \bar{\Omega} \times [0, T]$

$$v(x, t) = \inf \left\{ \sup_{y \in \bar{\Omega}} \{u_0(y) - L(y, 0; x, t)\}, \sup_{\substack{y \in \partial\Omega \\ 0 \leq s < t}} \{g(y, s) - L(y, s; x, t)\} \right\}$$

is the LSD[1] *solution to the problem*

$$\begin{cases} -\frac{\partial u}{\partial t} + H(-Du) = f(x) & \text{in } \ \Omega \times [0, T] \\ u(x, t) = g(x, t) & \text{on } \partial\Omega \times [0, T] \\ u(x, 0) = u_0(x) & \text{in } \quad \bar{\Omega}. \end{cases}$$

Formula (14) *is interesting because it generalizes to the MINPLUS case the following formula related to the heat transfer equation, see Quadrat (1995):*

$$u(x, t) = \int_{\bar{\Omega}} u_0(y) \, L(x, t; y, 0) \, dy + \int_0^t \int_{\partial\Omega} g(y, s) \, L(x, t; y, s) \, ds \, dy$$

Remark. In the special case where $f \equiv f_0$ constant and where Ω is convex, the following holds:

$$L(x, t; y, s) = f_0(t - s) + (t - s)\hat{H}\left(\frac{x - y}{t - s}\right).$$

If one considers the case where $\Omega = \mathbb{R}^N$ and $f \equiv 0$, then (14) becomes the Hopf (1965) and Lax (1957) formula:

$$u(x, t) = \operatorname*{Inf}_{y \in \mathbb{R}^N}\left(u_0(y) + t\,\hat{H}(\frac{x - y}{t}\right).$$

Let us recall a few special cases of this formula.

The equation:

$$\begin{cases} \dfrac{\partial u}{\partial t} + |Du| = 0 & \text{in } \mathbb{R}^N \times [0, T] \\ u(x, 0) = u_0(x) & \text{in } \mathbb{R}^N \end{cases}$$

has the usc solution:

$$u(x, t) = \inf_{|y - x| < t} u_0(y).$$

The equation:

$$\begin{cases} -\dfrac{\partial u}{\partial t} + |Du| = 0 & \text{in } \mathbb{R}^N \times [0, T] \\ u(x, 0) = u_0(x) & \text{in } \mathbb{R}^N \end{cases}$$

has the lsc solution:

$$v(x, t) = \sup_{|y - x| < t} u_0(y).$$

For any real $p > 0$, the equation:

$$\begin{cases} \dfrac{\partial u}{\partial t} + \dfrac{1}{p}|Du|^p = 0 & \text{in } \mathbb{R}^N \times [0, T] \\ u(x, 0) = u_0(x) & \text{in } \mathbb{R}^N \end{cases}$$

has the solution:

$$v(x, t) = \inf_y \left\{u_0(y) + \frac{|x - y|^p}{p\,t^{1/p}}\right\}.$$

In the case where $p = 2$, then the equation:

$$\begin{cases} \dfrac{\partial u}{\partial t} + \dfrac{1}{2}|Du|^2 = 0 & \text{in } \mathbb{R}^N \times [0, T] \\ u(x, 0) = u_0(x) & \text{in } \mathbb{R}^N \end{cases}$$

has the USD[1] solution:

$$u(x, t) = \inf_y \left\{u_0(y) + \frac{|x - y|^2}{2t}\right\}.$$

and the equation:

$$\begin{cases} -\dfrac{\partial u}{\partial t} + \dfrac{1}{2}|Du|^2 = 0 & \text{in } \mathbb{R}^N \times [0, T] \\ u(x, 0) = u_0(x) & \text{in } \mathbb{R}^N \end{cases}$$

has the LSD[1] solution:

$$v(x, t) = \sup_y \left\{ u_0(y) - \frac{|x - y|^2}{2t} \right\}.$$

7. MINMAX Analysis

We have shown in the previous sections how MINPLUS analysis, based on the "scalar product" $\inf_x (f(x) + g(x))$, is used to synthesize and extend a branch of nonlinear analysis (lsc analysis and convex analysis, epiconvergence and Mosco-epiconvergence, viscosity solutions). In this section we will show that in a similar way one can construct a MINMAX analysis based on the "scalar product" $\inf_x$ (max (f(x), g(x)) and thus extend the previous method to quasi-convex analysis, developed in particular by Crouzeix (1977), Volle (1985), and Elquortobi (1992).

In particular, we introduce the infmax linear transform (see Gondran 1996b) which plays a role analogous in MINMAX analysis to the Legendre–Fenchel transform in MINPLUS analysis. Except for Theorem 9, the proofs are similar to those of MINPLUS analysis.

7.1. *Inf-Solutions and Inf-Wavelets in MINMAX Analysis*

Let X be a Banach space. We will denote |.| is norm, X^* its topological dual and $< \cdot, \cdot >$ the scalar product in the duality X^*, X.

When X is a Hilbert Space, we identify X and X^*.

We refer to as a *proper function* a function f: $X \to \mathbb{R} \cup \{+\infty\}$, lower bounded and non identical to $+\infty$.

Definition 7.1.1. *For two proper functions* f *and* g, *the "MINMAX scalar product"* (f, g) *is defined as:*

$$(f, g) = \inf_{x \in X} \{\max (f(x), g(x))\}$$

In this whole section, we will denote $(\cdot, \cdot)$ *the Minmax scalar product.*

Definition 7.1.2. *For any family* ϕ *of functions:* $X \to \mathbb{R}$ *(test functions), we define the infmax-*ϕ*-equivalence of two functions* f *and* g *by the equalities:*

$$(f, \varphi) = (g, \varphi) \; \forall \varphi \in \phi$$

and the infmax-bi-conjugate of f *with respect to* ϕ, *denoted* $P_\phi f$, *as:*

$$P_\phi f(x) = \sup_{\varphi \in \phi} \{(f, \varphi) / \text{under the condition } \varphi(x) < (f, \varphi)\}.$$

Theorem 8. *The set of functions infmax-ϕ-equivalent to a proper function* f *has a smallest element (inf-solution) equal to* $P_\phi f$.

Proof. (see Gondran 1997) □

Consider a few interesting examples of test functions.

Example 7.1.3. If $\phi = \Delta$, the set of Dirac δ_y inf-functions, defined as $\delta_y(x) = -\infty$ if $x = y$, $+\infty$ otherwise, the *infmax-Δ-equivalence* corresponds to *pointwise equality*.

Example 7.1.4. If $\phi = \tilde{\Delta}$, the set of $\delta_{y\varepsilon}$ functions, defined as:

$$\delta_{y,\varepsilon}(x) = -\infty \quad \text{if} \quad |x - y| < \varepsilon, +\infty \quad \text{otherwise},$$

then $P_{\tilde{\Delta}} f = f_*$, the lsc closure of f, and two functions are *infmax-$\tilde{\Delta}$-equivalent* if and only if:

$$\forall \varepsilon > 0, \quad \forall x \in X: {}_\varepsilon f(x) = {}_\varepsilon g(x).$$

where we recall that ${}_\varepsilon f(x) = \inf\{f(y): |y - x| < \varepsilon\}$

Example 7.1.5. If $\phi = \tilde{C}$, the set of continuous and infcompact functions, then $P_{\tilde{C}} f = f_*$, and two functions are *infmax-$\tilde{C}$-equivalent* if and only if they have the same lsc closure.

We obtain the same result by taking the infcompact usc functions on X.

For any eigenfunction f, we refer to as *infmax-regularized* the function:

$$f_{\lambda,q} = \inf_{y \in X} \left(\max \left(f(y), \frac{1}{2\lambda} |y - x|^2 + q \right) \right)$$

Example 7.1.6. If $\phi = Q$, the set of quadratic functions of the form $\varphi_{\lambda,y,q}(x) = \frac{1}{2\lambda} |y - x|^2 + q$, then for any function f, $P_Q f = f_*$, and two functions f and g are *infmax-Q-equivalent* if and only if all the infmax-regularized functions are pointwise equal.

$$f_{\lambda,q}(x) = g_{\lambda,q}(x) \quad \forall \lambda > 0, \quad \forall q \in \mathbb{R}; \quad \forall x \in X.$$

Definition 7.1.7. *For any proper function* f, *we refer to as* infmax linear transform *the function* $\overset{\circ}{f}(p, q)$ *defined for any* $p \in X$ *and any* $q \in \mathbb{R}$ *as:*

$$\overset{\circ}{f}(p, q) = (f(x), \langle p, x \rangle + q).$$

This transform will play the same role in MINMAX analysis as the Legendre–Fenchel transform in MINPLUS analysis.

Theorem 9. *The quasi convex lsc closure of a function* f, *denoted* $f_{\circledast}$ *is equal to:*

$$f_{\circledast}(x) = \sup_{p \in X^*} \overset{\circ}{f}(p, -\langle p, x\rangle + m)$$

or again to:

$$f_{\circledast}(x) = \sup_{p \in X, q \in \mathbb{R}} \{(f, \langle p, .\rangle + q \quad \text{with} \quad \langle p, x\rangle + q < (f, \langle p, .\rangle + q)\}.$$

where m is a lower bound of f.

The proof (see Gondran 1997) is provided by using the Hahn-Banach theorem and by following a proof by Elquortobi (1992). It is shown that the result does not depend on m.

Example 7.1.8. If A is the set of continuous linear functions on X of the form $\langle p, x\rangle + q$, then for any proper function f, $P_A f = f_{\circledast}$ and two proper functions f a g are infmax-A-equivalent if they have the same infmax linear transform.

Now we have introduced in Sect. 3 the MINPLUS wavelet transforms: likewise we define the MINMAX wavelet transforms for the multi-resolution analysis of lsc functions as:

Definition 7.1.9. *The MINMAX wavelet transform of a function* f *is provided for any* $a \in \mathbb{R}_+, b \in \mathbb{R}^n$ *by:*

$$T_f(h; a, b) = \inf_{x \in \mathbb{R}^n}\left(\max\left(f(x), h\left(\frac{b-x}{a}\right)\right)\right) = \left(f(\cdot), h\left(\frac{b-\cdot}{a}\right)\right)$$

where h *is an inf-compact usc function which will be referred to as an "analyzing" function.*
Examples 7.1.4 *and* 7.1.6 *correspond respectively to* $h(x) = \delta_{0,1}(x)$ *and*

$$h(x) = \frac{1}{2}|x|^2 + q.$$

Theorem 8 *ensures the reconstruction formula of lsc* f *through the formula:*

$$f(x) = \sup_{a \in \mathbb{R}_+, b \in \mathbb{R}^n}\left\{T_f(h; a, b)/\text{under the condition } h\left(\frac{b-x}{a}\right) < T_f(a, b)\right\}.$$

7.2. Inf-Convergence in MINMAX Analysis

Definition 7.2.1. *For any family* ϕ *of test functions, we will say that a sequence of proper functions* f_n *semi infmax-*ϕ*-converges towards* f *if*

$$\liminf\,(f_n, \varphi) = (f, \varphi) \quad \forall \varphi \in \phi.$$

Definition 7.2.2. *A sequence of proper functions* f_n *infmax-*ϕ*-converges towards* f *if*

$$\lim(f_n, \varphi) = (f, \varphi) \quad \forall \varphi \in \phi.$$

In the case of Example 7.1.3, $\phi = \Delta$ and the *infmax-*Δ*-convergence* corresponds exactly to the simple convergence and f_n semi *infmax-*Δ*-converges* towards f if $f(x) = \liminf\ f_n(x)$.

Examples 7.1.4–7.1.6 correspond to one of the main convergences in nonlinear analysis, epiconvergence, and Example 7.1.8 corresponds to a new convergence in nonlinear analysis, comparable to the Mosco-epiconvergence, but for quasi-convex functions.

We will next take into consideration that the f_n sequences are *uniformly locally bounded.*

Theorem 10. *Any sequence of proper functions* f_n *semi infmax-*$\widetilde{C}$*-converges towards* $\underline{f}$*, the lsc envelope of* f_n *being defined by* $f_n(x) = \liminf\ f_n(y)$ *for* $y \to x$ *and* $n \to \infty$*. For any sequence* f_n *converging towards x, we have liminf* $f_n(x_n) \geq \underline{f}(x)$.

If a sequence of proper functions f_n *infmax-*$\widetilde{C}$*-converges towards* f*, then* $f_* = \underline{f}$.

Theorem 11. *In a Banach space, the epiconvergence implies the infmax-*$\widetilde{C}$*-convergence.*

Definition 7.2.3. *A sequence of lsc quasi convex proper functions* f_n *Mosco-epiconverges towards* f*, a quasi convex proper function if, in all points* $x \in X$*, we have:*

- *for any sequence* x_n *converging weakly towards* x, $\liminf f_n(x_n) \geq f(x)$*;*
- *there exists a sequence* x_n *converging strongly towards* x *such that limsup* $f_n(\bar{x}_n) \leq f(x)$.

Theorem 12. *In a Hilbert space, the Mosco-epiconvergence implies the infmax-*A*-convergence and we have the converse if the* f_{n*} *functions are equicoercive.*

Theorem 13. *In* $\mathbb{R}^N$*, the infmax-*$\tilde{\Delta}$*-convergence, the infmax-*$\widetilde{C}$*-convergence and the infmax-Q-convergence are identical to the epiconvergence.*

Theorems 12 and 13 are adaptations of results by Attouch (1984). Results closely related to these theorems can be found in Akian et al. (1994).

Remark 7.2.4. All of the above results also apply to the dioid ($\overline{\mathbb{R}}$, max, min) by replacing lsc with usc, quasi-convex with quasi-concave, inf with sup.

7.3. Explicit Solutions to Nonlinear PDEs in MINMAX Analysis

Let us consider the following Hamilton–Jacobi problem:

$$\begin{cases} \frac{\partial u}{\partial t} + H(u, Du) = 0 & \text{in } \Omega =]0, +\infty[\times \mathbb{R}^N \\ u(x, 0) = g(x) & \text{on } \mathbb{R}^N \end{cases} \tag{15}$$

where the Hamiltonian H(r, p) is continuous in r and p, nondecreasing in r for any $y \in \mathbb{R}^N$, sublinear in p for any $r \in \mathbb{R}$, and where g is Lipschitzian and bounded.

We then show, see Barron et al. (1996) that the only viscosity solution to problem (15) is the function $u(x, t) \in USD^1$ defined as:

$$u(x,t) = \min_{y\in\mathbb{R}^N}\left(\max\left(g(y), h\left(\frac{x-y}{t}\right)\right)\right) \tag{16}$$

where, for any $x \in \mathbb{R}^N, h(x) = \inf(r \in \mathbb{R}/\forall y \in \mathbb{R}^N$, we have $H(r, y) \geq \langle x, y\rangle)$ is the *conjugated function* of H. We verify that h is quasi-convex and lsc on $\mathbb{R}^N$ and that:

$$H(r, p) = \sup(\langle p, q\rangle \text{ such that } h(q) \leqq r).$$

In the case where g is quasi-convex continuous (instead of Lipschitzian and bounded), solution (16) is, in addition, quasi-convex, see Volle (1997).

Let us give a few examples of formula (16).

Definition 7.3.1. *Let us consider the problem:*

$$\frac{\partial u}{\partial t} + e^u\,|Du| = 0 \quad \text{in} \quad \Omega =]0, +\infty[\, \times \mathbb{R}$$
$$u(x, 0) = |x| \quad \text{on } \mathbb{R}.$$

We verify that $h(x) = \log|x|$, *then that the problem has the solution:*

$$u(x,t) = 0 \quad \text{if} \quad |x| \leq t$$
$$u(x,t) = y \quad \text{if} \quad |x| > t$$

where $y > 0$ *is the only solution to* $\frac{x}{t} = e^y + \frac{y}{t}$.

Definition 7.3.2. *Let us consider the problem:*

$$\begin{cases} \dfrac{\partial u}{\partial t} + u\,|Du| = 0 & \text{in } \Omega =]0, +\infty[\, \times \mathbb{R} \\ u(x, 0) = g(x) & \text{on } \mathbb{R} \end{cases}$$

We have $h(x) = |x|$ *and the problem allows the solution:*

$$u(x,t) = \min_{y\in\mathbb{R}}\left(\max g(y), \left|\frac{x-y}{t}\right|\right).$$

7.4. Eigenvalues and Eigenfunctions for Endomorphisms in MINMAX Analysis

An extension into the continuous field of results on eigenvalues and eigenvectors from the discrete case for the dioid $(\mathbb{R} \cup \{+\infty\}, \text{Min}, +)$ was carried out by Dudnikov and Samborskii (1989), see Exercise 3 of Chap. 6.

Here we will present the extension into the continuous field of results obtained on eigenvalues and eigenvectors from the discrete case for the dioid ($\overline{\mathbb{R}}$, Min, Max), see Gondran and Minoux (1997, 1998).

It will be seen (Theorem 14) that in this case one obtains a complete explicit characterization of the eigenvalues and eigenvectors of endomorphisms based on the dioid ($\overline{\mathbb{R}}$, Min, Max).

Let X be a real, reflexive Banach space. Let F be the set of functions: $X \to \overline{\mathbb{R}}$, bounded from below and inf-compact.

By endowing F with the laws $\oplus$ and $\otimes$ defined as:

$$(f \oplus g)(x) = \operatorname{Min}(f(x), g(x)) \qquad \forall f, g \in F, \quad \forall x \in X$$
$$\alpha \otimes f(x) = \operatorname{Max}(\alpha, f(x)) \ \forall f \in F, \quad \forall \alpha \in R, \qquad \forall x \in X$$

F has the structure of a semi-module, called a MinMax functional semi-module, on the dioid ($\overline{\mathbb{R}}$, Min, Max) with as neutral element the function h^{ε} defined as:

$$h^{\varepsilon}(x) = +\infty \quad \forall x \in X.$$

We then define the "scalar product" $\langle f, g\rangle$ of two functions f and g in the MinMax functional semi-module as:

$$\langle f, g\rangle = \operatorname*{Min}_{x \in X} \{\operatorname{Max}(f(x), g(x))\}.$$

Let Δ be the set of inf-compact functions A: $X \times X \to \overline{\mathbb{R}}$, satisfying the following conditions:

(i) there exists $\theta_A > -\infty$ such that $A(x, x) = \theta_A \ \forall x \in X$;
(ii) $A(x, y) \geq \theta_A \ \ \forall x, y \in X \times X$.

Conditions (i) and (ii) correspond in MinMax analysis to the classical concept of diagonal dominance in the sense of the order relation of the dioid $\left(\sum_{y \neq x} A(x, y) \leq A(x, x)\right)$.

We then define the image of $f \in F$, denoted Af, as:

$$Af(x) = \operatorname*{Min}_{y \in X} \{\operatorname{Max}\{A(x, y), f(y)\}\} \quad \forall x \in X.$$

We easily verify that the functional Af is bounded from below and inf-compact and that the mapping $f \to Af$ is "linear" in the dioid ($\overline{\mathbb{R}}$, Min, Max):

$$A(\alpha \otimes f \oplus \beta \otimes g) = \alpha \otimes Af \oplus \beta \otimes Ag \quad \forall f, g \in F, \ \forall \alpha, \beta \in \mathbb{R}.$$

Δ therefore corresponds to a set of endomorphisms on the MinMax functional semi-module. The product of two endomorphisms A and B of Δ is the endomorphism C, denoted AB, defined as:

$$C(x, y) = \operatorname*{Min}_{z \in X} \{\operatorname{Max}\{A(x, z), B(z, y)\}\} \quad \forall (x, y) \in X \times X.$$

One easily verifies that $C \in \Delta$ and that the product is associative.

As $A^2(x, y) \leq A(x, y)$, the sequence $A(x, y), A^2(x, y), \ldots, A^n(x, y)$ is bounded is monotone nonincreasing and bounded from below, and the endomorphism A^* can be defined as:

$$A^*(x, y) = \lim A^n(x, y).$$

We can then state:

Proposition 7.4.1.

$$(A^*)^2 = A^* = AA^* = A^*A.$$

Given a sub-semi-module H of F, we refer to as generator of H any mapping $G \in H^X$ (which with every $z \in X$ associates $G^z \in H \subset F$), such that, for every $\psi \in H$, there exists $\varphi \in F$, such that for any $x \in X$,

$$\psi(x) = \langle \varphi(\cdot), G^{\cdot}(x) \rangle = \operatorname*{Min}_{z \in X} \left\{ \operatorname{Max}\left(\varphi(z), G^z(x)\right) \right\}.$$

The following theorem provides a complete characterization of eigenvalues and eigenvectors for endomorphisms with a "dominant diagonal" in MinMax analysis.

Theorem 14. *(Gondran and Minoux (2007)*

Let $A \in \Delta$ be an endomorphism with a "dominant diagonal." Then any $\lambda > \theta_A$ is an eigenvalue of A, and for any $y \in X$, φ_λ^y defined as:

$$\varphi_\lambda^y(x) = \lambda \otimes A^*(x, y)\ \forall x \in X \text{ is a proper function for the eigenvalue } \lambda.$$

Let $G_\lambda = \bigcup_{y \in X} \{\varphi_\lambda^y\}$ be the set of distinct elements of $\{\varphi_\lambda^y / y \in X\}$, then G_λ is a generator of the MinMax functional semi-module of the set of eigenfunctions F_λ (eigen-semi-module) corresponding to the eigenvalue λ, i.e. every eigenfunction $f \in F_\lambda$ is written:

$$f(x) = \langle h(\cdot), \varphi_\lambda^{\cdot}(x) \rangle \quad \text{with} \quad h \in F.$$

Moreover, G_λ is the unique minimal generator of F_λ.

Proof. Here we only provide a sketch of proof, see Gondran and Minoux (2007) for a more detailed proof.

- First we show that for any λ, φ_λ^y is an eigenfunction.
  Indeed, $A\,\varphi_\lambda^y\,(x) = A(\lambda \otimes A^*(x, y)) = \lambda \otimes AA^*\,(x) = \lambda \otimes A^*\,(x, y) = \lambda\,\varphi_\lambda^y(x)$ taking into account Proposition 7.4.1 and the idempotency of $\otimes$.
- Then we show that for any $\lambda > \theta_A$ and $f \in F$, we have $f = \lambda \otimes f = Af = A^*f$.
  Indeed, $Af = \lambda \otimes f$ implies $\operatorname{Max}(A(x, x), f(x)) \geq \operatorname{Max}(f(x), \lambda)$ which, together with $\lambda > \theta_A$, makes it possible to conclude $f(x) \geq \lambda$ and therefore $f = \lambda \otimes f = Af \geq A^*f$. The reverse inequality $A^*f \geq f$ is obtained by showing that for any $\bar{x}$ and any $\varepsilon > 0$, we can obtain the inequality $A^*f(\bar{x}) + \varepsilon > f(\bar{x})$.

– We deduce from the above result that for any $f \in F_\lambda$ with $\lambda > \theta_A$ we have $f = A^*(\lambda \otimes f) = (\lambda \otimes A^*) f$, i.e.:

$$f(x) = \underset{y\in X}{\text{Min}} \left\{ \text{Max} \left\{ \varphi_\lambda^y(x), f(y) \right\} \right\} = \left\langle f(\cdot), \varphi_\lambda^{\cdot}(x) \right\rangle .$$

– The proof of the minimality and the uniqueness of G_λ is shown through contradiction by assuming that $\varphi_\lambda^y(x)$ decomposes into the form $\varphi_\lambda^y(x) = \underset{z\in Z}{\text{Min}}\{\text{Max}\{h(z),$ $\Psi^z(x)\}\}$ with $\Psi^z \in F_\lambda$ and $\Psi^z \neq \varphi_\lambda^y$.

First one has to show that there exists $z' \in Z$ such that $\varphi_\lambda^y \leq \Psi^{z'}$. It is the one for which $\varphi_\lambda^y = \text{Max}\,\{A^*(y,y), \lambda\} = \text{Max}\,\{\theta_A, \lambda\} = \lambda = \text{Max}\left\{h(z'), \Psi^{z'}(x)\right\}$; indeed, we then have $h(z') \leq \lambda$, $\Psi^{z'} = \lambda \otimes \Psi^{z'}$ and therefore:

$$\varphi_\lambda^y(x) \leq \text{Max}(h(z'), \Psi^{z'}(x)) \leq \text{Max}(\lambda, \Psi^{z'}(x)) = \Psi^{z'}(x).$$

Now, it just remains to show that $\Psi^{z'} \leq \varphi_\lambda^y$; indeed, we have:

$$\begin{aligned}\Psi^{z'}(x) &= \underset{u\in X}{\text{Min}} \left\{ \text{Max} \left\{ \varphi_\lambda^u(x), \Psi^{z'}(u) \right\} \right\} \leq \text{Max} \left\{ \varphi_\lambda^y(x), \Psi^{z'}(y) \right\} \leq \text{Max} \left\{ \varphi_\lambda^y(x), \lambda \right\} \\ &= \varphi_\lambda^y(x) \quad \square\end{aligned}$$

The uniqueness of the minimal generator is an interesting property because it will enable one to provide interpretations to the eigenfunctions of this generator, as was the case with matrices in Chap. 6, Sect. 6: see for example Gondran and Minoux (1998) where generating proper functions G_λ are identified with the aggregates of a of percolation process at threshold λ.

8. The Cramer Transform

Historically, the Cramer transform was introduced to study the theory of large deviations. Let us recall the principle.

Let X_i $(i = 1, 2, \ldots)$ be random independent variables of the same law and uniformly distributed. Then the sequence $S_n = \frac{1}{n} \sum_{i=1,n} X_i$ converges almost definitely towards the mean $\overline{X}$ of the X_i by the strong law of large numbers. The theory of large deviations will give us an estimation of the probability that S_n is close to x, for x different from $\overline{X}$ in O(1) in relation to n.

$$\text{Let } h(n; a, b) \equiv -\mathit{ln}\ \text{Prob}\,\{a < S_n < b\}.$$

The function $h(n + n'; a, b)$ is nonnegative and subadditive, i.e.:

$$h(n + n'; a, b) \leq h(n; a, b) + h(n'; a, b).$$

Indeed, the independence of the X_1 entails:

$$\text{Prob}\left\{a < \frac{1}{n+n'} \sum_{i=1,n+n'} x_i < b\right\}$$

$$\geq \text{Prob}\left\{a < \frac{1}{n} \sum_{i=1,n} X_i < b\right\} . \text{Prob}\left\{a < \frac{1}{n'} \sum_{i=n+1,n+n'} X_i < b\right\}$$

hence the announced relation.

A consequence of the subadditivity is that the limit

$$s(a,b) \equiv \lim_{n\to+\infty} \frac{h(n;a,b)}{n} = \inf_n \frac{h(n;a,b)}{n}$$

exists. Just take two integers n and n_0, do the Euclidian division $n = n_0 q + r$, apply the subadditvity and successively extend n and n_0 towards $+\infty$. We then set:

$$S(x) \equiv \inf_{a<x<b} s(a,b).$$

The function S(x) is positive or zero and it is easy to show that it is convex. The theorem of large deviations is then written:

$$\text{Prob}\left\{x \leq \frac{1}{n} \sum_{i=1,n} X_i < x + dx\right\} = e^{-nS(x)}\, dx. \tag{17}$$

As we will again see through heuristic reasoning, this theorem is used in statistical mechanics to reach the thermodynamic limit. In this context, S(x) is identified with entropy. For a precise proof, see for example Lanfort (1973).

Let $Z(\beta)$ be the Laplace transform of X.

$$Z(\beta) \equiv \left\langle e^{+\beta X}\right\rangle = \int e^{+\beta X} p(x)\, dx$$

where p(x) is the density of the probability of X.

We have $Z^n(\beta) = \left\langle e^{+\beta(X_1+\ldots+X_n)}\right\rangle$. Through the theorem of large deviations, for large n, the sum $X_1 + X_2 + \ldots X_n$ is near nx with the probability $\sim e^{-nS(x)}$. Thus the contribution to $Z^n(\beta)$ from the sums of nx is $\sim e^{n[\beta x - S(x)]}$. When one integrates with respect to all the x variables, the dominant contribution comes from the x which maximizes $\beta x - S(x)$, hence

$$Z^n(\beta) \sim e^{n(\beta x - S(x)).}$$

and

$$\ln Z(\beta) = \sup_\beta(\beta x - \ln Z(\beta)). \tag{18}$$

We obtain S(x) through the inverted Legendre transform

$$S(x) = \sup_\beta(\beta x - \ln\ Z(\beta)) \tag{19}$$

Definition 8.1. *The Cramer transform* C *is a function of* M, *the set of positive measures on* $E = \mathbb{R}^n$, *in* C_X, *the set of convex lsc proper functions, defined by* $C \equiv F \circ \log \circ L$ *where* L *is the Laplace transform and* F *the Fenchel transform.*

The Cramer transform possesses a large number of properties such as the transformation of the product of convolution into inf-convolution, see for example Azencott et al. (1978) and Akian (1995).

The use of this transformation in the field of partial differential equations (PDE's) is particularly interesting as the following theorem shows:

Theorem 15. *(Akian et al. (1995)*

The Cramer transform v *of the solution* u *to the PDE on* $E = \mathbb{R}$

$$-\frac{\partial u}{\partial t} + \hat{c}\left(-\frac{\partial}{\partial x}\right)(u) = 0, \quad u(0, \cdot) = \delta \tag{20}$$

(with $\hat{c} \in C_X$*) satisfies the Hamilton–Jacobi equation:*

$$+\frac{\partial v}{\partial t} + \hat{c}\left(\frac{\partial v}{\partial x}\right)(u) = 0, \quad v(0, \cdot) = \chi. \tag{21}$$

The latter equation is the HJB *equation of a problem of dynamic control* $x' = u$, *of instantaneous cost* c(u) *and of initial cost* χ.

Proof. The Laplace transform of u, denoted q, satisfies:

$$-\frac{\partial w}{\partial t}(t, \theta) + \hat{c}\,(\theta)\, q(t, \theta) = 0, \quad q(0, \cdot) = 1.$$

So $w = \log(q)$ satisfies:

$$-\frac{\partial w}{\partial t}(t, \theta) + \hat{c}(\theta) = 0, \quad w(0, \cdot) = 0$$

which yields $w(t, \theta) = \hat{c}(\theta)\, t$. As $\hat{c}$ is lsc convex, w is usc convex and can be considered as the Fenchel transform of a function v:

$$w(t, \theta) = \sup_x(\theta x - v(t, x)).$$

We deduce from the above $\theta = \frac{\partial v}{\partial x}$ and $\frac{\partial w}{\partial t} = -\frac{\partial v}{\partial t}$.

So v satisfies (21).

This equation is the HJB equation of a problem of control with the dynamic $x' = u$, of instantaneous cost c(u) and of initial cost χ, since $\hat{c}$ is the Fenchel transform of c and the HJB equation of the problem of control is:

$$-\frac{\partial v}{\partial t} + \min_u\left\{c(u) - u\frac{\partial v}{\partial x}\right\} = 0, \quad v(0, \cdot) = x. \quad \square$$

If $\hat{c}$ is time-independent, the optimal trajectories are straight lines with $v(x) = tc(x/t)$.

The solution to a linear PDE with constant coefficients is classically calculated with the Fourier transform: this is the case of (20) if $\hat{c}$ is a polynomial. The previous theorem shows that a first order nonlinear PDE with constant coefficients is isomorphic to a linear PDE with constant coefficients and therefore can be calculated explicitly. Such explicit solutions are known by the name of Hopf, Bardi and Evans formulas (1984), see also Sects. 6.2 and 7.3.

Example 8.2. Let us consider the HJB equation

$$\frac{\partial v}{\partial t} + \frac{1}{p}\left|\frac{\partial v}{\partial x}\right|^p = 0, \qquad v(0, x) = v_0(x).$$

We deduce from the above $w(t, \theta) = t\frac{1}{p}|\theta|^p$, then $v(x, t) = \frac{|x|^p}{p\, t^{1/p}}$, and finally

$$v(x, t) = v_0(x) \,\Box\, \frac{|x^p|}{p\, t^{1/p}} = \inf_y \left(v_0(y) + \frac{|x - y|^p}{p\, t^{1/p}} \right)$$

where $\Box$ corresponds to the inf-convolution on x.

Example 8.3. Let us consider the HJB equation

$$\frac{\partial v}{\partial t} + \frac{1}{2}\left(\frac{\partial v}{\partial x}\right)^2 + \frac{2}{3}\left(\frac{\partial v}{\partial x}\right)^{3/2} = 0, \qquad v(0, \cdot) = v_0(x).$$

We deduce from the above $w(t, \theta) = t\left(\frac{1}{2}\theta^2 + \frac{2}{3}|\theta|^{3/2}\right)$ and finally $v(t, x) = v_0(x) \,\Box\, \frac{x^2}{2t} \,\Box\, \frac{|x|\,3}{3t^2}$ where $\Box$ corresponds to the inf-convolution on x.

Exercises

Exercise 1. Proximity and duality in a Hilbert space

Let H be a real Hilbert space and $\Gamma_0(H)$ the set of functions with values in $]-\infty, +\infty]$ defined everywhere on H, convex, lower semi-continuous, not everywhere equal to $+\infty$.

(1) Show that for any $f \in \Gamma_0(H)$, $x \in H$, the function

$$z \to f(z) + \frac{1}{2}\|z - x\|^2$$

has a strict minimum.

We will denote $\tilde{f}(x)$ this minimum and $z^* = \mathrm{prox}_f\, z$ ("proximal point") the unique point where the minimum is reached:

$$\tilde{f}(x) = \min_z \left(f(z) + \frac{1}{2}\|z - x\|^2 \right), \quad z^* = \arg\min_z \left(f(z) + \frac{1}{2}\|z - x\|^2 \right).$$

(2) Determine the proximal point for an affine function $f(z) = (a, z) - \beta$ (where $a \in H, \beta \in \mathbb{R}$); for the characteristic function of a closed convex set C, $f(z) = \Psi_c(z) = 0$ if $z \in C, +\infty$ if $x \notin C$; for $f(z) = (a, z) - \beta + \Psi_c(z)$; for a quadratic function $f(z) = k||z||^2$.

(3) For any function $f \in \Gamma_0(H)$, we define its dual function

$$g(y) = \sup_{x \in H} \{(x, y) - f(x)\} \quad \forall y \in H$$

Show that $g \in \Gamma_0(H)$ and that

$$f(x) = \sup_{x \in H} \{(x, y) - g(x)\}.$$

If two points x and y are such that we have the equality

$$f(x) + g(x) = (x, y)$$

we say that they are conjugated with respect to the pair of dual function f and g. Show that y is a *subgradient* of f at the point x, i.e.:

$$f(u) \geq f(x) + (y, u - x)$$

and that the subdifferential (set of the subgradients) $\partial f(x)$ is a closed convex set. Show that $\inf_{x \in H} f(x) = -g(0)$ and $\arg\min_{x \in H} f(x) = \partial g(0)$.

(4) If $f \in \Gamma_0(H)$ and $g \in \Gamma_0(H)$ are two dual functions of each other; let x, y, z be three elements of H. Show that the following properties are equivalent:

(i) $z = x + y, f(x) + g(y) = (x, y)$
(ii) $x = \text{prox}_f\, z, y = \text{prox}_g\, x$

Deduce that for every z and z' in H, we have:

$$||\text{prox}_f z - \text{prox}_f z'|| \leq ||z - z'||$$

so that the application prox_f is continuou: $H \to H$ in the strong sense.

(5) Show that $\tilde{f}$ is convex differentiable (in the Fréchet sense) and that

$$\nabla \tilde{f}(x) = x - z^* = x - \text{prox}_f\, x$$

[*Indications*: Moreau (1965)].

Exercise 2. Properties of the Moreau-Yosida transform

Let $f: \mathbb{R}^n \to \mathbb{R} \cup \{+\infty\}$ be a convex, lsc function, finite in at least one point. We denote $\langle .,. \rangle$ the scalar product and $|.|$ the associated norm.

(1) For any $\lambda > 0$, we consider the function f_λ, the so-called Moreau-Yosida transform, defined on $\mathbb{R}^n$ by: $f_\lambda(x) = \inf_{y \in \mathbb{R}^n} \left\{f(y) + \frac{\lambda}{2}|x - y|^2\right\}$.

(a) Show that the inf of the previous definition is reached in a unique point x_λ.

(b) Write f_λ as the inf-convolution of two convex functions. Verify that the inf-convolution is exact and that f_λ is differentiable in any $x \in \mathbb{R}^n$ with

$$\nabla f_\lambda(x) = \lambda(x - x_\lambda) \in \partial f(x_\lambda).$$

(c) Show that:

$$I + \frac{1}{\lambda}\partial f \text{ is a surjective multi valued mapping: } \mathbb{R}^n \to \mathbb{R}^n;$$

$$\forall x \in \mathbb{R}^n, \left(I + \frac{1}{\lambda}\,\partial f\right)^{-1}(x) = x_\lambda$$

(2) Determine $f_\lambda(x)$ and x_λ for any $x \in \mathbb{R}^n$ in each of the following special cases:

(a) $f(x) = \langle a, x\rangle + b$, where $a \in \mathbb{R}^n$, $b \in \mathbb{R}$.
(b) f indicator function of a closed non empty convex set C of $\mathbb{R}^n$.
(c) $f(x) = \frac{1}{2}\langle Ax, x\rangle$ with A self-adjoint.

(3) Show that x_λ can be characterized by one or the other of the following conditions:

$$f(y) - f(x_\lambda) + \lambda\langle x_\lambda - x, y - x_\lambda\rangle \geq 0 \quad \forall y \in \mathbb{R}^n$$
$$f(y) - f(x_\lambda) + \lambda\langle y - x, y - x_\lambda\rangle \geq 0 \quad \forall y \in \mathbb{R}^n$$

(4) (a) Show that the mapping $x \to x_\lambda$ is Lipschitzian with Lipschitz constant 1.
(b) Show that the mapping $x \to \nabla f_\lambda(x) = \lambda(x - x_\lambda)$ is Lipschitzian with Lipschitz constant λ.
(c) Prove the inequality:

$$0 \leq f_\lambda(y) - f_\lambda(x) - \lambda\langle x - x_\lambda, y - x\rangle \leq \lambda|x - y|^2.$$

(5) What is the (Legendre–Fenchel) conjugate of

$$N_\lambda(x) = \frac{\lambda}{2}|x|^2?$$

Deduce the expression of the conjugate $\hat{f}_\lambda$ of f_λ.

$$\text{Compare } \inf_{x\in\mathbb{R}^n}\{f(x)\} \text{ with } \inf_{x\in\mathbb{R}^n}\{f_\lambda(x)\}.$$

(6) (a) Show that $f(x_\lambda) \leq f_\lambda(x) \leq f(x)$.
(b) Establish the equivalence of the following statements:
(i) x minimizes f on $\mathbb{R}^n$
(ii) x minimizes f_λ on $\mathbb{R}^n$
(iii) $x = x_\lambda$
(iv) $f(x) = f(x_\lambda)$
(v) $f(x) = f_\lambda(x_\lambda)$

(7) (a) Let $x \in \operatorname{dom} f := \{x | f(x) < +\infty\}$. Show that $x_\lambda \to x$ when $\lambda \to +\infty$. Deduce that $\{x \in \mathbb{R}^n | \partial f(x) \neq \emptyset\}$ is dense in dom f and that $f_\lambda(x) \to f(x)$ when $\lambda = +\infty$.
(b) Let $x \notin \operatorname{dom} f$. Show that $f_\lambda(x) \to +\infty$ when $\lambda \to +\infty$.

[*Indications*: see for example Hiriart-Urruty (1998).

(1) (b) One can use the classical result: if the inf-convolution $(f_1 \,\square\, f_2)$ of two lsc convex proper functions f_1 and f_2 is exact in x (i.e. if there exists x_1 and x_2 with $x = x_1 + x_2$ such that $(f_1 \,\square\, f_2)(x) = f_1(x_1) + f_2(x_2)$ then

$$\partial(f_1 \,\square\, f_2)(x) = \partial f_1(x_1) \cap \partial f_2(x_2).$$

(c) from the characterization of the point x_λ:

$$0 \in \partial(f + \frac{\lambda}{2}|\cdot - x|^2)(x_\lambda).$$

(2) (a) $f_\lambda(x) = \langle a, x\rangle + b - \frac{|a|^2}{\lambda}, x_\lambda = x - \frac{a}{\lambda}$.
(b) $f_\lambda(x) = \frac{\lambda}{2} d_c^2(x)$ where d_c is the functional distance to C.
(c) $x_\lambda = \left(I + \frac{A}{\lambda}\right)^{-1}(x), f_\lambda\,(x) = \frac{1}{2}\langle A_\lambda x, x\rangle$ with $A_\lambda = A\left(I + \frac{A}{\lambda}\right)^{-1}$.
(3) Optimality conditions for an objective function of the form f + g, f convex lsc and g convex differentiable.
(5) $\hat{N}_\lambda(p) = \frac{1}{2\lambda}|p|^2$.
Since $f_\lambda = f \,\square\, N_\lambda$, $\hat{f}_\lambda = \hat{f} + \hat{N}_\lambda$ and $\hat{f}_\lambda(p) = \hat{f}(p) + \frac{1}{2\lambda}|p|^2$.
In particular $\hat{f}_\lambda(0)(= -\inf\limits_{x\in\mathbb{R}^n} f_\lambda(x)) \equiv \hat{f}(0)(= -\inf\limits_{x\in\mathbb{R}^n} f(x))$.
(7) (a) First use an affine lower bound of f to prove that $|x_\lambda - x| \to 0$, then that $x_\lambda \in \operatorname{dom} \partial f = \{x | \partial f(x) \neq \emptyset\}$ to prove the density of dom ∂f in dom f.
(b) Can be proved by contradiction assuming "$\{f_\lambda\,(x)\}_\lambda$ bounded from above."]

Exercise 3. A few Legendre–Fenchel transforms

With every proper function $f: \mathbb{R}^n \to \mathbb{R} \cup \{+\infty\}$, we associate its Legendre–Fenchel transform $\hat{f}: \mathbb{R}^n \to \mathbb{R} \cup \{+\infty\}$ defined as:

$$y \in \mathbb{R}^n, \hat{f}(y) = \sup_{x\in\mathbb{R}^n} \{\langle x, y\rangle - f(x)\}.$$

(1) Check that $\hat{f}$ is convex lsc, and that the Fenchel inequality

$$\hat{f}(y) + f(x) \geq \langle x, y\rangle \quad \forall (x, y) \in \mathbb{R}^n \times \mathbb{R}^n$$

is satisfied with equality if and only if $y \in \partial f(x)$, where $\partial f(x)$ is the sub-differential of f in x.

Show that if f is 1-coercive on $\mathbb{R}^n$ (i.e. if $f(x)/||x|| \to +\infty$ when $||x|| \to +\infty$), then $\hat{f}$ is finite everywhere on $\mathbb{R}^n$. Show that if f is strictly convex, differentiable and 1-coercive on $\mathbb{R}^n$, then $\hat{f}$ is also finite everywhere, strictly convex, differentiable and 1-coercive on $\mathbb{R}^n$.

(2) Calculus of some Fenchel transforms.

- If I_A is the *indicator function* of a non empty subset A of $\mathbb{R}^n$ ($I_A(x) = 0$ if $x \in A$, $+\infty$ otherwise), show that

$$\hat{I}_A(y) = \sup_{x \in A} \langle y, x \rangle \quad \text{(support function A)}$$

- If A is a closed non empty subset of $\mathbb{R}^n$ and f_A the function defined as $f_A(x) = \frac{1}{2}||x||^2$ if $x \in A$, $+\infty$ otherwise, show that $\hat{f}_A(y) = \frac{1}{2}\left(||y||^2 - d_A^2(y)\right)$ where d_A designates the function distance to A.
- For $a > 0$ and $c \geq 0$, let us set
$f_{a,c}$: $x \in \mathbb{R} \to f_{a,c}(x) = -\sqrt{a^2 - (x-c)^2}$ if $|x - c| \leq a$, $+\infty$ therwise.
Calculate $\hat{f}_{a,c}$, then $f_{a_1,c_1} \square f_{a_2,c_2}$ where $\square$ corresponds to the inf-convolution.
- For given reals m and σ, let us set:
$q_{m,\sigma}$: $x \in \mathbb{R} \to q_{m,\sigma}(x) = \frac{1}{2}\left(\frac{x-m}{\sigma}\right)^2$ if $\sigma \neq 0$, $q_{m,0}(x) = \delta_m(x)$
(function of Dirac type centered in m)
Calculate $\hat{q}_{m,\sigma}$, then $q_{m_1,\sigma_1} \square\, q_{m_2,\sigma_2}$.
$f(x) = \frac{1}{2}x^2 = q_{0,1}(x)$ is its own Fenchel transform. Show that it is the only function with this property.

[*Indications*:

(1) See Exercise 1, question 3.
(2) $\hat{f}_{a,c}(y) = a\sqrt{1+y^2} + cy$, $f_{a_1,c_1} \square f_{a_2,c_2} = f_{a_1+a_2,c_1+c_2}$.

$$\hat{q}_{m,\sigma}(y) = \frac{1}{2}\sigma^2 y^2 + my,\ q_{m_1,\sigma_1} \square\, q_{m_2,\sigma_2} = q_{m_1+m_2,\sqrt{\sigma_1^2+\sigma_2^2}}.$$

$$f(x) + \hat{f}(y) \geq \langle x, y \rangle \text{ with } x = y, \text{ therefore if } f(x) \geq \frac{1}{2}||x||^2, \hat{f}(y) \leq \frac{1}{2}||y||^2.]$$

Exercise 4. "Proximal" algorithm
The task is to solve the problem

$$\text{(P) Minimize } f(x) \text{ with } x \in C = \{y / y \in \mathbb{R}^n;\ Ax \leq b\}$$

where f: $\mathbb{R}^n \to \mathbb{R}$ is a convex nonsmooth function, where A is a $m \times n$ real matrix, and $b \in \mathbb{R}^m$.

We assume that $C \neq \emptyset$ and that we either have bounded C (condition c_1) or $f(x) \to +\infty$ when $|x| \to +\infty$ (condition c_2).

Under each of these conditions, the problem (P) has a finite optimal solution.

We recall that the Moreau-Yosida regularization

$f_\lambda(x) = \underset{z\in R^n}{\mathrm{Min}}\left(f(z) + \frac{\lambda}{2}|z - x|^2\right)$, defined for any $\lambda > 0$ (see Exercise 2), is convex and continuously differentiable with gradient:

$$\nabla f_\lambda(x) = -\lambda(x_\lambda - x)$$

where

$$x_\lambda = \underset{z\in R^n}{\mathrm{argmin}}\left\{f(z) + \frac{\lambda}{2}|z - x|^2\right\}$$

(1) Consider the problem

$$(P_\lambda)\ \text{Minimize } f_\lambda(x) \text{ with } x \in \mathbb{R}^n.$$

Show that the values of the objective optimal function and that the optimal solutions of (P) and (P_λ) are the same.

The proximal point algorithm (Martinet 1970; Rockafellar 1976a,b) then consists in using (P_λ) to solve (P).

Proximal point Algorithm

Let x^k be the current point and $\lambda_k > 0$. Solve (exactly or approximately) the sub-problem

$$SP(x^k)\text{: } \underset{z\in C}{\mathrm{Minimiser}}\left(f(z) + \frac{\lambda_k}{2}|z - x^k|^2\right)$$

Take x^{k+1} as the (exact or approximate) solution to $SP(x^k)$.

Prove that if λ_k is chosen at each iteration such that $\lambda_\kappa \in [\lambda_{min}, \lambda_{max}]$ with $0 < \lambda_{min} < \lambda_{max} < +\infty$ and $0 < \lambda_k < +\infty$ and x^{k+1} is taken to be the exact solution to $SP(x^k)$, then the sequence $\{x^k\}$ converges toward an optimal solution to (P).

(2) SP (x^k) will be solved iteratively using tangential approximation of f based on the sub-gradients of f which are generated during the calculations. Thus tangential approximations $w_0(z), w_1(z), \ldots, w_i(z)$ of f will be successively constructed. The ith approximation $w_i(z)$ is constructed from the values of the function f and the sub-gradients obtained in the previous iterations at the points $z^0 = x^k, z^1, \ldots, z^i$ by:

$$w_i(z) = \underset{0\le j\le i-1}{\mathrm{Max}}\left\{f(z^j) + \left\langle g^j, z - z^j\right\rangle\right\}$$

where $z^j \in C$, $g^j \in \partial f(z^j)$.

For each stage i, the sub-problem $SP(x^k)$ is replaced with the following approximated sub-problem:

$$SP_i(x^k) = \underset{z\in C}{\mathrm{Min}}\left\{\psi_i(z, x^k) = w_i(z) + \frac{\lambda_k}{2}|z - x^k|^2\right\}$$

$SP_i(x^k)$ can be reformulated as the following quadratic problem with linear constraints:

$$Q_i(x^k) = \begin{cases} \underset{z,\eta}{\text{Min}}\, \eta + \frac{\lambda_k}{2}|z - x^k|^2 & \text{under the contraintes} \\ \eta \geq f(z^j) + \langle g^j, z - z^j \rangle \ (j = 0, 1, \ldots, i-1) \\ z \in C. \end{cases}$$

Let (z^i, η^i) be the unique optimal solution to $Q_i(x^k)$.
It is this z^i which is used to construct the next approximation:

$$w_{i+1}(z) = \max \left\{ w_i(z), f(z^i) + \left\langle g^i, z - z^i \right\rangle \right\}$$

Let z^* be the only optimal solution to $SP(x^k)$ and $\Phi^* = \Phi(z^*, x^k)$, where, $\forall z, \Phi(z, x^k) = f(z) + \frac{\lambda_k}{2}\left|z - x^k\right|^2$.
Show that:

(i) $z^i \to z^*$
(ii) $\Phi(z^i, x^k) \to \Phi^*$ and furthermore:

$$\forall i\colon \Phi(z^i, x^k) \geq \Phi^* + \frac{\lambda_k}{2}|z^i - z^*|^2.$$

(iii) $\psi_i\,(z^i, x^k) \to \Phi^*$ and furthermore:

$$\forall i\colon \psi_i(z^i, x^k) \leq \psi_{i+1}(z^{i+1}, x^k) \leq \Phi^* \leq \Phi(z^i, x^k).$$

(3) The procedure of approximate solution to $SP(x^k)$ will be stopped as soon as $\varepsilon_i = \Phi(z^i, x^k) - \psi_i(z^i, x^k) = f(z^i) - w_i(z^i)$ becomes sufficiently small (therefore when $\Phi^* - \psi_{i+1}\,(z^{i+1}, x^k)$ is small, see question 2 iii).

For ε, ε' two positive reals, we propose the proximal algorithm (see Dodu et al. 1994) whose current step k is the following:
Let x^k be the current point.

(a) Initialization.

$$Q_0(x^k) \text{ has solution } z^0 = x^k. \text{ Take } g^0 \in \partial f(z^0) \text{ and set } i = 1.$$

(b) Solve the problem $Q_i(x^k)$. Let (z^i, η^i) be the optimal solution, calculate $f(z^i)$ and $g^i \in \partial f(z^i)$.
Calculate $w_i(z^i) = \underset{0 \leq j \leq i-1}{\text{Max}} \left\{ f(z^j) + \langle g^j, z^i - z^j \rangle \right\}$ and $\varepsilon_i = f(z^i) - w_i(z^i)$
If $\varepsilon_i \leq \frac{\lambda_k}{2}|z^i - x^k|^2$, go to c)
If $\frac{\lambda_k}{2}|z^i - x^k|^2 < \varepsilon_i \leq \varepsilon$, END of iteration k and end of the proximal iterations.
If $\frac{\lambda_k}{2}|z_i - x^k|^2 < \varepsilon_i$ and $\varepsilon_i > \varepsilon$, then add to $Q_i(x^k)$ the new constraint $\eta \geq f(z^i) + \langle g^i, z - z^i \rangle$, set $i \leftarrow i + 1$ and go into b).
(c) If $|z^i - x^k| \leq \varepsilon'$, END of the proximal iterations. Otherwise terminate the current stage k by setting $x^{k+1} \leftarrow z^i$.

Show that:

(i) $f(z^i) \le f(x^k) + \varepsilon_i - \lambda_k |z^i - x^k|^2$.

(ii) If $\varepsilon > 0$, $\varepsilon' > 0$ and $\forall k$: $\lambda_k \in [\lambda_{min}, \lambda_{max}]$ with $\lambda_{min} > 0$, then the algorithm ends in a finite number of proximal iterations by satisfying one of the following two conditions (C1) or (C2):

$$\text{(C1)} \quad \frac{\lambda_k}{2} |z^i - x^k|^2 < \varepsilon_i \le \varepsilon$$

$$\text{(C2)} \quad |z^i - x^k| \le \varepsilon'.$$

In the case (C1), we have $|\nabla f_{\lambda_k}(x^k)| \le 2\sqrt{2\lambda_k \varepsilon}$;
In the case (C2), we have $|\nabla f_{\lambda_k}(x^k)| \le 2\lambda_k\, \varepsilon'$.
which shows that x^k is, in both cases, an approximate solution to (P_{λ_k}) and therefore to (P).

[*Indications*:

(1) See Martinet (1970) and Rockafellar (1976a,b). For further discussion concerning the case where x^{k+1} is not the exact optimal solution to SP(x^k), see Auslender (1984).

(2) (ii) follows from the strong convexity of $\Phi(z, x^k)$ (see Dodu et al. 1994); (i) and (iii) are well-known sproperties of the "cutting-plane method" Kelley (1960).

(3) See Dodu et al. (1994). One will also find there stopping rules to further improve the efficiency of this algorithm (the computational results reported show that it is one of the most efficient algorithms available for convex nonsmooth optimization to linear constraints).]

Exercise 5. Duality between probability and optimization

We refer to as *decision space* the triple $(\mathcal{U}, \mathcal{A}, \mathbb{K})$ where $\mathcal{U}$ is a topological space, $\mathcal{A}$ the set of open subsets of $\mathcal{U}$ and $\mathbb{K}$ a functional: $A \to \overline{\mathbb{R}}_+$ such that:

(i) $\mathbb{K}(\mathcal{U}) = 0$

(ii) $\mathbb{K}(\emptyset) = +\infty$

(iii) $\mathbb{K}(\cup A_n) = \inf_n\ \mathbb{K}(A_n)$ for any $A_n \in \mathcal{A}$.

The mapping $\mathbb{K}$ is called a *cost measure*.

A function c: $u \in \mathcal{U} \to c(u) \in \overline{\mathbb{R}}_+$ such that $\mathbb{K}(A) = \inf\limits_{u \in A} c(u)\ \forall A \subset \mathcal{U}$ is called a *cost density* of the cost measure $\mathbb{K}$.

(1) Let a real positive lsc function c such that $\inf\limits_{u} c(u) = 0$, then $\mathbb{K}(A) = \inf\limits_{u \in A} c(u)$ for any open set of $\mathcal{U}$ defines a cost measure. Show that, conversely, each cost measure defined on an open set of a Polish space (topological space with a countable base of open sets) has a unique minimal extension K_* to $\mathcal{P}(\mathcal{U})$ (the power set of **U**) having an lsc density c on $\mathcal{U}$ such that $\inf\limits_{u} c(u) = 0$. The most classical cost densities will be:

$$X_m(x) \equiv +\infty \text{ if } x \ne m, 0 \text{ if } x = m;$$

$$M^p_{m,\sigma} \equiv \frac{1}{p} \left\| \sigma^{-1}(x - m) \right\|^p \text{ with } p \ge 1 \text{ and } M^p_{m,0} \equiv X_m.$$

(2) By analogy with the random variables, we define a decision vector X on $(\mathcal{U}, \mathcal{A}, \mathbb{K})$ as a functional: $\mathcal{U} \to \mathbb{R}^n$. We use the term decision variable (DV).

This decision vector induces a cost measure $\mathbb{K}_X$ on $(\mathbb{R}^n, B)$ (B being the set of open set of $\mathbb{R}^n$) defined by $\mathbb{K}_X(A) = \mathbb{K}_*(X^{-1}(A)), \forall A \in B$. This cost measure $\mathbb{K}_X$ has an lsc density C_X.

We then define the *characteristic function* $\mathbb{F}(X)$ of a decision variable X as the Fenchel transform of the cost density C_X, $\mathbb{F}(X) \equiv F(C_X)$, and the optimum $\mathbb{O}(X)$ of the DV of X by $\mathbb{O}(X) \equiv \arg\min_X C_X(x)$ if the minimum exists. When the optimality is unique and in the vicinity of the optimum, this yields:

$$C_X(x) = \frac{1}{p}\left|\frac{x - \mathbb{O}(X)}{\sigma}\right|^p + \sigma(|x - \mathbb{O}(X)|^p).$$

Prove that if C_X is convex and has a unique minimum of order p, then we have:

$$(\mathbb{F}(X)'(0) = \mathbb{O}(X), \quad \mathbb{F}(X - \mathbb{O}(X)^p(0) = \Gamma(q)[S^p(X)]^q$$

with $\dfrac{1}{p} + \dfrac{1}{q} = 1$.

(3) By analogy with conditional probabilities, we define the *conditional excess cost* of taking the best decision in A knowing the best one taken in B by:

$$\mathbb{K}(A/B) \equiv \mathbb{K}(A \cap B) - K(B).$$

Two decision variables X and Y are said to be independent if:

$$C_{X,Y}(x, y) = C_X(x) + C_Y(y).$$

The conditional excess cost of X knowing Y is defined as:

$$C_{X,Y}(x, y) \equiv K_*(X = x/Y = y) = C_{X,Y}(x, y) - C_Y(y).$$

For two independent decision variables X and Y of order p and for every real k, show that we have:

$$\begin{aligned}
&C_{X+Y} = C_X \otimes C_Y, \mathbb{F}(X + Y) = \mathbb{F}(X) + \mathbb{F}(Y), [\mathbb{F}(kX)](\theta) = [F(X)](k\theta),\\
&\mathbb{O}(X + Y) = \mathbb{O}(X) + \mathbb{O}(Y), \mathbb{O}(kX) = k\mathbb{O}(X), \mathbb{S}^p(kX) = |k|\, \mathbb{S}^p(X),\\
&\mathbb{S}^p(X + Y)^q + [\mathbb{S}^p(X)]^q + [\mathbb{S}^p(Y)]^q
\end{aligned}$$

where $C_X \otimes C_Y$ is the inf-convolution of C_X and C_Y:

$$C_X \otimes C_Y(z) = \inf_{x,y}[C_X(x) + C_Y(y) \text{ with } x + y = z] \text{ and } \frac{1}{p} + \frac{1}{q} = 1.$$

[*Indications*: see Akian et al. (1992).]

Exercise 6. Characterization of the super and subdifferentials of a function on $\mathbb{R}$

Let Ω be an open set of $\mathbb{R}^n$ and $u \in C(\Omega)$. Show that:

(a) $p \in \partial^+ u(x)$ (resp. $p \in \partial^- u(x)$) if and only if there exists $\varphi \in C^1(\Omega)$ such that $D\varphi(x) = p$ and that $u - \varphi$ has a local maximum (resp. minimum) in x;
(b) $\partial^+ u(x)$ and $\partial^- u(x)$ are convex closed (possibly empty) subsets of $\mathbb{R}^n$;
(c) If u is differentiable in x, then $\{Du(x)\} = \partial^+ u(x) = \partial^- u(x)$;
(d) If for a x, we have both non empty $\partial^+ u(x)$ and $\partial^- u(x)$, then $\partial^+ u(x) = \partial^- u(x) = \{Du(x)\}$
(e) The sets $A^+ = \{x \in \Omega: \partial^+ u(x) \neq \emptyset\}$, $A^- = \{x \in \Omega: \partial^- u(x) \neq \emptyset\}$ are dense.

[*Indications*: See Crandall et al. (1984) and Barles (1994).

(a) Let $p \in \partial^+ u(x)$. Then there exists $\delta > 0$ and a continuous increasing function σ on $[0, +\infty[$ with $\sigma(0) = 0$ such that

$$u(y) \le u(x) + \langle p, y - x\rangle + \sigma(|y - x|)|y - x| \quad \forall y \in B(x, \delta)$$

where $B(x, \delta)$ is the ball of radius δ centered in x.
Then ρ, the function C^1 defined as

$$\rho(r) = \int_0^r \sigma(t)dt$$

verifies $\rho(0) = \rho'(0) = 0$ and $\rho(2r) \ge \sigma(r)\, r$.
We deduce from the above that the function

$$\varphi(y) = u(x) + \langle p, y - x\rangle + \rho(2|y - x|)$$

belongs to $C^1(R^n)$, verifies $D\varphi(x) = p$ and that $u - \varphi$ has a maximum in x in $B(x, \delta)$.

(b) Let $\bar{x} \in \Omega$ and let us consider the function $\varphi_\varepsilon(x) = \frac{1}{\varepsilon}|x - \bar{x}|^2$. For any $\varepsilon > 0$, $u - \varphi_\varepsilon$ has a maximum on $\overline{B} = B(\bar{x}, R)$ at the point x_ε. We deduce

$$|x_\varepsilon - \bar{x}|^2 \le 2\varepsilon \sup_{x \in \overline{B}} |u(x)|.$$

If ε is sufficiently small, x_ε is not on the boundary of $\overline{B}$ and, according to a) $D\varphi_\varepsilon(u_\varepsilon) = 2(x_\varepsilon - \bar{x})/\varepsilon \in \partial^+ u\,(x_\varepsilon)$. A^+ is therefore dense.]

Exercise 7. Marginal functions

Let us consider the function u(x) defined as:

$$u(x) = \inf_{b \in B} g(x, b)$$

where $g: \Omega \times B \to R$, $\Omega \subset R^n$ is an open set and B is a topological space. Functions of this type are often referred to as *marginal functions*.

The most classical example is the distance function to a set $S \subset R^n$,

$$d(x, S) = \inf_{s \in S} |x - s|.$$

Another example is the "regularized Yosida function."

$$f_\lambda(x) = \inf_{y \in \Omega} (u(y) + \frac{1}{2\lambda}|x - y|^2).$$

(1) It is assumed that g (x, B) is bounded for any $x \in \Omega$ and that $x \to g(x, b)$ is continuous in x uniformly with respect to $b \in B$; i.e. that

(a) $|g(x, b) - g(y, b) \leq w(|x - y|, R)$, for $|x|, |y| \leq R, b \in B$, for a module w. We denote M(x) the set (which may be empty):

$$M(x) = \arg\min_{b \in B} g(x, b) := \{b \in B : u(x) = g(x, b)\}.$$

It is non empty if B is compact.
Show that under the hypothesis (1), we have $u \in C\ (\Omega)$ and

$$\partial^+ u(x) \supseteq \partial_x^+ g(x, b) \text{ for any } b \in M(x), \text{ and}$$
$$\partial_x^- g(x, b) \supseteq \partial^- u(x).$$

(2) Assume now that g (·, b) is differentiable in x uniformly for b i.e. there exists for a module w_1 such that:

(b) $|g(x + h, b) - g(x, b) - \langle D_x\, g(x, b), b\rangle\ | \leq |h| w_1(|h|)$ for any $b \in B$ and small h.
We also assume that:
(c) $b \to D_x\, g(x, b)$ is continuous,
(d) $b \to g(x, b)$ is lsc. We further denote $Y(x) := \{D_x\, g(x, b) : b \in M\ (x)\}$.
Show that under the hypotheses (1), (2), (3), (4) and B compact, we have:
(e) $Y(x) \neq \emptyset$
(f) $\partial^+ u(x) = \overline{C}_0 Y(x)$
(g) $\partial^-\ u\ (x) = \begin{cases} \{y\} & \text{if} \quad Y(x) = y \\ \emptyset & \text{if} \quad Y(x) \text{ is not singleton} \end{cases}$
and therefore that $u \in SDS^1$.

(3) It is now assumed that B is compact, g continuous on $\Omega \times B$, differentiable with respect to x with D_x g continuous on $\Omega \times B$. Show that u is SDS^1 and locally Lipschitz ($u \in Lip_{loc}\ (\Omega)$) and that we still have (e), (f) and (g).

[*Indications*: see Bardi and Capuzzo-Dolcetta (1997), Chap. 2. (1) Lemma 2.11. (2) Proposition 2.13. (3) Proposition 4.4.]

Exercise 8. Crandall, Ishii and Lions lemma

Let $\theta \subset R^n$ be locally compact, $u \in LSC(\theta)$, $z \in \theta$ and $(p, x) \in \partial_2^- v(z)$.

Show that for any sequence $u_n \in LSC(\theta)$ which *epiconverges* toward u, there exists for any $\delta > 0$

$\hat{x}_m \in \theta$, $(p_m, x_m) \in \partial_2^- u_m(\hat{x}_n)$ such that

$(\hat{x}_n, u_n(\hat{x}_n), p_n, X_n)$ converge toward $(z, u(z), p, X - 4\delta I)$.

[*Indications*:

Taking $z = 0$, we will seek to minimize the function $a(x) - \langle p, x\rangle - \frac{1}{2}\langle Xx, x\rangle + 2\delta|x|^2$ in the compact $B_r = \{x \in \theta\colon |x| \le r\}$ where $r > 0$ is chosen in relation to $z = 0$ and δ. $\hat{x}_n$ will be the argument of this minimum. see Crandall et al. (1992).]

Exercise 9. Solutions of nonlinear PDE by separation of variables

(1) Consider the following HJB problem:

$$\frac{\partial u}{\partial t} + \frac{1}{2}|\nabla_u|^2 = 0 \text{ in}]0, +[\times\mathbb{R}^N$$
$$u(x, 0) = \max_{i\in I}\{a_i x + b_i\} \text{ on } \mathbb{R}^N$$

where $a_i \in \mathbb{R}^N$, $b_i \in \mathbb{R}$.

Show that the solution SDI[1] (and convex) is equal to

$$u(x, t) = \max_{i\in I}\{a_i x - \frac{1}{2}|a_i|^2 t + b_i\}$$

(2) What happens to this solution if we now consider the HJB problem:

$$\begin{aligned} &\frac{\partial u}{\partial t} + H(|\nabla_u|) = f(t) && \text{in }]0, +\infty[\times\mathbb{R}^N \\ &u(x, 0) = \max_{i\in I}\{a_i x + b_i\} && \text{on } \mathbb{R}^N \end{aligned}$$

where $a_i \in \mathbb{R}^N$, $b_i \in \mathbb{R}$, $f(t)$ increasing in t and $H(p)$ convex.

[*Indications*: (2) $u(x, t) = \max_{i\in I}\{a_i\, x + \int_0^t f(s)ds - H(a_i)\, t + b_i\}$.]

Chapter 8
Collected Examples of Monoids, (Pre)-Semirings and Dioids

The present chapter is intended as a catalogue of examples covering the various algebraic structures studied throughout this book.

The successive items to be found in this chapter are: in Sect. 1, examples of monoids with emphasis on canonically ordered monoids; in Sect. 2, examples of pre-semirings and pre-dioids which are not semirings; in Sect. 3, examples of semirings (and rings) which are not dioids; and in Sect. 4, examples of dioids.

Using the many examples stated in this chapter, a virtually unlimited number of other examples may, of course, be derived by homomorphism, but also by considering *matrices, polynomials, formal series* on these algebraic structures. Verification of the main properties enjoyed by the various examples stated here is either explicitly stated in the text of the chapter, or can be found in the previous chapters.

Throughout this chapter, for most of the examples considered, we provide references to the chapters (in bold character) and sections or subsections dealing with each particular example.

1. Monoids

The typology of monoids is recalled in Fig. 1 below.

The first level of the classification contains:

- Groups (see Sect. 1.2);
- Canonically ordered monoids (see Sect. 1.3–1.6);
- "General" monoids which belong to none of the previous categories (see Sect. 1.1).

We recall that, by virtue of Theorem 1 of chapter 1 (Sect. 3.4), a monoid cannot both be a group and be canonically ordered, the corresponding subclasses are therefore *disjoint*.

Table 1 recalls the main definitions concerning monoids.

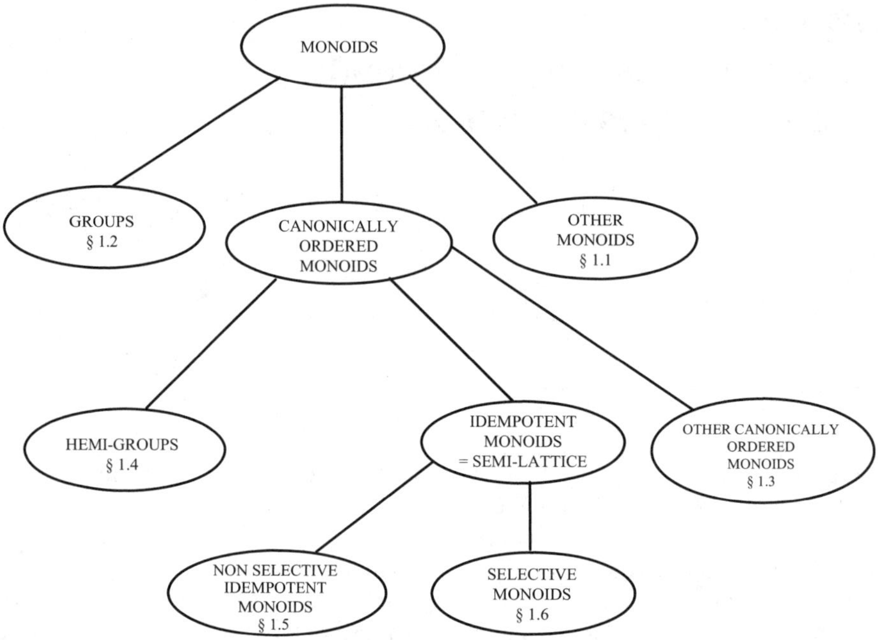

Fig. 1 Typology of monoids

Table 1 Basic terminology concerning monoids

Monoid	Set E endowed with an associative internal law ⊕
Cancellative monoid	Monoid in which ⊕ is cancellative every element is cancellative
Group	There exists a neutral element ε and every element of E has an inverse for ⊕
Canonically ordered monoid	The canonical preorder relation $\leq$ (defined as $a \leq b \Leftrightarrow \exists c$ such that $b = a \oplus c$) is an *order* relation
Idempotent monoid	⊕ is idempotent ($\forall a \in E \; a \oplus a = a$)
Selective monoid	⊕ is selective ($\forall a \in E, b \in E: \; a \oplus b = a$ or b)
Hemi-group	Every element is cancellative (property of hemi-group) and the canonical preorder relation is an order relation

1.1. General Monoids

The examples presented in this section are monoids which are neither *groups* (see Sect. 1.2), nor *canonically ordered monoids* (see Sects. 1.3–1.6).

After the presentation of each example, we indicate, whenever appropriate, the chapters and sections where the example is discussed (chapter number in bold followed by section numbers). If opportune, additional references are pointed out.

1.1.1. (Int($\mathbb{R}$), +)

We consider the set Int ($\mathbb{R}$) of the intervals of $\mathbb{R}$ of the form $a = [\underline{a}, \bar{a}]$ with $\underline{a}, \bar{a} \in \mathbb{R}$ and $\underline{a} \leq \bar{a}$. This set is endowed with the addition of intervals:

$$a \oplus b = [\underline{a} + \underline{b}, \bar{a} + \bar{b}]$$

The operation $\oplus$ is *associative, commutative* and has [0, 0] as neutral element.

If $\ell(a) = \bar{a} - \underline{a}$ is the length of the interval, this yields $\ell(a \oplus b) = \ell(a) + \ell(b)$. Since $\ell([0, 0]) = 0$, the only invertible elements are the intervals of length 0.

$(\mathrm{Int}(\mathbb{R}), \oplus)$ is not canonically ordered because $a \leq b$ and $b \leq a$ implies $\ell(a) = \ell(b)$, but not $a = b$; thus for example: $[1, 3] \oplus [1, 1] = [2, 4]$ and $[2, 4] \oplus [-1, -1] = [1, 3]$.

1.1.2. ($\mathcal{P}(\mathbb{R}^k)$, +), (Conv ($\mathbb{R}^k$), +), (Conv$_c$ ($\mathbb{R}^k$), +)

These are more general cases than Example 1.1.1. $\mathcal{P}(\mathbb{R}^k)$ is the power set of $\mathbb{R}^k$ (k integer ≥ 1).

The addition + of two subsets A and B of $\mathbb{R}^k$ (Minkowski addition) is defined as: $C = A + B$ with

$$C = \{x/x \in \mathbb{R}^k \quad \text{and} \quad x = a + b \quad \text{with} \quad a \in A, b \in B\}$$

This operation is *associative, commutative* and has neutral element $\{(0)^k\}$, the subset reduced to the zero vector of $\mathbb{R}^k$.

The set of convex subsets of $\mathbb{R}^k$, Conv($\mathbb{R}^k$), is a submonoid of $\mathcal{P}(\mathbb{R}^k)$ for the above-defined addition. To be convinced of this, just check that the addition operation on subsets is internal in Conv ($\mathbb{R}^k$).

The same is true for all the subsets of $\mathcal{P}(\mathbb{R}^k)$ for which addition is an internal operation. A typical example is Conv$_c$ ($\mathbb{R}^k$), the set of *compact convex* subsets of $\mathbb{R}^k$.

1.1.3. ($\mathcal{P}(\mathbb{R})$, .)

The set $\mathcal{P}(\mathbb{R})$ is the power set of the real line.

The law $\oplus$ is the product of two subsets of $\mathbb{R}$ defined as:

$$C = A\,.\,B = \{x/x = a\,.\,b \quad \text{with} \quad a \in A \quad \text{and} \quad b \in B\}$$

This operation is *associative, commutative* and has neutral element {1}, the subset reduced to the real number 1.

1.1.4. (Conv ($\mathbb{R}^{n+k}$), ⊕)

We consider the set of the convex subsets of $\mathbb{R}^{n+k}$ endowed with the *generalized sum* ⊕ defined for any A and B ∈ Conv ($\mathbb{R}^{n+k}$) by:

$$A \oplus B = \{(y, z_1 + z_2) \in \mathbb{R}^n \times \mathbb{R}^k / (y, z_1) \in A, (y, z_2) \in B\}.$$

One verifies that ⊕ is an *internal associative, commutative* operation having *neutral element* $\mathcal{P}(\mathbb{R}^n) \times \{(0)^k\}$.

[Reference: Rockafellar (1970), Chap. 3.]

1.1.5. $\left(\hat{\mathbb{R}}^k, \overset{(k)}{+}\right)$

The elements of $\hat{\mathbb{R}}^k$ are ordered k-tuples of real numbers of the form:

$$a = \begin{pmatrix} a^{(1)} \\ a^{(2)} \\ \vdots \\ a^{(k)} \end{pmatrix}$$

with ∀i: $a^{(i)} \in \hat{\mathbb{R}} = \mathbb{R} \cup \{+\infty\}$ and $a^{(1)} \le a^{(2)} \le \cdots \le a^{(k)}$

The operation $\overset{(k)}{+}$ is defined as:

$$c = \begin{pmatrix} c^{(1)} \\ c^{(2)} \\ \vdots \\ c^{(k)} \end{pmatrix} = a \overset{(k)}{+} b$$

where $c^{(1)}, c^{(2)}, \ldots c^{(k)}$ are the k smallest values considered in nondecreasing order out of the set:

$$Z = \{z/z = a^{(i)} + b^{(j)}, \quad 1 \le i \le k, \quad 1 \le j \le k\}$$

This operation is *commutative, associative* and has neutral element $\begin{pmatrix} 0 \\ 0 \\ \vdots \\ 0 \end{pmatrix}$.

This monoid *is not canonically ordered*, as illustrated by the following example:

Let us take k = 2, $a = \begin{pmatrix} 1 \\ 3 \end{pmatrix}$ $b = \begin{pmatrix} 3 \\ 5 \end{pmatrix}$

This yields:

$$\begin{pmatrix} 1 \\ 3 \end{pmatrix} \overset{(2)}{+} \begin{pmatrix} 2 \\ 6 \end{pmatrix} = \begin{pmatrix} 3 \\ 5 \end{pmatrix} \quad \text{therefore } a \le b$$

and

$$\begin{pmatrix} 3 \\ 5 \end{pmatrix} \overset{(2)}{+} \begin{pmatrix} -2 \\ 2 \end{pmatrix} = \begin{pmatrix} 1 \\ 3 \end{pmatrix} \quad \text{therefore} \quad b \le a$$

and yet $a \neq b$

[References: **4** (Sect. 6.8), **8** (Sect. 4.3.1)]

1.1.6. $\left(\hat{\mathbb{R}} + [0, \eta]^{(N)}, \overset{(\le\eta)}{+}\right)$

Let $\eta > 0$ be a given positive real number.

We consider the set E of sequences of real numbers of $\hat{\mathbb{R}} = \mathbb{R} \cup \{+\infty\}$ having variable finite length, and with extreme values differing by at most η.

For $a \in E$ we denote $\nu(a)$ the number of terms of the sequence a, and a can be written:

$$a = \left(a^{(1)}, a^{(2)}, \dots a^{\nu(a)}\right)$$

$$\text{with} \quad a^{(1)} = \underset{i=1\dots\nu(a)}{\text{Min}} \left\{a^{(i)}\right\} \qquad a^{\nu(a)} = \underset{i=1\dots\nu(a)}{\text{Max}} \left\{a^{(i)}\right\}$$

$$\text{and} \quad a^{\nu(a)} - a^{(1)} \le \eta$$

Thus E can be considered as a subset of $\hat{\mathbb{R}} + [o, \eta]^{(N)}$ where $[0, \eta]^{(N)}$ denotes the set of finite sequences of elements in the real interval $[0, \eta]$.

We consider on E the law $\overset{(\le\eta)}{+}$ defined as follows:

$$a = \left(a^{(1)}, a^{(2)}, \dots, a^{\nu(a)}\right)$$

$$b = \left(b^{(1)}, b^{(2)}, \dots, b^{\nu(b)}\right)$$

$$\text{then} \quad c = a \overset{(\le\eta)}{+} b = \left(c^{(1)}, c^{(2)}, \dots, c^{\nu(c)}\right)$$

with:

$$c^{(1)} = a^{(1)} + b^{(1)} = \underset{i=1\dots\nu(a)}{\text{Min}} \left\{a^{(i)}\right\} + \underset{i=1\dots\nu(b)}{\text{Min}} \left\{b^{(i)}\right\}$$

and where c is the sequence formed by all the values of the form $a^{(i)} + b^{(j)}$ ($i = 1, \dots \nu(a); j = 1, \dots \nu(b)$) such that:

$$c^{(1)} \le a^{(i)} + b^{(j)} \le c^{(1)} + \eta$$

This therefore implies $\nu(c) \le \nu(a) . \nu(b)$.

Example. For $\eta = 3$: $(2, 3, 5) \overset{(\le\eta)}{+} (1, 4) = (3, 4, 6, 6)$

One verifies that $\overset{(\le\eta)}{+}$ is *commutative, associative* and has neutral element $\varepsilon = (0)$ (the sequence formed by a single term equal to 0).

However $\overset{(\le\eta)}{+}$ is not idempotent as shown by the following example where $\eta = 2$: $(1, 2, 3) \overset{(\le\eta)}{+} (1, 2, 3) = (2, 3, 3, 4, 4, 4)$.

[References: **4** (Sect. 6;10), **8** (Sect. 4.3.2)]

1.1.7. Qualitative Multiplication

On the set of signs $E = \{+, -, 0\}$, we consider the product of signs $\otimes$ defined by the table:

$\otimes$	$+$	$-$	0
$+$	$+$	$-$	0
$-$	$-$	$+$	0
0	0	0	0

It is readily checked that $(E, \otimes)$ is a monoid with neutral element $+$.

E can be augmented with the indeterminate sign (denoted: ?) such that: $? \otimes + = ?$, $? \otimes - = ?$, $? \otimes 0 = 0$, $? \otimes ? = ?$.

We still have a monoid structure with neutral element $+$. The element denoted $+$ can be interpreted as representing the real interval $]0, +\infty[$, $-$ as representing $]-\infty, 0[$, ? as representing $]-\infty, +\infty[$, and 0 as representing the set $\{0\}$.

Observe that, contrary to qualitative addition (see Sect. 1.5.5), $\otimes$ is not idempotent and the monoid $(E, \otimes)$ is not canonically ordered.

[References: **1** (Sect. 3.4), **1** (Sect. 6.1.3)]

1.2. Groups

1.2.1. $(\mathbb{R}, +)$, $(\mathbb{Z}, +)$, $(\mathbb{C}, +)$

The set of real numbers (the set of signed integers, of complex numbers) endowed with ordinary addition. $+$ is *commutative, associative* and has as neutral element 0. Every element has an inverse for $+$.

1.2.2. $(\mathbb{R}^*, \times)$, $(\mathbb{Q}^*, \times)$, $(\mathbb{C}^*, \times)$

The set of nonzero real numbers (the set of nonzero rational numbers, nonzero complex numbers) endowed with ordinary multiplication $\times$ which is commutative, associative with neutral element 1. Every element has an inverse for $\times$.

1.2.3. $(\mathbb{R}\backslash\{1\}, a \oplus b = a + b - ab)$

The set $\mathbb{R}\backslash\{1\}$ endowed with the operation $\oplus$ defined as: $a \oplus b = a + b - ab$ is associative, commutative and has 0 as neutral element. Every element a is invertible: $a^{-1} = \dfrac{a}{a-1}$.

[Reference: **1** (Sect. 2.1)]

1.3. Canonically Ordered Monoids

The subclass of canonically ordered monoids includes:

- Hemi-groups (see Sect. 1.4);
- Idempotent monoids and selective monoids (see Sects. 1.5 and 1.6);
- Other canonically ordered monoids which are neither semi-groups nor idempotent monoids, which we introduce in the present section.

1.3.1. $\left(\hat{\mathbb{R}}^k, \mathrm{Min}_{(k)}\right)$

The elements of $\hat{\mathbb{R}}^k$ are ordered k-tuples of real numbers $\in \hat{\mathbb{R}}$.

$$\text{If} \quad a = \begin{pmatrix} a^{(1)} \\ a^{(2)} \\ \vdots \\ a^{(k)} \end{pmatrix} \in E \quad \text{this implies: } a^{(1)} \le a^{(2)} \le \cdots \le a^{(k)}$$

The law $\oplus$ on E is the operation $\mathrm{Min}_{(k)}$ defined for:

$$a = \begin{pmatrix} a^{(1)} \\ a^{(2)} \\ \vdots \\ a^{(k)} \end{pmatrix} \quad b = \begin{pmatrix} b^{(1)} \\ b^{(2)} \\ \vdots \\ b^{(k)} \end{pmatrix}$$

$$\text{as} \quad c = \begin{pmatrix} c^{(1)} \\ c^{(2)} \\ \vdots \\ c^{(k)} \end{pmatrix} = a \ \mathrm{Min}_{(k)} \ b$$

where $c^{(1)}\ c^{(2)}\ \cdots\ c^{(k)}$ are the k smallest values taken in the set

$$\left\{a^{(1)}, a^{(2)}, \dots a^{(k)}, b^{(1)}, b^{(2)}, \dots, b^{(k)}\right\}$$

$\oplus$ is commutative, associative and has a neutral element $\varepsilon = \begin{pmatrix} +\infty \\ +\infty \\ \vdots \\ +\infty \end{pmatrix}$

We observe that $\oplus$ is not idempotent but is "*k-idempotent*," that is to say:

$$\underbrace{\begin{pmatrix} a^{(1)} \\ \vdots \\ a^{(k)} \end{pmatrix} \oplus \cdots \oplus \begin{pmatrix} a^{(1)} \\ \vdots \\ a^{(k)} \end{pmatrix}}_{\text{k times}} = \begin{pmatrix} a^{(1)} \\ \vdots \\ a^{(1)} \end{pmatrix} = \underbrace{\begin{pmatrix} a^{(1)} \\ \vdots \\ a^{(k)} \end{pmatrix} \oplus \cdots \oplus \begin{pmatrix} a^{(1)} \\ \vdots \\ a^{(k)} \end{pmatrix}}_{\text{k+1 times}}$$

We deduce from the above that the canonical preorder relation $\propto$ is an *order* (see Chap. 1, Proposition 3.4.6).

[References: **4** (Sect. 4.6.8), **8** (Sect. 4.3.1)]

1.3.2. $(\hat{\mathbb{R}} + [0, \eta]^{(\mathbf{N})}, \mathbf{Min}_{(\leq\eta)})$

Let $\eta > 0$ be a given positive real number. As in Example 1.1.6, we consider the set $E = \hat{\mathbb{R}} + [0, \eta]^{(\mathbf{N})}$ whose elements are sequences of real numbers $\in \hat{\mathbb{R}}$ of finite (variable) length with extreme values which differ by η at most.

The law $\oplus$ on E is the operation $\text{Min}_{(\leq\eta)}$ defined as:

$$a = \left(a^{(1)}, a^{(2)}, \ldots a^{\nu(a)}\right), b = \left(b^{(1)}, b^{(2)}, \ldots b^{(\nu(b))}\right)$$

then:

$$c = a \ \text{Min}_{(\leq\eta)} b = \left(c^{(1)}, c^{(2)}, \ldots c^{\nu(c)}\right)$$

where c is the sequence formed by all the terms $a^{(i)}$ and $b^{(j)}$ satisfying:

$$a^{(i)} \leq z + \eta \quad \text{and} \quad b^{(j)} \leq z + \eta$$

with

$$z = \text{Min}\left\{\underset{i=1\ldots\nu(a)}{\text{Min}}\left\{a^{(i)}\right\};\ \underset{j=1\ldots\nu(b)}{\text{Min}}\left\{b^{(j)}\right\}\right\}$$

Observe that: $\nu(c) \leq \nu(a) + \nu(b)$

Example:

$$\begin{aligned} &\text{let} \quad \eta = 3 \quad \text{and} \quad a = (4, 6, 7), b = (2, 3, 5) \\ &\text{so} \quad c^{(1)} = 2 \quad \text{and} \quad c = (2, 3, 4, 5) \end{aligned}$$

$\oplus$ is *commutative, associative* and has a neutral element ε which is the element $(+\infty)$ (sequence formed by a single term equal to $+\infty$).

Observe that $\oplus$ is neither idempotent nor cancellative. Thus for example, for $\eta = 2$: $(1, 2, 3) \oplus (1, 2, 3) = (1, 1, 2, 2, 3, 3)$ (non idempotent)
and:

$$\begin{aligned} (1) \oplus (2) &= (1, 2) \\ (1) \oplus (2, 4) &= (1, 2) \end{aligned}$$

(non cancellative).

Let us show that the canonical preorder relation is an *order relation*. Therefore consider two elements $a \in E$, $b \in E$ such that there exists $u \in E$ and $v \in E$ satisfying:

$$\begin{aligned} &a \oplus u = b \quad \text{and} \quad a = b \oplus v \\ &\text{as} \quad b^{(1)} = \text{Min}\left\{a^{(1)}, u^{(1)}\right\} \quad \text{we have:} \quad b^{(1)} \leq a^{(1)} \\ &\text{As} \quad a^{(1)} = \text{Min}\left\{b^{(1)}, v^{(1)}\right\} \quad \text{we have:} \quad a^{(1)} \leq b^{(1)} \end{aligned}$$

hence we deduce $a^{(1)} = b^{(1)}$.

Now, given the equality $a^{(1)} = b^{(1)}$:

$$b = a \oplus u \Rightarrow \nu(b) \geq \nu(a)$$
$$\text{and} \quad a = b \oplus v \Rightarrow \nu(a) \geq \nu(b)$$

which implies: $\nu(a) = \nu(b) = k$

For the latter relation to be satisfied, it is then necessary that: $u^{(1)} > a^{(1)} + \eta$ from which we deduce that: $b = a \oplus u = a$. $(E, \oplus)$ is therefore a canonically ordered monoid. □

[References: **4** (Sect. 6.10), **8** (Sect. 4.3.2)]

1.3.3. Nilpotent t-Norm: ([0, 1], Max{0, a + b − 1})

On the set [0, 1], we define the operation $\otimes$ by: $a \otimes b = \max\{0, a + b - 1\}$.

$\otimes$ is associative, commutative and has 1 as neutral element. One verifies that it induces a structure of canonically ordered monoid. Indeed, $a \leq b$ and $b \leq a$ imply that there exists $c \in [0, 1]$ and $d \in [0, 1]$ such that:

$$\max\{0, a + c - 1\} = b \quad \text{and} \quad \max\{0, b + d - 1\} = a$$

and if $a \neq 0$ and $b \neq 0$, this yields $a + c - 1 = b$ and $b + d - 1 = a$, hence $c + d = 2$ and therefore $c = d = 1$, hence $a = b$.

One verifies that every element $a \neq 1$ is nilpotent: $\underbrace{a \otimes a \otimes \cdots \otimes a}_{\text{n times}} = 0$ as soon as $n \geq \dfrac{1}{1-a}$. This law $\otimes$ is referred to as a t-*norm*, (see Exercise 2, Chap. 1).

[References: **1** (Sect. 3.2.5), **1** (Exercise 2)]

1.3.4. Nilpotent t-conorm: ([0, 1], Min {a + b, 1})

On the set [0, 1], we define the operation $a \oplus b = \text{Min}\{a + b, 1\}$.

$\oplus$ is associative, commutative and has 0 as neutral element. One easily checks that it induces a canonically ordered structure. One verifies that every element $a \neq 0$ is nilpotent: $n\, a = 1$ as soon as $n > \frac{1}{a}$. This is referred to as a t-*conorm*, (see Exercise 2, Chap. 1).

[References: **1** (Sect. 3.2.5), **1** (Exercise 2)]

1.3.5. ([0, 1], a ⊕ b = a + b − ab)

On the set [0, 1], we define the operation $a \oplus b = a + b - ab$.

$\oplus$ is associative, commutative and has 0 as neutral element. One easily checks that it induces the structure of a canonically ordered monoid.

The antisymmetry of the canonical preorder relation is proved as follows. Assuming that both $a \geq b$ and $a \leq b$ hold, there exist $c \in [0, 1]$ and $d \in [0, 1]$ such that

$$a = b + c - bc \quad \text{and} \quad b = a + d - ad.$$

Eliminating b from these two relations yields:

$$(1 - a)[d(1 - c) + c] = 0$$

If $a \neq 1$, $d(1 - c) + c = 0$ implies $c = 0$ and $d = 0$, hence $a = b$.
If $a = 1$, then $b = 1$ and in this case, too, $a = b$.

1.3.6. Order of Magnitude Monoid

On the set E of pairs $(a, \alpha) \in (\mathbb{R}_+\backslash\{0\}) \times \mathbb{R}$, to which we add the pair $(0, +\infty)$, we define the operation $\oplus$ by:

$$(a, \alpha) \oplus (b, \beta) = (c, \min(\alpha, \beta))$$

with $c = a$ if $\alpha < \beta$, $c = b$ if $\beta < \alpha$, $c = a + b$ if $\alpha = \beta$.

$\oplus$ is associative, commutative and has $(0, +\infty)$ as neutral element and induces the structure of a canonically ordered monoid.

The elements (a, α) are in $1 - 1$ correspondence with the numbers of the form a ε^{α} when ε tends towards 0^+.

[Reference: **1** (Sect. 3.4)]

1.3.7. Nonstandard Number Monoid

On the set E of triples $(a, b, \alpha) \in (\mathbb{R}_+\backslash\{0\})^3$ to which we add the triple $(0, 0, +\infty)$, we define the operation $\oplus$ by:

$$(a_1, b_1, \alpha_1) \oplus (a_2, b_2, \alpha_2) = (a_1 + a_2, b, \min\{\alpha_1, \alpha_2\})$$

with $b = b_1$ if $\alpha_1 < \alpha_2$, $b = b_2$ if $\alpha_2 < \alpha_1$, $b = b_1 + b_2$ if $\alpha_1 = \alpha_2$.

This is a canonically ordered monoid, product of $(\mathbb{R}_+, +)$ with the order of magnitude monoid (Sect. 1.3.6). It is isomorphic to the set of nonstandard numbers of the form $a + b\varepsilon^{\alpha}$ endowed with ordinary addition when ε tends to 0^+.

[Reference: **1** (Sect. 6.1.6)]

1.3.8. Power Monoid

On the set E of pairs $(a, A) \in (\mathbb{R}_+\backslash\{0\})^2$ to which we add the pair $(0, 0)$, we define the operation $\oplus$ as

$$(a, A) \oplus (b, B) = (c, \max\{A, B\})$$

with $c = a$ if $A > B$, $c = b$ if $A < B$, $c = a + b$ if $A = B$.

This monoid is isomorphic to the one in Sect. 1.3.7 by setting $A = e^{-\alpha}$.

This canonically ordered monoid is, moreover, isomorphic to the set of numbers of the form a A^p when p tends to $+\infty$, endowed with ordinary addition.

[Reference: **1** (Sect. 3.4.4)]

1.4. Hemi-Groups

Such monoids are both *canonically ordered* and *cancellative* (see Chap. 1, Sect. 3.5).

1.4.1. $(\mathbb{R}_+, +)$, $(\mathbb{N}, +)$

The set of nonnegative real numbers (resp. set of natural numbers) endowed with ordinary addition $+$.

$+$ is *associative, commutative* and has neutral element 0. Every element is *cancellative* and the canonical preorder relation is an *order*.

A virtually unlimited number of hemi-groups can be deduced by isomorphism from $(\mathbb{R}_+, +)$. Thus, with every one-to-one correspondence φ of $M \subset \mathbb{R}$ in $\mathbb{R}_+$, one can associate the hemi-group $(M, \oplus)$, where $\oplus$ is defined as: $a \oplus b = \varphi^{-1}(\varphi(a) + \varphi(b))$.

Examples will be found in sections 1.4.5, 1.4.7, 1.4.8 below.

1.4.2. $(]0, 1], \times)$

The set of real numbers of the interval $]0, 1]$ endowed with ordinary multiplication.

$\times$ is *associative, commutative* and has neutral element 1. Every element is *cancellative* and the canonical preorder relation is an *order*.

Note that $(]0, 1], \times)$ and $(\mathbb{R}_+, +)$ are isomorphic through φ defined as: $\varphi(x) = -\ln(x)$.

1.4.3. $(\mathbb{R}_+\backslash\{0\}, \times)$, $(\mathbb{N}_*, \times)$

The set of strictly positive real numbers (resp. strictly positive natural numbers) endowed with ordinary multiplication. $\times$ is *associative, commutative* and has neutral element 1. Every element is *cancellative* and the canonical preorder relation is an *order*.

1.4.4. The Free Monoid

Let A be a set (called "alphabet") whose elements are called *letters*.

We take for E the set of finite sequences of elements of A which we call *words*, and we define the operation $\otimes$ as *concatenation* i.e.:

$$
\begin{aligned}
\text{if} \quad & m_1 \in E : m_1 = s_1 s_2 \ldots s_p \\
& m_2 \in E : m_2 = t_1 t_2 \ldots t_q \\
& m_1 \otimes m_2 = s_1 s_2 \ldots s_p t_1 t_2 \ldots t_q
\end{aligned}
$$

The set of words on A endowed with the operation of concatenation, denoted A^*, is called the *free monoid on* A.

The canonical preorder relation is an order:

$$m_1 \le m_2 \Leftrightarrow m_1 \text{is a } \textit{prefix} \text{ of } m_2.$$

The operation $\otimes$ is not commutative. Finally, we observe that in E, every element is right-cancellative and left-cancellative. Indeed:

$$m \otimes m' = m \otimes m'' \Rightarrow m' = m''$$

and

$$m' \otimes m = m'' \otimes m \Rightarrow m' = m''.$$

[References: **1** (Sect. 2.1.13), **1** (Sect. 2.3.2)]

1.4.5. $(\mathbb{R}_+\backslash\{1\}, a \oplus b = \dfrac{a+b}{1+ab})$, $(\mathbb{R}_+, a \oplus b = a\,(1+b^2)^{1/2} + b\,(1+a^2)^{1/2})$

The set $\mathbb{R}_+\backslash\{1\}$ endowed with the operation $\oplus$ defined as $a \oplus b = \dfrac{a+b}{1+ab}$ is a hemi-group (see Example 2.3.3, Chap. 1), because it is isomorphic to $(\mathbb{R}_+, +)$ by the hyperbolic tangent transform; we have indeed:

$$th(w_1 + w_2) = \frac{th(w_1) + th(w_2)}{1 + th(w_1) \cdot th(w_2)}.$$

The same is true for $\mathbb{R}_+$ endowed with the operation $\oplus$ defined as $a \oplus b = a\,(1 + b^2)^{1/2} + b\,(1 + a^2)^{1/2}$ because it is isomorphic to $(\mathbb{R}_+, +)$ by the hyperbolic sine transform.

[Reference: **1** (Exercise 1)]

1.4.6. (Int $(\mathbb{R}_+)$, +)

This is the set of intervals on $\mathbb{R}_+$, that is to say elements a of the form $a = [\underline{a}, \bar{a}]$ with $0 \le \underline{a} \le \bar{a}$ and endowed with the addition of intervals:

$$a \oplus b = [\underline{a} + \underline{b}, \bar{a} + \bar{b}].$$

As for Int $(\mathbb{R})$ (see Example 1.1.1), the operation $\oplus$ is associative, commutative and has [0, 0] as neutral element. But one easily checks that Int $(\mathbb{R}_+)$ is moreover canonically ordered by $\oplus$ and that each element (interval) is cancellative.

1.4.7. $(\mathbb{R}_+, a \oplus_p b = (a^p + b^p)^{1/p})$

This is the set of nonnegative reals endowed with the operation $\oplus_p$ (with $p \in \mathbb{R}_*$), defined as: $a \oplus_p b = (a^p + b^p)^{1/p}$.

This is a hemi-group which is isomorphic to $(\mathbb{R}_+, +)$. Moreover, it is observed that: $\lim\limits_{p\to+\infty} (a \oplus_p b) = \max\{a, b\}$.

[Reference: **1** (Sect. 3.2.5)]

1.4.8. $(\mathbb{R}, \mathbf{a} \oplus^{\mathbf{h}} \mathbf{b} = \mathbf{h} \ln (\mathbf{e}^{\mathbf{a}/\mathbf{h}} + \mathbf{e}^{\mathbf{b}/\mathbf{h}}))$

For any real number $h \neq 0$, $\mathbb{R}$ endowed with the operation $a \oplus^h b = h\, ln\, (e^{a/h} + e^{b/h})$ is a hemi-group which is isomorphic to $(\mathbb{R}_+, +)$.

When h tends to 0^+, this yields $\lim(a \oplus^h b) = \max\{a, b\}$, and when h tends to 0^-, we have $\lim(a \oplus^h b) = \min\{a, b\}$.

[Reference: **1** (Sect. 3.2.5)]

1.5. Idempotent Monoids (Semi-Lattices)

The subclass of idempotent monoids includes:

- Selective monoids (see Sect. 1.6);
- Other idempotent monoids which are not selective monoids and which we introduce in the present section.

We observe that, for all these examples, the operation $\oplus$ being idempotent, the canonical preorder relation is an *order*.

1.5.1. $(\mathbb{N}, \textbf{gcd})$

The set of natural numbers endowed with the gcd operation ("greatest common divisor") which associates gcd (a, b) with every pair of integers a, b.

This operation is *associative, commutative, idempotent* and has neutral element $\varepsilon = +\infty$, by viewing ε as the (infinite) product of all the prime numbers raised to the power $+\infty$.

1.5.2. $(\mathbb{N}, \textbf{lcm})$

The set of natural numbers endowed with the lcm operation ("least common multiple") which associates lcm(a, b) with every pair of integers a, b.

This operation is *associative, commutative, idempotent* and has as neutral element 1.

1.5.3. $(\mathcal{P}(\mathbf{X}), \cup)$

The power set of a given set X endowed with the *union* operation.

This operation is *associative, commutative, idempotent* and has as neutral element the empty subset $\emptyset$.

1.5.4. $(\mathcal{P}(\mathbf{X}), \cap)$

The power set of a given set X endowed with the operation *intersection*. This operation is *associative, commutative, idempotent* and has as neutral element the set X itself.

1.5.5. Qualitative Addition

The set E = {+, −, 0, ?} endowed with the idempotent operation ⊕ defined as

⊕	+	−	0	?
+	+	?	+	?
−	?	−	−	?
0	+	−	0	?
?	?	?	?	?

is a canonically ordered monoid, with neutral element 0 and absorbing element ?.

The following relations hold: ? $\geq$ + $\geq$ 0 and ? $\geq$ − $\geq$ 0, which can be illustrated by the following graphic representation:

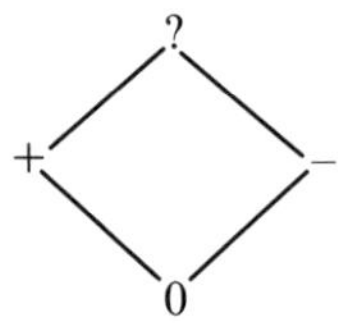

By identifying E, with $\{0, 1\}^2$, ? ≡ (1, 1), + ≡ (1, 0), − ≡ (0, 1) and 0 ≡ (0, 0), it is observed that (E, ⊕) is isomorphic to the square of the monoid ({ 0, 1}, max).

[Reference: **1** (Sect. 3.4.2)]

1.5.6. (Conv ($\mathbb{R}^n$), A⊕B = Conv (A∪B)), (Conv$_c$ ($\mathbb{R}^n$), A⊕B = Conv (A∪B))

We consider the set of the convex subsets of $\mathbb{R}^n$, conv ($\mathbb{R}^n$), endowed with the operation ⊕ defined as:

$$A \oplus B = \text{conv}(A \cup B)$$

where conv (X) denotes the convex hull of a subset X.

This operation is *associative, commutative, idempotent* and has as neutral element the empty subset Ø.

The set of *compact* convex sets of $\mathbb{R}^n$, Conv$_c$ ($\mathbb{R}^n$), is an idempotent sub-monoid of Conv ($\mathbb{R}^n$) for ⊕.

1.5.7. (Conv ($\mathbb{R}^n$), ∩), (Conv$_c$ ($\mathbb{R}^n$), ∩)

Here we consider the set of the convex subsets of $\mathbb{R}^n$, Conv ($\mathbb{R}^n$), endowed with the operation *intersection*. This operation is *associative, commutative*, idempotent and has as neutral element, the set $\mathbb{R}^n$ itself.

The set of *compact* convex sets of $\mathbb{R}^n$, Conv$_c$ ($\mathbb{R}^n$), is a idempotent sub-monoid of Conv ($\mathbb{R}^n$) for intersection.

1.5.8. (Int ($\mathbb{R}$), ∩)

This is the special case of (conv ($\mathbb{R}^n$), ∩) with $n = 1$, (see Sect. 1.5.7 above).

1.5.9. (Int ($\mathbb{R}$), Conv (A ∪ B))

This is the special case of (conv ($\mathbb{R}^n$), $A \oplus B = \text{conv}(A \cup B)$) with $n = 1$, (see Sect. 1.5.6).

Let us complete this presentation of idempotent monoids with examples involving vectors "monitored" by the first component.

1.5.10. ($\hat{\mathbb{R}} \times \mathcal{P}(\mathbb{R})$, Min)

$\mathcal{P}(\mathbb{R})$ denoting the power set of $\mathbb{R}$, we consider the set $\hat{\mathbb{R}} \times \mathcal{P}(\mathbb{R})$ of elements $a = (a_1, a_2)$ such that $a_1 \in \hat{\mathbb{R}}$ and $a_2 \in \mathcal{P}(\mathbb{R})$, endowed with the Min operation defined as follows:

$$\text{Min}(a, b) = \begin{cases} a & \text{if} \quad a_1 < b_1 \\ b & \text{if} \quad b_1 < a_1 \\ (a_1, a_2 \cup b_2) & \text{if} \quad a_1 = b_1 \end{cases}$$

This operation is associative, commutative, idempotent and has $(+\infty, \emptyset)$ as neutral element.

This type of monoid can be used to define a minimum operation for complex numbers.

Case 1. if $z_1 = x_1 + iy_1$ and $z_2 = x_2 + iy_2$ are two complex numbers, we will take

$$\text{Min}(z_1, z_2) = \begin{cases} z_1 & \text{if} \quad x_1 < x_2 \\ z_2 & \text{if} \quad x_2 < x_1 \\ \{z_1, z_2\} & \text{if} \quad x_1 = x_2 \end{cases}$$

where $\{z_1, z_2\}$ denotes the set of complex numbers with real component $x_1 = x_2$ and $y_1 \cup y_2$ as imaginary component.

Observe that we only keep a single value for the real component, but that the imaginary component can take *multiple values*.

Case 2. If z_1 and z_2 are two complex numbers ($|\,.\,|$ denoting the modulus), we will take:

$$\text{Min}(z_1, z_2) = \begin{cases} z_1 & \text{if} \quad |z_1| < |z_2| \\ z_2 & \text{if} \quad |z_2| < |z_1| \\ \{z_1, z_2\} & \text{if} \quad |z_1| = |z_2| \end{cases}$$

Observe that we only keep a single value for the modulus, but that the argument can take multiple values.

[References: this type of monoid appears to be useful for studying complex Hamilton–Jacobi equations arising in quantum mechanics (see Gondran 1999b, 2001a,b)].

1.6. Selective Monoids

For all the examples in this section, the operation $\oplus$ being selective, the canonical preorder relation is a *total order* (see Chap. 1, Proposition 3.4.7).

1.6.1. $(\hat{\mathbb{R}}, \mathrm{Min})$, $(\hat{\mathbb{Z}}, \mathrm{Min})$

The set of real numbers (resp. signed integers) endowed with the operation Minimum of two numbers.

This operation is *associative, commutative, selective* and has neutral element $+\infty$.

1.6.2. $(\check{\mathbb{R}}, \mathrm{Max})$, $(\check{\mathbb{Z}}, \mathrm{Max})$

The set of real numbers (resp. signed integers) endowed with the operation Maximum of two numbers.

This operation is *associative, commutative, selective* and has neutral element $-\infty$.

1.6.3. $(\mathbb{R}_+, \mathrm{Max})$, $(\mathbb{N}, \mathrm{Max})$

The set of nonnegative reals (resp. natural numbers) endowed with the operation "maximum of two numbers."

This operation is *associative, commutative, selective* and has neutral element 0.

1.6.4. $(\hat{\mathbb{R}}_+, \mathrm{Min})$, $(\hat{\mathbb{N}}, \mathrm{Min})$

The set of nonnegative reals (resp. of natural numbers) endowed with the operation "Minimum of two numbers." This operation is *associative, commutative, selective* and has as neutral element $+\infty$.

1.6.5. $(\hat{\mathbb{R}}^n$, Min-Lexicographic)

We consider the set of vectors of $\hat{\mathbb{R}}^n$ (totally) ordered by the lexicographic order:

$$\begin{aligned} a \underline{\propto} b \quad \text{if} \quad & a_1 < b_1 \\ \text{or} \quad & a_1 = b_1 \quad \text{and} \quad a_2 < b_2 \\ & \ldots\ldots \\ \text{or} \quad & a_1 = b_1 \quad \text{and} \quad a_2 = b_2, \ \ldots, \ a_{n-1} = b_{n-1}, \quad \text{and} \quad a_n < b_n \end{aligned}$$

The operation Min-lexicographic is associative, commutative and selective and has $(+\infty)^n$ neutral element.

This type of monoid can be used for example to define the minimum of complex numbers: $z_1 = x_1 + iy_1$, $z_2 = x_2 + iy_2$.

$$\mathrm{Min}(z_1, z_2) = \begin{cases} z_1 & \text{if} \quad x_1 < x_2 \quad \text{or if} \quad x_1 = x_2 \quad \text{and} \quad y_1 < y_2 \\ z_2 & \text{if} \quad x_2 < x_1 \quad \text{or if} \quad x_1 = x_2 \quad \text{and} \quad y_2 < y_1 \end{cases}$$

Table 2 Recapitulatory list of monoids

General Monoids	Properties of $\oplus$	Neutral element	Canonical preorder $\leq$	Additional properties and comments
$(\mathcal{P}(\mathbb{R}^k), +)$	Associative Commutative	$\{(0)^k\}$	Preorder	
$(\mathrm{Int}(\mathbb{R}), +)$	Associative Commutative	$[0, 0]$	Preorder	
$(\mathrm{Conv}(\mathbb{R}^k), +)$	Associative Commutative	$\{(0)^k\}$	Preorder	
$(\mathrm{Conv}(\mathbb{R}^{n+k}), \oplus)$	Associative Commutative	$\mathcal{P}(\mathbb{R}^n)\times \{(0)^k\}$	Preorder	
$(\mathcal{P}(\mathbb{R}), \cdot)$	Associative Commutative	$\{1\}$	Preorder	
$((\hat{\mathbb{R}}^k, \overset{(k)}{+})$	Associative Commutative	$(0)^k$	Preorder	
$\left(\hat{\mathbb{R}} + [0, \eta]^{(N)}, \overset{(\eta)}{+}\right)$	Associative Commutative	(0)	Preorder	
Qualitative multiplication	Associative Commutative	$+$	Preorder	
Groups				
$(\mathbb{R}, +), (\mathbb{Q}, +), (\mathbb{C}, +)$	Associative Commutative	0	Preorder	
$(\mathbb{R}^*, \times), (\mathbb{Q}^*, \times), (\mathbb{C}^*, \times)$	Associative Commutative	1	Preorder	
$(\mathbb{R}\backslash\{1\}, a + b - ab)$	Associative Commutative	0	Preorder	
Canonically Ordered Monoids				
$\left(\hat{\mathbb{R}}^k, \mathrm{Min}_{(k)}\right)$	Associative Commutative	$(+\infty)^k$	Order	
$\left(\hat{\mathbb{R}} + [0, \eta]^{(N)}, \mathrm{Min}_{(\eta)}\right)$	Associative Commutative	$(+\infty)$	Order	
$([0, 1], \min (a + b;\ 1))$	Associative Commutative	0	Order	Every element is nilpotent
$([0, 1], \max (0;\ a + b - 1))$	Associative Commutative	1	Order	Every element is nilpotent
$([0, 1], a + b - ab)$	Associative Commutative	0	Order	
Order of magnitude monoid	Associative Commutative	$(0, +\infty)$	Order	
Nonstandard number monoid	Associative Commutative	$(0, 0, +\infty)$	Order	
Power monoid	Associative Commutative	$(0, 0)$	Order	

Table 2 (continued)

Hemi-groups	Properties of $\oplus$	Neutral element	Canonical preorder $\leq$	Additional properties and comments
$(\mathbb{R}_+, +)$	Associative Commutative	0	Total order	
$(\mathbb{N}, +)$	Every element is cancellative			
$(]0, 1], \times)$	Associative Commutative Every element is cancellative	1	Total order	
$(\mathbb{N}_*, \times)$ $(\mathbb{R}_+\backslash\{0\}, \times)$	Associative Commutative Every element is cancellative	1	Total order	
Free monoid	Concatenation Associative Every element is cancellative	$\emptyset$ (empty word)	Order	Concatenation is not commutative
$(\mathbb{R}_+\backslash\{1\}, \frac{a+b}{1+ab})$	Associative Commutative Every element is cancellative	0	Total order	
$(\mathbb{R}_+, a(1+b^2)^{1/2} + b(1+a^2)^{1/2})$	Associative Commutative Every element is cancellative	0	Total order	
(Int $(\mathbb{R}_+), +)$	Associative Commutative Every element is cancellative	[0, 0]	Order	
$(\mathbb{R}_+, \oplus_p)$ $a \oplus_p b = (a^p + b^p)^{1/p}$	Associative Commutative Every element is cancellative	0	Total order	
$(\mathbb{R}_+, \oplus^h)$ $a \oplus^h b = h\ln(e^{a/h} + e^{b/h})$	Associative Commutative Every element is cancellative	$-\infty$ (if $h > 0$) $+\infty$ (if $h < 0$)	Total order	

Idempotent Monoids

$(\mathbb{N}, \text{gcd})$	Associative Commutative Idempotent	$+\infty$	Order	
$(\mathbb{N}, \text{lcm})$	Associative Commutative Idempotent	1	Order	
$(\mathcal{P}(X), \cup)$	Associative Commutative Idempotent	$\emptyset$	Order	Sup-semi lattice

Table 2 (continued)

Idempotent Monoids (*continued*)	Properties of $\oplus$	Neutral element	Canonical preorder $\leq$	Additional properties and comments
$(\mathcal{P}\ (X), \cap)$	Associative Commutative Idempotent	X	Order	Inf-semi-lattice
Qualititative addition	Associative Commutative Idempotent	0	Order	
$(\mathrm{Conv}(R^n), \mathrm{Conv}\ (A \cup B))$	Associative Commutative Idempotent	$\emptyset$	Order	
$(\mathrm{Conv}(\mathbb{R}^n), \cap)$	Associative Commutative Idempotent	$\mathbb{R}^n$	Total order	
$(\mathrm{Int}(\mathbb{R}), \cap)$	Associative Commutative Idempotent	$\mathbb{R}$	Total order	
$(\mathrm{Int}(\mathbb{R}), \mathrm{Conv}(A \cup B))$	Associative Commutative Idempotent	$\emptyset$	Total order	
$(\hat{\mathbb{R}} \times \mathcal{P}\ (R), \mathrm{Min})$	Associative Commutative Idempotent	$\{+\infty, \emptyset\}$	Order	
Selective Monoids				
$(\hat{\mathbb{R}}, \mathrm{Min})\ (\hat{\mathbb{Z}}, \mathrm{Min})$	Associative Commutative Selective	$+\infty$	Total order	
$(\check{\mathbb{R}}, \mathrm{Max})\ (\check{\mathbb{Z}}, \mathrm{Max})$	Associative Commutative Selective	$-\infty$	Total order	
$(\mathbb{R}_+, \mathrm{Max})\ (\mathbb{N}, \mathrm{Max})$	Associative Commutative Selective	0	Total order	
$(\hat{\mathbb{R}}_+, \mathrm{Min})\ (\hat{\mathbb{N}}, \mathrm{Min})$	Associative Commutative Selective	$+\infty$	Total order	
$(\hat{\mathbb{R}}^n, \text{Min-lexico})$	Associative Commutative Selective	$(+\infty)^n$	Total order	

2. Pre-Semirings and Pre-Dioids

Pre-semirings and pre-dioids are algebraic structures with two operations which do not enjoy all the properties of semirings and dioids. In this section we present a few typical examples of such structures. Table 3 recalls the basic definitions. Figure 2

Table 3 Definition of pre-semirings and pre-dioids

Pre-semiring $(E, \oplus, \otimes)$	A set E endowed with two internal laws of $\oplus$ and $\otimes$ where $\oplus$ is associative $\otimes$ is associative and right and/or left distributive with respect to $\oplus$
Pre-dioid $(E, \oplus, \otimes)$	Pre-semiring for which $(E, \oplus)$ is a canonically ordered monoid

shows the place of pre-semirings and pre-dioids in the typology. We limit ourselves to the case where there is at least right or left distributivity of $\otimes$ with respect to $\oplus$, see Sect. 4.1.2, Chap. 1.

2.1. *Right or Left Pre-Semirings and Pre-Dioids*

2.1.1. The Set of Mappings of a Commutative Monoid onto Itself

Let $(E, \dot{+})$ be a commutative monoid, and H the set of mappings $E \rightarrow E$. One endows H with the operations $\oplus$ and $\otimes$ defined as:

$$(f \oplus g)(a) = f(a) \dot{+} g(a)$$
$$(f \otimes g)(a) = g \circ f(a)$$

where $\circ$ denotes the usual composition of mappings. One verifies (see Example 4.1.3 of Chap. 1) that $(H, \oplus, \otimes)$ is a left pre-semiring.

[Reference: **1** (Sect. 4.1.3)]

2.1.2. Monotone Data Flow Algebra (Left Pre-Dioid)

Let $(L, \wedge)$ be an idempotent monoid. A *data flow* algebra is formed by a set F of *functions* defined on L and with value in L.

L being idempotent, we know (see Chap. 1 Sect. 3.4) that the canonical preorder relation $\leq$ is an order relation, and that, in this case, an equivalent definition of $\leq$ is:

$$a \leq b \Leftrightarrow a \wedge b = b$$

(see Chap. 1 Proposition 3.6.2).

We do not assume the existence of a neutral element for $\wedge$ but only the existence of a largest element Ω (necessarily unique), i.e. of an element such that:

$$\forall a \in L: a \wedge \Omega = \Omega$$

Observe that the monoid L endowed with $\wedge$ has the structure of a sup-semi-lattice.

Table 4 Recapitulatory list of main pre-semirings and pre-dioids

Right and left pre-semirings	Property of the monoid (E, $\oplus$)	Property of the monoid (E, $\otimes$)	Distributivity of $\otimes$ with respect to $\oplus$	Comments, additional properties
Endomorphisms of a commutative monoid (S, $\oplus$)	Commutative, neutral element: h^ε	Non commutative, neutral element = identity endomorphism	Right and left	Pre-semiring
Mappings of a monoid onto itself (S, $\oplus$)	Commutative, neutral element: h^ε	Non commutative neutral element = identity endomorphism	Left only	Pre-semiring
Product of a pre-dioid and a ring	Commutative		Right and left	Pre-semiring (neither pre-dioid nor semiring)
Monotone data flow algebra	Commutative idempotent	Non commutative $e = h^e$	Left only	Left pre-dioid (the canonical preorder is an order)
Pre-dioids				
($\mathbb{R}_+$, Max, +)	Commutative selective $\varepsilon = 0$	Commutative hemi-group	Right and left	The canonical preorder is a total order
($\mathbb{N}$, lcm, $\times$)	Commutative idempotent $\varepsilon = 1$	Commutative $e = 1$. Every element of $E\backslash\{0\}$ is cancellative	Right and left	The canonical preorder is an order. ε non absorbing for $\otimes$
(Int ($\mathbb{R}$), Conv ($A \cup B$), +)	Commutative idempotent $\varepsilon = [0, 0]$	Commutative $e = [0, 0]$	Right and left	The canonical preorder is an order. ε non-absorbing for $\otimes$

We now consider the set F of functions: L $\to$ L, endowed with the internal laws $\oplus$ and $\otimes$ defined as follows:

$$\forall f, g \in F, \forall a \in L:$$
$$(f \oplus g)(a) = f(a) \wedge g(a)$$
$$(f \otimes g)(a) = g \circ f(a)$$

where $\circ$ is the usual law of composition of mappings.

One verifies that $\oplus$ is commutative, associative and idempotent.

One verifies that $\otimes$ is associative, in general non commutative, and has a neutral element which is the identity mapping h^e (defined as: $\forall a$: h^e (a) = a).

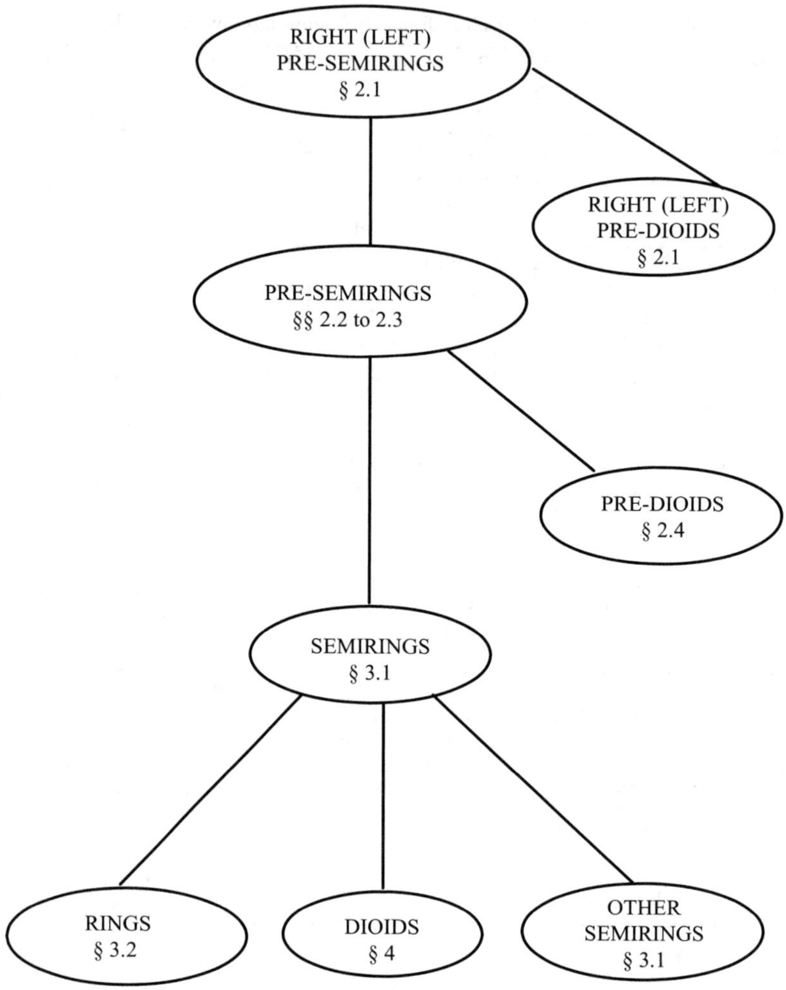

Fig. 2 Typology of pre-semirings and semirings

Moreover, the law $\otimes$ is *left distributive* with respect to $\oplus$ because:

$$\begin{aligned}\forall \mathrm{a} \in \mathrm{L}\colon \mathrm{h} \otimes (\mathrm{f} \oplus \mathrm{g})(\mathrm{a}) &= ((\mathrm{f} \oplus \mathrm{g}) \circ \mathrm{h})(\mathrm{a}) = (\mathrm{f} \oplus \mathrm{g})(\mathrm{h}(\mathrm{a}))\\ &= \mathrm{f} \circ \mathrm{h}(\mathrm{a}) \oplus \mathrm{g} \circ \mathrm{h}(\mathrm{a})\\ &= (\mathrm{h} \otimes \mathrm{f} \oplus \mathrm{h} \otimes \mathrm{g})(\mathrm{a})\end{aligned}$$

$(\mathrm{F}, \oplus, \otimes)$ is therefore a *left pre-dioid.*

In the so-called *monotone* data flow problems (see for example Kam and Ullman 1977), we assume moreover that F is the subset of monotone functions: $\mathrm{L} \to \mathrm{L}$, i.e. satisfying:

$$\forall \mathrm{a}, \mathrm{b} \in \mathrm{L}, \mathrm{a} \leq \mathrm{b} \;\Rightarrow\; \mathrm{f}(\mathrm{a}) \leq \mathrm{f}(\mathrm{b})$$

In this case, only the property of left distributivity of $\otimes$ with respect to $\oplus$ is satisfied and consequently the algebraic structure $(F, \oplus, \otimes)$ is only a *left pre-dioid*.

Remark. In the so-called *continuous* data flow problems (Kildall 1973), F is assumed to be the subset of functions satisfying the property of *endomorphism* on L:

$$\forall a, b \in L: \quad f(a \wedge b) = f(a) \wedge f(b)$$

(this property is known under the name of *continuity* in the literature devoted to data flow problems). In this case, the structure obtained corresponds to an algebra of endomorphisms of a monoid (see Sect. 2.2 below). ||

[References: The algebra of monotone data flow has been studied by many authors in the framework of Data Flow Analysis models of computer programs, and finds applications in the optimization of compilers, the verification of programs, and the transformation of programs (in particular, cancellation through the elimination of common subexpressions). Refer for example to Graham and Wegman (1976), Kam and Ullman (1976, 1977), and Tarjan (1981)].

2.2. Pre-Semiring of Endomorphisms of a Commutative Monoid

Let $(E, \dot{+})$ be a commutative monoid with neutral element ε and H the set of *endomorphisms of* E, that is to say the set of mappings: $h: E \to E$ such that:

$$\forall a, b \in E \quad h(a \dot{+} b) = h(a) \dot{+} h(b)$$

On H we define the operation $\oplus$ by:

$$\forall h, f \in H, \quad g = h \oplus f \quad \text{is such that} \quad g(a) = h(a) \dot{+} f(a) \quad \text{(for all a)}$$

Moreover, H is endowed with the law $\otimes$ defined as:

$$\forall h, f \in H: h \otimes f = f \circ h$$

where $\circ$ denotes the usual law of composition for mappings.

One verifies that $\oplus$ has a neutral element, namely the endomorphism $h^{\varepsilon}: E \to E$ defined as:

$$\forall a \in E \quad h^{\varepsilon}(a) = \varepsilon$$

One also verifies that $\otimes$ is right and left distributive with respect to $\oplus$ and with neutral element the identity endomorphism h^{e} defined as:

$$\forall a \in E: h^{e}(a) = a$$

Note that the property of right distributivity follows from the property of endomorphism:

$$[f \oplus g] \otimes h(a) = h[f(a) \oplus g(a)] = h(f(a)) \dot{+} h(g(a)) = [f \otimes h \oplus g \otimes h](a)$$

Also observe that the multiplication $\otimes$ defined above is not commutative in general. Moreover, without further assumption, the element h^{ε} *is not absorbing for* $\otimes$.

Indeed, we have that:

$$\forall h \in H \quad h \otimes h^{\varepsilon} = h^{\varepsilon}$$

(because, $\forall a \in E$: $h^{\varepsilon}(h(a)) = \varepsilon$) but we do not necessarily have $h^{\varepsilon} \otimes h = h^{\varepsilon}$ (indeed, $h(h^{\varepsilon}(a)) = h(\varepsilon)$ which, without an explicit assumption, has no reason to be equal to ε).

The structure $(H, \oplus, \otimes)$ is therefore a pre-semiring.

Remark. The above example can easily be generalized by no longer considering a single monoid $(E, \dot{+})$ but two monoids $(E, \dot{+})$ and $(F, \Box)$ and taking for H the set of homomorphisms: $E \to F$. ||

[References: **1** (Sect. 4.2.2), **4** (Sect. 4.4)

This algebraic structure was suggested independently by Kildall (1973) in the context of data flow analysis of programs (so-called "continuous" data flow problems) and by Minoux (1976) as a generalization of the path algebras of Carré et al. (1971) and Gondran (1974, 1975) in order to model complex and non-standard path-finding problems in graphs (for example finding the shortest path with time constraints, refer to Chap. 4 Sect. 4.4 of the present book).]

2.3. Pre-Semiring, Product of a Pre-Dioid and a Ring

We consider $E = E_1 \times E_2$ where: E_1, endowed with the operations $\oplus_1$ and $\otimes_1$ is a *pre-dioid*; and E_2, endowed with the operations $\oplus_2$ and $\otimes_2$ is a *ring*.

$(E_1, \oplus_1)$ is therefore a canonically ordered monoid whereas $(E_2, \oplus_2)$ is not canonically ordered.

The operations $\oplus$ and $\otimes$ on E are defined as the product operations:

$$\forall \begin{pmatrix} x_1 \\ x_2 \end{pmatrix} \in E \begin{pmatrix} y_1 \\ y_2 \end{pmatrix} \in E:$$

$$\begin{pmatrix} x_1 \\ x_2 \end{pmatrix} \oplus \begin{pmatrix} y_1 \\ y_2 \end{pmatrix} = \begin{pmatrix} x_1 \oplus_1 y_1 \\ x_2 \oplus_2 y_2 \end{pmatrix}$$

and:

$$\begin{pmatrix} x_1 \\ x_2 \end{pmatrix} \otimes \begin{pmatrix} y_1 \\ y_2 \end{pmatrix} = \begin{pmatrix} x_1 \otimes_1 y_1 \\ x_2 \otimes_2 y_2 \end{pmatrix}$$

As the operations $\oplus$ and $\otimes$ inherit basic properties of $(E_1, \oplus_1, \otimes_1)$ and $(E_2, \oplus_2, \otimes_2)$ one easily sees that:

$\oplus$ is commutative, associative
$\otimes$ is associative and right and left distributive with respect to $\oplus$.

However $(E, \oplus)$ is neither a canonically ordered monoid, nor a group.

The structure $(E, \oplus, \otimes)$, a product of a pre-dioid and a ring, is therefore the example of a *pre-semiring* which is *neither a pre-dioid nor a ring*.

A typical example of such a structure is the product of the pre-dioid $(\mathbb{R}_+, \text{Max}, +)$ (see Sect. 2.4) and the ring $(\mathbb{Z}, +, \times)$.

Another closely related example is that of the product of the pre-dioid $(\mathbb{R}_+, \text{Max}, +)$ by $(\mathbb{R}, +, \times)$ which is a field (therefore also a ring).

In the latter case, E is the set of pairs $\begin{pmatrix} x \\ x' \end{pmatrix}$ with $x \in \mathbb{R}_+$ and $x' \in \mathbb{R}$ endowed with the laws $\oplus$ and $\otimes$ defined as:

$$\begin{pmatrix} x \\ x' \end{pmatrix} \oplus \begin{pmatrix} y \\ y' \end{pmatrix} = \begin{pmatrix} \text{Max}\{x, y\} \\ x' + y' \end{pmatrix}$$

and

$$\begin{pmatrix} x \\ x' \end{pmatrix} \otimes \begin{pmatrix} y \\ y' \end{pmatrix} = \begin{pmatrix} x + y \\ x' \times y' \end{pmatrix}$$

2.4. Pre-Dioids

2.4.1. Pre-Dioid $(\mathbb{R}_+, \text{Max}, +)$

E is the set $\mathbb{R}_+$ of nonnegative reals.

$\oplus$ is the operation Max (maximum of two real numbers) with neutral element $\varepsilon = 0$
$\otimes$ is the operation + (sum of two real numbers) with unit element $e = 0$

One easily verifies the distributivity of $\otimes$ with respect to $\oplus$.

However, ε is not absorbing for $\otimes$ because:

$$\forall a \in E: a \otimes \varepsilon = \varepsilon \otimes a = a$$

$(\mathbb{R}_+, \text{Max}, +)$ is therefore a pre-semiring. As $(\mathbb{R}_+, \text{Max})$ is a canonically ordered monoid, it is a *pre-dioid*.

2.4.2. Pre-Dioid of the Natural Numbers Endowed with lcm and Product $(\mathbb{N}, \text{lcm}, \times)$

We take for E the set of natural numbers.

The law $\oplus$ is defined as:

$a \oplus b = \text{lcm}(a, b)$ (least common multiple)

This law is associative, commutative and idempotent, has neutral element $\varepsilon = 1$, and endows $\mathbb{N}$ with a structure of *canonically ordered monoid*.

The law $\otimes$ is associative, commutative, has unit element $e = 1$, and is distributive with respect to $\oplus$.

$(\mathbb{N}, \text{lcm}, \times)$ is therefore a pre-dioid.

However, observe that $\varepsilon = 1$ *is not absorbing for* $\otimes$ (indeed, for $a \in E$, $a \neq 1$, $a \times 1 = a \neq 1$). $(\mathbb{N}, \text{lcm}, \times)$ is therefore neither a semiring nor a dioid.

Let us observe that, if one replaces the lcm operation by the gcd operation we obtain $(\mathbb{N}, \text{gcd}, \times)$ which is a *dioid* (an idempotent-cancellative dioid actually) (see Sect. 4.7.4).

2.4.3. Pre-Dioid of Intervals (Int $(\mathbb{R})$, Conv $(A \cup B)$, $+$)

Let us take for E the set Int $(\mathbb{R})$ of the intervals of the real line $\mathbb{R}$ of the form $[\underline{a}, \bar{a}]$ with $\underline{a} \leq 0$ and $\bar{a} \geq 0$.

Let us define the operation $\oplus$ as the union of two intervals, in other words:

$$[\underline{a}, \bar{a}] \oplus [\underline{b}, \bar{b}] = [\text{Min}[\underline{a}, \underline{b}], \text{Max}[\bar{a}, \bar{b}]]$$

$\oplus$ is commutative and idempotent and has for neutral element the interval [0, 0]. It endows Int $(\mathbb{R})$ with a structure of *canonically ordered monoid.*

Moreover, let us define the operation $\otimes$ by:

$$[\underline{a}, \bar{a}] \otimes [\underline{b}, \bar{b}] = [\underline{a} + \underline{b}, \bar{a} + \bar{b}]$$

The operation $\otimes$ has for neutral element [0, 0].

The distributivity of $\otimes$ with respect to $\oplus$ is deduced from the obvious properties:

$$\text{Min}[\underline{a}, \underline{b}] + \underline{c} = \text{Min}[\underline{a} + \underline{c}, \underline{b} + \underline{c}]$$
$$\text{Max}[\bar{a}, \bar{b}] + \bar{c} = \text{Max}[\bar{a} + \bar{c}, \bar{b} + \bar{c}]$$

Remark. This example is one of the few cases where distributivity holds with intervals. This is not the case for (int $(\mathbb{R})$, Conv $(A \cup B)$, $\cap$) nor for (Int $(\mathbb{R})$, $\cap$, $+$). ||

However, $(E, \oplus, \otimes)$ *is not a semiring* because ε is not absorbing for $\otimes$. Indeed, for an arbitrary element $[\underline{a}, \bar{a}] \neq \varepsilon$ we have that $[\underline{a}, \bar{a}] \otimes [0, 0] = [\underline{a}, \bar{a}] \neq \varepsilon$.

The structure $[E, \oplus, \otimes]$ defined above is therefore a *pre-dioid* but it is not a semiring.

Note that this example illustrates the fact that assuming $\leq$ to be an order relation is not sufficient to guarantee the absorption property.

3. Semirings and Rings

The class of semirings includes:

- Rings (see the examples of Sect. 3.2);
- Dioids (dealt with in Sect. 4);
- Other semirings (see the examples of Sect. 3.1).

We recall that, since a monoid cannot both be a group and be canonically ordered, the subclass of rings and the subclass of dioids are *disjoint*.

Figure 2 indicates the place of semirings and rings in the typology.

Table 5 recalls the basic definitions concerning semirings and rings.

Table 5 Definitions of semirings and rings

Semiring (E, ⊕, ⊗)	(E, ⊕) monoid with neutral element ε; (E, ⊗) monoid with neutral element e ⊗ right and/or left distributive with respect to ⊕; ε is absorbing for ⊗ (i.e.: $\forall a \in E$, $a \otimes \varepsilon = \varepsilon \otimes a = \varepsilon$)
Ring (E, ⊕, ⊗)	Semiring such that (E, ⊗) is a group

3.1. General Semirings

These are semirings which are neither rings nor dioids.

3.1.1. Semiring, Product of a Dioid and of a Ring

We consider $E = E_1 \times E_2$ where:

E_1 endowed with the laws $\oplus_1$ and $\otimes_1$ is a dioid.

E_2 endowed with the laws $\oplus_2$ and $\otimes_2$ is a ring.

The laws ⊕ and ⊗ on E are defined as the product laws:

$$\forall \begin{pmatrix} x_1 \\ x_2 \end{pmatrix} \in E \begin{pmatrix} y_1 \\ y_2 \end{pmatrix} \in E :$$

$$\begin{pmatrix} x_1 \\ x_2 \end{pmatrix} \oplus \begin{pmatrix} y_1 \\ y_2 \end{pmatrix} = \begin{pmatrix} x_1 \oplus_1 y_1 \\ x_2 \oplus_2 y_2 \end{pmatrix}$$

$$\begin{pmatrix} x_1 \\ x_2 \end{pmatrix} \otimes \begin{pmatrix} y_1 \\ y_2 \end{pmatrix} = \begin{pmatrix} x_1 \otimes_1 y_1 \\ x_2 \otimes_2 y_2 \end{pmatrix}$$

The laws ⊕ and ⊗ inherit the basic properties of $(E_1, \oplus_1, \otimes_1)$ and $(E_2, \oplus_2, \otimes_2)$, one therefore easily checks that:

⊕ is communicative, associative and has neutral element $\varepsilon = \begin{pmatrix} \varepsilon_1 \\ \varepsilon_2 \end{pmatrix}$ where ε_1 (resp. ε_2) is the neutral element of E_1 for $\oplus_1$ (resp. E_2 for $\oplus_2$).

⊗ is associative and distributive (on the right and on the left) with respect to ⊕.

$\varepsilon = \begin{pmatrix} \varepsilon_1 \\ \varepsilon_2 \end{pmatrix}$ is absorbing for ⊗ .

We deduce from the above that the product structure (E, ⊕, ⊗) is a semiring. However, it is neither a dioid, nor a ring.

A typical example of the above is the product of the dioid $(\mathbb{R}_+ \cup \{+\infty\}, \text{Min}, +\}$ and the ring $(\mathbb{Z}, +, \times)$ with the laws ⊕ and ⊗ defined as:

$$\begin{pmatrix} x_1 \\ x_2 \end{pmatrix} \oplus \begin{pmatrix} y_1 \\ y_2 \end{pmatrix} = \begin{pmatrix} \text{Min}\{x_1, y_1\} \\ x_2 + y_2 \end{pmatrix}$$

$$\begin{pmatrix} x_1 \\ x_2 \end{pmatrix} \otimes \begin{pmatrix} y_1 \\ y_2 \end{pmatrix} = \begin{pmatrix} x_1 + y_1 \\ x_2 \times y_2 \end{pmatrix}$$

One verifies that $\varepsilon = \binom{+\infty}{0}$ is absorbing for ⊗.

[Reference: **1** (Sect. 5.4)]

3.1.2. Semiring of Signed Numbers

Let us consider the pair $(a, s) \in \mathbb{R} \times S$ where $S = \{+, -, 0, ?\}$ is the set of signs of qualitative algebra (see Chap. 1, Example 6.1.3 and Chap. 8, Sect. 4.5.3 for the presentation of this dioid).

With every real number a, we thus associate four signed numbers a^+, a^-, a° and $a^?$ corresponding respectively to: a obtained as the limit of a sequence of numbers $> a$ (a^+); of a sequence of numbers $< a$ (a^-); of a sequence of numbers all equal to a (a°); of a sequence of numbers only convergent towards a $(a^?)$.

We define the addition $\oplus$ of two signed numbers (a, s) and (b, σ) as: $(a, s) \oplus (b, \sigma) = (a + b, s \dot{+} \sigma)$ and the multiplication $\otimes$ by:

$(a, s) \otimes (b, \sigma) = (ab, (sg\ (a) \dot{\times} \sigma) \dot{+} (sg\ (b) \dot{\times} s) \dot{+} (s \dot{\times} \sigma))$ where $\dot{+}$ and $\dot{\times}$ are addition and the multiplication of qualitative algebra (see below Sect. 4.5.3) and sg (a) the sign of a (with the convention $sg\ (0) = 0$).

One verifies that $(\mathbb{R} \times S, \oplus, \otimes)$ is a semiring. It is not a dioid however, because the set $\mathbb{R} \times S$ is not canonically ordered by $\oplus$.

[References: **1** (Sect. 6.1.3), **8** (Sect. 4.5.3)]

3.2. Rings

3.2.1. Ring $(\mathbb{Z}, +, \times)$

The set of signed integers endowed with standard addition and multiplication.

3.2.2. Ring $(\mathbb{R}[X], +, \times)$

The set of polynomials with real coefficients of a real variable x endowed with the sum and the product of polynomials.

3.2.3. Ring $(M_n(\mathbb{R}), +, \times)$

The set of square $n \times n$ matrices with real entries endowed with the sum and product of matrices.

3.2.4. Ring $(\mathcal{P}(E), \Delta, \cap)$

The power set of a set E, endowed with the symmetric difference Δ $(A \Delta B = (A \cup B) \backslash (A \cap B)$ and the intersection $\cap$.

Table 6 Recapitulatory list of semirings and rings

	Property of the monoid (E, ⊕)	Property of the monoid (E, ⊗)	Distributivity of ⊗ with respect to ⊕	Comments, additional properties
Semiring product of a dioid and a ring ($E_1 \times E_2$, ⊕, ⊗)	Commutative neutral element $\varepsilon = \begin{pmatrix} \varepsilon_1 \\ \varepsilon_2 \end{pmatrix}$	Neutral element $e = \begin{pmatrix} e_1 \\ e_2 \end{pmatrix}$	Right and left	$\varepsilon = \begin{pmatrix} \varepsilon_1 \\ \varepsilon_2 \end{pmatrix}$ absorbing for ⊗
Ring ($\mathbb{Z}$, +, ×)	Group $\varepsilon = 0$	Commutative $e = 1$	Right and left	ε absorbing for ⊗
Ring of polynomials ($\mathbb{R}[X]$, +, ×)	Group $\varepsilon = 0$ (zero polynomial)	Commutative $e = 1$	Right and left	ε absorbing for ⊗
Ring of matrices n × n ($M_n(\mathbb{R})$, +, ×)	Group $\varepsilon = 0$ (zero matrix)	$e = I_n$ (n × n identity matrix)	Right and left	ε absorbing for ⊗
($\mathcal{P}(E)$, Δ, ∩)	Commutative group, $\varepsilon = \emptyset$	Commutative idempotent	Right and left	ε absorbing for ⊗

4. Dioids

The typology of dioids is recalled in Fig. 3 below. The second level of the classification contains:

- Symmetrizable dioids (see Sect. 4.4);
- Idempotent dioids (see Sect. 4.5–4.8);
- "General" dioids which do not belong to any of the previous categories (see Sect. 4.3);

Table 7 recalls the main definitions concerning dioids.

Before presenting examples of each of these classes, we first provide a few examples of right or left dioids (Sect. 4.1), then examples of the general class formed by the endomorphisms of a canonically ordered commutative monoid (Sect. 4.2).

4.1. *Right or Left Dioids*

4.1.1. Right Dioid and Shortest Path with Gains or Losses

On the set $E = \mathbb{R} \times \mathbb{R}_+ \backslash \{0\}$, we define the following operations ⊕ and ⊗:

$$(a, k) \oplus (a', k') = \begin{cases} (a, k) & \text{if} \quad \frac{a}{k} < \frac{a'}{k'} \quad \text{or if} \quad \frac{a}{k} = \frac{a'}{k'} \quad \text{and} \quad k \geq k' \\ (a', k') & \text{otherwise} \end{cases}$$

$$(a, k) \otimes (a', k') = (a + ka', kk')$$

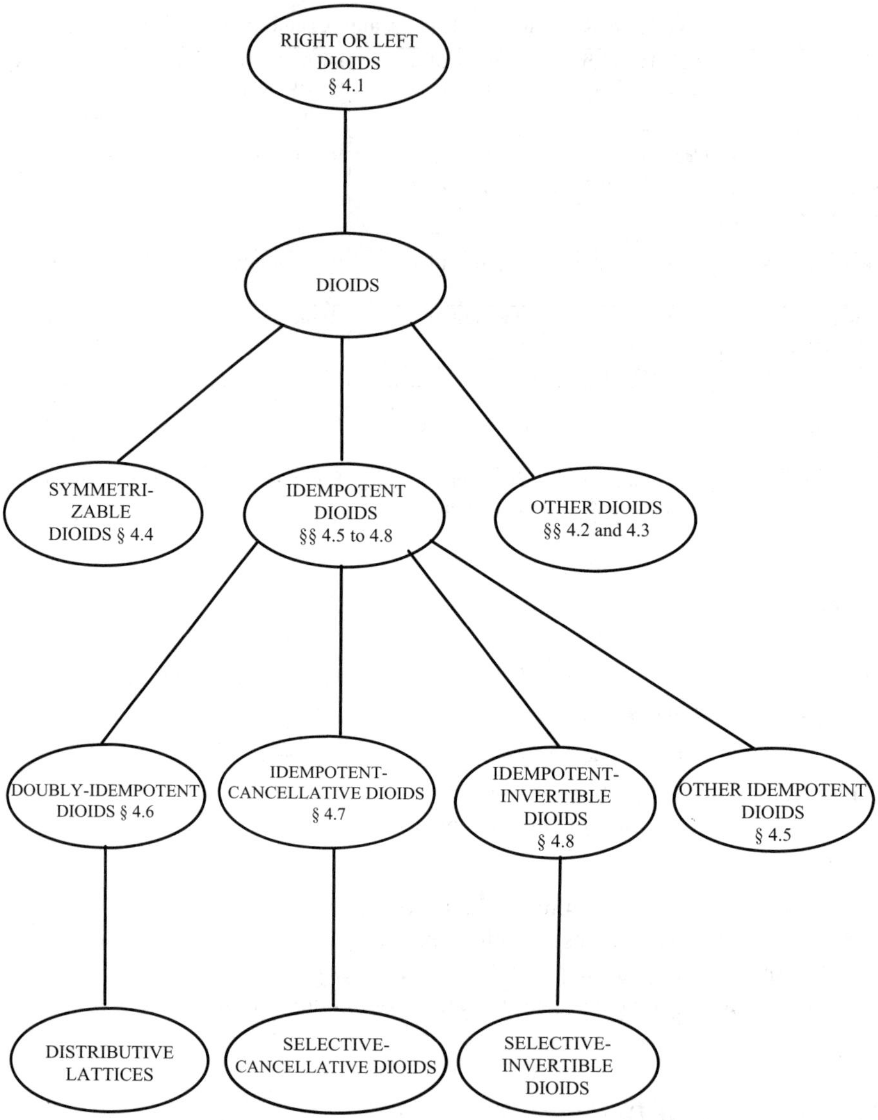

Fig. 3 Classification of dioids

$\oplus$ has for neutral element ε any element of the form $(+\infty, k)$ and $\otimes$ has neutral element $e = (0, 1)$.

One verifies (see Example 6.1.4 Chap. 1), that $(E, \oplus, \otimes)$ is a *right dioid*. This is the algebraic structure required to solve the shortest path problem with gains or losses.

[Reference: **4** (Exercise 4)]

Table 7 Basic terminology relating to dioids

Dioid (E, ⊕, ⊗)	(E, ⊕) is a canonically ordered monoid with neutral element ε. (E, ⊗) is a monoid with neutral element e ⊗ is right and/or left distributive with respect to ⊕ ε is absorbing for ⊗
Symmetrizable dioid	Dioid for which (E, ⊕) is a hemi-group
Idempotent dioid	Dioid with an idempotent law ⊕
Doubly idempotent dioid (distributive lattices)	Dioid with two idempotent laws ⊕ and ⊗
Idempotent-cancellative dioid	Idempotent dioid for which (E, ⊗) is a cancellative monoid
Selective-cancellative dioid	Idempotent-cancellative dioid with the ⊕ law selective
Idempotent-invertible dioid	Idempotent dioid for which (E, ⊗) is a group
Selective-invertible dioid	Idempotent-invertible dioid with the ⊕ law selective

4.1.2. A Right Dioid

On the set $E = \mathbb{R} \times \mathbb{R}_+$, we define the following operations ⊕ and ⊗:

$$(a, k) \oplus (a', k') = \begin{cases} (a, k) & \text{if} \quad a < a', \quad \text{or if} \quad a = a' \quad \text{and} \quad k > k' \\ \\ (a', k') & \text{otherwise} \end{cases}$$

$$(a, k) \otimes (a', k') = (a + ka', kk')$$

⊕ has for neutral element ε any element of the form $(+\infty, k)$ and ⊗ has neutral element $e = (0, 1)$.

4.1.3. Left Dioid of Semi-Cancellative Fractions

In automatic control, the study of the stability of a system through the analysis of transfer functions (Laplace transform of the input-output function) presents inconsistencies when the (rational) transfer function $\frac{b(s)}{a(s)}$ is simplified through polynomials with unstable poles (roots lying in the right half-plane, Re (s) > 0). To illustrate this, if we consider the example where $\frac{b(s)}{a(s)} = \frac{s-1}{(s-1)\,(s+1)}$, the system is unstabilizable; but since, on the ring of rational fractions, $\frac{s-1}{(s-1)(s+1)} = \frac{1}{s+1}$, one can rewrite $\frac{b(s)}{a(s)} = \frac{1}{s+1}$ and the system would be (incorrectly) considered as stabilizable if it were impossible to distinguish these two fractions.

To eliminate this risk of inconsistency, one must therefore prevent cancellation using unstable zeros and thus instead of the ring of rational fractions we have to consider the left dioid of *semi-cancellative fractions* (SCF) introduced by Bourlès (1994).

To obtain this, we consider the set of pairs (b(s), a(s)) where b(s) (resp. a(s)) is a polynomial in the ring $\mathbb{R}[s]$ of polynomials with real coefficients (resp. a polynomial in $\mathbb{R}^*[s] = \mathbb{R}[s]\backslash\{0\}$).

Let $P(\mathbb{C}_+)$ (resp. $P(\mathbb{C}_-)$ be the subset of $\mathbb{R}^*[s]$ formed by the polynomials having roots in the closed right half-plane $\mathbb{C}_+$ (resp. open left half-plane $\mathbb{C}_-$).

Any polynomial a (s) may be decomposed (in a unique way) into the form

$$a(s) = a^+ (s)\, a^- (s) \text{ with } a^+ (s) \in P(\mathbb{C}_+),\ a^- (s) \in P(\mathbb{C}_-).$$

On the pairs of $\mathbb{R}[s] \times \mathbb{R}^*[s]$, we define the equivalence relation $(b, a)_\sim (b', a')$ if and only if we have: $b (s)\, a' (s) = a (s)\, b' (s)$ and $a^+(s) = a'^+ (s)$.

The set of semi-cancellative fractions (SCF) will be the quotient set of the set of pairs $\mathbb{R}[s] \times \mathbb{R}^* [s]$ by this equivalence relation.

We denote [b, a] the class of (b, a) modulo $\sim$ ([b, a] is therefore a SCF).

We then define on $\mathbb{R}[s] \times \mathbb{R}^* [s]$ the operations $\oplus$ and $\otimes$ by:

(1) $(b, a) \oplus (b', a') = (b\alpha' + b'\alpha, p\alpha\alpha')$
where $p = \gcd (a, a')$, $a = p\alpha$, $a' = p\alpha'$ (hence $p\, \alpha\, \alpha' = \text{lcm}\, (a, a')$)

(2) $(b, a) \otimes (b', a') = (\beta b', a\alpha')$
where $\pi = \gcd(b, a') \cap P (\mathbb{C}_+)$, $b = \pi\beta$, $a' = \pi\alpha'$.

Proposition 1. *The operations* (1) *and* (2) *are compatible with the equivalence relation* $\sim$.

Proof. It is enough to check that if $(b_1, a_1)_\sim (b'_1, a'_1)$ and $(b_2, a_2)_\sim (b'_2, a'_2)$ we have:

$$(b_1, a_1) \oplus (b_2, a_2) \sim (b'_1, a'_1) \oplus (b'_2, a'_2)$$
$$(b_1, a_1) \otimes (b_2, a_2) \sim (b'_1, a'_1) \otimes (b'_2, a'_2). \quad \square$$

One can therefore define on the set SCF an addition and a multiplication by setting:

$$[b_1, a_1] \oplus [b_2, a_2] = [(b_1, a_1) \oplus (b_2, a_2)]$$
$$[b_1, a_1] \otimes [b_2, a_2] = [(b_1, a_1) \otimes (b_1, a_2)].$$

Proposition 2. *The set* SCF *endowed with the laws* $\oplus$ *and* $\otimes$ *is a left dioid.*

Proof. One verifies first of all that the addition $\oplus$ is associative and commutative (it corresponds to the connection of two systems in parallel). [0, 1] is the neutral element for the addition because $[b, a] \oplus [0, 1] = [b, a]$.

The multiplication $\otimes$ is associative (it corresponds to the connection of two systems in sequence). [1, 1] is the neutral element for the multiplication. But the multiplication is not commutative.

We observe that [0, 1], the neutral element of $\oplus$, is left-absorbing but not right-absorbing $\otimes$; indeed, we clearly have:

$$[0, 1] \otimes [b, a] = [0, a^-] = [0, 1] \quad \text{and}$$
$$[b, a] \otimes [0, 1] = [0, a] \neq [0, 1] \quad \text{if} \quad a \notin P (\mathbb{C}_+).$$

The multiplication is left distributive, because:

$$[b_1, a_1] \otimes ([b_2, a_2] \oplus [b_3, a_3]) = ([b_1, a_1] \otimes [b_2, a_2]) \oplus ([b_1, a_1] \otimes [b_3, a_3])$$

but not right distributive, as shown by the following example:

$$[(1, 1) - (1, s)][1, s - 1] = [1, s], \quad \text{but}$$
$$[1, 1][1, s - 1] - [1, s][1, s - 1] = [s - 1, s\,(s - 1)].$$

SCF is therefore clearly a left dioid. □

One can refer to Bourlès (1994) for a generalization of semi-cancellative fractions to general unitary commutative rings.

The following results then show that these SCF can indeed be used to rigorously study the stability of a system described by its transfer function.

Indeed, considering (b, a) and (r, s) two elements of $\mathbb{R}[s] \times \mathbb{R}^*[s]$, such that $a\,s + b\,r \neq 0$, then the SCF [c, d] where $(c, d) = (b\,s, a\,s + b\,r)$ only depends on the SCF [b, a] and [r, s] (and not on the particular pairs (b, a) and (r, s)).

We will refer to as a *negative feedback operator* the operator Γ : SCF × SCF → SCF defined as:

$$\Gamma\,([b, a], [r, s]) = [b\,s, a\,s + b\,r]$$

for all SCF [b, a] and [r, s] such that $a\,s + b\,r \neq 0$.

Then one can state:

Proposition 3. *(Bourlès 1994)*

Consider a SCF [b, a]*; then there exists a* SCF [r, s] *such that* Γ *(*[b, a], [r, s]*) is stable, if and only if* [b, a] *is stabilizable.*

[Reference: for more detail, see Bourlès 1994]

4.2. Dioid of Endomorphisms of a Canonically Ordered Commutative Monoid. Examples.

Let (E, ⊕) be a canonically ordered commutative monoid with neutral element ε.

As in Sect. 2.2. of this chapter, we consider the set H of endomorphisms on E verifying, ∀h ∈ H:

$$h\,(a \oplus b) = h\,(a) \oplus h\,(b) \quad \forall a, b \in E$$
$$h\,(\varepsilon) = \varepsilon$$

endowed with the laws ⊕ and ⊗ defined as: ∀h, g ∈ H

$$(h \oplus g)(a) = h\,(a) \oplus g\,(a) \quad \forall a \in E$$
$$(h \otimes g)(a) = g \circ h\,(a) \quad \forall a \in E$$

where ° is the law of composition of mappings. One verifies that (H, ⊕, ⊗) is a dioid (pre-semiring of Sect. 2.2 with the extra property $h\,(\varepsilon) = \varepsilon$, which guarantees that h^{ε} is absorbing).

This is a very important class of dioids, in particular for studying complex path-finding problems in graphs such as those corresponding to the following examples.

4.2.1. Dioid of Nondecreasing Functions

On the monoid $(E, \oplus)$ with $E = \hat{R} = \mathbb{R} \cup \{+\infty\}$, $\oplus = \min$ and $\varepsilon = +\infty$, we consider the set H of nondecreasing functions $h: E \to E$, with $h(t)$ tending to $+\infty$ when t tends to $+\infty$.

These functions satisfy the equations

$$h\,(\min(t, t')) = \min(h\,(t), h\,(t'))$$
$$h\,(+\infty) = +\infty$$

and the set $(H, \oplus, \otimes)$ is a dioid.

This dioid is the algebraic structure required to solve the shortest path problem with time dependent lengths on the arcs, see Example 6.2.1 of Chap. 1 and the generalized algorithms $1'$, $2'$, $2''$, $3'$ of Chap. 4, Sect. 4.4.

[References: Minoux (1976), **1** (Sect. 6.2), **4** (Sect. 4.4)]

4.2.2. A Dioid for the Shortest Path with Discounting Problem (Minoux 1976)

With each arc (i, j) of a graph G, we associate a length which will depend, in a path, on the number of arcs taken previously. For example, if we interpret the path-traversal along the arc (i, j) as the completion of an annual investment program, the cost of the arc (i, j) is $C_{ij}/(1+\tau)^t$ if t is the number of arcs previously traversed along by the path, that is to say the year of the expenditure C_{ij} (τ being the discounting rate).

We seek the shortest path with discounting from vertex 1 to the other vertices.

If T is the final period, we will take for S the set of vectors with $(T+1)$ components in $\mathbb{R}_+ \cup \{+\infty\}$. If $a = (a_0, a_1, \ldots, a_T)$ and $b = (b_0, b_1, \ldots, b_T)$, we will define $d = a \oplus b = (d_0, d_1, \ldots, d_T)$ by setting $d_t = \min(a_t, b_t)$, for $t = 0, 1, \ldots, T$. $\varepsilon = (+\infty, \ldots, +\infty)$. Thus to each arc (i, j) in G we let correspond the function $h_{ij}: S \to S$ defined as:

$h_{ij}(a) = b \quad$ with: $b_0 = +\infty$

$$b_t = a_{t-1} + \frac{C_{ij}}{(1+\tau)^{t-1}} \quad \text{for } t = 1, \ldots, T.$$

We observe that such endomorphisms are T-nilpotent (which guarantees the existence of the quasi-inverse $(h_{ij}{}^*)$ of the matrix of endomorphisms (h_{ij})).

Knowing the "state" $a = h^*_{1j}\,(0)$ of a vertex j, we can deduce the shortest path with discounting from vertex 1 to this vertex, whose value is equal to $\min\limits_{0 \le t \le T} (a_t)$.

The various generalized algorithms of Chap. 4, Sect. 4.4 can be applied to solve this problem.

[References: Minoux (1976), **1** (Sect. 6.2), **4** (Sect. 4.4)]

4.2.3. A Dioid for the Shortest Path Problem with Time Constraints

(Halpern and Priess 1974; Minoux 1976)

With each arc (i, j) of a graph G, we associate:

- A duration $d_{ij} > 0$ measuring the traversal time on arc (i, j),
- A set of intervals $V_{ij} \subset [0, +\infty[$ representing the set of instants at which departure is possible from i to j via arc (i, j);

With each node i we associate a set of intervals denoted $W_i \subset [0, +\infty[$ representing the set of instants at which parking at node i is allowed.

The problem is to find between two given vertices x and y the shortest path (in the sense of the traversal time) compatible with the temporal constraints induced by V_{ij} (on the arcs) and W_i (on the vertices).

We define the state E_i, of a vertex i as the set of possible instants of arrival at i when starting from the origin x. We denote S the set of states. An element of S will therefore be a set of intervals $\subset [0, +\infty[$. We define in S the operation $\oplus$ (union of two sets of intervals) as:

$$a \oplus b = \{t/t \in a \quad \text{or} \quad t \in b\} \; \forall a, b \in S.$$

The empty set $\emptyset$ is the neutral element of $\oplus$. We define the transition between i and j (i.e. the endomorphism h_{ij}) in several stages.

- If E_i corresponds to the set of possible time instants of arrival at i, then the set D_i of the possible instants of departure from i will be:

$$D_i = E_i \perp W_i$$

  where the operation $\perp$ is defined as follows:

$$\text{if} \quad E_i = \left\{[\alpha_1, \alpha'_1], [\alpha_2, \alpha'_2], \ldots, [\alpha_p, \alpha'_p]\right\}$$
$$W_i = \left\{[\beta_1, \beta'_1], [\beta_2, \beta'_2], \ldots, [\beta_p, \beta'_p]\right\}$$

  then:

$$D_i = [\gamma_1, \gamma'_1] \oplus [\gamma_2, \gamma'_2] \oplus \cdots \oplus [\gamma_p, \gamma'_p]$$

  with, for k from 1 to p:

$$[\gamma_k, \gamma'_k] = \begin{cases} [\alpha_k, \alpha'_k] & \text{if } \alpha'_k \notin W_i \\ [\alpha_k, \beta'_j] & \text{if } \alpha'_k \in [\beta_j, \beta'_j] \end{cases}$$

- The set of possible instants of departure from i towards j using arc (i, j) will then be:

$$D_i \cap V_{ij}$$

– Let us define on S an external operation T ("translation") by:

$$a \in S, \tau \in \mathbb{R}_+, \tau\ \mathrm{T}a = \{t + \tau / t \in a\}.$$

The set of possible instants of arrival at j from i using arc (i, j) will therefore be:

$$d_{ij}\ \mathrm{T}\ (D_i \cap V_{ij}).$$

– Thus, with each arc (i, j) of the graph, we associate an endomorphism φ_{ij}: $(S, \oplus) \to (S, \oplus)$:

$$\varphi_{ij}(E_i) = d_{ij}\ \mathrm{T}[(E_i \perp W_i) \cap V_{ij}].$$

Let us observe that φ_{ij} is entirely determined by the triple (W_i, V_{ij}, d_{ij}) but that the product of two such endomorphisms cannot be described by such a triple.

One can then show that the set of endomorphisms φ_{ij} is a p-nilpotent set which implies the existence of Φ^*, the quasi-inverse of the matrix of endomorphisms $\Phi = (\varphi_{ij})$.

The earliest time to reach y will then be the minimum element of $E_y = \Phi^*_{xy}\ (E_x)$.

This problem can be solved by applying one of the generalized algorithms from Chap. 4, Sect. 4.4, e.g. algorithm 3′ (generalized Dijkstra's algorithm).

[References: Minoux (1976), Halpern and Priess (1974), **4** (Sect. 4.4)]

4.3. General Dioids

4.3.1. K Shortest Path Dioid $(\hat{\mathbb{R}}^k, \mathrm{Min}_{(k)}, \overset{(k)}{+})$

The elements of E are ordered k-tuples of real numbers chosen in $\hat{\mathbb{R}} = \mathbb{R} \cup \{+\infty\}$.

The $\oplus$ law is the operation $\mathrm{Min}_{(k)}$ introduced in Sect. 1.3.1. It endows E with a structure of canonically ordered monoid, with neutral element $\varepsilon = \begin{pmatrix} +\infty \\ +\infty \\ \vdots \\ +\infty \end{pmatrix}$.

The $\otimes$ law is the operation $\overset{(k)}{+}$ introduced in Sect. 1.1.5. It has unit element $e = \begin{pmatrix} 0 \\ +\infty \\ \vdots \\ +\infty \end{pmatrix}$.

One verifies the distributivity of $\otimes$ with respect to $\oplus$, and the absorption property for ε.

$(E, \oplus, \otimes)$ thus defined is therefore clearly a dioid.

The use of this dioid as a model to state and solve the kth shortest path problem in a graph is discussed in Chap. 4 Sect. 6.8.

[References: **4** (Sect. 6.8), **8** (Sect. 1.1.5), **8** (Sect. 1.3.1)]

4.3.2. η-Optimal Path Dioid $\left(\hat{\mathbb{R}} + [0, \eta]^{(N)}, \text{Min}_{(\leq\eta)}, \overset{(\leq\eta)}{+}\right)$

$\eta > 0$ being a given nonnegative real number, the elements of E are finite (variable length) sequences of real numbers in $\hat{\mathbb{R}} = \mathbb{R} \cup \{+\infty\}$, with extreme values differing by η at most.

If $a \in E$, we denote ν (a) the number of terms of the sequence corresponding to a, and a can be written:

$$a = \left(a^{(1)}, a^{(2)}, \ldots a^{(\nu(a))}\right)$$
$$\text{with} \quad a^{(1)} = \underset{i=1\ldots\nu(a)}{\text{Min}} \left\{a^{(i)}\right\}$$
$$a^{\nu(a)} = \underset{i=1\ldots\nu(a)}{\text{Max}} \left\{a^{(i)}\right\}$$
$$\text{and:} \quad a^{\nu(a)} - a^{(1)} \leq \eta$$

The $\oplus$ law is the operation $\text{Min}_{(\leq\eta)}$ introduced in Sect. 1.3.2. It endows E with a structure of *canonically ordered monoid*, with neutral element $\varepsilon = (+\infty)$ (sequence formed by a single term equal to $+\infty$).

The $\otimes$ law is the operation $\overset{(\leq\eta)}{+}$ introduced in Sect. 1.1.6. It has unit element $e = (0)$ (the sequence formed by a single term equal to 0).

The distributivity of $\otimes$ with respect to $\oplus$ is verified by observing that if

$$(a \oplus b) \otimes c = u = \left(u^{(1)}, u^{(2)}, \ldots u^{\nu(u)}\right)$$
$$u^{(1)} = \text{Min}\left\{a^{(1)}, b^{(1)}\right\} + c^{(1)}$$

and u is the sequence of all the terms of the form:

$a^{(i)} + c^{(k)}$ and $b^{(j)} + c^{(k)}$ (for $i = 1, \ldots \nu(a)$; $j = 1, \ldots \nu(b)$; $k = 1, \ldots \nu(c)$) which do not exceed $u^{(1)} + \eta$. On the other hand, evaluating the expression $(a \otimes c) \oplus (b \otimes c)$ yields the same result.

Finally, if we agree to identify ε with any finite sequence of the form $(+\infty, +\infty, \ldots + \infty)$, the element ε is absorbing for $\otimes$ since

$$\forall a \in E, a \otimes \varepsilon = \varepsilon \otimes a = \varepsilon$$

Thus this yields, for example:

$$\begin{pmatrix} 2 \\ 3 \\ 5 \end{pmatrix} \otimes (+\infty) = \begin{pmatrix} +\infty \\ +\infty \\ +\infty \end{pmatrix} = \varepsilon$$

The structure $(E, \oplus, \otimes)$ thus defined is therefore a *dioid*.

The use of this dioid as a model to represent and solve η-optimal path problems in graphs is discussed in Chap. 4 Sect. 6.10.

[References: **4** (Sect. 6.10), **8** (Sect. 1.1.6), **8** (Sect. 1.3.2)]

4.3.3. Order of Magnitude Dioid (Semi-Field)

On the set E of pairs (a, α) with $a \in \mathbb{R}_+\backslash\{0\}$ and $\alpha \in \mathbb{R}$, to which we add the pair $(0, +\infty)$, we define the two laws $\oplus$ and $\otimes$ as:

$$(a, \alpha) \oplus (b, \beta) = (c, \min(\alpha, \beta))$$
$$\text{with} \quad c = a \quad \text{if} \quad \alpha < \beta, c = b \quad \text{if} \quad \alpha > \beta, c = a + b \quad \text{if} \quad \alpha = \beta,$$
$$(a, \alpha) \otimes (b, \beta) = (a\,b, \alpha + \beta).$$

One verifies that $(E, \oplus, \otimes)$ is a (non idempotent) dioid.

This dioid is isomorphic to the set of elements of the form $a\,\varepsilon^{\alpha}$ endowed with ordinary addition and multiplication when $\varepsilon > 0$ tends towards 0^+.

We obtain a dioid isomorphic to the previous one by setting $A = e^{-\alpha}$ and by considering the set E of pairs $(a, A) \in (\mathbb{R}_+\backslash\{0\})^2$ to which we append the pair (0, 0); this set is endowed with the laws $\oplus$ and $\otimes$ defined as:

$$(a, A) \oplus (b, B) = (c, \text{Max}(A, B))$$
$$\text{with} \quad c = a \quad \text{if} \quad A > B, c = b \quad \text{if} \quad A < B, c = a + b \quad \text{if} \quad A = B,$$
$$(a, A) \otimes (b, B) = (ab, AB)$$

Moreover, the elements (a, A) of this dioid are in $1 - 1$ correspondence with the elements of the form $a\,A^p$ endowed with ordinary addition and multiplication when p tends towards $+\infty$.

One can interpret (a, A) as the coding of an asymptotic expansion of the form $a\,A^p + 0(A^p)$ when $p \to +\infty$.

The latter dioid was introduced by Finkelstein and Roytberg (1993) to calculate the asymptotic expansion of probability distribution functions in the study of biopolymers. It was also used by Akian et al. (1998) for the calculus of the eigenvalues of a matrix with entries of the form $\exp(-a_{ij}/\varepsilon)$ where ε is a small positive parameter.

[References: **1** (Sect. 6.1.5), **8** (Sect. 1.3.6)]

4.3.4. Nonstandard Number Dioid (Semi-Field)

On the set E of triples $(a, b, \alpha) \in (\mathbb{R}_+\backslash\{0\})^3$ to which we add the triples $(0, 0, +\infty)$ and $(1, 0, +\infty)$, we define the two laws $\oplus$ and $\otimes$ by:

$$(a_1, b_1, \alpha_1) \oplus (a_2, b_2, \alpha_2) = (a_1 + a_2, b, \min(\alpha_1, \alpha_2))$$
$$\text{with} \quad b = b_1 \quad \text{if} \quad \alpha_1 < \alpha_2, b = b_2 \quad \text{if} \quad \alpha_1 > \alpha_2, b = b_1 + b_2 \quad \text{if} \quad \alpha_1 = \alpha_2,$$
$$(a_1, b_1, \alpha_1) \otimes (a_2, b_2, \alpha_2) = (a_1\,a_2, b, \min(\alpha_1, \alpha_2))$$
$$\text{with} \quad b = a_2\,b_1 \quad \text{if} \quad \alpha_1 < \alpha_2, b = a_1\,b_2 \quad \text{if} \quad \alpha_1 > \alpha_2,$$
$$b = a_1\,b_2 + a_2\,b_1 \quad \text{if} \quad \alpha_1 = \alpha_2.$$

One verifies that (E, $\oplus$, $\otimes$) is a dioid. This dioid is isomorphic to the set of nonstandard numbers of the form $a + b\,\varepsilon^{\alpha}$, $(a > 0, b > 0)$, endowed with ordinary addition and multiplication, when $\varepsilon > 0$ tends towards 0^+.

This nonstandard number dioid can be related to the field ${}^{\rho}\mathbb{R}$ of nonstandard numbers introduced by Robinson (1973).

These are the numbers one can represent by the series

$$x = a_0 + a_1\,\rho^{\nu 1} + a_2\,\rho^{\nu 2} + \cdots (0 < \nu_1 < \nu_2 < \cdots)$$

where $a_0, a_1, a_2 \ldots$ are classical real numbers and where ρ is an infinitesimal.

x represents the sequence of the real numbers

$$x_{\varepsilon} = a_0 + a_1\,\varepsilon^{\nu 1} + a_2\,\varepsilon^{\nu 2} + \cdots (0 < \nu_1 < \nu_2 < \cdots)$$

when $\varepsilon > 0$ tends towards 0^+ (one can thus identify ρ with 0^+).

[References: **1** (Sect. 6.1.6), **8** (Sect. 1.3.7)]

4.4. Symmetrizable Dioids

4.4.1. Dioid ($\mathbb{R}_+$, +, x)

The set E is $\mathbb{R}_+$, the set of nonnegative reals.

$\oplus$ is ordinary addition with neutral element 0;
$\otimes$ is ordinary multiplication with unit element 1.
($\mathbb{R}_+$, +) is a hemi-group (see Chap. 1, Sect. 3.5)
($\mathbb{R}_+\backslash\{0\}$, $\times$) is a hemi-group.

The canonical preorder relation is a (total) order: it is the usual order on the nonnegative real numbers.

[Reference: **1** (Sect. 6.3.2)]

4.4.2. Dioid of Intervals of $\mathbb{R}_+$ with the Operations Sum and Product (Int ($\mathbb{R}_+$), +, x)

Consider the set E of all the real intervals of the form $[\underline{a}, \bar{a}]$ with $0 \le \underline{a} \le \bar{a}$.

The operation $\oplus$ is defined as:

$$[\underline{a}, \bar{a}] \oplus [\underline{b}, \bar{b}] = [\underline{a} + \underline{b}, \bar{a} + \bar{b}]$$

which has neutral element $\varepsilon = [0, 0]$. In E, every element is regular for $\oplus$.

The operation $\otimes$ is defined as:

$$[\underline{a}, \bar{a}] \otimes [\underline{b}, \bar{b}] = [\underline{a}\,\underline{b}, \bar{a}\,\bar{b}]$$

which has neutral element $e = [1, 1]$.

One easily verifies:

- The commutativity and the associativity of $\oplus$
- The commutativity and the associativity of $\otimes$
- The right and left distributivity of $\otimes$ with respect to $\oplus$.

ε is absorbing for $\otimes$ because, $\forall[\underline{a}, \bar{a}]$:

$$[\underline{a}, \bar{a}] \otimes [0, 0] = [0, 0] = \varepsilon.$$

The canonical preorder relation is defined as:

$$[\underline{a}, \bar{a}] \le [\underline{b}, \bar{b}] \Leftrightarrow \text{ there exists } [\underline{c}, \bar{c}] \text{ such that: } [\underline{b}, \bar{b}] = [\underline{a} + \underline{c}, \bar{a} + \bar{c}]$$

which is equivalent to the conditions:

$$\begin{cases} \underline{a} \le \underline{b} \\ \bar{a} \le \bar{b} \\ \bar{b} - \bar{a} \ge \underline{b} - \underline{a} \end{cases}$$

To verify that it is an order relation, let us assume that $[\underline{a}, \bar{a}] \le [\underline{b}, \bar{b}]$ and $[\underline{b}, \bar{b}] \le [\underline{a}, \bar{a}]$.

From the above we deduce:

$$\begin{cases} \underline{b} \ge \underline{a} \\ \bar{b} \le \bar{a} \\ \bar{b} - \bar{a} \ge \underline{b} - \underline{a} \end{cases} \quad \text{and} \quad \begin{cases} \underline{a} \ge \underline{b} \\ \bar{a} \ge \bar{b} \\ \bar{a} - \bar{b} \ge \underline{a} - \underline{b} \end{cases}$$

This implies: $\underline{a} = \underline{b}$ and $\bar{a} = \bar{b}$ therefore $[\underline{a}, \bar{a}] = [\underline{b}, \bar{b}]$ which proves the property and $[E, \oplus, \otimes]$ is indeed a dioid.

4.4.3. $(\mathbb{R}_+, \oplus_p, \otimes_p)$

This is the set of nonnegative reals endowed with the operation $\oplus_p$ (with $p \in \mathbb{R}_*$) defined as $a \oplus_p b = (a^p + b^p)^{1/p}$ and the operation $\otimes_p$ defined as $a \otimes_p b = (a^p \cdot b^p)^{1/p} = a \cdot b$ ($\otimes_p$ is therefore the standard multiplication).

When p tends towards $+\infty$ (resp. $-\infty$), $(\mathbb{R}_+, \oplus_p, \otimes_p)$ "tends" towards the dioid $(\mathbb{R}_+, \text{Max}, \times)$ (resp. $\mathbb{R}_+, \text{Min}, \times$). (see Sect. 4.8.2).

[References: **1** (Sect. 3.2.5) **8** (Sect. 1.4.7)]

4.4.4. $(\mathbb{R}_+, \oplus^h, \otimes^h)$

This is the set of nonnegative reals endowed with the operation $\oplus^h$ (with $h \in \mathbb{R}_*$) defined as $a \oplus^h b = h\, \ell n\, (e^{a/h} + e^{b/h})$ and the operation $\otimes^h$ defined as $a \otimes^h b = h\, \ell n\, (e^{a/h}. e^{b/h}) = a + b$ ($\otimes^h$ is thus the usual addition).

When h tends to 0^+ (resp. 0^-), $(\mathbb{R}_+, \oplus^h, \otimes^h)$ "tends" towards the dioid $(\mathbb{R}_+, \text{Max}, +)$ (resp. $\mathbb{R}_+, \text{Min}, +$)

[References: **1** (Sect. 3.2.5) **8** (Sect. 1.4.8)]

4.4.5. $(L^2 (\mathbb{R}^n)^+, +, *)$

We consider the set of functions: $\mathbb{R}^n \to \mathbb{R}_+$, square integrable, endowed with the addition of functions and the convolution product. This is a symmetrizable dioid.

4.5. Idempotent Dioids

The subclass of idempotent dioids includes (see Fig. 3):

- Doubly idempotent dioids (see Sect. 4.6);
- Idempotent-cancellative dioids (see Sect. 4.7);
- Idempotent-invertible dioids (see Sect. 4.8);
- Other idempotent dioids which do not belong to any of the previous categories, and which are dealt with in the present section.

4.5.1. Dioid $(\mathcal{P}(\mathbb{R}^n), \cup, +)$

$n \in N$ being a given integer, we consider $E = \mathcal{P}(\mathbb{R}^n)$ the power set of $\mathbb{R}^n$. The law $\oplus$ is defined as the *union* of two subsets of $\mathbb{R}^n$ (neutral element $\emptyset$).
The law $\otimes$ is taken as the *sum* of two subsets of $\mathbb{R}^n$ defined as:

$$A \subset \mathbb{R}^n, B \subset \mathbb{R}^n$$
$$A + B = \{z/z \in \mathbb{R}^n, z = x + y \quad \text{with} \quad x \in A, y \in B\}$$

The neutral element of $\otimes$ is the subset reduced to the zero vector $(0, 0, \ldots 0)$ of $\mathbb{R}^n$.

One easily verifies that $(\mathcal{P}(\mathbb{R}^n), \cup, +)$ possesses all the properties of an *idempotent dioid.*

4.5.2. Dioid $\mathcal{P}(\mathbb{R})$ Endowed with the Union and the Product $(\mathcal{P}(\mathbb{R}), \cup, .)$

The set E is the power set of $\mathbb{R}$.
The law $\oplus$ is the *union of* two subsets of $\mathbb{R}$ (neutral element $\emptyset$).
The law $\otimes$ is the *product* of two subsets of $\mathbb{R}$ defined as:

$$A \subset \mathbb{R}, B \subset \mathbb{R}:$$
$$A \cdot B = \{z/z \in \mathbb{R}, z = x \cdot y \quad \text{with} \quad x \in A, y \in B\}$$

The neutral element of $\otimes$ is the subset $\{1\}$.

One easily verifies that $(\mathcal{P}(\mathbb{R}), \cup, .)$ possesses all the properties of an *idempotent dioid.*

4.5.3. Qualitative Algebra

On the set of signs, augmented with the indeterminate ?, S = {+, −, 0, ?}, we consider the qualitative addition ⊕ and qualitative multiplication ⊗ defined in the following tables (see Sects. 1.5.5 and 1.1.7):

⊕	+	−	0	?
+	+	?	+	?
−	?	−	−	?
0	+	−	0	?
?	?	?	?	?

⊗	+	−	0	?
+	+	−	0	?
−	−	+	0	?
0	0	0	0	0
?	?	?	0	?

As a result, an idempotent dioid is obtained.

[References: **1** (Sect. 3.4.2), **1** (Sect. 3.4.3), **1** (Sect. 6.1.3)]

4.5.4. $(\mathrm{Conv}(\mathbb{R}^k), \mathrm{Conv}(A \cup B), +)$, $(\mathrm{Conv}_c(\mathbb{R}^k), \mathrm{Conv}(A \cup B), +)$

We consider $\mathrm{Conv}(\mathbb{R}^k)$, the set of convex subsets of $\mathbb{R}^k$, endowed with the two following operations (see Sects. 1.5.6 and 1.1.2):

$$A \oplus B = \mathrm{Conv}(A \cup B)$$
$$A \otimes B = A + B$$

where conv(X) denotes the convex hull of X and A + B the vector sum of A and B. (see Sect. 1.1.2).

To show that this is a dioid, the non trivial property of the *distributivity* of ⊗ with respect to ⊕ remains to be proven, i.e.:

$$\mathrm{conv}(A \cup B) + C = \mathrm{Conv}[(A + C) \cup (B + C)].$$

Let us first of all check the following property:

V convex ⇒ for any subset U, $\mathrm{Conv}(U + V) = \mathrm{Conv}(U) + V$.

Indeed, $\mathrm{Conv}(U)+V$ is a convex set containing $U+V$, therefore $\mathrm{Conv}(U+V) \subset \mathrm{Conv}(U) + V$.

Conversely, every element w of $\mathrm{Conv}(U) + V$ is written $w = \sum_{i=1}^{n} \alpha_i\, u_i + v$ with $\alpha_i \geq 0$, $\sum_{i=1}^{n} \alpha_i = 1$, $u_i \in U$, $v \in V$, hence $w = \sum_{i=1}^{n} \alpha_i\, (u_i + v) \in \mathrm{Conv}\,(U + V)$, which proves the expected property.

This therefore yields $\mathrm{Conv}(A \cup B) + C = \mathrm{Conv}((A \cup B) + C) = \mathrm{Conv}[(A + C) \cup (B + C)]$, which proves distributivity.

$\mathrm{Conv}(\mathbb{R}^k)$ the set formed by all convex subsets of $\mathbb{R}^k$, is therefore an idempotent dioid.

The set $\mathrm{Conv}_c(\mathbb{R}^k)$ of *compact convex* subsets of $\mathbb{R}^k$ forms a subdioid of $\mathrm{Conv}(\mathbb{R}^k)$.

Let us show that this dioid is *idempotent-cancellative*. Indeed, let the *support* mapping, which characterizes a closed convex subset, be defined as:

$$\delta_A(p) = \sup_{x \in A} \langle p, x \rangle \quad \text{for} \quad p \in \mathbb{R}^k$$

and which, in view of the compactness of A, is finite. One verifies that $\delta_{A+B} = \delta_A + \delta_B$. If $A + B = C$, B is the unique convex subset having support function $\delta_B = \delta_C - \delta_A$. It follows that the application $B \to A + B$ is injective, and therefore the dioid $\text{Conv}_c(\mathbb{R}^k)$ is idempotent-cancellative.

[References: Rockafellar (1970), and **8** (Sect. 1.1.2), **8** (Sect. 1.5.6)]

4.5.5. Dioid of Relations

Let E be a set and $\mathcal{R}$ (E) the set of binary relations on E. One verifies that $\mathcal{R}$ (E), endowed with the two following laws:

$$R \oplus R': x\,(R \oplus R')\,y \text{ iff } x\,R\,y \quad \text{or} \quad xR'y,$$
$$R \otimes R': x\,(R \otimes R')\,y \text{ iff } \exists\, z \in E \quad \text{with} \quad x\,R\,z \quad \text{and} \quad z\,R'\,y$$

is an idempotent dioid.

The canonical order relation corresponds to: $R \leq R'$ if and only if x Ry implies x R' y, i.e. iff R is *finer* than R'.

4.5.6. Dioid of Mappings with Max and Convolution $\left(\overline{\mathbb{R}}_{\max}^{\mathbb{R}^n}\right)$

The set of mappings: $\mathbb{R}^n \to \overline{\mathbb{R}}$, endowed with the operation "pointwise maximum" ($(h \oplus g)(x) = \max\{(h(x), g(x))\}$) and with the sup-convolution product defined as:

$$(f \otimes g)(x) = \sup_{y \in R^n} \{f(x - y) + g(y)\}$$

(where we agree to set $(+\infty) + (-\infty) = -\infty$) is a complete idempotent dioid (refer to Chap. 1 Sect. 6.1.8 for the definition of a complete dioid). The neutral element for the product is the function e defined as: $e(x) = +\infty$ if $x \neq 0$ and $e(0) = 0$.

4.5.7. Dioid ($\hat{\mathbb{R}} \times \mathcal{P}(\mathbb{R})$, Min, +)

This is an extension of the "Min-Plus" dioid (see Sect. 4.8.3 below) to sets of vectors "monitored" by the first component.

E is the set $\hat{\mathbb{R}} \times \mathcal{P}(\mathbb{R})$; its elements are in the form $a = (a_1, a_2)$ with $a_1 \in \hat{\mathbb{R}}$ and $a_2 \in \mathcal{P}(\mathbb{R})$.

$\oplus$ is the Min operation defined on $\hat{\mathbb{R}} \times \mathcal{P}(\mathbb{R})$ in Sect. 1.5.10:

$$\mathrm{Min}(a, b) = \begin{cases} a & \text{if} \quad a_1 < b_1 \\ b & \text{if} \quad a_1 > b_1 \\ (a_1, a_2 \cup b_2) & \text{if} \quad a_1 = b_1 \end{cases}$$

$\otimes$ is the operation defined as:

$$a \otimes b = \begin{cases} a_1 + b_1 \\ a_2 + b_2 \end{cases}$$

($a_2 + b_2$ denotes the set of reals in the form $\alpha + \beta$, for all values of $\alpha \in a_2, \beta \in b_2$)

$\otimes$ is associative, commutative and endows E with a *monoid* structure with unit element $e = (0, 0)$. One also verifies the right and left distributivity of $\otimes$ with respect to $\oplus$.

Finally, $\varepsilon = (+\infty, \emptyset)$ is absorbing ($\forall a \in E$: $a \otimes \varepsilon = \varepsilon \otimes a = \varepsilon$).

$(E, \oplus, \otimes)$ thus defined is an idempotent dioid. This is an extension to dimension 2 of the "Min-Plus" dioid.

Finally, let us mention two interesting specializations of the above in connection with complex numbers.

4.5.8. Dioid ($\mathbb{C}_R$, Min, +)

E is the set of elements of the form $a = a_1 + i\, a_2$ with $a_1 \in \hat{\mathbb{R}}$ and $a_2 \in \mathcal{P}(\mathbb{R})$. An element a of E is therefore a set of complex numbers all having the same real component a_1. $\oplus$ and $\otimes$ are then defined as in Sect. 4.5.7.

4.5.9. Dioid ($\mathbb{C}_\rho$, Min, x)

E is the set of elements of the form $a = \rho(a)\, e^{i\,\theta(a)}$ with $\rho(a) \in \hat{\mathbb{R}}_+$ and $\theta(a) \in \mathcal{P}(\mathbb{R})$. An element a of E is therefore a set of complex numbers all having the same modulus ρ.

$\oplus$ is the Min operation defined on $\hat{\mathbb{R}}_+ \times \mathcal{P}(\mathbb{R})$ by:

$$\mathrm{Min}(a, b) = \begin{cases} a & \text{if} \quad \rho(a) < \rho(b) \\ b & \text{if} \quad \rho(b) < \rho(a) \\ (\rho(a), \theta(a) \cup \theta(b)) & \text{if} \quad \rho(a) = \rho(b) \end{cases}$$

$\otimes$ is the "multiplication" operation defined on $\hat{\mathbb{R}}_+ \times \mathcal{P}(\mathbb{R})$ by:

$\rho(a \otimes b) = \rho(a)\, \rho(b)$

$\theta(a \otimes b) = \theta(a) + \theta(b)$ (Minkowski sum of the sets $\theta(a)$ and $\theta(b)$).

($\theta(a) + \theta(b)$ is the set of real numbers of the form $\alpha + \beta$, α running through $\theta(a)$ and β running through $\theta(b)$).

[The two previous idempotent dioids appear to be relevant to the so-called complex Min-Plus analysis which has application in the study of complex Hamilton–Jacobi equations in quantum mechanics (see Gondran 1999b).]

4.6. Doubly Idempotent Dioids, Distributive Lattices

4.6.1. Dioid ({0, 1} , Max, Min): Boole Algebra

E is the set $\{0, 1\}$ endowed with the operations $\oplus$ and $\otimes$ defined as:

$$a \oplus b = \text{Max}\{a, b\}$$
$$a \otimes b = \text{Min}\{a, b\}$$

$\varepsilon = 1$ is the neutral element of $\oplus$ and $e = 0$ the neutral element of $\otimes$.

We thus have that $E = \{\varepsilon, e\}$.

One easily checks all the properties necessary for $(E, \oplus, \otimes)$ to enjoy a dioid structure. In particular, the canonical preorder relation is an order relation because $\oplus$ is idempotent (see Chap. 1, Sect. 3.4).

Observe that the Boole algebra $(E, \oplus, \otimes)$ as defined above is an ordered set in which any pair a, b of elements has an upper bound given by:

$$\sup(a, b) = a \oplus b$$

and a lower bound given by

$$\inf(a, b) = a \otimes b$$

This is therefore a *distributive lattice*. As, moreover, there is a finite number of elements, this is a complete lattice.

[Reference: see e.g. Birkhoff, 1979]

4.6.2. Dioids (ℕ, lcm, gcd) and (ℕ, gcd, lcm)

Here we take $E = \mathbb{N}$, the set of natural numbers.

The lcm operation on $\mathbb{N}$ is associative, commutative, idempotent and has as neutral element 1. Similarly, the gcd operation on $\mathbb{N}$ is associative, commutative and idempotent. It has as neutral element $+\infty$ by viewing this element as the infinite product of all the prime numbers raised to the power $+\infty$. One easily verifies the distributivity of lcm with respect to gcd and of gcd with respect to lcm.

Finally, 1 is absorbing for gcd in the structure $(\mathbb{N}, \text{lcm}, \text{gcd})$ and $+\infty$ is absorbing for lcm in the structure $(\mathbb{N}, \text{gcd}, \text{lcm})$.

Each of these structures is therefore a *doubly idempotent dioid*. Moreover, since, Proposition 6.5.7 of Chap. 1 holds, these are *distributive lattices*.

4.6.3. Dioids $\left(\overline{\mathbb{R}}, \text{Max}, \text{Min}\right)$, $\left(\overline{\mathbb{R}}, \text{Min}, \text{Max}\right)$

We take as our basic set E the set $\overline{\mathbb{R}}$ of real numbers augmented with the elements $\varepsilon = -\infty$ and $e = +\infty$, the operations $\oplus$ and $\otimes$ being defined as:

$$a \oplus b = \text{Max}\{a, b\}$$
$$a \otimes b = \text{Min}\{a, b\}$$

The element ε is absorbing because, $\forall a \in E$, $a \otimes \varepsilon = \text{Min}\{a, -\infty) = -\infty = \varepsilon$.

One easily verifies that (E, $\oplus$, $\otimes$) enjoys all the properties of a dioid. In particular the operation $\oplus$ being idempotent, the canonical preorder relation is an *order* (it is the usual total order on the set of real numbers).

Moreover, E can also be considered as a *lattice* taking as the upper bound of two arbitrary elements a and b $\sup(a, b) = a \oplus b = \text{Max}\{a, b\}$ and as the lower bound of two arbitrary elements a and b

$$\inf(a, b) = a \otimes b = \text{Min}\{a, b\}$$

One indeed verifies that all the axioms of the algebraic definition of a distributive lattice are satisfied: idempotence, commutativity, the associativity of each law, distributivity, as well as the absorption properties:

$$a \oplus (a \otimes b) = a$$
$$a \otimes (a \oplus b) = a$$

One recognizes a doubly idempotent dioid structure. By exchanging the Max and Min operations, we obtain the lattice-dioid $\left(\overline{\mathbb{R}}, \text{Min}, \text{Max}\right)$ which has similar properties.

The $(\overline{\mathbb{R}}, \text{Max}, \text{Min})$ structure defined above is the one underlying the algebraic formulation of the maximal capacity path problem in a graph or equivalently (see Gondran and Minoux 1995) the problem of finding a path with *maximal inf-section*.

Similarly, the structure $(\overline{\mathbb{R}}, \text{Min}, \text{Max})$ is the one underlying the algebraic formulation of the minimal sup-section path problem.

[References: **1** (Sect. 6.5), **4** (Sect. 6.3)]

4.6.4. Lattice of the Power Set of a Given Set $(\mathcal{P}(X), \cup, \cap)$, $(\mathcal{P}(X), \cap, \cup)$

X being a given set, we take $E = \mathcal{P}(X)$, the power set of X. The law $\oplus$ is the union of two subsets (neutral element: $\emptyset$) and the law $\otimes$ is the intersection of two subsets (neutral element: X). One easily verifies that $(\mathcal{P}(X), \cup, \cap)$ satisfies all the properties of a *distributive lattice*. The same is true for $(\mathcal{P}(X), \cap, \cup)$.

4.7. Idempotent-Cancellative and Selective-Cancellative Dioids

4.7.1. Dioids $(\hat{\mathbb{R}}_+, \text{Min}, +)$, $(\hat{\mathbb{N}}, \text{Min}, +)$

The set E is the set of nonnegative reals augmented with $+\infty$.

$\oplus$ is the Min operation (minimum of two real numbers). It is associative, commutative, selective, and has neutral element $\varepsilon = +\infty$.

$\otimes$ is the operation + (sum of two real numbers). It is associative, commutative, and has unit element $e = 0$. It endows the set E with a *hemi-group* structure (see Sect. 1.4.1.).

One easily verifies the right and left distributivity of $\otimes$ with respect to $\oplus$. Finally $\varepsilon = +\infty$ is absorbing for $\otimes$ ($\forall a \in E, a \otimes \varepsilon = \varepsilon \otimes a = \varepsilon$).

$(\mathbb{R}_+ \cup \{+\infty\}, \text{Min}, +)$ is therefore a *selective-cancellative dioid.*

$(\hat{\mathbb{N}}, \text{Min}, +)$ is a subdioid of $(\hat{\mathbb{R}}, \text{Min}, +)$. It is sometimes referred to as the "tropical dioid" and has been used to solve some problems of enumeration in language theory (see for example Simon 1994).

[Reference: **1** (Sect. 6.6.4)]

4.7.2. Regular Language Dioid

A being a set of letters ("alphabet"), the free monoid (sets of words formed by a finite number of letters on A) is denoted A^* (see Sect. 1.4.4). It includes the empty word, denoted ε.

Every subset (whether finite or infinite) of A^*, $L \subset A^*$, is called a *language* on A. We denote $\mathcal{L} = \mathcal{P}(A^*)$ the set of all the languages on A.

The sum of two languages $L_1 \oplus L_2$ is defined as the set union of the words of L_1 and the words of L_2.

The product of two languages $L_1 \otimes L_2$ is the set of words formed by the concatenation of a word m_1 of L_1 and a word m_2 of L_2 (in this order).

The two following dioids are relevant to language theory:

$(\mathcal{P}_{\text{finite}}(A^*), \oplus, \otimes)$, the dioid of finite languages (finite subsets of A^*) formed on the alphabet A.

$(\mathcal{P}(A^*), \oplus, \otimes)$ the dioid of arbitrary languages formed on the alphabet A.

$(\mathcal{P}(A^*), \oplus, \otimes)$ has the empty language $\emptyset$ as neutral element of $\oplus$ and $\{\varepsilon\}$ as neutral element for $\otimes$. It has $(\mathcal{P}_{\text{finite}}(A^*), \oplus, \otimes)$ as subdioid.

In the case where the alphabet is finite, $(\mathcal{P}_{\text{finite}}(A^*), \oplus, \otimes)$ corresponds to the dioid of the polynomials on A, and $(\mathcal{P}(A^*), \oplus, \otimes)$ to the dioid of the formal series on A.

The structure defined above is that of regular languages (see e.g. Salomaa 1969). However, we observe that in addition to the dioid structure, the axioms of regular languages include the so-called closure operation denoted *.

[Reference: **1** (Sect. 6.6.2)]

4.7.3. Dioid of Integer Sequences with Min and Sum Operations $(\mathbb{N}^{\mathbb{N}}, \text{Min}, +)$

Here we take $E = \mathbb{N}^{\mathbb{N}}$ as the set of (infinite) sequences of nonnegative integers.

The operation $\oplus$ is the term by term minimum operation of two sequences of integers, in other words for two sequences $a = (a_i)_{i \in \mathbb{N}}$ and $b = (b_i)_{i \in \mathbb{N}}$, we set:

$$a \oplus b = c = (c_i)_{i \in \mathbb{N}} \quad \text{with} \quad c_i = \text{Min}\{a_i, b_i\} \ (\forall i).$$

The law $\oplus$ is associative, commutative and idempotent and its neutral element is the sequence $\varepsilon = (\varepsilon_i)_{i \in \mathbb{N}}$ with, $\forall i \in \mathbb{N}$: $\varepsilon_i = +\infty$.

The operation $\otimes$ is the term by term sum operation of two sequences of integers, in other words for $a = (a_i)_{i \in \mathbb{N}}$ and $b = (b_i)_{i \in \mathbb{N}}$, we set:

$$a \oplus b = c = (c_i)_{i \in \mathbb{N}} \quad \text{with} \quad c_i = a_i + b_i (\forall i).$$

The operation $\otimes$ is associative, commutative and its unit element is the zero sequence $e = (e_i)_{i \in \mathbb{N}}$ with $\forall i: \ e_i = 0$.

We observe that every element of E is cancellative for $\oplus$. Moreover, one easily verifies that $\otimes$ is distributive with respect to $\oplus$ and that ε is absorbing for $\otimes$.

Consequently $(\mathbb{N}^{\mathbb{N}}, \text{Min}, +)$ is an idempotent-*cancellative dioid.*

4.7.4. Dioid $(\mathbb{N}^*, \text{gcd}, \times)$

We take $E = \mathbb{N}^* = \mathbb{N}\backslash\{0\}$ as the set of strictly positive integers.

The operation $\oplus$ is the gcd of two numbers. This operation is associative, commutative and idempotent. The neutral element of $\oplus$ is $\varepsilon = +\infty$, by viewing ε as the (infinite) product of all the prime numbers, raised to the power $+\infty$.

The operation $\otimes$ is the product of two integers (neutral element $e = 1$). We observe that every element of $\mathbb{N}^*$ is cancellative for $\otimes$.

One also verifies that $\otimes$ is distributive with respect to $\oplus$ and that ε is absorbing for $\otimes$.

$(\mathbb{N}^*, \text{gcd}, \times)$ is therefore an idempotent *dioid.*

The dioid $(\mathbb{N}^*, \text{gcd}, \times)$ is isomorphic to the dioid $(\mathbb{N}^{\mathbb{N}}, \text{Min}, +)$ (see Sect. 4.7.3)

Indeed, if we denote $p_1, p_2 \ldots$ the (infinite) sequence of prime numbers, every integer $n \in \mathbb{N}$ may be decomposed into prime factors and expressed in the form:

$n = \prod_{i=1}^{q} p_i^{n_i}$ (q is the rank of the largest prime number not greater than $\sqrt{n}$).

One therefore sees that with every integer $n \in \mathbb{N}$ one can associate a sequence:

$$(n) = (n_1, n_2, \ldots n_{||}) \ \in \mathbb{N}^{\mathbb{N}}$$

If $n \in \mathbb{N}$ is defined as the sequence (n) and $m \in \mathbb{N}$ by the sequence (m), the gcd of n and m is defined as the sequence (r) such that:

$$\forall i \in \mathbb{N} : r_i = \min\{n_i, m_i\}$$

We denote:

$$(r) = (n) \oplus (m)$$

Similarly, the product $n \times m$ is defined as the sequence (s) such that:

$$\forall i: s_i = n_i + m_i$$

The gcd and $\times$ operations on $\mathbb{N}$ therefore clearly correspond to the operations Min and + of the dioid $(\mathbb{N}^{\mathbb{N}}, \text{Min}, +)$.

Finally, let us observe that, if one replaces the gcd operation with the lcm operation, we obtain $(\mathbb{N}, \text{lcm}, \times)$ which is only a *pre-dioid* (the property of absorption not being satisfied) see Sect. 2.4.2.

4.8. Idempotent-Invertible and Selective-Invertible Dioids

4.8.1. Dioid $(\hat{\mathbb{R}}^n, \text{Min}, +)$

The set E is $\hat{\mathbb{R}}^n$ and we take $\oplus$ as the Min operation (component-wise minimum, of two vectors of $\mathbb{R}^n$). It is associative, commutative, and has neutral element $\varepsilon = \begin{pmatrix} +\infty \\ +\infty \\ \vdots \\ +\infty \end{pmatrix}$.

This operation is *idempotent* but *not selective.*

$\otimes$ is the operation + (sum of two vectors of $\mathbb{R}^n$). It is associative, commutative, and has unit element $e = \begin{pmatrix} 0 \\ 0 \\ \vdots \\ 0 \end{pmatrix}$. It endows E with a *group* structure.

One easily verifies the right and left distributivity of $\otimes$ with respect to $\oplus$ and the absorption property of ε.

$(\hat{\mathbb{R}}^n, \text{Min}, +)$ is therefore an *idempotent-invertible dioid.*

4.8.2. Dioid $(\mathbb{R}_+, \text{Max}, \times)$

E is the set $\mathbb{R}_+$ of nonnegative reals.

$\oplus$ is the operation Max (maximum of two real numbers) with neutral element $\varepsilon = 0$.

$\otimes$ is the operation $\times$ (product of two real numbers) with unit element $e = 1$.

$\oplus$ is selective and $\otimes$ endows $\mathbb{R}_+ \backslash \{0\}$ with a group structure.

It is therefore a selective-invertible dioid.

The dioid $(\mathbb{R}_+, \text{Max}, \times)$ is isomorphic to the dioid $(\mathbb{R} \cup \{-\infty\}, \text{Max}, +)$ through the one-to-one correspondence: $x \to e^x$.

We observe that if we take $\otimes$ to be the addition of real numbers instead of the multiplication of real numbers, we obtain the structure $(\mathbb{R}_+, \text{Max}, +)$ which is a *pre-dioid* (see Sect. 2.4.1).

4.8.3. "Min-Plus" Dioid $(\hat{\mathbb{R}}, \text{Min}, +)$

E is the set $\hat{\mathbb{R}} = \mathbb{R} \cup \{+\infty\}$

$\oplus$ is the Min operation (minimum of two real numbers).

$\oplus$ is associative, commutative, selective and has neutral element $\varepsilon = +\infty$.

$\otimes$ is the operation + (sum of two real numbers).

$\otimes$ is associative, commutative and endows E with a *group* structure with unit element $e = 0$.

One also verifies the right and left distributivity of $\otimes$ with respect to $\oplus$.

Finally $\varepsilon = +\infty$ is absorbing ($\forall a \in E$: $a \otimes \varepsilon = \varepsilon \otimes a = \varepsilon$).

(E, $\oplus$, $\otimes$) thus defined is therefore a *selective-invertible dioid.*

The MIN-PLUS dioid is the algebraic structure underlying the shortest path problem in a graph (see for example Gondran and Minoux 1995, Chap. 3).

[References: **4** (Sect. 2), **4** (Sect. 6.5)]

4.8.4. "Max-Plus" dioid ($\check{\mathbb{R}}$, Max, +)

E is the set $\check{\mathbb{R}} = \mathbb{R} \cup \{-\infty\}$

$\oplus$ is the Min operation (minimum of two real numbers) with neutral element $\varepsilon = -\infty$.

$\otimes$ is the operation + (sum of two real numbers).

This dioid is isomorphic to the MIN-PLUS dioid.

The MAX-PLUS dioid is the appropriate structure to express in algebraic terms the behavior of some discrete event dynamic systems and thus generalize many classical results in automatic control of linear systems.

[References: **6** (Sect. 7); Baccelli et al. 1992; Gaubert 1992, 1995a,b.]

4.8.5. Dioid ($\hat{\mathbb{R}}^2$, Min-lexico, +), ($\hat{\mathbb{R}}^n$, Min-lexico, +)

E is the set of vectors of $\hat{\mathbb{R}}^2$ (totally) ordered by the lexicographic order.

$\oplus$ is the *lexicographic minimum* on $\hat{\mathbb{R}}^2$ defined as:

$$a \oplus b = \begin{cases} a & \text{if} \quad a_1 < b_1 \quad \text{or if} \quad a_1 = b_1 \quad \text{and} \quad a_2 < b_2 \\ b & \text{if} \quad b_1 < a_1 \quad \text{or if} \quad a_1 = b_1 \quad \text{and} \quad b_2 \le a_2 \end{cases}$$

$\otimes$ is the component-wise addition:

$$a \otimes b = \begin{pmatrix} a_1 + b_1 \\ a_2 + b_2 \end{pmatrix}$$

One verifies that (E, $\oplus$, $\otimes$) is a selective-invertible dioid. This is another possible extension to dimension 2 of the "Min-Plus" dioid.

More generally, for arbitrary $n > 2$ the dioid ($\hat{\mathbb{R}}^n$, Min-lexico, +) would be defined in a similar way.

	Property of the monoid $(E, \oplus)$	Property of the monoid $(E, \otimes)$	Distributivity of $\otimes$ with respect to $\oplus$	Comments, additional properties
General dioids				
Dioid of the k shortest paths (see Sect. 4.3.1.)	Commutative, neutral element ε	Commutative, unit element e	Right and left	
Dioid of the η-optimal path (see Sect. 4.3.2)	Commutative, neutral element ε	Commutative, unit element e	Right and left	
Order of magnitude dioid	Commutative, neutral element $(0, +\infty)$	Commutative, unit element $(1, 0)$	Right and left	
Nonstandard number dioid	Commutative, neutral element $(0, 0, +\infty)$	Commutative, unit element $(1, 0, +\infty)$	Right and left	
Symmetrizable dioids				
$(\mathbb{R}_+, +, \times)$	Commutative, neutral element $\varepsilon = 0$	Commutative, unit element $e = 1$	Right and left	Every element is cancellative for $\oplus$
Intervals in $\mathbb{R}_+(\text{Int}\ (\mathbb{R}_+), +, \times)$	Commutative, $\varepsilon = [0, 0]$	Commutative, $e = [1, 1]$	Right and left	Every element is cancellative for $\oplus$
Idempotent dioids				
$(\mathcal{P}(\mathbb{R}^n), \cup, +)$	Commutative, idempotent, $\varepsilon = \emptyset$	Commutative, $e = \begin{pmatrix} +\infty \\ +\infty \\ \vdots \\ +\infty \end{pmatrix}$	Right and left	
$(\mathcal{P}(\mathbb{R}), \cup, \cdot)$	Commutative, idempotent, $\varepsilon = \emptyset$	Commutative, $e = \{1\}$	Right and left	

	Property of the monoid (E, $\oplus$)	Property of the monoid (E, $\otimes$)	Distributivity of $\otimes$ with respect to $\oplus$	Comments, additional properties
Qualitative algebra	Commutative, idempotent, $\varepsilon = 0$	Commutative, e = +	Right and left	
$(\mathrm{Conv}(\mathbb{R}^k), \cup, +)$	Commutative, neutral element $\emptyset$	Commutative, unit element {0}	Right and left	
Dioid of relations	Commutative	Commutative	Right and left	

Doubly idempotent dioids

({0,1} , Max, Min) Boole algebra	Commutative, idempotent, $\varepsilon = 0$	Commutative, idempotent, e = 1	Right and left	(distributive lattice)
$(\mathbb{N}, \mathrm{lcm}, \mathrm{gcd})$ $(\mathbb{N}, \mathrm{gcd}, \mathrm{lcm})$	Commutative, idempotent, $\varepsilon = 1$ $(\varepsilon = +\infty)$	Commutative, idempotent, $e = +\infty$ $(e = 1)$	Right and left	(distributive lattice)
$(\overline{\mathbb{R}}^n, \mathrm{Max}, \mathrm{Min})$ $(\overline{\mathbb{R}}^n, \mathrm{Min}, \mathrm{Max})$	Commutative, idempotent, $\varepsilon = -\infty$ $(\varepsilon = +\infty)$	Commutative, idempotent, $e = +\infty$ $(e = -\infty)$	Right and left	(distributive lattice)
$(\mathcal{P}(X), \cup, \cap)$	Commutative, idempotent, $\varepsilon = \emptyset$	Commutative, idempotent, e = X	Right and left	(distributive lattice)

Idempotent-cancellative and selective-cancellative dioids

$(\overline{\mathbb{R}}^n_+, \mathrm{Min}, +)$	Commutative, selective, $\varepsilon = +\infty$	Commutative, e = 0	Right and left	Every element is cancellative for $\otimes$
Regular languages	Commutative, idempotent, $\varepsilon = \emptyset$	$e = L_{\emptyset}$	Right and left	Every element is cancellative on the right and on the left for $\otimes$

	Property of the monoid (E, ⊕)	Property of the monoid (E, ⊗)	Distributivity of ⊗ with respect to ⊕	Comments, additional properties
($\mathbb{N}^{\mathbb{N}}$, Min, +)	Commutative, idempotent, $\varepsilon = (+\infty, +\infty \ldots)$	Commutative, $e = 1$	Right and left	Every element is cancellative for ⊗
($\mathbb{N}^*$, gcd, ×)	Commutative, idempotent, $\varepsilon = +\infty$	Commutative, $e = 1$	Right and left	Every element is cancellative for ⊗

Idempotent-invertible and selective-invertible dioids

($\mathbb{R}_+$, Max, ×)	Commutative, selective, $\varepsilon = 0$	Abelian group, $e = 1$	Right and left	Every element has an inverse for ⊗
($\hat{\mathbb{R}}$, Min, +)	Commutative, selective, $\varepsilon = +\infty$	Abelian group, $e = 0$	Right and left	Every element has an inverse for ⊗
"Max-Plus" dioid ($\mathbb{R} \cup \{-\infty\}$, Max, +)	Commutative, selective, $\varepsilon = -\infty$	Abelien group, $e = 0$	Right and left	Every element has an inverse for ⊗
($\overline{\mathbb{R}}^n$, Min, +)	commutative, idempotent, $\varepsilon = \begin{pmatrix} +\infty \\ +\infty \\ \vdots \\ +\infty \end{pmatrix}$	commutative, $e = \begin{pmatrix} 0 \\ 0 \\ \vdots \\ 0 \end{pmatrix}$	Right and left	Every element has an inverse for ⊗
($\mathbb{R}^n$, Min-lexico, +)	Commutative, selective, $\varepsilon = \begin{pmatrix} +\infty \\ +\infty \\ \vdots \\ +\infty \end{pmatrix}$	Group, $e = \begin{pmatrix} 0 \\ 0 \\ \vdots \\ 0 \end{pmatrix}$	Right and left	Every element has an inverse for ⊗

References

Abdali K.S., Saunders D.B. (1975), Transitive closure and related semiring properties via eliminants, in Backhouse R.C., Carré B.A.

Ahuja R.K., Magnanti T.L., Orlin J.B. (eds.) (1993), *Network Flows*, Prentice Hall, Englewood Cliffs, NJ.

Akian M. (1999), "Densities of idempotent measures and large deviations", *Transactions of the American Mathematical Society*, 351, 11, pp. 4515–4543.

Akian M., Quadrat J., Viot M. (1994), "Bellman processes", in Cohen G., Quadrat J.P. (eds.), *LNC in Control and Information Sciences no. 199: 11th International Conference on Analysis and Optimization of Systems*, Springer Verlag, Berlin.

Akian M., Bapat R., Gaubert S. (1998), "Asymptotics of the Perron eigenvalue and eigenvector using max-algebra", *C.R. Acad. Sci. Paris*, t. 327, série I, pp. 927–932.

Akian M., Quadrat J.P., Viot M. (1998), "Duality between probability and optimisation", in J. Gunawardana (éd.) *Idempotency*, Cambridge University Press, pp. 331–353.

Aleksandrov A.D. (1939), "Almost everywhere existence of the second differential of a convex function and some properties of convex functions", *Leningrad University Annals (Mathematical Series)*, 37, pp. 3–35 (in russian).

Atkin A.O.L., Boros E., Cechlárová K., Peled U.N. (1997), Powers of circulants in bottleneck algebra, *Linear Algebra and its Applications*, 258, pp. 137–148.

Attouch H. (1984), *Variational Convergences for Functions and Operators*, Pitman, London, 1984.

Attouch H., Thera M. (1993), Convergences en analyse multivoque et variationnelle, *Matapli*, no. 36, Smai, 1993.

Attouch H., Wets R.J.B. (1986), Isometries of the Legendre-Fenchel transform, *Transactions of the American Mathematical Society*, vol. 296, pp. 33–60.

Attouch H., Aubin J.-P., Clarke F., Ekeland I. (1989), *Analyse non linéaire. Contributions en l'honneur de J.J. Moreau*, Gauthiers-Villars, Bordas, Paris.

Aubin J.P., Frankowska H. (1990), *Set-Valued Analysis*, Birkhaüser, Borton, 1990.

Auslender A. (1984), "A descent method for nondifferentiable convex programs", *Comptes rendus Acad. Sci. Paris*, t. 298, série 1, no. 20, pp. 529–532.

Azencott R., Guivarc'h Y., Gundy R.F. (1978), Ecole d'été de Saint-FLour 8, *Lectures Notes in Math*, Springer Verlag, Berlin, 1978.

Baccelli F., Cohen G., Olsder G.J., Quadrat J.P. (1992), *Synchronization and Linearity. An Algebra for Discrete Event Systems*, Wiley, London, 489 pp.

Backhouse R.C., Carré B.A. (1975), Regular algebra applied to path-finding problems, *J. Inst. Math. Appl.*, 15, pp. 161–186.

Bako A. (1974), Solution of the multiterminal minimal path problem in a network with gains, in A. Prekopa (ed.), *Progress in Operations Research.*

Bapat R.B. (1995), Permanents, max algebra and optimal assignment, *Linear Algebra and its Applications*, 226–228, pp. 73–86.

Bardi M., Capuzzo-Dolcetta I. (1997), *Optimal Control and Viscosity Solutions of Hamilton-Jacobi-Bellman Equations*, Birkhäuser.

Bardi M., Evans L.F. (1984), "On Hopf's formulas for solutions of Hamilton–Jacobi equations", *Nonlinear Analysis Theory, Methods and Applications*, vol. 8, no. 11, pp. 1373–1381.

Barles G. (1994), *Solutions de viscosité des équations de Hamilton–Jacobi*, Mathématiques et Applications, Springer Verlag, Berlin, 1994.

Beasley L.B., Pullman N.J. (1988), Semiring rank versus column rank, *Linear Algebra and its Applications*, 101, pp. 33–48.

Bellman R.E. (1958), On a routing problem, *Quart. Appl. Math.* 16, pp. 87–90.

Benzaken C. (1968), Structures algébriques des cheminements: pseudo-treillis, gerbiers de carré nul, in G. Biorci (ed.), *Network and Switching Theory*, Academic Press, NY, pp. 40–57.

Benzecri J.P. (1974), *L'Analyse des Données*: Tome 1: *Taxinomie*, Dunod, Paris.

Berge C. (1958), *La Théorie des graphes et ses applications*, Dunod Paris.

Berge C. (1959), *Espaces topologiques, fonctions multivoques*, Dunod, Paris.

Berge C. (1970), *Graphes et Hypergraphes*, Dunod Paris.

Berge C. (1972), Balanced Matrices, *Mathematical* Programming 2, 1, pp. 19–31.

Betrema J. (1982), Topologies sur des espaces ordonnés, *R.A.I.R.O. Informatique Théorique*, vol. 16, no. 2, pp. 165–182.

Birkhoff G. (1979), Lattice Theory, *Am. Math. Soc. Colloquium Publications*, vol. 25, 418 pp.

Blyth T., Janowitz M. (1972), *Residuation Theory*, Oxford, Pergamon Press.

Borchardt C.W. (1860), Uber eine der Interpolation entsprechende Darstellung der Eliminations-Resultante, *J. Reine Angew.Math*, 57, pp. 111–121.

Bourlès H. (1994), Semi-cancellable fractions in system theory, *IEEE Trans. Automatic Control*, vol. 39, no. 10, pp. 2148–2153.

Brams G.W. (1983), *Réseaux de Petri: Théorie et Pratique*, Masson, Paris.

Brenier Y. (1989), Un algorithme rapide pour le calcul de la transformée de Legendre-Fenchel discrète, *C.R.A.S.*, no. 308, série I, pp. 587–589.

Brualdi R.A., Ryser H.J. (1991), *Combinatorial Matrix Theory*, Cambridge University Press, 367 pp.

Bunks C., Chancelier J.P., Delebecque F., Gomez C. (ed.), Goursat M., Nikoukhah R., Steer S. (1999), *Engineering and Scientific Computing with Scilab*, Birkhäuser.

Burkard R.E. (1977), An algebraic approach to assignment problems, *Math. Program.*, 12, pp. 318–327.

Burkard R.E., Cuninghame-Green R.A., Zimmermann U. (1984), Algebraic and combinatorial methods in operations research, *Ann. Discrete Math.* 19, North-Holland, Amsterdam, 381 pp.

Butkovic P., Cechlárová K., Szabó P. (1987), Strong linear independence in bottleneck algebra, *Linear Algebra and its Applications*. 94, pp. 133–155.

Cao Z.Q., Kim K.H., Roush, F.W. (1984), *Incline Algebra and Applications*, Ellis Horwood, Chichester, Wiley, New York.

Carré B.A. (1971), An algebra for network routing problems, *J. Inst. Math. Appl.*, 7, pp. 273–294.

Cartier P., Foata D. (1969), *Problèmes Combinatoires de Commutation et Réarrangement*. Lecture Notes, in *Mathematics*, 85, Springer Verlag, Berlin.

Cayley A. (1854), Sur les déterminantes gauches, *Crelle's J.*, 38, pp. 93–96.

Cechlárová K. (1992), Eigenvectors in Bottleneck Algebra, *Linear Algebra and its Applications* 175, pp. 63–73.

Cechlárová K. (1995), Unique solvability of max-min fuzzy equations and strong regularity of matrices over fuzzy algebra, *Fuzzy Sets and Systems*, vol. 75, 2, pp. 165–177.

Cechlárová K. (1996), On the powers of matrices in bottleneck/fuzzy algebra, *Linear Algebra Appl.* 246, pp. 97–112.

Cechlárová K. (2003), Powers of matrices over distributive lattices – a review, *Fuzzy Sets and Systems*, vol. 138, 3, pp. 627–641.

Cechlárová K. and Plávka J. (1996), Linear independence in bottleneck algebras, *Fuzzy Sets and Systems*, vol. 77, 3, pp. 337–348.

Chaiken S. (1982), A combinatorial proof of the all- minors matrix-tree theorem, *S.I.A.M. J. Algebraic and Discrete Methods* 3, pp. 319–329.

Chang, C. C. (1958), Algebraic analysis of many valued logics, *Trans. AMS* 93, pp. 74–80.

Charnes A., Raike W.M. (1966), One-pass algorithms for some generalized network problems, *Oper. Research* 14, pp. 914–923.

Chen W.K. (1976), *Applied Graph Theory: Graphs and Electrical Networks*, North-Holland, Amsterdam.

Choquet G. (1953), "Theory of capacities", *Annales de l'Institut Fourier* 5, pp. 131–295.

Chrétienne P. (1983), *Les Réseaux de Petri Temporisés*, Thèse, Université Paris 6.

Clifford A.H., Preston G.B. (1961), *The Algebraic Theory of Semigroups*, vol. 1, Amer. Math. Soc.

Clifford A.H., Preston G.B. (1961), *The Algebraic Theory of Semigroups*, vol. 2, Amer. Math. Soc.

Cochet-Terrasson J., Cohen G., Gaubert S., Mc Gettrick M., Quadrat J.P. (1998), "Numerical computation of spectral elements in max-plus algebra", IFAC conference on system structure and control, Nantes, France.

Cohen G., Quadrat J.P. (1994) (eds.), 11th International Conference on Analysis and Optimization of Systems, Discrete Event Systems, Number 199, in *Lect. Notes in Control and Inf. Sci.*, Springer Verlag, Berlin.

Cohen G., Dubois D., Quadrat J.P., Viot M. (1983), *Analyse du comportement périodique de systèmes de production par la théorie des dioïdes*, Rapport Technique 191, INRIA, Rocquencourt.

Cohen G., Dubois D., Quadrat J.P., Viot M. (1985), A linear-system theoretic view of discrete event processes and its use for performance evaluation in manufacturing, *IEEE Trans. Autom. Control AC-30*, 3, pp. 210–220.

Cohen G., Moller P., Quadrat J.P., Viot M. (1989), Algebraic tools for the performance evaluation of discrete event systems, *I.E.E.E. Proceedings, Special Issue on Discrete Event Systems*, 77, no. 1.

Collomb P., Gondran M. (1977), Un algorithme efficace pour l'arbre de classification, *R.A.I.R.O. Recherche Opérationnelle*, vol. 1, pp. 31–49.

Crandall M.G., Lions P.L. (1983), Viscosity solutions of Hamilton–Jacobi equations, *Trans. Am. Math. Soc.* 277, 1983.

Crandall M.G., Evans L.C., Lions P.L. (1984), Some properties of viscosity solutions in Hamilton–Jacobi equations, *Trans Am. Math. Soc.*, vol. 282, no. 2, pp. 487–502.

Crandall M.G., Ishii H., Lions P.L. (1992), User's guide to viscosity solutions of second order partial differential equations, *Bull. Am. Math. Soc.*, vol. 27, no. 1, pp. 1–67.

Crouzeix J.-P. (1977), *Contribution à l'étude des fonctions quasiconvexes*, Thèse de Doctorat, Université de Clermont-Ferrand II (France), 1977.

Cruon R., Herve P.H. (1965), Quelques problèmes relatifs à une structure algébrique et à son application au problème central de l'ordonnancement, *Revue Fr. Rech. Op.* 34, pp. 3–19.

Cuninghame-Green, R.A. (1960), Process synchronisation in a steelworks – a problem of feasibility, in: Banbury and Maitland (eds.), *Proc. 2nd Int. Conf. on Operational Research*, English University Press, pp. 323–328.

Cuninghame-Green R.A. (1962), Describing industrial processes with inference and approximating their steady state behaviour, *Operl. Res. Quart.* 13, pp. 95–100.

Cuninghame-Green R.A. (1976), Projections in a minimax algebra, *Math. Progr.* 10, pp. 111–123.

Cuninghame-Green R.A. (1979), Minimax algebra, lecture notes, in *Economics and Mathematical Systems* 166, Springer, Berlin, Heidelberg, New York.

Cuninghame-Green R.A. (1991), Minimax algebra and applications, *Fuzzy Sets and Systems*, vol. 41, 3, pp. 251–267.

Cuninghame-Green R.A. (1991), Algebraic realization of discrete dynamical systems, *Proceedings of the 1991 IFAC Workshop on Discrete Event System Theory and Applications in Manufacturing and Social Phenomena*, International Academic Publishers, pp. 11–15.

Cuninghame-Green R.A. (1995), Maxpolynomial equations, *Fuzzy Sets and Systems*, vol. 75, 2, pp. 179–187.

Cuninghame-Green R.A., Cechlárová K. (1995), Residuation in fuzzy algebra and some applications, *Fuzzy Sets and Systems*, 71, pp. 227–239.

Cuninghame-Green R.A., Meijer P.F.J. (1980), An algebra for piecewise-linear minimax problems, *Discrete Appl. Math.* 2, pp. 267–294.

Dal Maso G. (1993), *An Introduction to* Γ*-convergence*, Birkhäuser.

Dantzig G.B. (1966), all shortest routes in a graph, in *Théorie des Graphes*, Proceedings Int. Symp. Rome, Italy, Dunod, Paris, pp. 91–92.

Dantzig G.B., Blattner W.O., Rao M.R. (1967), Finding a cycle in a graph with minimum cost-to-time ratio with application to a ship routing problem, in *Théorie des graphes*, proceedings int. symp. Rome, Italy, Dunod, Paris, pp. 209–214.

Dautray R., Lions J.L. (1985), *Analyse mathématique et calcul numérique pour les sciences et les techniques*, tome 3, Masson, 1985.

Defays D. (1978), Analyse hiérarchique des Préférences et généralisations de la transitivité, *Math. Sc. Hum.*, no. 61, pp. 5–27.

De Giorgi E. (1975), Sulla converginza di alcune successioni di integrali del tipo dell' area, *Rend. Math.* (6), 8, 1975, pp. 277–294.

De Giorgi E., Marino A., Tosques M. (1980), Problemi di evoluzione in spazi metrici e curve di massina pendenza, *Rend. Classe Sci. Fis. Mat. Nat. Accad. Naz. Lincei* 68, pp. 180–187.

Derigs U. (1979), Duality and admissible transformations in combinatorial optimization, *Zeitschrift für Oper. Res.*, 23, pp. 251–267.

Devauchelle-Gach B. (1991), *Diagnostic mécanique des fatigues sur les structures soumises à des vibrations en ambiance de travail*, Thèse de doctorat, Paris 9 Dauphine.

Dijkstra E.W. (1959), A Note on two problems in connection with graphs, *Numerische Mathematik* 1, pp. 269–271.

Di Nola A., Lettieri A., Perfilieva I. and Novák V. (2007), Algebraic Analysis of Fuzzy Systems, *Fuzzy Sets and Systems*, vol. 158, 1, 1, pp. 1–22.

Di Nola A., Gerla B., Algebras of Lukasiewicz's logic and their semirings reducts, in *Proc. Conf. on Idempotent Mathematics and Mathematical Physics*, to appear.

Dodu J.C., Eve T., Minoux M. (1994), "Implementation of a proximal algorithm for linearly constrained nonsmooth optimization problems and computational results", *Numerical Algorithms*, 6, pp. 245–273.

Dubois D. (1987), "Algèbres exotiques et ensembles flous". Séminaire "Algèbres Exotiques et Systèmes à Evénements Discrets", 3–4 juin 1987, CNET, Issy-les-Moulineaux.

Dubois D., Prade H. (1980), *Fuzzy Sets and Systems. Theory and Application*. Academic Press, New York.

Dubois D., Prade H. (1987), *Théorie des Possibilités. Application à la Représentation des Connaissances en Informatique*, Masson, Paris (2e éd.).

Dubois D., Prade H. (1991), Fuzzy sets in approximate reasoning – part1: inference with possibility distributions, *Fuzzy Sets and Systems*, 40, pp. 143–202.

Dubois D., Stecke K.E. (1990), Dynamic analysis of repetitive decision – free discrete event processes: the algebra of timed marked graphs and algorithmic issues, *Ann. Oper. Res.* 26, pp. 151–193.

Dubreil P., Dubreil-Jacotin M.L. (1964), *Leçons d'Algèbre Moderne*, Dunod, Paris, France.

Dudnikov P.L., Samborskii S.N. (1989), Spectra of endomorphisms of semimodules over semirings with an idempotent operation, *Soviet Math. Doki.*, vol. 40, no. 2, pp. 363–366.

Eilenberg S. (1974), *Automata, Languages and Machines*, Academic Press, London.

Ekeland I., Teman R. (1974), *Analyse convexe et problèmes variationnels*, Dunod, Gauthier-Villars, Paris.

Elquortobi A. (1992), Inf convolution quasi-convexe des fonctionnelles positives, *Recherche opérationnelle/Operations research*, vol. 26, no. 4, 1992, pp. 301–311.

Fan Z.T. (2000), On the convergence of a fuzzy matrix in the sense of triangular norms, *Fuzzy Sets and Systems* 109, pp. 409–417.

Fenchel W. (1949), On the conjugate convex functions, *Canadian J. Math.*, vol. 1, pp. 73–77.

Finkelstein A.V., Roytberg M.A. (1993), "Computation of biopolymers: a general approach to different problems", *Bio Systems*, 30, pp. 1–20.

Fliess M. (1973), "Automates et séries rationnelles non commutatives", *Automata Languages Programming* (M. NIVAT éd.), North Holland.

Floyd R.W. (1962), Algorithm 97: Shortest Path, *Comm. A.C.M.* 5, pp. 345.

Foata D. (1965), *Etude algébrique de certains problèmes d'Analyse Combinatoire et du Calcul des Probabilités*, Publ. Inst. Statist. Univ. Paris, 14.

Foata D. (1980), A combinatorial proof of Jacobi's identity, *Annals Discrete Maths* 6, pp. 125–135.

Ford L.R. Jr (1956), Network flow theory, *The Rand Corporation*, pp. 293.

Ford L.R., Fulkerson D.R. (1962), *Flows in Networks*, Princeton University Press, Princeton.

Frobenius G. (1912), "Über Matrizen aus nicht negativen Elementen", *S.B. Deutsch. Akad. Wiss. Berlin. Math.-Nat.* K1, pp. 456–477.

Gantmacher F.R. (1966), *Théorie des matrices*, tome 2, Dunod.

Gaubert S. (1992), *Théorie linéaire des systèmes dans les dioïdes*, Thèse de Doctorat, Ecole des Mines, Paris.

Gaubert S. (1994), "Rational series over dioids and discrete event systems", 11th Int. conf. on analysis and optimization of systems: Discrete Event Systems, Number 199, in *Lect. Notes in Control and Inf. Sci.*, Springer Verlag.

Gaubert S. (1995a), Performance evaluation of (Max, +) Automata, *IEEE Trans. Automatic Control* 40 (12), pp. 2014–2025.

Gaubert S. (1995b), Resource optimization and (Min, +) Spectral Theory, *IEEE Trans. Automatic Control* 40 (12), pp. 1931–1934.

Gaubert S. (1998), Communication orale à la 26e Ecole de Printemps on Theoretical Computer Science, Noirmoutier.

Gaubert S., Gunawardena J. (1998), "The Duality Theorem for Min-Max Functions", *C.R. Acad. Sci.* 326, pp. 43–48.

Gaubert S., Mairesse J. (1998), Modeling and analysis of timed petri nets using heaps of pieces, *IEEE Trans. Automatic Control*, à paraître.

Gaubert S., Butkovic P., Cuninghame-Green R. (1998), Minimal (Max, +) Realizations of Convex Sequences, *SIAM Journal Control Optim.* 36, 1, pp. 137–147.

Gavalec M. (1997), Computing matrix period in max-min algebra, *Discrete Appl. Math.* 75, pp. 63–70.

Gavalec M. (2002), Monotone eigenspace structure in max-min algebra, *Linear Algebra Appl.* 345, pp. 149–167.

Ghouila-Houri A. (1962), Caractérisation des matrices totalement unimodulaires, *C.R.A.S.*, Paris, tome 254, pp. 1192.

Giffler B. (1963), Scheduling general production systems using schedule algebra, *Naval Research Logistics Quarterly*, vol. 10, no. 3.

Golan J.S. (1999), *Semirings and Their Applications*, Dordrecht, Kluwer Academic Publishers.

Gondran M. (1974), Algèbre linéaire et Cheminement dans un Graphe, *R.A.I.R.O.*, vol. 1, pp. 77–99.

Gondran M. (1975), L'algorithme glouton dans les algèbres de chemins, *Bulletin de la Direction Etudes et Recherches*, EDF, Série C, 1, pp. 25–32.

Gondran M. (1975a), path algebra and algorithms, in *Combinatorial Programming: Methods and Applications* (B. Roy Ed.) D. Reidel Publish Co., pp. 137–148.

Gondran M. (1976a), Valeurs Propres et Vecteurs Propres en Classification Hiérarchique, *R.A.I.R.O. Informatique Théorique*, 10, 3, pp. 39–46.

Gondran M. (1976b), *Des algèbres de chemins très générales: les semi-anneaux à gauche et à droite*, Note interne EDF 2194/02, unpublished.

Gondran M. (1977), Eigenvalues and eigenvectors in hierarchical classification, in: Barra, J.R., et al. (eds.), *Recent Developments in Statistics*, North Holland, Amsterdam, pp. 775–781.

Gondran M. (1979), Les Eléments p-réguliers dans les Dioïdes, *Discrete Mathematics* 25, pp. 33–39.

Gondran M. (1995b), Valeurs Propres et Vecteurs Propres en Analyse des Préférences ou Trois solutions du paradoxe de Condorcet, Note EDF HI-00/95–001.

Gondran M. (1996), Analyse MINPLUS, *C.R.A.S.* Paris, t. 323, série 1, pp. 371–375.

Gondran M. (1996b), "Analyse MINMAX", *C.R. Acad. Sci. Paris*, 323, série I, pp. 1249–1252.

Gondran M. (1998), Inf-sup-ondelettes et analyse multirésolution. Note interne EDF DER HI 1997.

Gondran M. (1999a), Analyse MINMAX: infsolutions, infondelettes, infconvergences et analyse quasi-convexe. Note EDF DER HI 1999.

Gondran M. (1999b), Convergences de fonctions à valeurs dans $\mathbb{R}^k$ et analyse Min-Plus complexe. *C.R.A.S.*, Paris, t. 329, série I, pp. 783–788.

Gondran M. (2001a), "Calcul des variations complexes et solutions explicites d'équations d'Hamilton-Jacobi complexes", *C.R. Acad. Sci. Paris*, t. 32, série I, pp. 677–680.

Gondran M. (2001b), "Processus complexe stochastique non standard en mécanique quantique", *C.R. Acad. Sci. Paris*, série I, pp. 593–598.

Gondran M., Minoux M. (1977), Valeurs propres et Vecteurs propres dans les Dioïdes et leur Interprétation en Théorie des Graphes, *Bull. Dir. Et. Rech. EDF*, Série C, pp. 24–41.

Gondran M., Minoux M. (1978), L'indépendance linéaire dans les Dioïdes, *Bull. Dir. Et. Rech. EDF*, Série C, no. 1, pp. 67–90.

Gondran M., Minoux M. (1979), *Graphes et Algorithmes*, Eyrolles, Paris, 3e édition revue et augmentée 1995.

Gondran M., Minoux M. (1979), Valeurs propres et Vecteurs propres en Théorie des graphes, Colloques Internationaux, CNRS Paris 1978, pp. 181–183. *Lecture Notes in Econom. and Math. Systems* 166, Springer Verlag, Berlin.

Gondran M., Minoux M. (1984), Linear Algebra in Dioïds. A survey of recent results, in *Algebraic and Combinatorial Methods in Operations Research* (Burkard R.E., Cuninghame-Green R.A., Zimmermann U. Editors), *Annals of Discrete Mathematics* 19, pp. 147–164.

Gondran M., Minoux M. (1996), Valeurs propres et fonctionnelles propres d'endomorphismes à diagonale dominante en analyse MIN-MAX, *C.R.A.S.*, Paris, t. 325, série 1, pp. 1287–1290.

Gondran M., Minoux M. (1998), Eigenvalues and Eigen-Functionals of Diagonally Dominant Endomorphisms in MIN-MAX Analysis, *Linear Algebra and its Appl.*, 282, pp. 47–61.

Gondran M., Minoux M. (2007), "Dioids and Semirings: Links to Fuzzy sets and other Applications". *Fuzzy sets and Systems*, 158, 12, pp. 1273–1294.

Grabisch M. (1995), On equivalence classes of fuzzy connectives. The case of fuzzy integrals, *IEEE Trans. on Fuzzy Systems*, vol. 3, no. 1, pp. 96–109.

Grabisch M. (2003), The symmetric Sugeno integral, *Fuzzy Sets and Systems*, 139, pp. 473–490.

Grabisch M. (2004), "The Möbius function on symmetric ordered structures and its application to capacities on finite sets", *Discrete Math.*, 287, 1–3, pp. 17–34.

Grabisch M., Sugeno (1992), M., Multiattribute classification using fuzzy integral, *1st IEEE Int. Conf. on Fuzzy System*, pp. 47–54.

Grabisch M., De Baets, B., Fodor, J. (2004), The quest for rings on bipolar scales, *Int. J. Uncertainty, Fuzziness and Knowledge-Based Systems*, 12, 4, pp. 499–512.

Graham S.L., Wegman M. (1976), A fast and usually linear algorithm for global flow analysis, *J.l ACM*, 23, 1, pp. 172–202.

Grossmann A., Morlet J. (1984), "Decomposition of Hardy functions into square integrable wavelets of constant shape", *SIAM J. Math.*, Amal. 15, pp. 723, 1984.

Gunawardena J. (1994), "Min-max functions", *Discrete Event Dynamic Systems*, 4, pp. 377–406.

Gunawardena J. (1998), *Idempotency*, Cambridge University Press.

Guo S.Z., Wang P.Z., Di Nola A., Sesa S. (1988), Further contributions to the study of finite fuzzy relation equations, *Fuzzy Sets and Systems* 26, pp. 93–104.

Halpern J., Priess I. (1974), Shortest paths with time constraints on movement and parking, *Networks*, 4, pp. 241–253.

Hammer P.L., Rudeanu S. (1968), *Boolean Methods in Operations Research*, Springer Verlag.

Han S.C., Li H.X. (2004), Invertible incline matrices and Cramer's rule over inclines, *Linear Algebra and Its Applications* 389, pp. 121–138.

Hebisch U., Weinert H.J. (1998), Semirings. Algebraic theory and application in computer science, *World Scientific*, 361 pp.

Hille E., Phillips R.S. (1957), *Functional Analysis and Semi-groups*, American Mathematical Society, vol. XXXI.

Hiriart-Urruty J.B. (1998), *Optimisation et analyse convexe*, PUF, Paris.

Hoffman A.J. (1963), On Abstract dual Linear Programs, *Naval Research Logistics Quarterly* 10, pp. 369–373.

Höhle U.H., Klement E.P. (1995), Commutative Residuated 1-monoïds, in *Non-classical Logics and Their Applications to Fuzzy Subsets. A Handbook of the Mathematical Foundations of Fuzzy Set Theory*, Dordrecht, Kluwer, pp. 53–106.

Hopf E. (1965), "Generalized solutions of non linear equations of first-order", *J. Math. Mech.*, 14, pp. 951–973.

Hu T.C. (1961), The maximum capacity route problem, *Oper. Res.* 9, pp. 898–900.

Hu T.C. (1967), Revised matrix algorithms for shortest paths, *S.I.A.M. J. Appl. Math.*, 15, 1, pp. 207–218.

Jaffard S. (1989), "Exposants de Hölder en des points donnés et coefficients d'ondelettes", *C.R. Acad. Sci. Paris*, 308, série I, pp. 79–81.

Johnson, S.C. (1967), Hierarchical clustering schemes, *Psychometrica*, 32, pp. 241–243.

Jun-Sheng Duan J.-S. (2004), The transitive closure, Convergence of powers and adjoint of generalized fuzzy matrices, *Fuzzy Sets and Systems*, vol. 145, 2, pp. 301–311.

Kailath T. (1980), *Linear Systems*, Prentice Hall, Englewood Cliffs.

Kam J.B., Ullman J.D. (1976), Global data flow analysis and iterative algorithms, *Journal ACM* 23, 1, pp. 158–171.

Kam J.B., Ullman J.D. (1977), Monotone data flow analysis frameworks, *Acta Informatica* 7, pp. 305–317.

Karp R.M. (1978), A characterization of the minimum cycle mean in a digraph, *Discrete Math.* 23, pp. 309–311.

Kaufmann A., Malgrange Y. (1963), Recherche des chemins et circuits hamiltoniens d'un graphe, *R.I.R.O.*, 26, pp. 61–73.

Kelley J.E. (1960), "The cutting-plane method for solving convex programs", *J.l SIAM*, 8, pp. 703–712.

Kildall G.A. (1973), A Unified approach to program optimization, *Conference Rec ACM Symp. on Principles of Programming Languages*, Boston, MA, pp. 10–21.

Kim K.H., Roush F.W. (1995), Inclines of algebraic structures, *Fuzzy Sets and Systems*, vol. 72, 2, pp. 189–196.

Kim K.H., Roush F.W. (1980), Generalized fuzzy matrices, *Fuzzy Sets and Systems* 4, pp. 293–315.

Kleene S.C. (1956), "Representation of Events in nerve nets and finite automata" in *Automata Studies* (Shannon & Mc Carthy Editors), Princeton University Press, pp. 3–40.

Klement E.P., Mesiar R., Pap E. (2000), *Triangular Norms*, Dordrecht, Kluwer.

Kolokol'tsov V.N., Maslov V. (1989), Idempotent Analysis as a tool of control theory and optimal synthesis, *Funktsional'nyi Analiz i Ego Prilozhemiya*, vol. 23, no. 1, pp. 1–14, 1989.

Kolokol'tsov V.N., Maslov V. (1997), *Idempotent Analysis and its Applications*, Kluwer Acad. Publ.

Kuntzmann J. (1972), *Théorie des Réseaux*, Dunod, Paris.

Kwakernaak H., Sivan R. (1972), *Linear Optimal Control Systems*, Wiley-Interscience, New York.

Lallement G. (1979), *Semigroups and Combinatorial Applications*, J. Wiley & Sons.

Laue H. (1988), A Graph Theoretic proof of the fundamental trace identity, *Discrete Math* 69, pp. 197–198.

Lawler E.L. (1967), Optimal Cycles in doubly weighted directed linear graphs, in *Théorie des Graphes*, Proc. Int. Symp., Rome, Italy, (1966), Dunod, Paris, pp. 209–213.

Lax P.D. (1957), "Hyperbolic systems of conservation laws II", *Comm. on Pure and Applied Math.*, 10, pp. 537–566.

Lehmann D.J. (1977), Algebraic structures for transitive closure, *Theor. Comput. Sc.*, 4, pp. 59–76.

Lesin S.A., Samborskii S.N. (1992). "Spectra of compact endomorphisms", in *Idempotent Analysis*, Maslov and Samborskii Editors, Advances in Soviet Mathematics, vol. 13, American Mathematical Society, pp. 103–118.

Li Jian-Xin (1990), The smallest solution of max-min fuzzy equations, *Fuzzy Sets and Systems* 41, 317–327.

Lions P.L. (1982), *Generalized solutions of Hamilton-Jacobi Equations*, Pitman, London.

MacMahon P.A. (1915), *Combinatory Analysis*, Cambridge University Press et Chelsea Publishing Co., New York, 1960.

Mallat S. (1989), A theory of multiresolution signal decomposition: the wavelet representation, *IEEE transaction on Pattern Analysis and Machine Intelligence*, vol. 2, no. 7.

Martinet B. (1970), "Régularisation d'inéquations variationnelles par approximations successives", *RAIRO*, 4, pp. 154–159.

Maslov V. (1987a), *Méthodes opérationnelles*, Editions Mir., Moscou, 1987.

Maslov V. (1987b), "On a New Principle of Superposition for Optimization Problems", *Russian Math Surveys*, 42 (3), pp. 43–54.

Maslov V., Samborski S.N. (1992), *Idempotent Analysis*, Advances in Soviet Mathematics, vol. 13, American Mathematical Society.

Mc Lane S., Birkhoff G. (1970), "Algèbre", Gauthier-Villars, Paris.

Menger K. (1942), "Statistical metrics", *Proc. Nat. Acad. Sci. USA*, 28; pp. 535–537.

Meyer Y. (1992), *Les Ondelettes: algorithmes et applications*, A. Colin.

Minieka E., Shier D.R. (1973), A note on an algebra for the k best routes in a network, *J. Inst. Math. Appl.*, 11, pp. 145–149.

Minoux M. (1975), Plus courts chemins avec contraintes, *Annales des Télécommunications* 30, no. 11–12, pp. 383–394.

Minoux M. (1976), Structures algébriques généralisées des problèmes de cheminement dans les graphes: Théorèmes, Algorithmes et Applications, *R.A.I.R.O. Recherche Opérationnelle*, vol. 10, no. 6, pp. 33–62.

Minoux M. (1977), Generalized path algebras, in *Surveys of Mathematical Programming* (A. PREKOPA Editor) Publishing House of the Hungarian Academy of Sciences, pp. 359–364.

Minoux M. (1982), *Linear Dependence and Independence in Lattice Dioïds*, Note interne C.N.E.T. non publiée.

Minoux M. (1997), Bideterminants, arborescences and extension of the matrix-tree theorem to semirings, *Discrete Mathematics*, 171, pp. 191–200.

Minoux M. (1998a), A generalization of the all minors matrix tree theorem to semirings. *Discrete Mathematics* 199, pp. 139–150.

Minoux M. (1998b), *Propriétés combinatoires des matrices sur les (pré-)semi-anneaux*, Rapport de Recherche LIP6 1998/050, Université Paris 6.

Minoux M. (1998c), *Résolution de Systèmes linéaires dans les semi-anneaux et les dioïdes*, Rapport de Recherche LIP6 1998/051, Université Paris 6.

Minoux M. (1998d), *Algèbre linéaire dans les semi-anneaux et les dioïdes*, Rapport de Recherche LIP6 1998/052, Université Paris 6.

Minoux M. (2001), Extensions of Mac Mahon's master theorem to pre-semi-rings, LinearAlgebra and Appl., 338, pp. 19–26.

Moller P. (1987), *Notions de rang dans les dioïdes vectoriels*, Actes de la Conférence "Algèbres exotiques et systèmes à événements discrets" C.N.E.T., 3–4 juin 1987.

Moreau J.J. (1965), "Proximité et dualité dans un espace hilbertien", *Bull. Soc. Math. France*, 93, pp. 273–299.

Moreau J.J. (1970), Inf-convolution. Convexité des fonctions numériques, *J. Math. Pures Appl.*, 49, pp. 109–154.

Morgan, *Geometric Measure Theory*.

Morlet J. (1983), "Sampling Theory and Wave Propagation" in: Chen C.H., Acoustic Signal/Image Processing and Recognition, no. 1 in NATO ASi Series. Springer Verlag, pp. 233–261.

Nachbin L. (1976), *Topology and Order*, Van Nostrand, Princeton.

Olsder G.J., Roos C. (1988), Cramer and Cayley–Hamilton in the Max-Algebra, *Linear Algebra and its Applications*, 101, pp. 87–108.

Orlin J.B. (1978), Line Digraphs, Arborescences and Theorems of Tutte and Knuth, *J. Combin. Theory*, B, 25, pp. 187–198.

Pap E. (1995), *Null-Additive Set Functions*, Kluwer, Dordrecht.

Perny P., Spanjaard O., Weng P. (2005), "Algebraic Markov decision processes", *Proceedings IJCAI*.

Perrin D., Pin J.E. (1995), "Semigroups and Automata on infinite words", in *Semi-groups, Formal Languages and Groups*, (J. Fountain Editor), Kluwer, Dordrecht, pp. 49–72.

Perron O. (1907). "Über Matrizen", *Math. Ann.*, vol. 64, pp. 248–263.

Perterson J.L. (1981), *Petri Net Theory and the Modeling of Systems*, Prentice Hall, Englewood Cliffs, NJ, 290 pp.

Peteanu V. (1967), An algebra of the optimal path in networks, *Mathematica*, vol. 9, no. 2, pp. 335–342.

Quadrat J.P. (1990), *Théorèmes asymptotiques en programmation dynamique*, C.R.A.S., 311, série I, pp. 745–748.

Quadrat J.P., Max-Plus working group (1997), "Min-plus linearity and statistical mechanics", *Markov Processes and Related Fields* 3, 4, pp. 565–587.

Reutenauer C., Straubing H. (1984), Inversion of matrices over a commutative semiring, *J.Algebra*, 88, no. 2, pp. 350–360.

Robert F., Ferland J. (1968), Généralisation de l'algorithme de Warshall, *R.I.R.O.* 7, pp. 71–85.

Robinson A. (1973), "Function theory of some nonarchimedean fields", *The American Mathematical Monthly*, vol. 80, pp. 87–109.

Rockafellar R.T. (1970), *Convex Analysis*, Princeton University Press, Princeton.

Rockafellar R.T. (1976a), "Monotone operators and the proximal point algorithm", *SIAM J. ControlOptimiz.*, 14, pp. 877–898.

Rockafellar R.T. (1976b), "Augmented lagrangians and applications of the proximal point algorithm in convex programming", *Math. Oper. Res.* 1, pp. 97–116.

Rote G. (1990), Path problems in graphs, *Comput. Suppl.* 7, pp. 155–189.

Roy B. (1975), Chemins et Circuits: Enumérations et optimisation, in *Combinatorial Programming: Methods and Applications* D. Reidel, Boston, pp. 105–136.

Rutherford D.E. (1964), The Cayley–Hamilton theorem for semirings, *Proc. R. Soc. Edinburgh Sec.* A, 66, pp. 211–215.

Salomaa A. (1969), *Theory of Automata*, Pergamon Press, Oxford.

Samborski S.N. (1994), time discrete and continuous control problems convergence of value functions. In 11th Inter. Conf. on Analysis and Optimization of Systems. L.N. in Control and Inf. Sc., no. 199, Springer Verlag, 1994, pp. 297–301.

Schweizer B., Sklar A. (1983), *Probabilistic Metric Spaces*, North Holland, Amsterdam.

Shimbel A. (1954), Structure in communication nets, *Proc. Symp. on Information Networks*, Polytechnic Institute of Brooklyn, pp. 119–203.

Simon I. (1994), "On semigroups of matrices over the tropical semiring", *Informatique Théorique et Appl.*, vol. 28, no. 3–4, pp. 277–294.

Spanjaard O. (2003), *Exploitation de préférences non classiques dans les Problèmes Combinatoires. Modèles et Algorithmes pour les Graphes*, Doctoral dissertation, Université Paris Dauphine, France.

Straubing H. (1983), A combinatorial proof of the Cayley–Hamilton theorem, *Discrete Mathematics* 43, pp. 273–279.

Sugeno M. (1974), *Theory of Fuzzy Integrals and Applications*, Ph.D. dissertation, Tokyo Institute of Technology.

Sugeno M. (1977). "Fuzzy measures and fuzzy integrals. A survey", in Gupta, Saridis, Gaines (eds.), *Fuzzy Automata and Decision Processes*, pp. 89–102.

Tarjan R.E. (1981), A unified approach to path problems, *J. A.C.M.*, vol 28, no. 3, pp. 577–593.

Tomescu I. (1966), Sur les méthodes matricielles dans la théorie des réseaux, *C.R. Acad. Sci. Paris*, Tome 263, pp. 826–829.

Tricot C. (1993), *Courbes et dimension fractale*, Springer Verlag.

Tricot C., Quiniou J.-F., Wehbi D., Roques-Carmes C., Dubuc B. (1988), Evaluation de la dimension fractale d'un graphe, *Revue Phys. Appl. 23*, pp. 111–124.

Tutte W.T. (1948), The dissection of equilateral triangles into equilateral Triangles, *Proc. Cambridge Philos. Soc.*, 44, pp. 463–482.

Volle M. (1985), Conjugaison par tranches, *Ann. Mat. Pura Appl.* (IV), 1985, 139, pp. 279–312.

Wagneur E. (1991), Moduloïds and Pseudomodules 1. Dimension Theory, *Discrete Mathematics* 98, pp. 57–73.

Wang Z., Klir, G.J. (1972), *Fuzzy Measure Theory*, PlenumPress.

Warshall S. (1962), A theorem on boolean matrices, *J. A.C.M.* 9, pp. 11.

Wongseelashote A. (1976), An algebra for determining all path values in a network with application to k-shortest-paths problems, *Networks*, 6, pp. 307–334.

Wongseelashote A. (1979), Semirings and path spaces, *Discrete Mathematics* 26, pp. 55–78.

Yoeli M. (1961), A note on a generalization of Boolean matrix theory, *Amer. Math. Monthly*, 68, pp. 552–557.

Zeilberger D. (1985), A combinatorial approach to matrix algebra, *Discrete Mathematics* 56, pp. 61–72.

Zimmermann U. (1981), Linear and combinatorial optimization in ordered algebraic structures, *Annals of Discrete Mathematics* 10.

Index